MEMBRANE BIOREACTOR PROCESSES

PRINCIPLES AND APPLICATIONS

Advances in Water and Wastewater Transport and Treatment

A SERIES

Series Editor
Amy J. Forsgren

Xylem, Sweden

Membrane Bioreactor Processes: Principles and Applications
Seong-Hoon Yoon

Wastewater Treatment: Occurrence and Fate of Polycyclic Aromatic Hydrocarbons (PAHs)
Amy J. Forsgren

Harmful Algae Blooms in Drinking Water: Removal of Cyanobacterial Cells and Toxins
Harold W. Walker

ADDITIONAL VOLUMES IN PREPARATION

MEMBRANE BIOREACTOR PROCESSES

PRINCIPLES AND APPLICATIONS

Seong-Hoon Yoon

CRC Press
Taylor & Francis Group
Boca Raton London New York

CRC Press is an imprint of the
Taylor & Francis Group, an **informa** business

CRC Press
Taylor & Francis Group
6000 Broken Sound Parkway NW, Suite 300
Boca Raton, FL 33487-2742

First issued in paperback 2020

ISBN 13: 978-0-367-57567-0 (pbk)
ISBN 13: 978-1-4822-5583-6 (hbk)

Visit the Taylor & Francis Web site at
http://www.taylorandfrancis.com

and the CRC Press Web site at
http://www.crcpress.com

To my wife, Mihyun Choi

Contents

Preface

Membrane bioreactor (MBR) processes are one of the most successful commercial applications of membrane technology along with membrane-based desalination, water clarification, process solutions, etc. The knowledge base of MBR has expanded rapidly as the surging number of research papers suggests in recent decades. Alongside, however, the gaps between the leading edge academic research and the industrial practices have widened. Many key aspects of the innovations made in industry are not a subject of open discussion, and as a result, those have only limited influences on academic researches. Meanwhile, the main stream academic researches have evolved following their own direction. Due to the diverging interests, MBR technology is not fully described in the traditional scientific literature.

Hoping to fill the gap, I started a personal project of launching an MBR blog in 2011, i.e., http://www.onlinembr.info. I tried to explain the practical aspects of MBR technology based on scientific principles as much as possible. The principles of commercial products/processes were also described based on the knowledge obtained from relevant patent applications, conference proceedings, and personal analyses. The open-source knowledge sharing through the website turned out to be a great success, having more than 200 unique visitors a day globally, almost equally distributed among the Americas, Europe, and Asia. The questions and comments I received were a great source of new information and inspiration. With numerous encouragements from audiences, I took courage to author this book. The contents in the blog were thoroughly reviewed and modified with additional fillers to make the contents more rigorous and complete.

This book is written with an emphasis on the principles behind the scene instead of simply introducing the way MBR works. Readers will learn the theoretical and practical backgrounds of current practices involved with membrane module design, biological and membrane system design, system optimization, and system operation. Chapters 1 and 2 are about the fundamentals of membrane filtration relevant to MBR. Chapter 2 is specifically about the principles of submerged membranes. Readers will learn why submerged and cross-flow membranes should be operated at the conditions they are operated today and how the filtration performance can be improved. In the next two chapters (Chapters 4 and 5), readers will also learn about the biological principles that are closely related to the membrane performance. MBR design principles are discussed in Chapter 6. All other MBR-related topics are discussed in Chapter 7. Covering the available knowledge as much as possible, I hope this book gives a more holistic view on MBR to audiences.

I am indebted greatly to the professors, engineers, and colleagues who helped directly or indirectly in preparing this book. I would like to thank, in particular, professors Chung-Hak Lee (Seoul National University, Korea), Icktae Yeom (Sungkyunkwan University, Korea), and Hyungsoo Kim (Sungkyunkwan University, Korea) for the inspirations and comments they gave me in the last two decades. I also thank all of my colleagues in Nalco, a company of Ecolab, for their support. In particular, I appreciate John Collins and Jelte Lanting for sharing with me their more than 30 years of expertise in water treatment with me. I profited greatly from the discussions with Youngseck Hong (GE), Sangho Lee (Kookmin University, Korea), Hyung Hoon (CH2M Hill), Yeomin Yoon (University of South Carolina, USA), and Hosang Lee (Asahi Kasei). I also thank the companies and publishers that provided me with figures and pictures. I would also like to gratefully acknowledge the use of the incredibly vast and versatile knowledge platform provided by Google Inc., including search engine, cloud storage, email, picture editing, etc. Furthermore, I thank Taylor & Francis/CRC Press, in particular, series editor Amy Forsgren, editor Irma Britton, and coordinator Hayley Ruggieri, for their great care and dedication in preparing this book. I also want to thank Ms. Amor Nanas of Manila Typesetting Company for her excellent and meticulous works.

Finally, I thank my family, Mihyun Choi, Seungjoon, Seungmin, and Inkyung, for their love, moral support, and patience.

Seong-Hoon Yoon
Naperville, Illinois

Author

Seong-Hoon Yoon is a senior staff engineer of Nalco, an Ecolab company, located in Naperville, Illinois, where he has served since 2001. He received his BS, MS, and PhD degrees in chemical technology (currently department of chemical and biological engineering) from Seoul National University in 1991, 1994, and 1998, respectively. Dr. Yoon was a research engineer at the LG group in Seoul, Korea, before joining Nalco. Throughout his career, Dr. Yoon has been dedicated to advancing water treatment technology with an emphasis on membrane separation. He is a recipient of Nalco Chairman's award in 2007. His research interest includes water reuse and recycle; gas transfer membranes; chemical, biological, and physical water treatment; and information communication technologies for remote monitoring and control. Dr. Yoon has published more than 50 research papers and reports and holds more than 10 U.S. patents. He has professional engineer's licenses in chemical engineering (USA) and environmental engineering (Korea).

1 Principle of Membrane Filtration

1.1 CLASSIFICATION OF FILTRATION

1.1.1 SURFACE FILTRATION AND DEPTH FILTRATION

Depending on where the particle rejection occurs, filtration can be classified as surface filtration or depth filtration, as illustrated in Figure 1.1.

In surface filtration, particles are rejected by the filter surface. This type of filtration is also called absolute filtration because any particle larger than the pore size is rejected. Because particles accumulate in two-dimensional spaces on the filter surface, the particle-holding capacity of the surface filter is generally low. To control the particle deposition, deliberate antifouling measures are typically required by generating shear stress on the filter surface. Various methods are available, for example, high crossflow velocity, aeration, rotating/vibrating filter assembly, turbulence generation using spacers in the feed channel, rotating turbulence generator, feed flow pulsing, and others, but their use is restricted by the economics depending on the application. In surface filtration, particles are not lost to the depth of the filter unlike in depth filtration, which is a beneficial characteristic for recovering valuable materials from the feed water. All microfiltration (MF)/ultrafiltration (UF)/ reverse osmosis (RO) membranes, whereas stainless steel screens, and strainers are examples of surface filters. The vast majority of commercial membranes have an asymmetric structure that consists of a thin skin layer with small pores and a thick support layer with large pores. Whereas the skin layer separates particles from water, the support layer provides mechanical strength to the membrane without causing significant filtration resistance.

Depth filters consist of a matrix of randomly oriented fibers or particles with tortuous water paths. Particles intrude into a tortuous maze and are captured by the filter medium by charge and van der Waals interactions or physical obstructions. Because particle removal relies on random collision with the filter matrix followed by adhesion and entrapment, some particles can pass the filter before they are captured. As a consequence, filtrate quality is not as good as that of surface filters in general, but the filter can hold a large amount of particles in the three-dimensional filter medium. Therefore, depth filters are typically used at dead-end filtration modes without a crossflow on the filter surface. At this filtration condition, most particles contained in the feed water are captured by the filter instead of being scoured. Surface filtration can simultaneously occur if large particles are captured by the filter surface and the particle entrance to the filter medium is interrupted by the cake layer. Cartridge filters, sand filters, multimedia filters, and most air filters fall into this category.

1.1.2 CROSSFLOW FILTRATION AND DEAD-END FILTRATION

Surface filtration can be divided into crossflow filtration and dead-end filtration depending on the existence of turbulence on the filter surface, as shown in Figure 1.2. Generally, the crossflow filtration mode is used for water with large amount of total suspended solids (TSS), whereas dead-end filtration mode is for low TSS water.

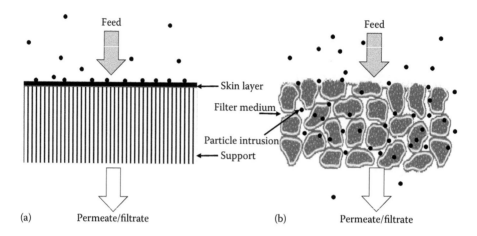

FIGURE 1.1 Surface filtration and depth filtration: (a) surface filtration by skin layer and (b) depth filtration by filter medium.

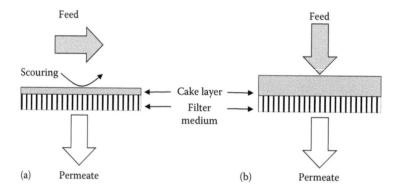

FIGURE 1.2 Crossflow filtration and dead-end filtration: (a) crossflow mode and (b) dead-end mode.

The term "crossflow" originated from the fact that the feed moves parallel to the filter surface and permeates flows perpendicular to the feed flow. In crossflow filtration, the cake layer depth can be controlled by the scouring effect but extra energy is required to move the feed water fast enough. Crossflow filtration can work more efficiently with surface filters than with depth filters because the crossflow can efficiently control solids deposition on the smooth filter surface. If the scouring effect and particle deposition are balanced, the cake layer hardly grows for an extended period. Crossflow filtration is particularly effective when the feed water carries a large amount of foulants such as suspended solids and macromolecules. All membrane bioreactor (MBR) processes and most wastewater filtrations rely on crossflow filtration. Stirred cell filtration (Figure 5.23) is also one form of crossflow filtration with an internal concentrate recycle.

In dead-end filtration, the feed is pushed through the filter medium without crossflow. Both surface filters and depth filters can be operated under this mode. Because dead-end filtration is not sustainable without removing the accumulated solids, the filter should either be replaced or backwashed periodically. Dead-end filtration can be energy-efficient due to the lack of crossflow, but prone to performance losses by solids accumulation. Therefore, dead-end filtration modes are primarily for water with low TSS. Most depth filters and some surface filters are operated in dead-end mode. Membrane applications for low TSS water are often practiced under dead-end modes, for example, surface water filtration, pretreatment for seawater RO, and tertiary filtration.

1.2 FILTRATION THEORY

1.2.1 CONCENTRATION POLARIZATION

When feed water permeates through the membrane, the solutes contained in the water are rejected by the membrane. Those rejected solutes accumulate near the membrane surface and form a high concentration zone called the concentration polarization (CP) layer (Porter 1972). The particles in the CP layer can diffuse back to the bulk by diffusion, if they are not fixed in the gel/cake layer. The CP phenomenon fundamentally limits the filtration performance by increasing the filtration resistance near the membrane surface. The concentration profile illustrated in Figure 1.3 settles at the equilibrium point, where the convective particle transport toward the membrane surface and the diffusive particle back-transport to the bulk are balanced. The corresponding equation can be written as Equation 1.1, where the convective particle transport, JC, in the left is equated to the diffusive particle back-transport in the right. The minus sign is required to reflect the opposite directions of those two effects.

$$JC = -D_{eff} \frac{dC}{dx} \tag{1.1}$$

where
 J water flux at steady state (m/s)
 C particle concentration (mg/L)
 x distance from the membrane surface (m)
 D_{eff} effective diffusion coefficient of solute (m^2/s)

Equation 1.1 can be integrated using the boundary conditions at steady state, that are, ($x = 0$, $C = C_G$) and ($x = d$, $C = C_B$), where δ is a boundary layer thickness in meters, C_G and C_B are particle concentrations in gel layer and in bulk in mg/L, respectively.

$$J_{SS} = -\frac{D_{eff}}{\delta} \ln\left(\frac{C_B}{C_G}\right) \tag{1.2}$$

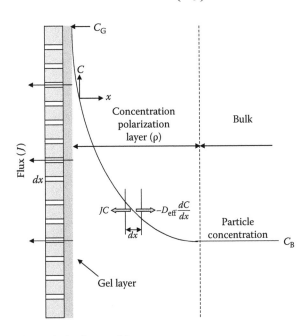

FIGURE 1.3 Concentration polarization model.

According to the above equation, steady state flux, J_{SS}, is inversely proportional to boundary layer thickness, δ (m), and is proportional to the effective diffusion coefficient, D_{eff} (m²/s). In fact, by increasing crossflow velocity on the membrane surface, thinner boundary layer and higher effective diffusivity can be achieved simultaneously (Bian et al. 2000). As will be discussed in Section 1.2.5.3, effective diffusivity increases when particle collisions are encouraged in a high shear field. The prelogarithmic factor, D_{eff}/δ, in Equation 1.2 can be expressed as mass transfer coefficient, k (m/s). As can be expected from the equation, small particles with high diffusivities are subject to a higher steady state water flux (J_{SS}) than large particles. The high water velocity on the membrane surface reduces boundary layer thickness (δ), which eventually results in high fluxes.

Equation 1.3 can be used to estimate gel layer concentration (C_G) for the small particles that do not form a cake layer in the CP layer, for example, emulsified oils, latex and paint particles, milk, albumin, gelatin, and others. In one study, 5% emulsified semisynthetic cutting oil was concentrated using a flat sheet polysulfone (PSU) membrane with a molecular weight cut-off (MWCO) of 100 kDa at 0.5 bar. Flow rate was maintained at 1.5 m/s to generate turbulence on membrane surface (Reynolds number = 8000). As shown in Figure 1.4, flux (J) can be plotted against $\ln(C_B)$ to obtain the interception with the x axis. In this figure, the oil concentration in the gel layer is estimated at 37% from the interception. This means that the flux (J) becomes zero if the oil concentration in the bulk becomes 37% according to the equation. The mass transfer coefficient, k (m/s), is obtained at 6.5×10^{-5} m/s from the slope.

$$J_{SS} = -\frac{D_{eff}}{\delta}[\ln(C_B) - \ln(C_G)] \tag{1.3}$$

Despite the popular use of the CP model to rationalize ultrafiltration, the model does not apply for many macromolecular colloidal and particulates solutions (Baker 2004). Flux is often too high to be rationalized by reasonable diffusivities and boundary layer thicknesses. It is due to the enhanced diffusivity at the crossflow filtration condition beyond the level predicted by the thermodynamic diffusivity. In fact, particles hit each other in the shear field and more collisions occur from the left side of the particles than from the right side (Figure 1.3) because of the particle concentration gradient in the CP layer. This is the so-called shear-induced diffusion, and its net effect is the enhanced particle movement away from the membrane surface toward the bulk. As a consequence,

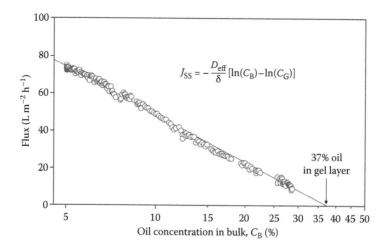

FIGURE 1.4 Flux decline as a function of bulk concentration (C_B) in ultrafiltration when 5% emulsified semisynthetic cutting oil is filtered by a PSU membrane with a MWCO of 100 kDa. (Modified from Um, M.-J. et al., *Water Res.* 35(17):4095–4101, 2001.)

the effective diffusivities in crossflow filtration tend to be much higher than the thermodynamic diffusivities. As particles become larger, more frequent collisions occur and the gap between the effective diffusivity and the thermodynamic diffusivity widens. In addition to shear-induced diffusion, particles tend to migrate to the space somewhere in between the membrane surface and the center of the flow channel to minimize the kinetic energy they carry. This phenomenon is called "inertial lift," which causes a tubular pinch effect, as will be discussed in Section 1.2.5.4. Therefore, to explain the flux reasonably, the effective diffusion coefficient, D_{eff}, should include not only the thermodynamic effect but also the hydrodynamic effect on particle back-transport. The particle back-transport theories are discussed in more detail in Section 1.2.5.

1.2.2 Resistance in Series Model

The relation between transmembrane pressure (TMP) and flux can be described using a simple resistance in series model as shown in Equation 1.4. This equation is fundamentally the same as other equations used to model heat and mass transfer, electrical conduction, air/water flow through pipelines, etc., where the flux/flow/current is proportional to the driving force and inversely proportional to the resistance.

$$J = \frac{\Delta P_T}{\mu(R_m + R_c + R_f)} \tag{1.4}$$

where

J water flux (m/s)
ΔP_T transmembrane pressure (Pa or kg/m/s^2)
μ viscosity of permeate (kg/m/s or cP, 1.00×10^{-3} for water at 20°C)
R_m membrane resistance (/m)
R_c cake resistance (/m)
R_f irreversible fouling resistance (/m)

The three resistances in the equation are operationally defined and can be measured experimentally. R_m is measured by filtering clean water through new membrane assuming R_f and R_c are zero. Because J, ΔP_T, and μ are known, R_m can be calculated using Equation 1.4. Similarly, $R_m + R_c + R_f$ is calculated from the operating data collected during the filtration. Subsequently, $R_c + R_f$ is calculated by subtracting R_m from $R_m + R_c + R_f$. After removing the cake layer from the membrane using a water jet, filtration is performed with clean water to obtain $R_m + R_f$. By subtracting R_m from $R_m + R_f$, R_f is obtained. Finally, by subtracting R_m and R_f from $R_m + R_f + R_c$, R_c is obtained. R_f is often included in R_c because R_f is typically much less than R_c in MF/UF, where particles rarely penetrate into the membrane causing irreversible fouling (Choo and Lee 1996).

The resistance in series model is often a useful tool to determine the major cause of membrane fouling by breaking down filtration resistances. For instance, the high R_c may indicate that a more vigorous membrane scouring is required to reduce cake layer formation. The high R_f may suggest a strong interaction between the membrane surface and the foulants contained in the feed water or that the pore sizes are too large to reject the foulants (or both). However, the resistant components are sensitive to the experimental condition/procedure used to measure them, especially when membranes are used to filter activated sludge (Chang et al. 2009). In one instance, both R_c and R_f vary depending on how thoroughly the cake layer is removed. In other instances, the residual cake layer that was not removed by the water jet can be broken down to smaller debris by a circulation pump during the clean water filtration and cause substantial resistances during the measurement of $R_m + R_f$ causing overestimated R_f. Extra care must be taken when the resistances from two different experimental sets are compared. In addition, when the resistance in series model is applied for hollow fiber membranes, the measured R_m is always overestimated because the effective ΔP_T is always lower than the apparent ΔP_T as a result of the internal pressure drop. Therefore, the R_m of hollow fiber membranes measured by the resistance in series model cannot be directly compared with the R_m of flat sheet membranes.

1.2.3 CAKE LAYER COMPACTION

1.2.3.1 Mechanism

The cake layer formed on the membrane surface acts as a filtration resistance that increases TMP at a constant flux mode or decreases flux at a constant pressure mode. It not only grows thicker over time because of the particles/solutes carried by the convective flow toward the membrane but it also becomes more compact, causing a gradual increase of filtration resistance. It has been considered that cake compaction proceeds with the further deposition of small particles in the void spaces of the cake layer, transformation of particle configuration, cake layer collapse, etc. Understanding the mechanism of cake layer compaction is crucial to understanding the cause of performance loss in MF/UF. It is also crucial to come up with the optimum module design and the associated operating methods, which can be achieved by minimizing the cake layer formation and its compaction.

The pressure loss through the cake layer can be calculated using the Carmen–Kozeny equation given as Equation 1.5. According to this equation (McCabe et al. 2005), the pressure loss through the cake layer, ΔP, increases as specific surface area, S, increases and cake porosity, ε, decreases if water flux, J, remains constant. S is also known as the surface area per volume and is inversely correlated with the particle size. By definition, the dynamic pressure loss through the cake layer is equivalent to the force pressing a unit area of cake layer and both have the same unit (i.e., N/m²). Therefore, the higher the dynamic pressure loss, the stronger the force pressing the cake layer.

$$\Delta P = \frac{5\,\mu S^2 (1-\varepsilon)^2 J}{\varepsilon^3}\,\Delta\ell \tag{1.5}$$

where

J	water flux based on cake surface area (m/s)
ΔP	pressure drop in cake layer (Pa or kg/m/s²)
S	specific surface area (/m)
ε	porosity of cake layer (–)
μ	viscosity of liquid permeating through cake layer (kg/m/s)
ℓ	depth of cake layer (m)

A cumulative compression effect exists toward the bottom of the cake layer because the force compressing one sublayer is transferred to the next lower sublayer, as illustrated in Figure 1.5 (Tiller 1953). As shown in Figure 1.5, the pressure drop in the first layer ($n = 1$) causes a downward force pressing the next layer ($n = 2$). Again, the sum of the downward forces from the first and the second layers is transferred to the third layer ($n = 3$) and so on. Finally, the sum of the forces from the first layer to the second from the last layer presses the last layer in the bottom ($n = n$). Therefore, the forces acting on each sublayer gradually increase toward the bottom of the cake layer. If the cake layer is compressible (or collapsible), cake layer compaction starts from the bottom and propagates upward.

It is also important to understand that TMP does not directly affect the extent of cake layer compaction contrary to popular perception. Only the dynamic pressure differential across the depth of the cake layer can compress the cake layer which is caused by permeate flow through the cake layer. If there is no permeate flow, cake layer compaction does not occur no matter how high the TMP is. Therefore, the cake layer compaction is more directly correlated with flux rather than TMP. In fact, reverse osmosis (RO) runs at a much higher TMP than MF/UF but the force pressing the cake

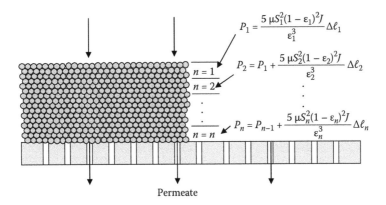

$$P_1 = \frac{5 \mu S_1^2 (1 - \varepsilon_1)^2 J}{\varepsilon_1^3} \Delta \ell_1$$

$$P_2 = P_1 + \frac{5 \mu S_2^2 (1 - \varepsilon_2)^2 J}{\varepsilon_2^3} \Delta \ell_2$$

$$P_n = P_{n-1} + \frac{5 \mu S_n^2 (1 - \varepsilon_n)^2 J}{\varepsilon_n^3} \Delta \ell_n$$

Permeate

FIGURE 1.5 Cumulative nature of the compressing cake layer, where the top sublayer is least compressed whereas the bottom sublayer is most compressed.

layer can be lower because of its low flux if the cake layer composition in both types of filtration are identical.

The macromolecules/colloids transported to the internal void spaces of the cake layer by permeate can also play a role in cake layer compaction. According to the porosity effect term, that is, $(1 - \varepsilon)^2/\varepsilon^3$ in the Carmen–Kozeny equation shown in Equation 1.5, cake layer resistance increases gradually as the cake layer porosity decreases, but it abruptly increases once the cake layer porosity falls below the threshold (~0.2). The sudden increase of cake layer resistance at least partially explains the phenomenon called "sudden TMP rise" observed in submerged membrane processes as discussed in Section 1.3.2.3.

1.2.3.2 Structure of Cake Layer

The cumulative nature of the hydrodynamic forces acting on cake layer compaction has been confirmed empirically. Figure 1.6 shows a cross-section of a cake layer formed on a UF membrane with a MWCO of 7000 Da at 150 L/m²/hr (LMH). It is apparent that the 760 nm polystyrene particles are most densely packed in the bottom of the cake layer whereas its packing density decreases toward the top (Tarabara et al. 2004).

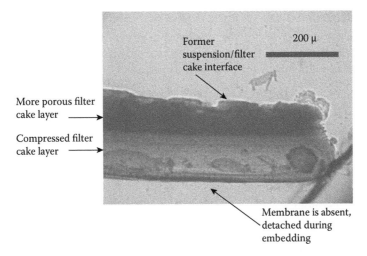

FIGURE 1.6 Filter cake cross-section imaged in transmitted light, where the diameter of the polystyrene particles = 760 nm; ionic strength = 1 mM; flat sheet PSU membranes with MWCO of 7000 Da at 150 LMH. (From Tarabara, V. et al., *J. Membrane Sci.* 241:65–78, 2004.)

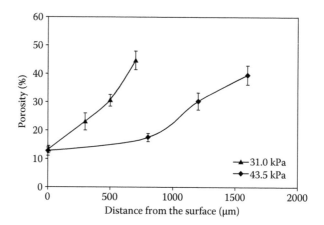

FIGURE 1.7 Cake layer area porosity distribution measured by analysis of optical microscopic image of cake layer cross-section. (From Gao, W.J. et al., *J Memb Sci.* 374(1–2):110–120, 2011.)

Similar observations were also made in a laboratory-scale anaerobic MBR (AnMBR) equipped with flat sheet submerged membranes treating whitewater from a thermomechanical pulping process (Gao et al. 2011). The cake layer was taken from the membrane after the filtration experiment and was cryogenically cut to expose the vertical structure. The pictures taken through an optical microscope were analyzed to obtain the vertical porosity distribution of the cake layer. As shown in Figure 1.7, porosity was lowest near the membrane surface at around 11%, and it increased gradually toward the top layer of the cake. It was also observed that cake layer porosity declined when high flux caused high TMP (43.5 kPa).

Although the trends of the porosities in the figure are valid, the porosities obtained based on image analyses should not be considered absolute values in most cases. It is because porosity can vary widely depending on the threshold used to distinguish void spaces from the solid constituents of cake layer in the image. Therefore, comparing the porosities from two different studies is generally not valid. In fact, the cake layer porosities observed in anaerobic MBR (AnMBR) are drastically lower than those obtained in aerobic MBR as shown in shown in Figures 1.7 and 3.24, but it does not necessarily suggest that cake layer porosities are lower in AnMBR than in aerobic MBR.

1.2.3.3 Model Equation

If there is no crossflow and all the particles contained in the feed deposit, the cake resistance, R_c, in Equation 1.4 can be expressed as Equation 1.6, where a proportionality coefficient called specific cake resistance (α) is multiplied by the cake layer thickness, cV/A.

$$R_c = \alpha \frac{cV}{A} \tag{1.6}$$

where
   R_c cake resistance (/m)
   α specific cake resistance (m/kg)
   c particle concentration (kg/m^3)
   V filtrate volume (m^3)
   A membrane surface area (m^2)

The above equation is valid for the rigid particles that do not undergo compaction because a fixed specific cake resistance, α, is used. However, when soft particles are filtered, an additional equation is required to reflect the changes in α depending on TMP. For instance, if yeast cells are filtered with a PSU membrane with 0.45 µm pores, the cake resistance, R_c, increases as TMP increases (McCarthy et al. 1998). The correlation between α and TMP was expressed as Equation 1.7. In this equation, α increases proportionally to the TMP at the TMP range of 30 to 500 kPa as a consequence of the cake layer compaction.

$$\alpha = \alpha_0(1 + k_c \Delta P_T) \tag{1.7}$$

where
  α_0 specific cake resistance when ΔP_T is zero
  k_c proportionality constant
ΔP_T TMP

1.2.3.4 Self-Acceleration of Cake Layer Compaction under Constant Flux Mode

When filtration is performed at a flux fixed at a low level, TMP increases very slowly in the early phase of the filtration cycle, but it increases abruptly at some point in time. This is the so-called "sudden TMP rise (or jump)." There are many different theories explaining the sudden TMP increase, which will be discussed in Section 1.3.2.1, but the self-accelerating nature of cake layer compaction seems to be the most significant factor.

As discussed in Section 1.2.3.1, cake layer compaction is triggered from the bottom of the cake layer, where the force compressing the cake layer is the largest due to the cumulative nature of the dynamic pressure loss in the cake layer. Once the compression force reaches a threshold at the bottom of the cake layer, it starts to collapse and filtration resistance increases. As a consequence, TMP must increase to compensate the flux loss under the constant flux mode. The increasing TMP is a direct consequence of the increasing pressure loss through the cake layer, which further increases the force compressing the cake layer. Once the filtration enters this vicious circle, TMP starts to increase abruptly (Park et al. 2006).

There are three major factors affecting cake layer compaction: (1) rigidity of the solids that form the cake layer, (2) pressure drop across the cake layer, and (3) exposure time. Because solids with biological origins are soft in MBR, the shape of the cake layer can change easily, reflecting the forces acting on them. The macromolecules/colloids transported by permeate continuously deposit in the void spaces of the cake layer and contribute to the triggering of the initial cake layer collapse by increasing pressure loss. As a result, even at low TMP such as less than 10 kPa, cake layer compaction eventually occurs and rapid TMP increase is triggered. In MBR, the threshold TMP that can cause cake layer compression was suggested to be 4.9 to 7.8 kPa, below which no cake layer compression occurs in the short term (Poorasgari et al. 2015).

1.2.4 Dynamic Membrane

It has been observed that pore sizes are not a critical factor affecting the permeate quality in terms of chemical oxygen demand (COD), total organic carbon (TOC), turbidity, etc., in most porous membrane processes. When MF and UF membranes with various pore sizes were used to filter anaerobic digester broth, there were no noticeable differences in permeate quality in terms of COD (Imasaka et al. 1989). No differences in permeate COD were observed in alcohol distillery wastewater filtration using tubular ceramic membranes with various pore sizes (Yoon 1994). Very small or no differences have been observed even in virus removal efficiencies between MF and UF membranes (Hirani et al. 2010). In addition, no meaningful differences have been observed in flux, regardless of the membrane permeability or pore size, if the pure water permeability is above a certain threshold and the filtration is performed with a cake layer on the membrane surface. According to a survey of 24 MBR plants using various crossflow membranes with tubular or plate and frame configurations, the design flux mostly ranged from 60 to 80 LMH regardless of the pore size, for example, 40 kDa (equivalent of <0.01 µm), 0.08 µm and

0.15 μm (Larrea et al. 2014). The same is also true for submerged membranes, where design flux is not a function of pore size. The design and operating fluxes are indeed nearly not affected by pore size or permeability among the well-established membranes despite the drastically different pore sizes in the range of 0.03 to 0.45 μm, although the pure-water permeability varies widely depending on the membrane.

The insensitivity of permeate quality and flux to the pore size and the membrane permeability is due to the cake layer formed in the beginning of the filtration cycle. Because particles in feed water see only the cake layer, initial membrane surface is not very influential to the filtration. The effective pore size of the cake layer is decided by the particles in feed water as shown in Figure 1.8. The cake layer formed on the membrane surface is called "dynamic membrane" that can be removed during membrane cleaning. It can typically reject most colloids such as single cell bacteria and viruses in MBR and surface water filtration, but it may not efficiently reject small biomolecules perhaps smaller than 100 to 300 kDa because the cake layer can provide large enough spaces to pass them depending on the compactness of it.

Dynamic membranes make pore size determination tricky. When membrane pore size is measured using surrogate particles such as latex or metal oxides particles, the concentration of surrogate particles in the test solution must be maintained as low as possible whereas the liquid velocity on the membrane surface is maintained fast enough to prevent cake layer formation. If a cake layer (or dynamic membrane) is formed, smaller particles than membrane pore can be rejected by the cake layer and the pore size will be underestimated.

Figure 1.9 illustrates the effect of dynamic membrane formation on permeate quality. In the experiment, 0.45 μm flat sheet submerged membranes (Yuasa Co., Japan) were used to filter 1.0 g/L

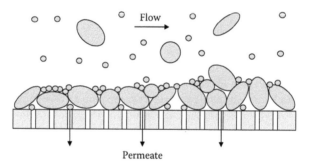

FIGURE 1.8   Dynamic membrane formation by particles existing in retentate.

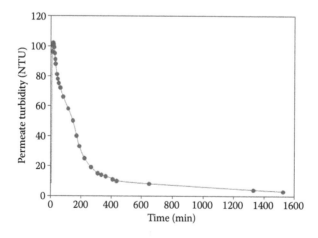

FIGURE 1.9   Time curve of permeate turbidity when 0.05 μm silica particles (1.0 g/L) are filtered by a flat sheet immersed membrane with 0.45 μm nominal pore size.

silica particles with a nominal size of 0.05 μm. In the beginning, the permeate turbidity was measured at around 100 nephelometric turbidity unit (NTU), but it gradually decreased to less than 10 within 6 h. Meanwhile, TMP increased from less than 5 kPa to more than 55 kPa at a constant flux of 25 LMH. These experimental results clearly demonstrate that the cake layer enables the membranes with a 0.45 μm nominal pore size to reject 0.05 μm particles at high TMP.

In many MF/UF, permeate quality is rarely affected by the membrane integrity due to the dynamic membranes unless the damage is catastrophic. If a pinhole is formed on a membrane, water flow rate to the damaged area immensely increases because the water flow is proportional to the fourth power of the pore diameter under the same pressure. Proportionally more solids are transported to the area and gradually plug up the pinhole, forming a dense cake layer that plays as a dynamic membrane. In composite hollow fiber membranes with laminated membrane layers on braids, the membrane film can be peeled off by mechanical and chemical damages. However, moderate damages are self-healed and hardly affect permeate quality because the large pores on fabric braids are easily plugged up by solids. Flat sheet submerged membrane modules typically have a woven or nonwoven fabric layer underneath the membrane and hence moderate damages on the membrane hardly deteriorate permeate quality. If pinholes are developed on the membrane layer, virus removal can be affected to some extent but it can potentially recover depending on the condition. In fact, in a study using RO membrane with pinhole damages, virus removal efficiency was recovered to near original levels (Mi et al. 2004). Similarly, if hollow fibers are cut during the filtration, lumens are eventually plugged up by solids and the effect on permeate quality is temporary.

In certain applications, a cake layer with a high permeability is formed at the beginning of the filtration cycle and it can act as a barrier against the deposition of finer particles. For instance, when yeast cells were added to bovine serum albumin (BSA) solution, the initially formed yeast cell layer on the crossflow MF membrane reduced flux decline by forming a dynamic membrane on the membrane surface (Güell et al. 1999). It is believed that the initially formed yeast layer reduced the BSA adsorption/deposition on the membrane surface while providing a relatively porous cake layer. It was also found that the loosely bound yeast layer on the membrane was easier to clean than the tightly bound BSA in the control experiment. A similar flux-enhancing effect was observed when BSA was filtered by membranes precoated with yeast cells (Arora and Davis 1994).

In some rare events, flux can be higher at lower crossflow velocities. It might be because the cake layer formed at lower crossflow velocity consists of more large particles, which makes the cake layer more porous. In fact, disproportionally more large particles deposit at low crossflow velocities according to the particle back-transport theory discussed in Section 1.2.5. On the contrary, at a high crossflow velocity, less particle deposition occurs on the membrane surface, but the hydraulic resistance of the cake layer can be greater because it consists of preferentially small particles under those conditions. In a pilot study, somewhat lower flux was observed at higher crossflow velocities when tubular ceramic membranes were used to filter the anaerobically digested sludge (Imasaka et al. 1993). A similar observation was made when a mixture of two different particles were filtered (Foley et al. 1995). However, these can be considered rare events that do not occur in typical filtration conditions. In general, higher crossflow velocities result in higher fluxes in most filtration processes.

When feed water is clear with low suspended solids content, cake layer formation is negligible and the membrane becomes the actual filtration barrier. Under these rare conditions, which are difficult to find in practical filtration, rejection efficiency is directly affected by the particle concentration in the CP layer for a given membrane. As crossflow velocity increases, the particle concentration in the CP layer decreases and the boundary layer becomes thinner. As a result, particle concentrations in the permeate also decrease. In a filtration study using bacteriophage $Q\beta$ and MS2 (Urase et al. 1993; Herath et al. 2000), the rejection efficiency of tubular ceramic membrane (Al_2O_3) increased as crossflow velocity increased, as shown in Figure 1.10. On the contrary, to some extent, a cake layer is always formed in MF and UF and acts as a dynamic membrane. As crossflow velocity increases, steady-state flux increases due to the thinning cake layer but the rejection efficiency can be affected somewhat negatively.

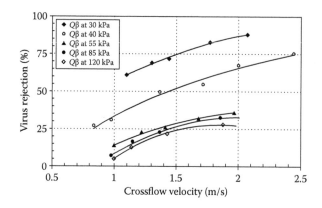

FIGURE 1.10 Bacteriophage $Q\beta$ rejection as a function of crossflow velocity at constant flux levels. (From Herath, G. et al., *J. Memb Sci.* 169(2):175–183, 2000.)

1.2.5 PARTICLE BACK-TRANSPORT

1.2.5.1 Overview

When particles are moving near the membrane surface, the drag forces associated with the axial (tangential) and the lateral (permeation) components of flow tend to carry particles along the streamline. At the same time, particles are subject to a number of different forces and effects that cause them to cross the streamline. Thermodynamic (or Brownian) diffusion, inertial lift, van der Waals attractions, charge repulsion, shear-induced diffusion, gravity settling, etc., are some of the known mechanisms as shown in Figure 1.11. Because the gravity settling effect is negligible compared with other effects, it can be neglected. The effective particle deposition velocity can be estimated from the difference between the permeation velocity (or flux) and the overall back-transport velocity.

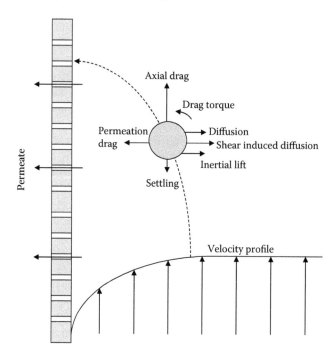

FIGURE 1.11 Forces and effects acting on a charged spherical particle suspended in a viscous fluid in laminar flow in the proximity of a flat porous surface.

The back-transport phenomenon has not been well theorized for the open channel system with a two-phase flow such as submerged membrane systems, but it has been extensively studied for closed channel systems with a single-phase flow such as tubular and plate and frame membranes. Although the magnitudes of back-transport velocities are different in open and closed channels, the basic principles must remain identical. It is worthwhile to review the back-transport theories developed for closed channel systems to understand the basic principles of submerged membrane filtration processes.

1.2.5.2 Brownian Diffusion

As described in Section 1.2.1, particles/solutes tend to accumulate near the membrane surface due to convective flow (or permeate flow) toward the membrane surface. At steady state, particle accumulation is balanced by diffusive particle back-transport to the bulk. The apparent back-transport velocity at steady state, v_D, can be expressed as Equation 1.8 (Zydney and Colton 1986; Wiesner and Chellam 1992). This equation is similar to Equation 1.2, where the prelogarithmic term in Equation 1.8 corresponds to the mass transfer coefficient, k (or D_{eff}/δ), in Equation 1.2.

$$v_D \propto \left(D_B^2 \gamma_w\right)^{\frac{1}{3}} \ln \frac{C_W}{C_B} \tag{1.8}$$

where
$\quad v_D$ diffusion induced back-transport velocity (m/s)
$\quad D_B$ Brownian diffusivity of particles (m²/s)
$\quad \gamma_w$ shear rate at the membrane surface (/s)
$\quad C_W$ particle concentration at the membrane surface (mg/L)
$\quad C_B$ particle concentration in bulk (mg/L)

The diffusivity, D_B, is calculated by the Stokes–Einstein equation as follows. This equation is valid only for small spherical molecules. It suggests that the diffusivity is proportional to the temperature and inversely proportional to the molecular size.

$$D_B = \frac{k_B T}{6\pi\mu r_p} \tag{1.9}$$

where
$\quad k_B$ Boltzmann constant (J/K)
$\quad T$ absolute temperature (K)
$\quad \mu$ fluid viscosity (kg/m/s)
$\quad r_p$ particle radium (m)

1.2.5.3 Shear-Induced Diffusion

Shear-induced diffusion is commonly observed when the viscosity of a concentrated rigid particle suspension is measured using a Couette viscometer, as shown in Figure 1.12a. When the bob (A) is spinning in the silicon oil suspended with rigid spheres at a very high concentration (45 v/v%), initially high viscosity decreases over time until it settles at a steady state, as shown in Figure 1.12b (Gadala-Maria and Acrivos 1980; Leighton and Acrivos 1987). This unusual observation was explained by shear-induced diffusion. In fact, the particles in the narrow spaces between the bob and the cup are exposed to much stronger shear fields than those in the center of the cup. Thus, the particles in the high shear field have many more chances to collide with each other and their effective diffusivity increases. Because the effective diffusivity is greater in the spaces between the bob

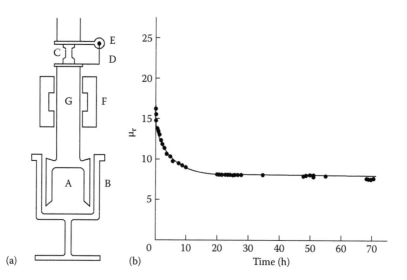

(a)   (b)   Time (h)

FIGURE 1.12 (a) Couette viscometer; (b) relative viscosity. Couette viscometer and the behavior of concentrated rigid particle suspension: (a) A, bob; B, cup; C, torsion bar; D, torsion arm; E, transducer; F, air bearing; G, torsion shaft. (b) Suspension concentration, 45% v/v; shear rate, 24/s; polystyrene spheres with 40 to 50 nm diameter. (From Leighton, D., and Acrivos, A. *J Fluid Mech.* 181:415–439, 1987.)

and the cup, more particles diffuse out from the space than those that diffuse into the space. The low particle concentration reduces the chance of transferring momentum from the spinning bob to the random particles in the liquid, which leads to low measurements of viscosity. The particle movement out of the in-between spaces increase as spinning frequency increases.

Similarly, shear-induced diffusion occurs in crossflow membrane filtration especially in the CP layer. Because the shear rate (not velocity) is the greatest on the membrane surface, shear-induced particle diffusivity is the greatest in the same place. In addition, the particle concentrations are the highest on the membrane surface, which increases the collision frequencies on the left side of the particles in Figure 1.3. The existence of the shear-induced diffusion makes the effective particle diffusivity greater than the Brownian diffusivity based on no-collision assumption (Eckstein et al. 1977). The shear-induced diffusivity, D_S, is expressed as the following (Zydney and Colton 1986):

$$D_S = 0.03\, r_P^2 \gamma_w \tag{1.10}$$

1.2.5.4 Inertial Lift

When liquid moves near the membrane surface, a velocity gradient forms, as shown in Figure 1.11. The liquid velocity is zero at the interface and maximum at the center of the flow channel. Because the liquid in the right side of the particle moves faster than that in the left side, particles spin counterclockwise in the figure. Particle spinning is the most vigorous near the membrane surface, where the velocity gradient is the steepest. On the contrary, in the center of the channel, the particles do not spin at all, but it moves fastest along the stream line of the convective flow.

In terms of the energy carried by the particle, the particles in the center carry the largest amount of kinetic energy in the form of velocity. On the contrary, the particles travelling near the membrane surface carry the largest amount of kinetic energy in the form of spin. A middle ground exists between the two extremes, where the total kinetic energy of the particles is at a minimum. Once the particles happen to enter into the lowest energy zone by Brownian or shear-induced diffusion, etc., they tend to stay in that area. As a result, the particle concentration near the channel surface tends to decline and this phenomenon is called "inertial lift." The "tubular pinch effect" is a consequence of the inertial

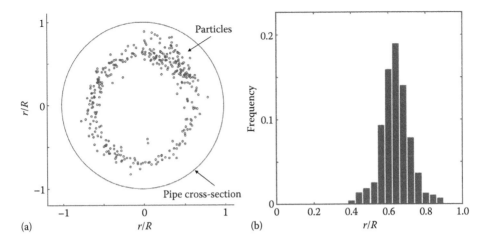

FIGURE 1.13 Experimental result demonstrating "tubular pinch effect" in the fluid flowing in a pipe at Reynolds number 67 and the ratio of channel and particle diameters is 9. (a) Particle distribution over a cross-section and (b) the corresponding histogram showing the probability as a function of dimensionless distance from the center of the pipe (r/R). (From Matas, J.-P. et al., *J Fluid Mech.* 515:171–195, 2004.)

lift and has been experimentally verified by monitoring the cross-sectional particle distribution in a pipe filled with a flowing medium. As shown in Figure 1.13, neutrally buoyant particles flowing in a pipe form a donut-shaped particle cloud somewhere in between the wall and the center. As flow velocity increases (or Reynolds number increases), particles tend to be off from the ring and are dispersed more evenly in the pipe due to turbulence, but the tendency of staying away from the wall remains.

Mathematically, the particle Reynolds number should not be negligible to have a significant inertial lift effect so that the nonlinear inertial terms in the Navier–Stokes equations play a role (Belfort et al. 1994). The lift velocity of a rigid, neutrally buoyant, freely rotating particle in a plain Poiseuille flow, v_L, is given by

$$v_L \propto r_p^3 U_m^2 \tag{1.11}$$

where
  v_L lift velocity (m/s)
  r_P particle radius (m)
  U_m maximum flow velocity in channel (m/s)

1.2.5.5 Total Back-Transport Velocity and Critical Flux

The particle back-transport induced by Brownian diffusion is thermodynamically motivated and is effective whenever a concentration gradient exists. The resulting particle back-transport is weaker for larger particles due to the smaller Brownian diffusivity. On the other hand, the particle back-transport induced by shear-induced diffusion and the lateral migration is kinetically motivated and is effective only when fluid movement causes a velocity gradient. The resulting particle back-transports are stronger for large particles than for small particles due to the higher chances of collision and the larger inertia moment. It is not completely rigorous to linearly add up those three different back-transport effects with different origins, but it might be worthwhile to do so to obtain a rough idea of the total back-transport velocity that the particles in moving fluid experience.

The back-transport velocity curves for the three different effects are plotted in Figure 1.14 for the condition. It was assumed that average flow velocity (U_m) was 0.24 m/s, channel height was 3×10^{-3} m in plate and frame module, and particle density (ρ) was 5745 kg/m³. The total back-transport velocity is the lowest for particles that were approximately 0.5 μm diameter. If liquid velocity

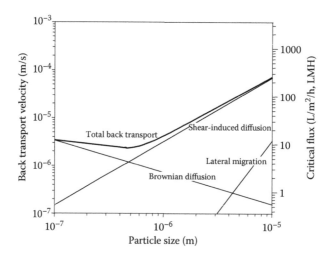

FIGURE 1.14 Back-transport velocity of iron oxide particles as a function of particle size in a slit channel. $U_m = 0.24$ m/s, $T = 298$ K, $\mu = 0.001$ kg/m/s, channel height $= 3 \times 10^{-3}$ m, $\rho = 5745$ kg/m³, and $k_B = 1.38 \times 10^{-23}$ J/K. (From Yoon, S.-H. et al., *J Memb Sci.* 161(1–2):7–20, 1999.)

increases, all the lines move up and the total back-transport velocity increases. In the meantime, the size of the particle that has the lowest back-transport velocity decreases because the lines for shear-induced diffusion and lateral migration move up more than the line for Brownian diffusion.

One important consequence of particle back-transport is that particle deposition does not occur until the convective flow toward the membrane surface exceeds the back-transport velocity. Here, the convective flow velocity is equivalent to the flux by definition. Therefore, the maximum flux obtainable without particle deposition is equivalent to the particle back-transport velocity. The critical flux is defined as the maximum flux obtainable without particle deposition, as will be discussed in Section 1.2.6. For instance, the back-transport velocity of a 1-μm particle is 4.2×10^{-6} m/s in Figure 1.14, and it corresponds to the critical flux of 15 LMH. Therefore, the particle will not deposit on the membrane unless the flux exceeds 15 LMH.

1.2.5.6 Effect on Cake Layer Structure

Particle back-transport causes a stratification of cake layer with large particles on the bottom and small particles on top. This phenomenon is more apparent in the constant pressure mode, where flux declines gradually over time. Under declining flux condition, large particles with high back-transport velocities can mainly deposit at the beginning of a filtration cycle because the convective flow velocity toward the membrane can exceed the back-transport velocity. However, the deposition of large particles becomes scarcer toward the end of the filtration cycle, where the slower water permeation at the low flux condition fails to exceed the back-transport velocity.

The probability of particle deposition (P_d) can be calculated by dividing the effective particle deposition velocity ($J - v_T$) by the convective flow caused by flux as expressed in Equation 1.12, where the higher the total back-transport velocity is, the lower the probability of deposition is.

$$P_d = \frac{J - v_T}{J} \qquad (1.12)$$

where

    P_d probability of particle deposition (–)
    J flux (m/s or LMH)
    v_T total back-transport velocity (m/s)

Figure 1.15 shows the size distributions of the particles that deposit on the membrane surface as a function of flux. The areas underneath the curves represent the relative amount of particles that deposit. It is apparent that fewer particles deposit at low fluxes. Simultaneously, the average size of the particles depositing on the membrane also decline as flux decreases. As a consequence of the gradual change of the particle sizes among the deposits, the cake layer becomes stratified in constant pressure mode, as illustrated in Figure 1.16. The bottom part of the cake layer consists of large and small particles, but the top consists of mainly small particles. On the contrary, cake layer stratification does not occur in constant flux mode, where the convective flow that brings particles down to the membrane is maintained constant. However, as discussed in Section 1.2.3, the cake layer still appears stratified as a consequence of more severe cake layer compaction in the bottom of the cake layer especially when soft particles are filtered.

Conceptually, cake layer resistance can be reduced by manipulating the TMP at the beginning of the filtration cycle. The deposition of large particles can be induced by allowing a high flux temporarily in the beginning and use the cake layer as a barrier for finer particle deposition under a lower flux. In addition, if the large particles in the bottom of the cake layer are less prone to the cake layer collapse than small particles, they can delay cake layer compaction. This mechanism may explain the reasons why the quick initial cake layer formation delayed flux loss in the long run when anaerobically digested wastewater was filtered by tubular ceramic membranes (Imasaka et al. 1993).

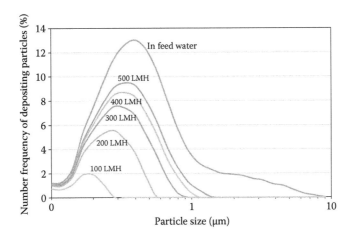

FIGURE 1.15 Size distribution of the particles that deposit on the membrane surface as a function of flux under the conditions described in the caption of Figure 1.14, in which the area below the curve represents the amount of particles deposit.

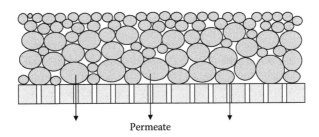

FIGURE 1.16 Side view of the cake layer formed under constant pressure mode, where less large particles deposit as flux decreases toward the end of the filtration cycle.

1.2.6   Critical Flux

1.2.6.1   Definition

The original definition of critical flux is the maximum flux obtainable with no foulant deposition on the membrane surface. At or below the critical flux, membrane permeability remains identical to the clean membrane's. However, it turns out that such a condition does not exist in practical filtration, where feed water with various foulants with various properties is filtered. Due to the nonexistence of rigorous critical flux in practical condition, the critical flux was split into two categories, namely, strong form and weak form (Field et al. 1995). The definition of the strong form is the same as the original definition of critical flux, and the weak form was defined as the maximum flux obtainable with the initial adsorption of macromolecules or colloids. In fact, particles/macromolecules adsorb on membrane surface as soon as the feed water comes into contact with membrane due to electrostatic and van der Waals interactions (Le-Clech et al. 2006). Once a thin adsorption layer is formed on the membrane surface, further adsorption may not occur as long as the operating flux is at or below the critical level. Under these conditions, membrane permeability would be nearly stable but somewhat lower than clean membrane's due to the thin adsorption layer.

In MBR, even the weak form critical flux turned out to be nonexistent. The mixed liquor of MBR contains various macromolecules with various charges, functional groups, sizes, etc. Some of the components can interact with the membrane surface as soon as the mixed liquor comes into contact with the membrane. In addition, some macromolecules with very low back-transport velocity not only continue to deposit at any flux condition (Zhang et al. 2006) but also chemically interact with the cake layer initially formed. As a consequence, membrane fouling can occur gradually even at extremely low flux and thus the true weak form critical flux does not exist in MBR.

Therefore, the term, "sustainable flux" is often used instead of critical flux to indicate the flux that lasts a considerable length of time without causing significant membrane fouling (Fane et al. 2002). The definition of sustainable flux is fuzzy because neither a standard protocol nor a consensus on the minimum duration of the sustainable flux exists. However, it can be considered that the sustainable flux is the maximum flux obtainable without frequent membrane cleaning that can interrupt efficient system operations under the given condition. For this criterion, the maximum flux sustainable for a month or more without chemical cleaning might be qualified for the sustainable flux in MBR. The minimum duration of the sustainable flux can be shorter for other membrane processes if more frequent membrane cleaning is acceptable.

1.2.6.2   Measurement

As discussed in the previous section, particle and macromolecule deposition occurs continuously regardless of the flux in the practical filtration condition in MBR. The true critical flux does not exist whether it is a strong form or a weak form at the condition, but the maximum flux obtained without a significant permeability loss in a short period, for example, 5 to 20 min, is often called a critical flux.

Flux stepping methods are used to measure the critical flux starting from the low flux in MBR. After tracing the TMP at a fixed low flux for 5 to 20 min under regular filtration conditions, filtration is stopped for 1 to 5 min to relax the membrane while aeration is performed normally. Filtration is resumed at a slightly higher flux by 3 to 10 LMH for 5 to 20 min. This filtration and relaxation cycle is repeated until a significant TMP increase is observed. The highest flux that does not cause a significant TMP increase during the filtration cycle is determined as a critical flux.

Figure 1.17 shows a critical flux test result obtained from a full-scale municipal MBR plant with an average daily flow (ADF) of 2300 m³/day. In this study, 12 min on and 3 min off cycles were repeated while elevating the flux step by step. As shown in the figure, TMP increased slowly at a rate of 2.0 kPa/h at 25 LMH, but the TMP increasing rate did not change much in the next high flux, that is, 34 LMH. This trend continued until flux reached 51 LMH, where the TMP increasing rate was 4.2 kPa/h. However, when flux became 60 LMH, TMP increasing rate suddenly jumped to 8.7 kPa/h. In this case, the critical flux was determined to be between 51 and 60 LMH. It is

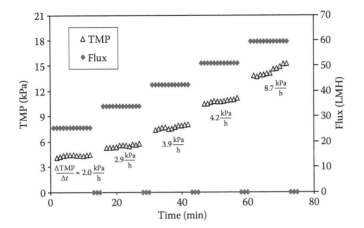

FIGURE 1.17 A critical flux test with flat sheet membranes (Kubota Co., Japan), influent BOD, 200 ppm; design flow rate, 2300 m³/day; MLSS, 12,000 mg/L; water temperature, 13°C; operational mode, 12 min suction and 3 min relaxation. (Data from Yoon, S.-H., and Collins, J.H., *Desalination* 191:52–61, 2006.)

noticeable that the measured critical fluxes are somewhat arbitrary because the extent of TMP increase in each flux level can vary depending on filtration duration, relaxation time, air flow rate, mixed liquor condition, initial membrane used, etc. Therefore, care must be taken when two critical fluxes from two different sources are compared.

1.2.7 EFFECT OF TMP ON FLUX

1.2.7.1 Theory

If the flux remains below a critical level, no significant membrane fouling occurs because particles are back-transported away from the membrane surface faster than they are dragged by the convective flow. When the flux is below the critical flux, it is linearly proportional to TMP, as shown by the pressure-controlled region in Figure 1.18. If the TMP increases beyond the pressure-controlled region, particles/solutes start to deposit on the membrane surface and create additional permeation barriers. Therefore, the TMP–flux curve departs from the linear line and the filtration enters into

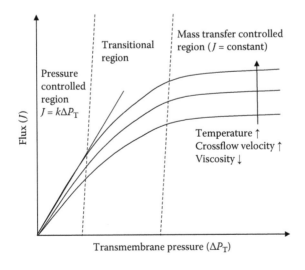

FIGURE 1.18 Conceptual diagram of the relation between TMP and flux in short-term filtration.

a transitional region. In this region, the specific flux gain for incremental TMP increase becomes smaller as TMP increases.

If TMP increases further beyond the transitional area, cake layer compaction starts to occur and the filtration enters into a mass transfer–controlled region. The pressure loss through cake layer acts as a force compressing the cake layer as explained in Section 1.2.3. In this region, the incremental gain in flux by increasing TMP can be completely offset by the increasing specific cake resistance. Therefore, flux remains almost constant regardless of TMP. Because flux is decided by the properties of water, such as diffusivity and viscosity rather than TMP, this region is called the mass transfer–controlled region. In this region, the flux can even decrease as TMP increases depending on the extent of the cake layer compaction. TMP is not a factor influencing flux in the pressure-controlled region, but high temperature and high crossflow velocity can increase the flux by lowering water viscosity and controlling cake layer depth, respectively.

The insensitivity of flux to TMP in the pressure-controlled region stems from the self-limiting nature of flux, as illustrated in Figure 1.19. When filtration is in a pseudo-steady state and flux stays nearly constant under constant pressure, increasing TMP causes an immediate flux increase. The increased flux not only brings more particles onto the cake layer but also expedites cake layer compaction following greater pressure loss through the cake layer, which acts as a force pressing the cake layer. Consequently, flux decreases rapidly until the new steady state is reached with a more compact and thicker cake layer (Fane 2007). If membrane permeability is increased by modifying the membrane chemistry and the manufacturing method, the high membrane permeability may allow a higher initial flux at the same TMP. However, the high flux causes more foulant deposition and in turn quicker cake layer compaction. As a consequence, the operating flux tends to converge to a certain level depending on the nature of the application despite the high membrane permeability. Therefore, the membrane permeabilities are rarely a factor determining the operating flux in porous membrane filtration as long as they are above a certain level.

Figure 1.20 shows flux and TMP profiles observed in chicory juice filtration using a disc membrane module with a rotating turbulent promoter. When a UF membrane with 100 kDa MWCO is equipped, the flux linearly increases as TMP increases at a low TMP range, but it enters into the

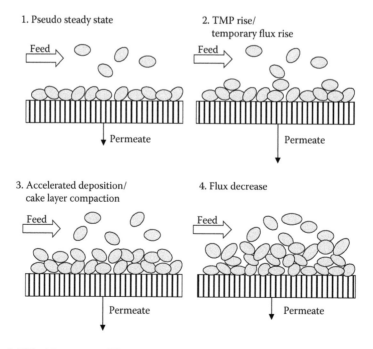

FIGURE 1.19 Self-limiting nature of flux.

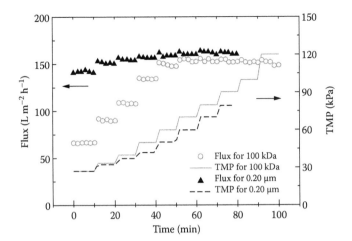

FIGURE 1.20   Flux profiles in TMP stepping experiment for 100 kDa and 0.2 μm membranes using a fixed disc membrane with rotating turbulence promoter at 1000 rpm in chicory juice filtration. (From Luo, J. et al., *J Memb Sci.* 435:120–129, 2013.)

mass transfer–controlled region after it reaches approximately 150 LMH at a TMP of 50 to 60 kPa. In the meantime, when MF membrane with 0.2 μm pores is equipped, flux is high even at a low TMP due to the high membrane permeability. Because the filtration is already in the mass transfer–controlled region at low TMP, only marginal flux increases are realized by increasing TMP. It is noteworthy that the maximum fluxes at the mass transfer–controlled region are nearly the same for MF and UF membranes despite the large difference in membrane permeability. It is because flux is determined mainly by the cake layer resistance rather than membrane resistance in the mass transfer–controlled region.

1.2.7.2   Where Do Submerged and Sidestream Membranes Stand in the Curve?
Submerged membranes run at the transitional region in the early phase of the filtration cycle with a moderate cake resistance. Flux can be increased by increasing vacuum pressure because cake resistance is not too large compared with the membrane resistance. Under these conditions, membrane fouling occurs very slowly, but particle/macromolecule deposition is not completely avoidable. The thin cake layer eventually collapses and void spaces in the cake layer are plugged up by the intrusion of very small particles. The loss of the cake layer porosity triggers the sudden TMP increase, as will be discussed in Section 1.3.2.1. As a consequence, filtration enters the mass transfer region as the filtration cycle proceeds to the end, where the membrane permeability is not a significant factor affecting the membrane performance.

The sidestream membranes running under crossflow mode at a positive pressure are operated largely at a mass transfer–controlled region with the possible exception of the small area near the concentrate exit. Although the membrane needs only a very low TMP to obtain the design flux at less than 0.1 to 0.2 bar, feed side TMP must be maintained more than an order of magnitude higher (3–6 bar) to obtain a sufficient crossflow velocity, for example, 1 to 4 m/s, in the membrane channel. As a result, the excess pressure applied to the channel entrance causes a severe cake layer compaction and the corresponding permeability loss. The extent of the cake layer compression decreases toward the exit of the membrane channel due to the lower TMP.

1.2.8   Effect of Membrane Resistance on Operating Flux

In general, the resistance (or permeability) of the membrane itself is not a major factor dictating the operating flux in crossflow filtration as discussed in the previous section. It is simply because membrane resistance takes only a small portion of the total filtration resistance in practical filtration condition. Instead, cake/fouling resistance, as defined in Section 1.2.2, is responsible for the vast majority of the filtration resistance. Figure 1.21 shows an example of filtration resistance breakdown in a tubular membrane running at 100 LMH, assuming 5.5 bar and 0.5 bar as TMP in the inlet and outlet, respectively. Membrane resistance takes only less than 2% and 20% of the total filtration resistance in the inlet and outlet of the membrane, respectively.

In submerged membrane filtration, membrane resistance takes a relatively larger portion of the total filtration resistance when it is compared with the crossflow filtration running under positive pressure. Because submerged membranes are operated at a much lower flux, therefore less compact cake layers are formed. Nonetheless, cake resistance is greater than membrane resistance with a potential exception in the very beginning of a filtration cycle. In addition to cake resistance, hollow fiber–based submerged membranes suffer from internal pressure loss caused by the permeate flow in the narrow lumen. As will be discussed in Section 1.2.6, the high permeability of the hollow fiber membrane does not always lead to high flux due to the performance loss by the internal pressure loss. Some degree of imbalanced filtration along the fiber is inevitable due to the TMP gradient along the hollow fiber. The imbalanced filtration expedites membrane fouling near the permeate exit by enhancing particle deposition and cake layer compaction. As a result, the initially high flux near the permeate exit is self-corrected to a lower level. The loss of flux near the permeate exit is compensated by the increased flux in the upstream. Eventually, the membrane fouling propagates to the upstream while TMP increases. If a membrane with high permeability is used at a high flux, the combined effect of the self-limiting flux behavior and the internal pressure loss expedites the membrane fouling and brings the flux back to the original level. Therefore, the membranes with high permeability can be better used to save energy by reducing TMP at the same flux instead of being used to obtain a higher flux in general.

In submerged membrane processes, however, only insignificant savings in energy costs can be realized by reducing TMP because the TMP is low from the beginning (e.g., <0.3 bar). For instance, according to the energy breakdown discussed in Section 6.15.6, only approximately 7% of the total plant energy consumption goes to the permeate pump in the MBR equipped with a submerged membrane. Moreover, this 7% includes the energy required to overcome not only the membrane resistance but also the cake layer resistance and the head pressure in the permeate discharge pipe. If the true energy required to overcome the membrane resistance, it should be much less than 1% of the total plant energy consumption. Even if the membrane itself does not have resistance at all, the expected maximum energy savings is well below 1%. Therefore, manufacturing highly permeable membranes is not the primary goal for submerged membrane processes as long as the membrane

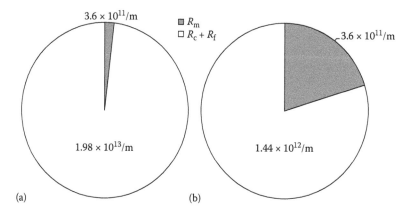

FIGURE 1.21 Breakdown of filtration resistance in tubular membrane process assuming inlet pressure 5.5 bar, outlet pressure 0.5 bar, average flux 100 LMH, and membrane flux 1000 LMH at 1 bar: (a) at entrance and (b) at exit.

has a reasonable permeability that does not cause a bottleneck during filtration. Meanwhile, high membrane permeability is not a determining factor for membrane performance in tubular membrane processes because cake layer resistance takes a vast majority of the total filtration resistance.

In contrast with the porous membrane, membrane permeability is an important factor affecting the performance of nonporous membranes such as RO and nanofiltration (NF) membranes because a large portion of the driving force (or TMP) is used to overcome membrane resistance. For example, if clean water flux was 35 LMH at 15 bar with a RO membrane, the operating flux can be 25 LMH at the same TMP in surface (or brackish) water filtration. In this case, the resistance caused by the membrane itself takes approximately 70% of the total resistance. Therefore, the same flux can be obtained at lower TMP if RO membrane permeability can be increased. In fact, by improving membrane permeability, the operating pressure is reduced from 15 bar to 10 bar to 6 bar while maintaining the same flux in recent decades. Conceptually, the high membrane permeability could be used to obtain higher fluxes in nonporous membrane processes to save capital costs and increase water recovery. But, at the elevated flux, salt concentration increases in the CP layer and the net driving force decreases due to the increased osmotic pressure. Simultaneously, the elevated salt concentration in the CP layer can expedite scale formation. The higher flux also expedites organic deposition and makes the cake layer more compact. Due to the osmotic pressure effect and the expedited fouling, the initially high flux decreases; thereby the initial gains can be wiped out. As a result, highly permeable RO membranes are used to lower the TMP rather than to increase the flux.

1.2.9 EFFECT OF TEMPERATURE ON FLUX

If all other variables in Equation 1.4 are constant, flux is inversely proportional to the viscosity of the permeate passing through the membrane pores in porous membrane processes. Because the MBR permeate contains little dissolved matter, its viscosity is virtually the same as that of clean water. The temperature effect on flux can be eliminated by normalizing the observed flux against the reference temperature. Typically, flux is corrected against 20°C using the following equations. It is assumed that the temperature effect is limited to the permeate passage through membrane pores and the flux is inversely proportional to the permeate viscosity following the Hagen–Poiseuille equation.

$$\mu_T = -9.802 \times 10^{-6}\, T^3 + 1.130 \times 10^{-3}\, T^2 - 5.793 \times 10^{-2}\, T + 1.785 \tag{1.13}$$

$$J_{T_0} = \frac{\mu_T}{\mu_{T_0}} J_T \tag{1.14}$$

where
- μ_T water viscosity at current temperature (cP or 10^{-3} kg/m/s)
- μ_{T_0} water viscosity at reference temperature (cP or 10^{-3} kg/m/s)
- J flux observed (LMH or gfd)
- J_{T_0} flux at reference temperature (LMH or gfd)
- T water temperature (°C)
- T_0 reference temperature (°C)

A more compact equation is also available as Equation 1.15.

$$J_{T_0} = J \left[\frac{42.5 + T_0}{42.5 + T} \right]^{1.5} \tag{1.15}$$

Although the above normalization equations are widely used to trend the flux (or permeability) in MF and UF processes, they suffer from a significant limitation. The above equations only count on the direct effect of permeate viscosity on flux, but temperature also affects flux indirectly by altering feed (or mixed liquor) viscosity, biological activity, dissolved oxygen concentration, and even microbial metabolism. For instance, at low temperature, high mixed liquor viscosity hampers turbulence on membrane surface and increases particle deposition. The increasing particle deposition in turn increases fouling resistance (R_c in Equation 1.4), which decreases flux more than expected based on the permeate viscosity. Because the above equation does not count on the effect of feed water characteristics and hydrodynamics, substantial errors are developed as the operating temperature deviates from the reference temperature. Therefore, the equation is meaningful only in a narrow range around the reference point. For instance, when temperature decreases from 30°C to 5°C in a full-scale MBR plant, the sustainable average daily flow (ADF) declines by 63% from 2650 m³/day to 984 m³/day (GE Water 2011). In contrast, viscosity declines by only 48% in the same temperature range.

1.2.10 FILTRATION THEORY FOR NONPOROUS MEMBRANE

NF and RO membranes share a large part of the filtration theory with MF and UF, but they are distinguished by a few unique filtration mechanisms due to their nonporous nature. In nonporous membranes, polymeric chains are randomly aligned and tangled with each other as well as with the tiny intermolecular spaces among them. Only the species smaller than the intermolecular spaces can enter into the membrane and this size exclusion mechanism is common for both nonporous and porous membranes. Meanwhile, charged species hardly enter into the intermolecular spaces even if they are small enough due to the elevated chemical potential in the intermolecular spaces. The rejected species on the membrane surface form a CP layer that increases the effective ionic concentration on the membrane surface beyond the concentration in the bulk. The difference in the ionic concentrations in between the CP layer and the permeate causes osmotic pressure that counteracts TMP.

The inability of ions to pass through nonporous membranes is explained by charge interaction. According to Coulomb's law, as shown in Equation 1.16, the force acting between two charged bodies is proportional to the magnitude of the charges (q_1 and q_2), but is inversely proportional to the dielectric permittivity of solvent ($\varepsilon_0 \varepsilon_r$) and the square of the distance (r^2). Here, dielectric permittivity is the measure of the ability of the solvent shielding the electrical charge of solutes by surrounding them. Molecular size and mobility, dipolar polarizability, molecular conformation, etc., are the factors affecting dielectric permittivity. The relative dielectric permittivity (ε_r) is a ratio of the dielectric permittivity of solvent/material to the permittivity of vacuum (ε_0). Because water has exceptionally high ε_r at around 78, it can effectively shield the charges of solutes. As a result, NaCl can easily dissolve into the water as Na^+ and Cl^- because water molecules shield the charges and prevent them from combining. On the contrary, only a negligible amount of NaCl dissolves in olive oils with ε_r of around 3, where the species with opposite charges strongly attract each other.

In nonporous membranes, small charge-neutral species such as water molecules can be forced to dissolve into the intermolecular spaces of polymeric chains using hydraulic pressure, as illustrated in Figure 1.22. Meanwhile, dissolving ionic species into the same spaces is much harder because the energy state of ions is disproportionally higher in polymers than water. As a consequence, the low dielectric permittivity of the membrane acts as a permeation barrier for charged species relative to water molecules. For instance, the relative electric permittivity of polyamide is only approximately 3, which is only a fraction of the permittivity of water (i.e., 78).

The rejection of charged species is further enhanced by hydration. In water molecules, the electron clouds are partially shifted toward the oxygen molecule due to the large difference in electronegativities. Thus, oxygen atoms have partial negative charges whereas hydrogen atoms have partial positive charges. In addition, two hydrogen atoms and one oxygen atom are not aligned linearly but are connected in a V-shape with a 104.5° angle; thereby water molecules retain a strong net dipole moment. The small water molecule with a strong dipole moment can effectively surround ionic

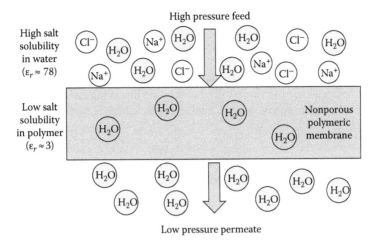

FIGURE 1.22 Ion rejection mechanism in RO membrane.

species to counter the charge and increase the effective ion radius. The enlarged ions are less likely to penetrate into the polymeric membrane due to the size exclusion mechanism. Divalent ions are rejected at much higher efficiencies than monovalent ions because they tend to be hydrated by more water molecules and hence form larger clusters.

This preferential permeation of water through nonporous membrane has been modeled in various ways depending on the viewpoint and the purpose. The solution diffusion model, preferential sorption-capillary flow model, surface force-pore flow model, and Donnon exclusion model are some examples. Detailed reviews on such models are found in Rautenbach and Albrecht (1989). It is beyond the scope of this book, but the important consequences of the unique separation mechanisms of nonporous membranes are briefly listed as follows:

- Water and solute passages rely on their own solubility and dissolution kinetics in polymeric membranes rather independently. Although water can dissolve faster or slower in polymeric membranes as TMP changes, salt dissolution rates are less affected by TMP. As a result, the apparent ion rejection efficiencies tend to increase as water flux increases. Temperature tends to affect the solute activities more than the water activity so that solute rejection efficiency tends to decrease as temperatures increase.
- Dynamic membrane (or cake layer) improves particle rejection efficiency in MF/UF, but the same concept does not apply for NF/RO. It is because the cake layer is not capable of rejecting the small species rejected by NF/RO. In addition, the cake layer increases the effective ionic concentration on the membrane surface by interfering with ion back-diffusion into the bulk. Due to the increased ion concentration on the membrane surface, more ions can pass through the membrane under the presence of the cake layer.
- Divalent ions are even less soluble than monovalent ions in a medium with low electric permittivity due to their stronger charge interaction. Therefore, rejection efficiency is much higher for divalent ions than for monovalent ions.
- Li^+ is smaller than Na^+, but rejection efficiencies are nearly comparable because rejection does not rely on the size exclusion mechanism.
- High hydraulic pressure is required to push water molecules through the small intermolecular spaces of a polymer. Cake layer resistance is only a minor portion of the total filtration resistance in typical conditions. Therefore, membranes with a high permeability are beneficial in reducing the operating pressure and saving the energy costs unlike in porous membrane filtration (MF and UF).

- Charge-neutral organic molecules can pass nonporous polymeric membranes relatively easily due to the lack of charge. Rejection efficiencies of small charge-neutral molecules, for example, methanol, ethanol, propanol, and acetone are known to be poor despite the larger sizes than inorganic ions.
- Rejection efficiency of the molecules with varying charges is dependent on pH. Ammonia rejection efficiencies are high at low pH, but they are low at high pH because the molecules lose charges. On the contrary, organic acid rejection is high at high pH and low at low pH.

$$F = \frac{1}{4\pi\varepsilon_0\varepsilon_r} \frac{q_1 q_2}{r^2} \qquad (1.16)$$

where
F force between two charged particles (N)
ε_0 absolute dielectric permittivity in vacuum ($C^2/m^2/N$)
ε_r relative dielectric permittivity (–)
q_1, q_2 charges of particle (C)
r distance between two particles (m)

1.3 MEMBRANE FOULING

1.3.1 CONSTANT FLUX VERSUS CONSTANT PRESSURE

All submerged membranes are operated under constant flux mode, where the permeate is drawn at a fixed rate using permeate pumps. As membranes are fouled, the vacuum pressure increases to compensate for the permeability loss. On the contrary, housed membranes are operated either at constant pressure mode or at constant flux mode depending on the application and the module structure. For example, most tubular or plate and frame membrane systems are operated at constant pressure modes whereas some housed hollow fiber membranes are operated under constant flux modes.

Operating membranes at constant flux mode is inherently advantageous from the perspective of membrane fouling control and energy efficiency. This is because expedited membrane fouling can be easily averted by avoiding excessively high flux in the constant flux mode by controlling the permeate pump output. On the contrary, under constant pressure mode, not only is cake layer formation expedited at high flux but also the cake layer becomes more compact as discussed in Sections 1.2.3 and 1.2.6. In this filtration mode, flux is very high at the beginning of the filtration but declines quickly. Initial TMP is set at a level that allows enough permeate flow rate after the initial sharp flux decline. Due to the high flux in the beginning, particles in water are dragged toward the membrane surface much faster than they are back-transported; thereby more particles deposit per permeate volume than in constant flux modes. In addition, the cake layer is quickly compressed at the very beginning of the filtration. If the particles are soft and compressible, cake layer compaction becomes more significant and the flux declines faster.

Figure 1.23 illustrates the relation between TMP and flux over the course of filtration in constant flux mode, constant pressure mode, and semiconstant pressure mode. In constant flux mode (Figure 1.23a), initial flux is kept low and maintained at the same level whereas TMP is allowed to increase. As will be discussed in Section 1.3.2.1, TMP increase is moderate in the beginning but it accelerates toward the end of the filtration cycle. In constant pressure mode (Figure 1.23b), initial flux is high due to high TMP but it decreases rapidly before it stabilizes. To minimize the rapid membrane fouling, low TMP is applied in the beginning and it is increased stepwise in semiconstant pressure mode (Figure 1.23c). In this modified constant pressure mode, TMP is adjusted manually or automatically to obtain sufficient permeate flow.

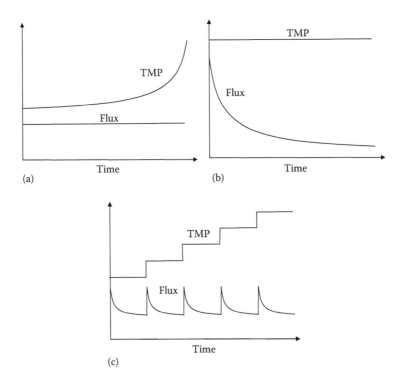

FIGURE 1.23 Conceptual diagram of three different modes of membrane operation: (a) constant flux mode, (b) constant TMP mode, and (c) semiconstant TMP mode.

1.3.2 MEMBRANE FOULING ROADMAP

1.3.2.1 Constant Flux Mode

The course of membrane fouling under constant flux modes in submerged membrane filtration can be split into the following three stages (Cho and Fane 2002; Zhang et al. 2006):

i. *Stage 1: Initial adsorption*

Most feed waters in MF/UF contain various contaminants with various characteristics, but the MBR mixed liquor contains particularly diverse contaminants originating from the wastewater, microorganisms, and chemicals added to the system. An initial short-term increase of TMP occurs due to the adsorption of macromolecules on the membrane surface, for example, soluble microbial products (SMP) and extracellular polymeric substances (EPS). Because the adsorption is primarily driven by chemical and physical interactions between membranes and macromolecules, the initial membrane fouling occurs even at zero flux as soon as the membrane comes into contact with mixed liquor. Adsorptive fouling might be less with more hydrophilic membranes because the bound water molecules on the membrane surface interfere with macromolecules directly contacting the membrane surface.

In Figure 1.24, the initial TMP is approximately 5 kPa, whereas the flux was 25.7 LMH, which can be translated to a permeability of approximately 5 LMH/kPa. Because the initial permeability of the membrane in clean water is approximately 15 LMH/kPa, the initial permeability loss in the very beginning of filtration cycle is approximately 67%.

ii. *Stage 2: Slow TMP increase*

A long-term increase of TMP occurs mainly due to the continuous deposition of EPS and SMP. In addition, the gradual compaction of the deposit layer also contributes to the TMP

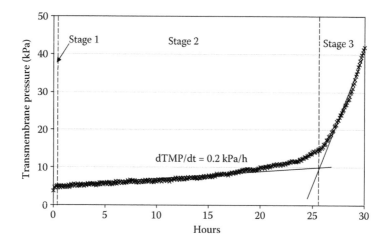

FIGURE 1.24 Typical TMP rising pattern at 15 LMH in submerged membrane filtration with Yuasa's flat sheet submerged membrane. MLSS = 12 g/L, 20°C.

increase. Because EPS and SMP have very low critical flux, they continuously deposit on the membrane surface even at a low flux. The TMP increase can be either linear or weakly exponential at this stage. In the figure, TMP increases at a rate of 0.2 kPa/h.

iii. *Stage 3: Sudden TMP increase (or jump)*

A sudden TMP increase can hardly be explained by sudden changes in operational parameters such as feed properties, rapid cell lyses, and other environmental changes because it also occurs in laboratory-scale experiments, of which operational parameters are rigorously controlled. The sudden TMP increase can be best explained by cake layer compaction (Chang et al. 2006a; Park et al. 2006; Fane 2007). Once pressure loss through the cake layer reaches a critical level, cake layer compaction starts from the bottom of the cake layer, where the force pressing (or squeezing) the cake layer is the strongest due to the cumulative nature of the downward compression force, as explained in Section 1.2.3. As cake layer compaction proceeds, TMP must increase to compensate for the permeability loss under constant flux mode. This in turn accelerates cake layer compaction by increasing the compression force. As a result, once TMP reaches a threshold, TMP increase accelerates by itself, as shown in Figure 1.24. Many other theories on sudden TMP increases exist and those are discussed in Section 1.3.2.3.

1.3.2.2 Constant Pressure Mode

Tubular and plate and frame modules that run under crossflow mode at positive pressure fit in this category. TMP is increased stepwise in typical processes whenever permeate flow rate falls below the target level and the filtration cycle ends when permeate flow is not sufficient at the highest TMP allowed. In this filtration mode, sudden performance drop is not apparent unlike in the constant flux mode simply because the performance drop occurs at the very beginning of the filtration cycle before it is even noticed. The flux behaviors in the two different operating modes appear different but the underlying principles are identical except for the time scale. The course of membrane fouling can be split into the following three stages:

i. *Stage 1: Rapid cake layer formation and compaction*

Due to the high TMP and flux in sidestream membrane process, initial flux is much higher than the critical flux. As a consequence, most of the particles contained in the feed are

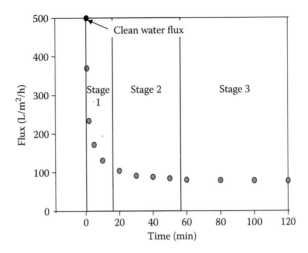

FIGURE 1.25   Typical flux curve of tubular membrane at a crossflow filtration mode under positive pressure.

deposited on the membrane when the particle-free permeate is lost through the membrane. Due to the high flux, cake layer compaction starts immediately after the filtration starts, as shown in Figure 1.25. This stage lasts for only a short period and the rapid permeability loss in the very early phase of the filtration is invisible in ordinary systems with ordinary monitoring equipment.

ii. *Stage 2: Slow cake layer growth with continuous cake layer compaction*
As filtration proceeds, filtration resistance increases with a growing and compacting cake layer. The particle deposition rate decreases partly because the decreasing flux causes less particle migration toward the membrane surface and partly because the flux approaches the critical flux, as discussed in Section 1.2.6. As a consequence, cake layer growth rate slows down after the initial rapid growth in stage 1. Other consequences of the lowering flux is that cake layer compaction slows down according to the cake layer compaction theory discussed in Section 1.2.3. It is noticeable that the permeability loss slows down as permeability declines in constant pressure mode. On the contrary, permeability loss accelerates as permeability declines in constant flux mode as discussed in the previous section.

iii. *Stage 3: Pseudo-steady state*
Once the flux decreases to a sufficiently low level, particle deposition becomes scarce because particle back-transport velocity matches the particle deposition velocity. In this stage, the cake layer grows very slowly and the permeability stays at a fairly stable level. However, gradual cake layer compaction still occurs, especially at the bottom of the cake layer, causing a long-term gradual permeability loss. In addition, fine particles or macromolecules carried by permeate continue to deposit in the void spaces in the cake layer, compromising permeability. If TMP is increased to compensate for the permeability loss, filtration goes through a new subcycle from stage 1.

1.3.2.3   Sudden TMP Increase

The sudden TMP increase (or jump) is commonly observed in submerged membrane filtration operated at a constant flux mode. As discussed in Section 1.3.2.1, the dominant cause of the sudden TMP increase is a vicious circle between cake layer compaction and increasing TMP (Chang et al. 2006a; Park et al. 2006; Fane 2007), but there are other explanations that may also contribute to the phenomenon.

- *Pore loss (or blocking) model*—If some pores are plugged by deposits, more water has to pass through the other pores remaining open at a constant flux mode. In the beginning of the filtration cycle, the lost water permeation from the plugged pores can be easily compensated by the large number of open pores. As pore plugging proceeds, however, the loss of open pores causes increasingly higher fluxes to a smaller number of intact pores. Once the local flux near the open pores exceeds the critical flux, particles can easily deposit and plug up pores. Thereby, filtration resistance abruptly increases at a certain point, which directly translates to TMP increase (Ognier et al. 2004; Ye et al. 2006).
- *Area loss model*—This model is based on a similar idea as the pore loss model. It is based on the assumption that membrane fouling does not occur evenly across the membrane surface, but it occurs area by area due to the inhomogenosity of the membrane surface and the scouring effect. If one area is fouled first, the flux in the clean areas must increase to maintain a constant average flux. Based on the same logic used for the pore loss model, flux increase in clean areas accelerates as clean areas become more scarce; thereby membrane fouling occurs increasingly faster. Eventually, membrane fouling propagates across the membrane surface once it exceeds a critical level (Cho and Fane 2002). This area loss model is particularly plausible in large-scale MBR plants, where diffuser fouling can cause a severe membrane fouling in the affected areas. The lost permeate flow from the fouled membrane cassette must be compensated by the cleaner membrane cassettes, which can trigger accelerated TMP increases across the system.
- *Percolation model*—In membrane filtration, the permeate carries a small amount of macromolecules or colloids through not only the cake layer but also through the membrane. A portion of such contaminants can deposit in the void spaces of cake layers by van der Waals interaction, electrostatic interaction, molecular hindrance, etc. It gradually reduces the cake layer's porosity and increases the cake layer's resistance. Once the cake porosity reaches a threshold at approximately 0.2, the cake resistance calculated by Equation 1.5 increases faster and in turn TMP increases accordingly to maintain constant flux (Hermanowicz 2004). The increase of TMP also expedites the cake layer compaction that again contributes to the sudden TMP increase.
- *Osmotic pressure model*—Osmotic pressure occurs when solute concentrations are different on the two sides of a semipermeable membrane that allows easier passage of the solvent. Solvent in the diluted side tends to permeate to the concentrated side to equalize the concentrations on both sides of the membrane. By definition, the minimum hydraulic pressure required to stop the solvent movement is equivalent to the osmotic pressure. According to the van't Hoff equation, osmotic pressure is proportional to the differences in the number density of solute molecules across the semipermeable membrane. In MF/UF, ionic concentrations are nearly identical in both sides of the membrane due to the free passage of ions through the membrane. However, the concentrations of macromolecules in the feed and the permeate can be slightly different because the cake layer partially rejects macromolecules. The macromolecular concentration on the membrane surface can be further enhanced by the hindered back-diffusion in the void spaces in the cake layer (Fane 2009). The accumulated macromolecules in the cake layer can cause osmotic pressure that diminishes the effective TMP. To maintain a constant flux, TMP must increase to counteract the osmotic pressure. Although the osmotic pressure theory is plausible, the significance of such an effect is yet to be proven because the free macromolecule concentration in the void spaces of the cake layer cannot be high enough to produce significant osmotic pressure. Meanwhile, the counterions in the void spaces of the cake layer, which balances the charges of the cake layer constituents, have been proposed as a source of osmotic pressure (Chen et al. 2012b; Lin et al. 2014), but it is not certain that the semifixed ions generate a noticeable osmotic pressure. No reliable empirical data that supports the osmotic pressure model is available.

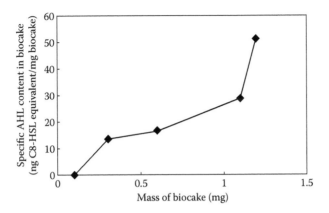

FIGURE 1.26 Specific AHL level in cake layer as a function of the amount of biological cake. (From Yeon, K.-M. et al., *Environ Sci Technol.* 43(19):7403–7409, 2009b.)

- *Quorum-induced biofouling and cake layer compaction*—Microorganisms change metabolic activity/state as a whole group depending on environmental conditions. It has been known that microorganisms use small molecules called *quorum* to communicate with each other. The biological system of stimulus and response is called *quorum sensing*. Quorums are based on various chemistries, but *N*-acyl homoserine lactone (AHL) has been identified as the main quorum related to cake layer compaction (Lee et al. 2009; Yeon et al. 2009a). In fact, it was found that AHL concentrations surge when sudden TMP increases occur due to cake layer compaction. As shown in Figure 1.26, specific AHL concentration in the cake layer (*y* axis) increases as the cake layer builds up (*x* axis). When sudden TMP increase occurred, specific ALH content jumped from approximately 30 ng AHL/mg cake layer to approximately 50 ng AHL/mg cake layer. However, it is not clear whether the increasing compression force on the cake layer stimulates ALH secretion or the increasing AHL secretion expedites cake layer compaction. Based on this finding, a quorum quenching technology has been developed to break down the molecules that trigger the membrane fouling, as discussed in Section 5.4.4.1 (Yeon et al. 2009b).

1.4 EFFECT OF MEMBRANE SURFACE PROPERTY ON FLUX

1.4.1 STREAMING ZETA POTENTIAL OF MEMBRANE SURFACE

1.4.1.1 Definition and Measurement

If a charged surface comes into contact with an electrolyte solution, counterions surround the charged surface according to Stern's model (Shaw 1969). Due to strong electrostatic interaction, some counterions firmly adsorb onto the surface to form a Stern layer. The remainder of the counterions are distributed outside the Stern layer and are free to move in and out of the diffuse layer, as shown in Figure 1.27. The counterion profile in the diffuse layer is decided at the equilibrium among electrostatic interaction, random Brownian motion, convective flow velocity, etc. Electric potential is greatest on the surface and it decreases rapidly on the Stern plane. Outside the Stern plane, electrical potential decreases slowly until it reaches zero in the bulk.

To measure the Stern potential accurately, all the excess counterions in the diffuse layer must be removed by exposing the measuring surface to a very strong shear field. However, this method not only does not guarantee the complete removal of all counterions from the diffuse layer but it may also remove some adsorbed counterions. Therefore, electrokinetic zeta potential is measured at a moderately shearing condition. The measured zeta potentials are dependent on the experimental

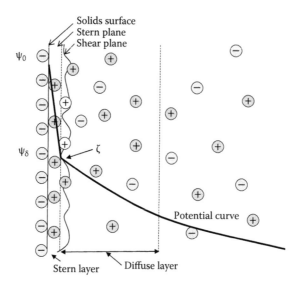

FIGURE 1.27 Electric double layer according to Stern's model.

condition because the location of the shear plane varies depending on the shear rate. Temperature, ionic strength, electrolytes, shear stress on surface, etc., are the factors affecting zeta potential. Therefore, comparing zeta potentials from different experiments needs to be done cautiously.

The streaming zeta potential of the surface is measured based on the voltage that develops along the charged surface while an electrolyte solution is moving over it (Figure 1.28). The movement of counterions is somewhat hindered by the surface charges whereas the co-ions move freely over the surface. The extent of the charge separation between the upstream and the downstream is influenced by the surface charge, flow velocity, concentration of ions, and the fluid properties. The zeta potential is calculated based on the electrical potential differential across the surface differential and other experimental parameters employed. In the membrane, zeta potential can be measured for either the membrane surface or the pore surface depending on the location of the electrodes as shown in the figure.

The streaming zeta potential of the pore wall is measured while filtering electrolytes under pressure (Nyström et al. 1994). Likewise, the streaming zeta potential of the membrane surface is measured while electrolyte is flowing over the membrane surface (Elimelech and Childress 1996).

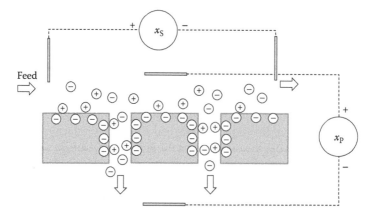

FIGURE 1.28 Streaming zeta potential of membrane surface (x_S) and pore wall (x_P). Voltage difference is measured upstream and downstream of the surface while electrolyte solution is flowing.

The relationship between the streaming potential and the zeta potential is given by the well-known Helmholtz–Smoluchowski equation.

$$\xi = \frac{\mu \lambda \Delta E}{\varepsilon_r \varepsilon_0 \Delta P} \tag{1.17}$$

where
   ξ zeta potential (V)
   ΔE streaming potential (V)
   ΔP pressure difference across the channel (kg/m/s^2)
   μ viscosity of the solution (kg/m/s)
   ε_r relative permittivity of the solution (–)
   ε_0 electric permittivity of vacuum (F/m)
   λ conductivity of polyelectrolyte solution (S/m)

1.4.1.2   Effect of Zeta Potential on Membrane Fouling

Conceptually, the membranes with negative zeta potential repulse negatively charged particles and macromolecules. Many references exist supporting this notion. However, those observations were mostly made in laboratory conditions, where charge repulsion effect was highlighted among other effects by carefully designing the experimental condition, for example, simple feed composition without complex organics and high shearing of membrane surface to prevent cake layer formation. In practical filtration conditions, however, the effect of membrane zeta potential is mostly not apparent mainly because membranes are coated by macromolecules and particles in the early phase of the filtration. In addition, zeta potential effect can be easily buried by many other factors affecting membrane fouling.

The zeta potential effect on membrane fouling was demonstrated in a controlled environment using yeast cells (*Saccharomyces cerevisiae*) and six different membranes with different zeta potentials between −5.5 and −19.7 mV (Kang et al. 2004). The streaming zeta potential of the yeast cells was measured at −8.7 mV under the experimental conditions employed. It was observed that yeast deposition rate decreased as the magnitude of the membrane zeta potential increased. The linear correlation factor between the membrane zeta potential and the yeast cell deposition rate was found to be 0.69. However, it should be noted that this experiment was performed under a rigorously controlled environment, which can maximize the charge repulsion effect. The yeast cells were cleaned twice with deionized water to remove contaminants and particle deposition was observed using an optical microscope only until a monolayer of yeast cells was formed under a low crossflow velocity (2.5 cm/s). In these conditions, yeast cells directly face the membrane surface and are exposed to a direct charge interaction with the membrane. Meanwhile, Brownian diffusion is very weak for the large yeast cells (5–10 μm) and shear-induced diffusion is also negligible under the exceptionally low crossflow velocity. As a result, the charge repulsion is the dominant mechanism that plays a role in particle back-transport. If the filtration had been performed beyond the monolayer formation, the zeta potential effect would have been less apparent because the suspended yeasts only interact with the other yeast cells already deposited in the cake layer.

When a polyvinyliden difuoride (PVDF) membrane surface (Durapore® from Millipore, Watford, UK) was modified with 2-acrylamido-2-methyl-1-propanesulfonic acid (AMPSA), 2-hydroxyethyl methacrylate (HEMA), and quaternized 2-(dimethylamino) ethyl methacrylate (qDMAEM), the modified membranes with higher hydrophilicity were less susceptible to fouling by *Escherichia coli* compared with the original membranes with less hydrophilic surfaces (Kochkodan et al. 2006). In the same study, chemically neutral surfaces were easier to clean compared with charged surfaces due to the lack of charge attraction between membrane surfaces and *E. coli* cells. However, this result is not completely relevant for practical membrane filtration including MBR because the experiment was performed with pure *E. coli* washed three times with deionized water, and the filtration was performed in very different conditions using a stirred cell at 300 rpm. In addition, membrane cleaning was performed by operating the stirred cell for 5 min with deionized water without cleaning chemicals, thus the adsorbed foulants might not be removed completely. In MBR, various particles and biopolymers exist at high concentration and strong oxidants (hypochlorite) are used at high concentration (e.g., 1000–5000 mg/L) as cleaning agents that could not only oxidize foulants but also modify the membrane surface charges, as reported by Levitsky et al. (2011).

In MBR, macromolecules with various chemical and physical properties exist in a mixed liquor. As soon as the mixed liquor comes into contact with membrane, initial coating (or fouling) occurs even before filtration starts. In fact, to some extent, the initial membrane fouling is inevitable regardless of the flux in MBR, as explained in Section 1.2.6. In addition, vigorous aeration is regularly performed at 20 to 60 cm/s of upflow liquid velocity in a typical aeration condition (Yamanoi and Kageyama 2010). Because the membrane surface is not directly visible to membrane foulants in mixed liquor due to the initial coating/deposit layer, charge repulsion between the original membrane surface and the foulants is hardly a dominant factor affecting membrane fouling.

The adsorption of solutes in feed water makes the actual streaming membrane zeta potential different from the initial zeta potential of new membranes. When four different membranes with different surface zeta potentials (−40 to −10 mV) were used to filter 1% whey proteins at pH 8 to 10, the surface zeta potential tended to converge to approximately −30 mV regardless of the initial zeta potential (Lawrence et al. 2006). In other experiments, a nanofiltration membrane with a positive zeta potential (+12 mV) immediately turned negative (−9 mV) at neutral pH when the membrane came into contact with an 18.5 mg/L humic acids solution (Yoon et al. 1998). The surface charge reversal was attributed to humic acid adsorption onto the membrane surface. In other experiments, streaming zeta potential became more negative when negatively charged membranes were exposed to humic acids (Elimelech and Childress 1996). When six different RO and NF membranes were compared, no correlations between initial membrane zeta potential and membrane fouling rate were observed (Hobbs et al. 2000), which suggested that initial membrane surface property was hardly influential to cake layer growth because the foulants in the feed water see only the cake layer formed on the membrane surface.

1.4.2 CONTACT ANGLE

1.4.2.1 Theory

In general, hydrophilic membranes are perceived to be less prone to fouling in water treatment. It is attributed to the hydration of membrane surface by water molecules that prevent membrane foulants from directly coming into contact with the membrane. In fact, hydrophilic membranes have polar or charged functional groups to which polar water molecules gather to counter the surface polarity. For hydrophobic membranes, on the contrary, weak van der Waals forces between membranes and water molecules are not enough to hold water molecules on the membrane surface. The lack of bound water molecules means that the chances are high that foulants will be adsorbed on the membrane surface in hydrophobic membranes (Liu et al. 2011).

Contact angle is the most commonly used parameter to quantify membranes' hydrophilicity or hydrophobicity. The higher the contact angle is, the higher the hydrophobicity is. The contact angle must be less than 90° to allow spontaneous water intrusion to the pore without extra pressure, as shown in Figure 1.29. If the contact angle is greater than 90°, extra pressure is required to initiate the water permeation or the membrane must be wetted or soaked in the liquid with lower surface tension, for example, water/alcohol mixtures and surfactant solutions. Because membrane wetting becomes impractical as the system capacity grows, all the MF/UF membranes available for bulk water treatment at the commercial scale are manufactured as hydrophilic to avoid the costly prewetting process. In some literature, the contact angle is defined as the angle on the opposite side of θ in the figure. In those cases, the contact angle should be larger than 90° for spontaneous pore wetting.

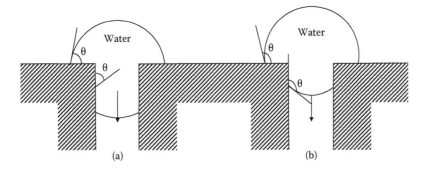

FIGURE 1.29 The effect of equilibrium contact angle, θ, on the pore wetting phenomenon: (a) voluntary wetting, θ < 90° and (b) nonvoluntary wetting, θ > 90°.

In practical situations, however, actual membrane surface is neither perfectly smooth nor uniform at the molecular level. Even if the nominal contact angle is less than 90°, it may take a substantial time to wet all the pores because the actual contact angle can vary area by area. On the contrary, even if contact angle is larger than 90°, pores will eventually be wet due to various reasons such as defects on membrane surface, water vapor condensation in the pores, depletion of the air contained in the pore by diffusion to the bulk water, declining contact angle due to organic and inorganic fouling of pore entrance, etc. Gradual pore wetting of the membrane is one of the major challenges when porous hydrophobic membranes are used for applications that require rigorous pore dryness, for example, membrane distillation, membrane-assisted biofilm reactor, hydrogen-based hollow-fiber membrane biofilm reactor, and membrane humidifier (Franken et al. 1987; Sachtler and Schmidt 1989; Nerenberg and Rittmann 2004; Lackner et al. 2010).

The membrane contact angle is influenced not only by the major polymeric constituent but also by many other factors. It widely varies depending on the properties of chemical additives such as plasticizer, solvents used to dissolve the polymer, crystallinity of polymer, manufacturing method and condition, surface morphologies, etc. For instance, pure PVDF typically has a near neutral or slightly hydrophobic surface, if the surface is smooth, as shown in Table 1.1, but the contact angle can even be increased to 148° using a proper alcohol as coagulant at a proper condition using a wet immersion method (Kuo et al. 2008). In addition, measuring the contact angle of MF/UF membranes with high accuracy is tricky due to the spontaneous water permeation and the surface

TABLE 1.1

Contact Angle of Various Materials

Pure Material	Contact Angle (Approximate)
Ordinary glass	20
Anodized aluminium	60
Polyvinyl alcohol (PVA)	60
Polystyrene	66
Nylon6,6	68
Polysulfone (PSU)	70
PMMA	74
Nylon	79
PVC	86
PVDF	89
Polyethylene (PE)	96
Polypropylene (PP)	102
PTFE	109

inhomogenosities in terms of surface roughness and pore size. As a result, the contact angles found in the literature for the same membrane can vary significantly depending on the experimental conditions employed.

Another aspect of contact angle is that it is not fixed but variable over time due to oxidative membrane cleaning using NaOCl, H_2O_2, etc. (Levisky et al. 2011). When PVDF membranes were cleaned with NaOCl, the contact angle decreased from 92.2° to 75°–80° depending on the extent of chlorine exposure. Contact angle decrease was more significant with polyethersulfone (PES) membrane, the contact angle of which decreased from 100.6° to 63°–76°. On the other hand, the membranes with a hydrophilic coating layer, for example, polyethylene glycol (PEG), polyvinyl alcohol (PVA), and polymethyl methacrylate (PMMA), may gradually lose hydrophilicity after oxidative chemical cleaning and physical abrasion.

Liquid entry pressure (LEP), which is the lowest pressure that triggers water passage through hydrophobic membrane, can be calculated using Equation 1.18.

$$\text{LEP} = \frac{4B\gamma \cos(180 - \theta)}{d} \tag{1.18}$$

where
  LEP liquid entry pressure (Pa)
    B capillary constant (–)
    d defect diameter (m)
    γ surface tension at the air–liquid interface (0.0728 N/m at 20°C)
    θ liquid–membrane contact angle (degree)

1.4.2.2 Effect of Contact Angle on Membrane Fouling

It has been widely perceived that hydrophilic membranes are less susceptible to organic fouling than hydrophobic counterparts in water treatment. The low fouling is attributed to the lower hydrophobic–hydrophobic interaction between the membrane and naturally occurring organic materials (Kabsch-Korbutowicz et al. 1999; Ho and Zydney 2006; Kang et al. 2004). When two homemade PVDF membranes with different contact angles (77° vs. 101°) were used to filter activated sludge, the hydrophilic membrane with smaller contact angles resulted in almost 100% higher critical flux, although those two membranes had identical pore sizes (0.1 μm) and porosities (15%). It was postulated that more macromolecules were adsorbed on the hydrophobic membrane (van der Marel et al. 2010). On the contrary, the results presented by other researchers have shown that surface hydrophilicity alone was not necessarily correlated with membrane fouling propensity (Knoell et al. 1999; Vrijenhoek et al. 2001; Hobbs et al. 2006).

It is very hard to explain the discrepancies without knowing the exact experimental conditions. However, it is noteworthy that the positive contact angle (or hydrophilic) effects were observed mostly in the laboratory environment. In fact, in the experiment mentioned above (van der Marel et al. 2010), a membrane with high contact angle (101°) was used as a hydrophobic membrane that required artificial pore wetting. However, no known commercial MF/UF membranes aimed for bulk water treatment have such hydrophobic surfaces as shown in Table 1.2, where the contact angle of the most hydrophobic MF/UF membrane is 82°. In the meantime, the contact angles of 20 commonly used commercial RO/NF membranes range from 38° to 73° (Norberg et al. 2006). In any case, using membranes with less than 90° contact angle is not practical in large-scale applications due to the costly wetting process that requires extra capital and operating costs. The contact angle effect on membrane performance is rarely noticeable in practical membrane filtration, perhaps because all commercial membranes are hydrophilic enough.

In one laboratory study performed with yeast cells as a foulant, the effect of contact angle on flux was apparent from the very beginning of the filtration cycle until a monolayer of yeast cells had

TABLE 1.2

Contact Angle of Commercial Membranes

Material	Manufacturer	Contact Angle
PVDF (UF)	Toray	82[a]
PVDF (UF)	Sterlitech	92.2[b]
Polyethersulfone (UF)	Sterlitech	100.6[b]
Chlorinated PE (MF)	Kubota	Not measurable[a]
PSU (UF)	Microdyn	67.6[c]
Regenerated cellulose (UF)	Microdyn	54.8[c]
Aromatic polyaramide (UF)	Microdyn	66.2[c]
Polyester, PETE (MF)	Osmonics	66[d]
Polycarbonate, PCTE (MF)	Osmonics	66[d]
PTFE	Sumitomo	49[d]
CA (UF)	Osmonics	55
MCE (MF, GSWP)	Millipore	49[e]

[a] From van der Marel, P. et al., *J Membrane Sci* 348:66–74, 2010.

[b] From Levisky, I. et al., *J Membrane Sci* 377:306–213, 2011.

[c] From Kabsch-Korbutowicz, M. et al., *Desalination* 126:179–185, 1999.

[d] From Choi, J.-H., Ng, H.-Y., Influence of membrane material on performance of a submerged membrane bioreactor. *Proceedings of the IWA Conference.* May 15–17, 2007. Harrogate, UK.

[e] From Clarke, A. et al., *Langmuir* 18:2980–2984, 2002.

formed (Kang et al. 2006). In that experiment, particle deposition on a translucent polycarbonate flat sheet membrane was directly monitored by an optical microscope through the membrane from the permeate side, which was a technique called direct observation through the membrane (DOTM) demonstrated by Li et al. (1998). Due to the limitation of the optical microscope, only a first monolayer formation could be observed. In fact, it is not hard to imagine that the interaction between yeast cells and bare membrane surface is a crucial factor affecting the first monolayer formation on the membrane. However, once the monolayer is formed, the direct interaction between membrane and foulant becomes less important. The surface of the membrane is completely screened by the first monolayer; thereby the benefits of membrane hydrophilicity become obscure in the long run (Le-Clech et al. 2006).

It has been reported anecdotally that hydrophilic membranes are easier to clean. If cleaning was performed under mild conditions using a water jet and base/acid, foulants may be more easily detached from the hydrophilic surfaces due to their weak hydrophobic interaction. The easily removable cake layer might be beneficial when membranes should be cleaned only in mild conditions. However, the benefit of using ultrahydrophilic membranes is rather fuzzy in MBR because strong oxidants–based chemical cleaning is routinely practiced and all the commercially available membranes are at least hydrophilic to some degree. In fact, cleaning MF/UFs is typically performed under very harsh conditions using strong oxidants such as 1000 to 5000 mg/L of NaOCl, and hence, the most organic molecules adsorbed on the membrane surface are degraded to smaller molecules and disengaged from the membrane regardless of their affinity on the membrane surface.

1.4.3 Surface Roughness

It is intuitive that rough surfaces would more easily catch particles compared with smooth surfaces. Rough surfaces not only provide large contact areas for the particles in the valleys but also have areas with slow flow, as illustrated in Figure 1.30. The large contact area provided by the curved

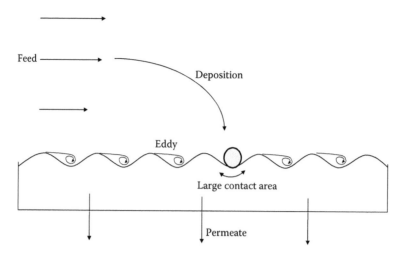

FIGURE 1.30   Particle deposition on rough membrane surfaces.

surfaces increase the chance of permanent particle settling by enhancing van der Waals forces, charge interactions, dipole–dipole interactions, etc. In addition, the low shear rates in the valleys help particle adhesion while reducing the chance of the particles being detached. The surface roughness can play a major role until the cake layer grows enough to bury the rough surface below the thick cake layer.

Surface roughness is measured by atomic force microscopy (AFM), which is also called scanning force microscopy. The AFM consists of a cantilever with a sharp tip (probe) used to scan the specimen surface with a resolution of a fraction of a nanometer. This device relies on electron tunneling in a narrow gap between the tip and the sample. Electrons jump from one side to the other through the vacuum when the two objects are close enough. The electric current flowing between the tip and the sample through the vacuum is related to the distance between them. Finally, the three-dimensional surface morphology is mapped based on the current in each pixel of the surface using a computer.

The effect of surface roughness, zeta potential, and contact angle on membrane fouling was investigated using two NF and two RO membranes with different surface characteristics (Vrijenhoek et al. 2000, 2001). The surface roughness was defined as the arithmetic average of the absolute surface height deviations from the center plane based on AFM images. Root mean square roughness was also defined as the standard deviation of the average roughness. The surface area difference represents the percentage increase of the three-dimensional surface area over the two-dimensional surface area (Hobbs et al. 2006). The four different membranes were used to filter a 200 mg/L silica solution at 51 LMH and the relative flux losses were monitored. As shown in Figure 1.31, the relative flux loss linearly increased as the surface roughness increased. In contrast, membrane zeta potential and contact angle were not correlated with the relative flux loss $(1 - J/J_0)$ at all. Similar observations were made when six different NF and RO membranes were compared in other studies (Hobbs et al. 2000, 2006). In these studies, it was found that the flux declines were very well correlated with the ratio of three-dimensional surface area and two-dimensional surface area, which was a form of surface roughness, when contaminated ground water was filtered.

In another study, yeast cell deposition rates were monitored using a technique called DOTM (Kang et al. 2006). When six different MF and UF membranes were compared, the initial cell deposition rate was best correlated with surface roughness with a linear correlation factor of 0.79. Zeta potential was also positively correlated with the cell deposition rate with a linear correlation factor of 0.69. However, it should be noted that DOTM is valid for the first monolayer formation due to the limitation of the technique as discussed in Section 1.4.2.2, and the observation is relevant only for the very initial filtration cycle.

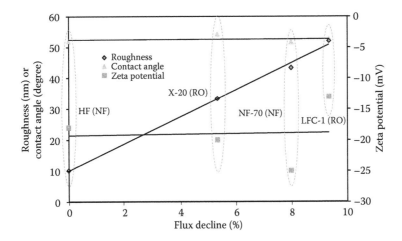

FIGURE 1.31 Effect of surface roughness, contact angle, and zeta potential on membrane fouling by 200 mg/L 100 nm colloidal silica (or flux decline) at 51 LMH: LFC-1 (Hydranautics, Oceanside, CA), NF-70 (Dow Filmtec, Minneapolis, MN), X-20 (Trisep, Goleta, CA), HL (GE Osmonics, Minnetonka, MN), J = flux with colloidal silica, J_0 = clean water flux. (Data from Vrijenhoek, E.M. et al., *J Memb Sci.* 188:115–128, 2001.)

1.4.4 PORE SIZE AND DISTRIBUTION

It has been postulated that the particles that closely fit the membrane pores efficiently plug the pores and reduce flux (Imasaka et al. 1989; Chang et al. 1994). It has been also assumed that particles that are smaller than the pore size can easily penetrate into the pores and cause pore blockings by being captured in the tortuous water channels. Although those theories seem plausible, a thorough survey failed to prove the existence of such an effect in practice (Le-Clech et al. 2006). The obscure pore and particle size effect can be attributed to the complex and changing nature of the feed water composition. In fact, the particles and molecules in the feed can change their shapes and characteristics by undergoing physical, chemical, and biological reactions before they meet with the membrane. Hence, finding out the correlation between feed water characteristics and membrane fouling is inherently difficult. Moreover, it is particularly difficult to find any correlation in MBR because mixed liquor contains the molecules/particles that originated not only from raw wastewater but also from biological metabolisms. The fluctuating nature of mixed liquor quality is another barrier for the correlation. In conclusion, no universally applicable trends have been identified with respect to the pore size effect on flux in MBR, although membranes with optimum pore sizes may exist for mixed liquors with a specific size distribution in theory.

The membranes with large pores tend to have rougher surfaces. Hence, membranes with large pore sizes may foul faster than those with small pores because they are more prone to catching particles from the mixed liquor, as discussed in Section 1.4.3. In practical filtration conditions, however, the pore size effect varies study by study, depending on the specific condition employed. For instance, if crossflow velocity is high enough to prevent significant cake layer formation, membranes with larger pore sizes likely show greater fluxes because they likely have greater permeabilities. If crossflow velocity is in a medium range, membranes with larger pore size may attract more deposit due to the high initial flux and the potentially rougher surfaces. As a result, the flux may appear inversely related with pore size. If crossflow velocity is low, a substantial amount of cake layer will be formed regardless of pore size. This leads to the cake filtration condition, where the membrane itself is not a factor affecting the flux. If membrane permeability is extremely low, membrane permeability itself can be a limiting factor in obtaining high flux. In this case, flux may increase as pore size increases because membranes with larger pore size tend to have higher permeabilities.

Pore size distribution may play a role in accelerated membrane fouling in certain cases. According to the Hagen–Poiseuille equation shown below, flow rate is proportional to the fourth power of the pore radius under a constant pressure condition. Thus, five times and sixteen times greater liquid flow rates are expected when pore size increases by 50% and 100%, respectively. Apart from the pore size distribution, pores may be more densely populated in some local areas due to manufacturing defects. Consequently, the permeability of some local areas can be higher than other areas, as shown in Figure 1.32. In the figure, zone 2 has higher flux than zone 1 due to the existence of a few big pores. Under these circumstances, zone 2 can be fouled quicker than zone 1 especially when the local flux exceeds the critical flux. In constant flux mode, the flux in zone 1 must increase to maintain a constant average flux. It in turn causes accelerated membrane fouling in the zone. The imbalanced local membrane fouling triggered by the inhomogeneous surface properties accelerates membrane performance loss. To minimize the fouling triggered by the membrane properties, membrane surface should be homogeneous with a uniform porosity and a narrow pore size distribution.

$$Q = \frac{\pi r^4 \Delta P_T}{8\,\mu L} \tag{1.19}$$

where

      Q flow rate (m³/s)
      r pore radius (m)
  ΔP_T transmembrane pressure (Pa or kg/m/s²)
      μ viscosity of liquid in pore (cP or kg/m/s)
      L length of pore (m)

The potential issue of localized high flux has been fundamentally solved by recently developed patterned membranes based on lithographic technology originally developed for microelectronics manufacturing. A master mold with the desired pattern is fabricated on a silicon wafer by photolithography following the methods used in semiconductor manufacturing. Replica molds are prepared by casting polydimethylsiloxane, PMMA, and others on the master mold. Finally, nanopatterned membranes with the desired surface morphology are produced by casting polymer solutions on the replica mold (Won et al. 2012). This technique allows for the manufacture of membranes with the exact surface contour desired for specific purposes, for example, pyramid-patterned or prism-patterned membranes. Moreover, membranes with the exact surface morphologies can also

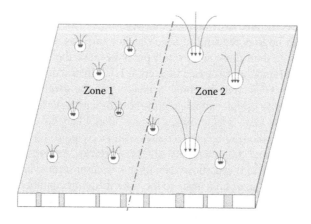

FIGURE 1.32 Local flux variation caused by porosity and pore size variation.

be made by articulating the mold. The surface of the membranes marketed by fluXXion B.V. had pores with identical size and shape distributed perfectly evenly (Fluxxion 2007). These membranes were produced as flat sheet membranes attached on both sides of discs that are stacked up in cylindrical pressure vessels. The most apparent application of this membrane is for the sterilization of juice, milk, and others due to the very tight pore size distributions that do not allow any bacterium to pass through the membrane. However, the economic feasibility of patterned membranes has not been verified because many other competing technologies exist.

1.4.5 INTERNAL PORE CONNECTIVITY

Track-etched membranes are made by the bombardment of fission fragments on polymer film to make narrow tracks in the film followed by etching the tracks with chemicals. As a result, pore size distribution is narrow and all the pores pass the membrane straight through with no internal connectivities. If any pore is plugged in the middle, it is permanently lost because there are no alternative channels to pass.

The internal connectivity of membrane pores has been proposed as a potential factor affecting membrane performance. All the membranes commonly used in MBR have a highly interconnected pore structure formed by random alignments of polymer clusters. The water that enters into a pore has multiple channels to go through until it completely passes the membrane skin. If one channel is plugged by particles or macromolecules, the water can still pass through alternative channels. This mechanism potentially plays a significant role when internal pore blocking is a major cause of membrane performance loss (Ho and Zydney 2006). Despite this plausibility, the significance of the internal pore connectivity effect in practical membrane filtration conditions is questionable because the vast majority of particles smaller than the pore are rejected by the cake layer and the cake resistance takes the vast majority of total resistance in the conditions commonly found in MF/UF.

1.4.6 LIMITATION OF THE THEORY

As discussed in this chapter, numerous different factors have been identified and validated as contributors to membrane fouling. However, membrane fouling occurs as a result of complex interactions among those factors, thereby determining the significance of each factor in the practical filtration condition is intrinsically hard. It is partly because of the complexities of the interconnectivities of such factors and partly because field conditions are different from the rigorous conditions employed to verify the validity of each factor. As a result, it is hard to predict the filtration performance with high accuracy without performing actual filtration.

Membrane fouling phenomena were popularly explained based on the interactions between particles and membrane surfaces, for example, charge interaction and hydrophobic interaction in the 1980s and 1990s. This notion was the basis of numerous attempts at membrane surface modification implementing specific surface properties to reduce membrane fouling. Noticeable progresses have been made in RO by implementing a fouling-resistant layer, which makes the surface smoother and reduces the attraction with organic foulants. On the other hand, despite the numerous different attempts to modify the surface chemistry, no significant progress has been made in MF and UF beyond the level achieved by the optimized polymer blending and manufacturing processes. As a result, the significance of particle–membrane interaction has dwindled over time especially in porous membrane filtration, for example, MF and UF.

The low efficacy of surface property modification in porous membranes is attributed to the dominance of cake layer resistance in filtration resistance. Therefore, more efforts have been made in the industry to optimize the hydrodynamic conditions around the membrane by modifying the module configuration and the dimension. It has been proven that the factors affecting hydrodynamic conditions on the membrane surface have a direct effect on membrane performance, for example, air–liquid velocity, direct air bubble contact, and random hollow fiber movements. As a consequence, submerged modules, rotating disc modules, vibrating modules, various backwashing methods, and others, have been developed as will be discussed in Chapter 2.

1.5 MEMBRANE CHEMISTRY AND MANUFACTURING

1.5.1 MEMBRANE STRUCTURE

Depending on the pore sizes along the depth of the membrane, membranes can be classified as symmetric or asymmetric. The terms used to classify filters are summarized in Table 1.3. Symmetric membranes have the same pore sizes throughout the depth of the entire membrane (Figure 1.35b), but asymmetric membranes have a tighter skin layer on the feed side of the membrane and a loose support layer (Figure 1.33a). Membranes with symmetric structures are adopted mainly for MF membranes in solid–liquid separation because the thick filtration layer imposes high filtration resistances. Asymmetric membranes have an advantage in permeability because the thin skin layer minimizes filtration resistance. Virtually all tight membranes are based on asymmetric structures, but some MF membranes have symmetric structures with large pores. The membrane skin layer of asymmetric membranes can be further classified as isotropic or anisotropic depending on the homogeneity of the pores. If pore sizes are identical in the skin layer regardless of the depth, it is called isotropic. Otherwise, it is called anisotropic. Anisotropic skin structures are advantageous in terms of filtration resistance, but may be prone to physical and chemical damages. The support layer of asymmetric membranes can be further classified as sponge type or finger type depending on the structure (Figure 1.33b and 1.33c). The membranes with a sponge type support layer might be less prone to integrity issues due to the lower chance of having defects on the surface by the extended finger structure through the skin layer. In addition, if TMP is high, the sponge type structure may be more resistant to compaction and subsequent permeability loss. If the materials of the skin layer and the support layer are different, the membranes are called composite membranes (Figure 1.33d). For example, in RO and NF membranes, an active separation layer is cast on the UF membrane that is again cast on nonwoven fabric. Some submerged hollow fibers and flat sheet membranes used for MBR have a dual layer structure with an active membrane layer laminated or cast on fabric supports. Screens are referred to the filtration barriers with large pores, perhaps with 50 μm or larger sizes and are not considered as membranes by convention. Depth filters are the filters that capture particles in the body of the filtration medium, as discussed in Section 1.1.1.

TABLE 1.3
Terms Used to Classify Filters

Classification	Terms	Remark
Surface filter	Symmetric	Pore sizes are identical throughout the membrane depth. No distinction between skin and support layers (Figure 1.35b)
	Asymmetric	Skin and support layer are distinctive, but they are based on the same chemistry. Support layer structure is further classified as finger-type and sponge-type depending on the configuration (Figure 1.35a)
	Isotropic	Pore sizes are identical in the vertical depth of the skin layer
	Anisotropic	Pore sizes are the smallest on the surface and gradually increase into the depth of the skin layer
	Composite	Skin and support layers are made of different materials. Skin layer can be either isotropic or anisotropic. Example: RO/NF membranes and dual layer hollow fiber membranes (Figure 1.33d)
	Screen	Gap, perforated, or mesh screens, single layer woven fabrics, etc.
Depth filter	Nonwoven	Fibers are randomly overlapped. Most air filters, many cartridge filters
	Filtration column	Sand filter, multimedia filter, etc.

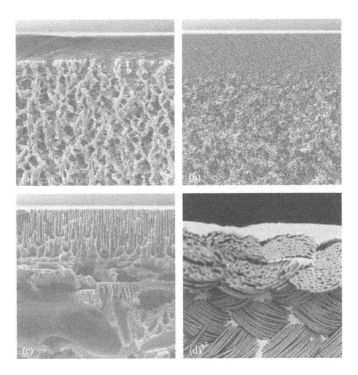

FIGURE 1.33 Cross-section of various membranes: (a) isotropic asymmetric membrane with microporous support (10 kDa Ultracel® PLC, regenerated cellulose); (b) isotropic asymmetric (or anisotropic) membrane with sponge type support layer (10 kDa Biomax®); (c) anisotropic asymmetric membrane with finger type support layer (Traditional PES 10); (d) composite hollow fiber membrane. (a–c: Courtesy of EMD Millipore Corp.)

1.5.2 MEMBRANE CHEMISTRY

PVDF had not been a popular membrane material until the 1980s, not only due to the difficulties in making membranes with good and consistent qualities but also due to its high prices compared with other polyolefins, such as PE and PP. However, as manufacturing technology improves and the demand for the chlorine-tolerant membranes increases, PVDF has become one of the most popular membrane materials for surface water filtration, tertiary filtration, and MBR (Liu et al. 2011). The low glass transition temperature (T_g) of PVDF at approximately −35°C makes it flexible at room temperature, which is an essential property required for submerged membranes to take advantage of random membrane movement. The high flexibility makes it more suitable for submerged membranes than PSU and PES. The pore sizes of PVDF membranes span from loose UF to MF, for example, 0.01 μm or larger. The membranes made of PVDF are hydrophobic and pore wetting is required using a liquid with a low surface tension, for example, alcohols, acetone, and surfactant. Because onsite pore wetting is highly impractical in the field, especially for large-scale systems, contact angle is reduced by doping the membrane materials using hydrophilic additives. PVDF membranes are compatible with 1000 to 5000 mg/L NaOCl or higher with a lifetime chlorine tolerance of around 500,000 ppm·h (Fenu et al. 2012) although the chlorine tolerance varies widely depending on the methods and additives used in the manufacturing process. The PVDF membrane itself is tolerant at high temperatures, for example, 80°C to 100°C, but most commercial modules are rated for 40°C or lower due to the low temperature tolerance of other module parts. Likewise, PVDF itself is compatible with all pH ranges, but commercial membranes may have a limit, for example, an operating pH of 5 to 9.5 and cleaning pH of 2.0 to 10.5 for ZW500® of GE, due to the doping materials used to modify the membrane properties. Microza® (Asahi Kasei), Zeeweed® (GE), Memcor® (Evoqua, formerly Siemens), Sterapore SADF™ (Mitsubishi Rayon), Airlift™ (Norit), Neosep™ (Veolia), Membray® (Toray), etc., are some of the commercial modules based on PVDF chemistry.

Polyolefins such as polyethylene (PE) and polypropylene (PP) have been used widely for MF membranes for decades. These materials have low T_g at approximately −120°C and −20°C, respectively, which provide excellent flexibility that is essential for submerged hollow fiber membranes (Section 3.4.2). Traditionally, heating and stretching methods were used with or without solvents. Lamellar microcystallites are induced by aligning the polymer chains by applying high shear stress when extruding PE/PP. The film is then stretched by 50% to 300% at just below the melting temperature. Under this stress, the amorphous phases among the crystallites deform and form slit-like pores (Nunes and Peinemann 2001). Despite the high intrinsic chlorine tolerance, the membranes based on these materials may not be as chlorine-tolerant as PVDF membranes depending on the crystallinities and the doping/coating materials. Polyolefin-based membranes are fabricated mostly as MF rather than UF with pore sizes larger than 0.1 μm. Polyolefins must be doped with hydrophilic additives such as surfactants and PVA, or surface-treated to make the membrane surface hydrophilic so that spontaneous pore wettings occur when membranes come into contact with water. Although intrinsic tolerances are much higher, pH and temperature ranges allowed for commercial membranes are generally 2 to 13 and 5°C to 40°C, respectively, due to the additives or coating materials used or the materials used for the module fabrication. The first submerged membrane module with horizontally mounted hollow fiber membranes was developed in the early 1990s by Mitsubishi Rayon (Sterapore LFB™) and was based on PE. Submerged hollow fiber membranes from Econity Inc. are also based on PE and is claimed to be highly chlorine tolerant with a concentration time (CT) value of 1,000,000 ppm·hr. Kubota's flat sheet membranes are known to be chlorinated polyethylene (or PVC).

PSU and PES have been used for many decades to fabricate MF and UF. PSU and PES have high glass transition temperature (T_g) at 195°C and 230°C, respectively. These materials are intrinsically stiff at room temperature but suitable for high-temperature applications, if modules are fabricated with the proper materials. Both materials are soluble in chloroform and dimethylformamide and are easily applied for conventional membrane manufacturing processes based on nonsolvent-induced phase separation (NIPS) process. However, the high solubility to solvents limits the PSU and PES only for the feed water without solvent contents. They are also hydrophobic in nature and do not allow spontaneous wetting when fabricated as a membrane. The manufactured membranes are often not allowed to dry completely and are often treated by hydrophobic agents with low vapor pressure such as glycerin. The hydrophilicity of the PSU/PES membrane can be increased by mixing sulfonated PSU to nonsulfonated PSU. Polyvinylpyrrolidone (PVP) is another material commonly used to improve hydrophilicity (Nunes and Peinemann 2001). PSU and PES are suitable for making UF membranes especially with low MWCO, for example, 10,000 Da or lower. It has been commonly used for beverage/bottled water filtrations to remove pyrogens, which may require occasional steam sterilization at 121°C. However, PSU is known to be unsuitable for submerged membranes due to the rigidity that makes the membrane brittle in turbulent conditions. PES can be fabricated as submerged membranes by doping it with additives. Puron® hollow fiber membranes (KMS) and BIO-CEL® flat sheet membranes (Microdyn-Nadir) are known to be based on PES chemistry. The temperature and pH ranges allowed for those membranes fabricated for MBR are similar to those for PVDF membranes in general, that is, 5°C to 40°C and pH 2 to 12. Chlorine tolerance of PES membranes are known to be high at 250,000 to 500,000 ppm·h, depending on the manufacturing methods (Arkhangelsky et al. 2007; KMS 2013).

Polyacrylonitrile (PAN) is a durable hydrophilic material with high chemical resistance against hydrolysis and oxidation. PAN has high T_g at 200°C and is stiff at room temperature, which makes it hard to use for submerged hollow fiber membranes. However, if properly fabricated as a membrane with qualified glue, o-ring, etc., it can be used at near boiling point and it can be steam-sterilized at 121°C. PAN based UF membranes with a MWCO of 5000 to 100,000 Da are suitable for the application recovering enzymes from water due to its suitability for steam sterilization. Therefore, it is used mainly for biotechnology and food industries. PAN membranes are less permeable and less stable at high pH (above 12–13) and high oxidizer concentrations compared with polyolefins and PVDF. Chlorine tolerances are known to be 120,000 to 200,000 ppm·h. As a result, PAN is not popularly used for large-scale water filtrations, where aggressive membrane cleanings at high oxidizer concentrations are essential.

1.5.3 Porous Membrane Manufacturing

Various methods are used to manufacture membranes depending on membrane materials, desired properties, existence of solvents, economic feasibilities, etc. The base material is the principal factor determining the membrane property but the manufacturing method is also a crucial factor affecting pore sizes and structures, pore size distributions, hydrophilicities, porosities, mechanical and chemical strength of the membrane, etc. A vast majority of the porous membranes used for water treatment are manufactured by phase inversion–based methods suitable for producing asymmetric membranes. The methods based on heating and stretching are also used for some polyolefin and polytetrafluoroethylene (PTFE) membranes. It is not commonly used for water treatment membranes, but track etching and nano-patterning of membrane surfaces are also used to obtain specific surface morphologies.

Most membrane manufacturing processes can be classified as either NIPS, a wet casting method, or thermally induced phase separation (TIPS), a dry method, depending on the driving force used to induce phase inversion. However, the differences between NIPS and TIPS have been fading with the use of solvents in TIPS to increase the porosity (Baker 2004). As a result of the use of solvents, the temperature of polymer melt in TIPS has gone down to 70°C to 100°C. In addition, the melt spun fibers in TIPS are often cooled and precipitated by being submerged in a water bath similar to NIPS processes.

In NIPS, which is also called diffusion-induced phase separation (DIPS), polymer solutions are made by dissolving polymer in a solvent, and are cast on woven or nonwoven supports or glass plates. By immersing the film in nonsolvents, the solvent exchange process starts and the polymer coagulation accelerates. Because solvent leaves from the film–nonsolvent interface, polymer concentration is the highest in the interface. The thin dense layer in the surface becomes a skin layer whereas the rest becomes a support layer, which is a characteristic feature of asymmetric membranes. To enhance the densification of the skin layer and obtain desired properties, a partial evaporation of the solvent can be allowed in a controlled environment before the film is immersed into nonsolvent.

Virtually any change in chemical and operating condition affects the membrane properties. Due to the complex interaction among polymers, solvents, nonsolvents, drying time, process temperatures, etc., the consequence of changing one variable is not always reasonably predictable. For instance, PVP has been used to avoid macrovoids in PES membranes (Wienk et al. 1996), but it increases macrovoids in PSU membranes (Yeo et al. 2000). Therefore, finding the optimum membrane manufacturing condition that allows the desired membrane properties involves testing nearly infinite combinations of chemicals and operating conditions. In addition, acceptable quality control is highly challenging because that requires maintaining not only all manufacturing conditions but also the qualities of all raw materials rigorously.

Various additives with various molecular weights can be added to the polymer dope to improve the membrane properties and the operability of the casting process, for example, PEG, PVP, propionic acids, surfactants, etc. (Figure 1.34). PEG is used as a pore-forming agent that affects pore size and porosity by modifying the thermodynamics and kinetics of the phase inversion process (Guillen et al. 2011). However, PEG can either increase or decrease the pore sizes depending on the polymers, solvents, nonsolvent, and its own molecular weight. As briefly discussed in the previous paragraph, PVP can also turn out to either increase or decrease the void spaces depending on its condition. Surfactants can affect the membrane structure by affecting the interfacial properties between nonsolvent and polymer solutions. Solvent–nonsolvent exchange can be expedited with surfactant and macrovoids can be formed. Propionic acids can be added to either a polymer dope or coagulation bath to modify the phase separation process (Fritzsche et al. 1990).

The interactions between solvent and nonsolvent also play a role in determining membrane structure. For example, if the miscibility of the nonsolvent and solvent is high, solvent is quickly deprived from the membrane, leaving large pores. This leads to a finger-type support structure. If

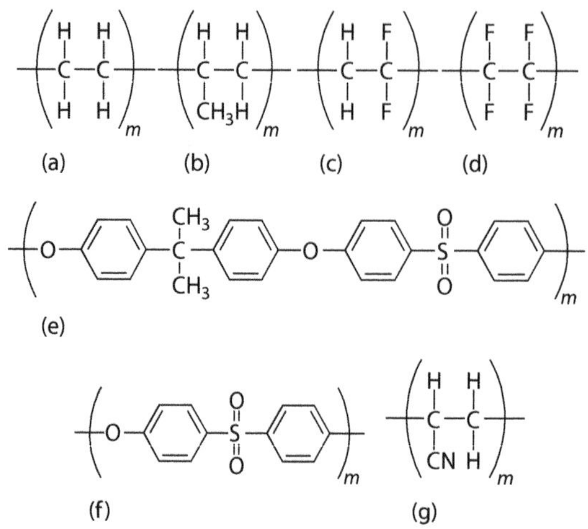

FIGURE 1.34 Some of the common membrane chemistries for MF and UF: (a) PE, (b) PP, (c) PVDF, (d) PTFE, (e) PSU, (f) PES, and (g) PAN.

the miscibility is low, slow solvent exchange occurs and the support structure becomes sponge-type with small pores (Guillen et al. 2011). Figure 1.35 shows commercial membranes with finger-type support and sponge-type support.

In NIPS process producing PVDF membranes, the homogeneous polymer solutions are prepared by dissolving PVDF powders at approximately 60°C in the solvent, which consists of one or more of dimethylformamide, dimethyl sulfoxide, tetrahydrofuran, butanone, N-methyl pyrrolidone, etc. To make the membrane hydrophilic, PEG, PVP, sulfonated polystyrene, PVA, PMMA, various alcohols and surfactants, etc., can be added to the polymer solution. The polymer solution is cast on substrates such as polyester fabric to provide mechanical strength to the final product. Drying time can be given at a specified temperature, humidity, solvent vapor pressure, etc., before it is immersed in coagulation bath. Water, ethanol, and i-propanol or their combinations are used as nonsolvents in the coagulation bath.

Traditionally, TIPS has been used to fabricate membranes from the polymers that do not dissolve in solvent easily, for example, PE and PP. However, the advantages of TIPS with regard to less wastewater production and potentially lower production costs allow it to be applied for other materials such as PVDF, PMMA, and PSU. In TIPS processes, polymers may be dissolved by heat with optionally a small amount of solvent with additives. After casting the polymer solution, it is cooled down at a controlled rate to induce phase separation. Solvent (or diluent) is allowed to evaporate at a specific rate in a defined atmosphere to obtain the desired membrane properties. After

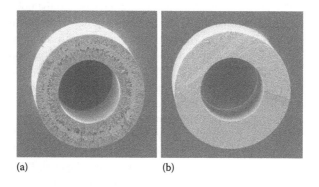

FIGURE 1.35 Cross-section of hollow fiber membranes: (a) finger-type support layer and (b) sponge-type support layer. (Courtesy of Asahi Kasei Corp.)

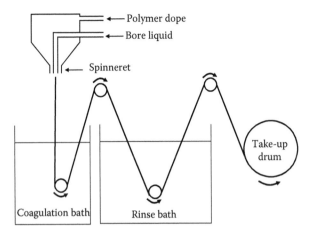

FIGURE 1.36   Hollow fiber spinning process based on NIPS.

the polymer is solidified by crystallization or glass transition, the remaining solvent is extracted by solvent exchange and the residual extractant in the membrane is usually evaporated to yield a microporous structure (Matsuyama et al. 2002; Ishigami et al. 2014). The selection of solvent (or diluent) is crucial in controlling the crystallization process that in turn determines the membrane morphologies and properties with respect to pore size, strength, flux, etc. (Liu et al. 2011).

The conditions developed for asymmetric flat sheet membrane manufacturing cannot be directly applied for hollow fiber spinning processes. Solvent exchange and coagulation occur on both sides of the wall in hollow fibers and the bore liquid properties and temperature play a role. In addition, polymer dopes for hollow fibers should be more viscous and elastic to maintain proper tension during spinning. Figure 1.36 illustrates the hollow fiber spinning process. A homogeneous polymer dope is prepared by mixing polymers, solvents, and additives followed by degassing. The dope and the bore liquid are fed to the spinneret by precision metering pumps. The extruded materials are immersed in the coagulation bath filled with nonsolvents to induce phase separation. Optionally, the extruded materials are exposed to an atmosphere for a certain period to evaporate solvents before being immersed in the coagulation bath. After coagulation, the solvent and nonsolvent in the hollow fiber are removed in a rinse bath before being taken up by a bobbin. Hollow fiber spinning can be as fast as up to 1 m/s. All tanks and baths are temperature controlled and humidity, solvent vapor pressure, etc., in the atmosphere are also controlled. The membrane property is affected by nearly every parameter involved in the spinning process such as composition and temperature of the polymer dope, geometry of the spinneret, spinning speed, composition and temperature of nonsolvent, spinning room temperature and atmospheric composition, tension on hollow fiber and the take-up speed, bore liquid temperature and composition, etc. (Baker 2004).

1.6   WHY IS INCREASING FLUX HARD IN MEMBRANE PROCESS?

1.6.1   History of Flux

In recent decades, tremendous progress has been made in membrane technology with regard to specific energy consumption, membrane life, module fabrication, etc. However, the operating flux has been nearly stagnant despite all this progress (Lesjean et al. 2011). When it comes to submerged membrane, it is apparent that membrane manufacturers have focused more on reducing souring air demand by modifying module/frame design, improving module design to reduce capital and installation costs, improving membrane life, etc., instead of focusing on improving flux.

After the submerged membrane apparatus was invented by F. Tajima and T. Yamamoto in 1986 for radioactive substances treatment (Tajima and Yamamoto 1986), the same concept was adapted by K. Yamamoto for wastewater treatment in 1989 (Yamamoto et al. 1989). Soon after, submerged membrane modules were commercialized by Mitsubishi Rayon Co. using 0.1 µm polyethylene hollow fibers with small inner diameters (<0.3 mm), which allowed 10 to 15 LMH in municipal wastewater treatment. After the commercialization of submerged flat sheet membranes by Kubota Co., sustainable flux reached 20 to 25 LMH in the early 1990s. Nearly simultaneously, a similar flux was realized for hollow fiber membranes after dual layer hollow fibers with bigger inner diameters (~0.9 mm) were developed by GE (formerly Zenon Environmental Inc.). It is believed that the lower internal pressure drop is the major reason for the higher flux realized by Kubota and GE as discussed in Section 3.4.3. Since then, design flux has hardly improved for submerged membranes for MBR. Later in the middle of 2000s, Pentair (formerly Norit) developed an aerated tubular membrane system called Airlift™. It is more costly to set up compared with submerged membranes, but design flux had increased to 40 to 60 LMH with better membrane scouring by the combined effect of forced mixed liquor circulation and aeration through the vertically mounted tubular membranes.

Crossflow tubular membranes have been used for many decades for MBR since the first MBR based on tubular membrane technology was demonstrated in South Africa in the 1960s. New membrane materials have been introduced for longer service life while accommodating challenging water sources. Module design has also been improved to lower capital and operating costs, etc. Despite this progress, there have been no fundamental changes in flux. The flux seems to be decided by crossflow velocity and the nature of the feed water. According to a survey, the flux of a tubular membrane running under positive pressure still remains mostly at 60 to 80 LMH (Larrea et al. 2014).

In reverse osmosis of surface water, the standard operating pressure of 15 bar (or ~225 psi) has been reduced down to 10 bar (or ~150 psi) with the development of low pressure membranes, and again to 7 bar (or ~100 psi) with ultralow pressure membranes during the last two decades. In order to lower the operating pressure, membrane chemistry has been modified to allow faster water molecule transport through polymeric chains. Simultaneously, untrathin skin layer with less than 1 µm depth was introduced to minimize the filtration resistance. The ultrathin skin layer is coated on the PSU based UF membrane with 40–50 µm depth, which is casted on polyester based nonwoven backing materials. Due to the ultrathin skin layer, low operating pressure is achieved by improving membrane permeability without losing too much rejection efficiency. Low fouling RO membranes have also been developed in the recent decade by coating with polyvinyl alcohol (PVA), which reduces surface roughness and the charge interaction with foulants (Myung et al. 2005). Nonetheless, operating flux remains nearly identical at 15 to 25 LMH (or 9–15 gfd) in typical applications.

1.6.2 Self-Limiting Nature of Flux

High TMP or high membrane permeability does not guarantee long-term high flux in MF/UF due to the self-limiting nature of flux. Flux may increase temporarily, but it expedites not only the foulants' deposition but also cake layer compaction. Consequently, the initially high flux will decrease to a new pseudo-steady state similar to or even lower than the original flux depending on the extent of cake layer compaction, as illustrated in Figure 1.19 and discussed in Sections 1.2.3.1 and 1.2.7.1. This self-limiting nature of flux in MF/UF processes is also the reason that mass transfer–controlled region exists in Figure 1.18.

The flux of submerged membranes can be increased to some extent by performing more intensive aeration. However, intensive aeration is an effective means to mitigate particulate deposition rather than the attachment of macromolecules. Therefore, soluble macromolecules still deposit on membranes, forming a gel/cake layer. The high flux makes the gel layer denser and accelerates the membrane performance loss. Tubular membranes run under mass transfer–controlled region. High TMP results in higher flux only for a short period before it returns near the original flux. To obtain a high steady-state flux, membrane fouling must be better controlled by improving hydrodynamic condition on the membrane surface by increasing crossflow velocity or using turbulence promoters, but the overall economic feasibility does not always allow us to adopt such options.

1.6.3 ECONOMIC CONSTRAINTS ON FLUX

Design flux can be increased by increasing scouring air flow rate in submerged membranes or cross-flow velocity in tubular membranes. Capital costs can be partially saved due to the smaller membrane surface areas required to treat the same flow rate. Overall, capital cost savings can be realized with the extra expense of operating costs. Therefore, there exists an optimum flux with which total capital and operating costs become the minimum. This relation is illustrated in Figure 1.37. According to the figure, when initial flux is low, large capital cost savings can be realized with a small increase of flux. However, saving capital cost becomes increasingly harder as flux increases because the incremental savings in membrane area declines. With respect to the operating cost, a relatively large flux gain is possible with a small increase of energy input (or operating costs) at a low flux range, but it becomes increasingly harder as flux increases. Overall, the total cost becomes the lowest somewhere in the middle. Conceptually, most commercial membrane processes target the lowest overall costs, although the actual optimum point can vary depending on the market environment.

The optimum flux can increase if the capital cost line shifts upward or if the operating cost line shifts downward in the figure. Alternatively, optimum flux can increase if a new module design allows improved membrane scouring with the same energy demand. Apart from the economic constraints, high design flux increases the operational risks during peak flow even if the new modules have an improved membrane scouring mechanism. In fact, with a high design flux, a small number of membrane modules must deal with the additional flow compared with the system designed for a lower flux, thereby the incremental flux increase is greater during peak flow. With a small number of membrane modules, membranes are prone to fouling due to the disproportionally increasing particle deposition and cake layer compaction at high flux.

There are numerous examples of engineering compromise for better economics. One example is that the speed of passenger jets has hardly improved for several decades. Although there are no apparent obstacles in the sky that can limit speed, unlike for cars on the ground, the time required to fly from one continent to another remains nearly identical despite the great progress made in jet engine and flight design. It is simply because increasing the speed beyond the current level is not economical due to the aerodynamic limitations around the wings. Because the drag force is proportional to the "square" of the plane speed, less fuel efficiency is inevitable at higher speed. In addition, the resistance abruptly increases as the jet approaches sonic speeds. As a result, passenger jets servicing inter continents fly at a compromised speed of 70% to 80% of the sonic speed in high altitudes. For better overall economics, focusing on improving the turbine efficiency and the aerodynamics is more rewarding than focusing on improving the speed.

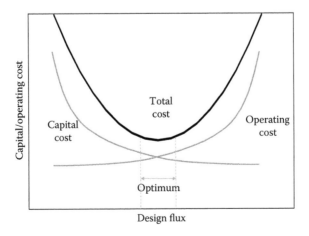

FIGURE 1.37 Conceptual diagram showing how optimum design flux is decided, where the x axis and y axis are in arbitrary scale.

2 Membrane Process

2.1 CLASSIFICATION OF MEMBRANE PROCESS

2.1.1 Depending on Pore Size

There are four different types of membranes depending on pore size and the molecular weight of the solute they can reject, as illustrated in Figure 2.1.

- *Microfiltration (MF)*—The membranes with a nominal pore size of 0.1 μm to 1 μm are typically called microfiltration membranes, although there are no universally accepted criteria. A vast majority of membranes have pore size distributions except for track-etched and patterned membranes. Nominal pore sizes are determined based on the smallest particles that a membrane can reject at 90% to 99% efficiency (depending on the manufacturer). It can be also determined based on pore size distribution curve, where the pore size that falls in the top 1% to 10% is called the nominal pore size. Pore size distribution is typically measured by a porometer, which measures gas flow rate through the membrane of which pores are prefilled (or wetted) with the liquids of known surface tension (Kim et al. 1994). While gas pressure is rising gradually, the largest pores are opened up first and the smaller pores follow sequentially in the order of pore sizes because large pores have a lower capillary pressure than small pores. The pore size distribution is calculated mathematically by analyzing the relation between the gas permeation rate and the gas pressure. In a practical membrane filtration occurring with a cake layer formed on the membrane surface, the cake layer acts as a dynamic membrane. As a result, membranes can reject much smaller particles than their pore size suggests. The quality of MF and UF permeates may be indistinguishable from each other in most applications in raw water and wastewater treatment once the cake layer is stabilized. Because cake layers dominate the membrane permeability in a vast majority of MF and UF applications, there are no substantial differences in operating pressures and other design parameters between MF and UF with few exceptions. MF and UF typically run at low pressure (e.g., 6 bar or below) whereas NF and RO run at relatively high pressure (e.g., 6 bar or higher).
- *Ultrafiltration (UF)*—Membranes with pores that are smaller than 0.1 μm are typically called UF membrane in the industry. UF membrane's pore size can be as small as a few nanometers, sufficient to reject macromolecules but not sufficient to reject inorganic ions. If pore sizes are smaller than 0.03 μm, they are often expressed as molecular weight cutoff (MWCO) determined by filtering surrogate molecules with known molecular weights. The approximate relation between pore size and MWCO is plotted in Figure 2.2 based on the approximate diameters of various proteins. Because the molecular weight is not the only factor determining the molecular diameter, MWCO should be considered only a rough guideline of pore size. In fact, the effective diameter of organic molecules vary depending not only on the conformation of them at varying pH, ionic strength, temperature, etc., but also on the molecular configuration. In addition, the actual size of the molecules rejected by membranes is highly affected by the characteristics of the cake layer formed on the membrane surface just like in MF. Hence, the actual MWCO observed can be significantly different from the nominal MWCO found in membrane specifications depending on the

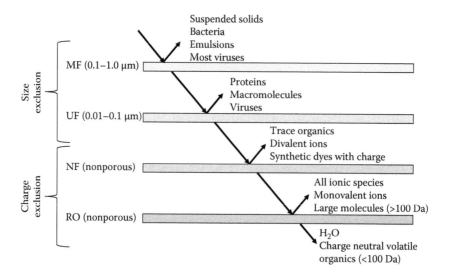

FIGURE 2.1   Conceptual diagram of the main particles rejected by different membranes.

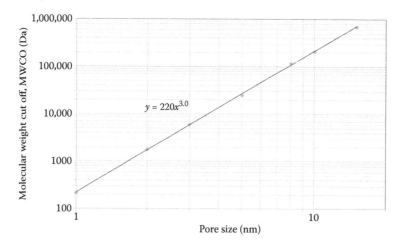

FIGURE 2.2   An approximate relation between pore size and molecular weight cutoff. (Data from Schoichet, M.S. and Sefton, M.V., *Immunoisolation*. In *Handbook of Biomaterials Evaluation*, eds. von Recum, A.F., Boca Raton, FL: CRC Press, 1999.)

operating conditions. Although the permeate qualities of MF and UF are indistinguishable in water treatment, some of the tight UF membranes with low MWCO are potentially better in recovering small proteins in food and pharmaceutical applications because cake layer may not be able to reject very small proteins. Additionally, the tight UF membranes provide an extra assurance of virus removal, if the pore sizes are smaller than the smallest virus, that is, 20–30 nm, although the actual virus removal efficiency may or may not be noticeably different from that of MF or loose UF membranes, as discussed in Section 2.6.4.1.

- *Nanofiltration (NF)*—This process was called nanofiltration to distinguish loose RO from tight RO. Although the name gives an impression that it has nanometer-sized pores, it does not have visible pores through electron microscopy. The major rejection mechanism of NF is charge exclusion rather than size exclusion, just like in RO. It can typically reject divalent ions at high efficiencies (e.g., 70%–99%) while rejecting monovalent ions at substantially lower efficiencies (e.g., 30%–80%). The operating pressure of NF is much lower than that of RO due to the high membrane permeability and the low osmotic pressure across the membrane; therefore, NF can be an economical option that can replace RO when the presence of monovalent ions in permeate is not a concern, for example, water recycling for cooling towers or surface/ground water treatment to produce potable water. With respect to MWCO, NF does not have clear-cut upper and lower borders with UF and RO, and hence NF is defined somewhat differently depending on the manufacturer.

- *Reverse osmosis (RO)*—RO membranes are nonporous and hence no pores are identified by scanning electron microscopy (SEM). The high hydraulic pressure on the feed side forces water molecules to dissolve in the polymeric membrane. Passing the spaces in between polymeric chains, water molecules emerge from the permeate side, where the hydraulic pressure is low. Similarly, impurities in the feed water undergo the same process to pass the membrane, but they tend not to dissolve in polymeric membrane, as discussed in detail in Section 1.2.10. Charged ions hardly permeate through the membrane, but small charge-neutral molecules pass the membrane easily depending on molecular size and electrical polarity. For instance, the rejection efficiency of lithium ions (6.9 g/mol) is much higher than that of much larger ethanol molecules (46.1 g/mol) due to the charge of lithium ion. It has been known that the rejection efficiencies of small charge-neutral organics such as methanol, ethanol, acetone, propanol, etc., are typically very low at a few tenths of a percentage.

2.1.2 MEMBRANE AND MODULE CONFIGURATION

There are three basic membrane configurations, as summarized in Table 2.1, but those are further split to the following six categories depending on how the membranes are fabricated as module.

TABLE 2.1
Summary of Membrane and Module Configurations

Membrane Configuration	Module Configuration or Operating Method	Driving Force	Pore Size	Common Applications	Example
Flat sheet (FS)	Plate and frame (PF)	Pressure	MF/UF	MM, WW, EDI	Pall DT™, Electrocell (EC)
	Submerged (no pressure vessel)	Vacuum	MF/UF	MBR	Kubota, Toray Membray, Microdyn Bio-Cel®, Pure Envitech®
	Spiral wound (SW)	Pressure	UF/RO	DS	DowFilmtec™, Hydranautics ESPA®, Woongjin CSM®
	Static discs with shearing paddle	Vacuum Pressure	MF/UF/NF/RO	MM, WW	BKT FMX™, Grundfos BioBooster™
	Rotating discs	Vacuum	UF	MBR	Huber VRM®
Hollow fiber (HF)	Contained in pressure vessels	Pressure	MF/UF/RO	WT, TT, MM	Asahi Microza®, Toyobo Hollowsep®, GE ZW1500
	Submerged (no pressure vessel)	Vacuum	MF/UF	WT, MBR	GE ZW500, Asahi Microza®, Mitsubishi Sterapore™, EconityCF
Tubular (TB)	Pressure filtration	Pressure	MF/UF	WW, MM	Koch Abcor®, Duraflow
	Vacuum filtration with bubbling	Vacuum	MF/UF	MBR	Norit Airlift™

Note: DS, desalination; ED, electrodialysis; MM, manufacturing; TT, tertiary treatment; WT, surface water treatment; WW, wastewater treatment.

- *Plate and frame (PF)*—PF-type modules are used mainly for small-scale applications that deal with difficult waters with high membrane fouling potential. This module's structure is inevitably complex, which necessitates the sealing of every corner and side of the flat sheet membranes under pressure, as shown in Figure 2.3. A large number of flat membrane assemblies are stacked up to increase the packing density. A high pressure loss is inevitable to maintain the high flow rate through the long and narrow channels with abrupt turns. The compact module structure enables easy steam sterilization. Due to the high capital and operating costs, PF is primarily suitable for applications that either create a large value, for example, food manufacturing, protein recovery, and pharmaceutical manufacturing, or for applications that do not have competing technologies that satisfy the process/regulatory requirement.
- *Spiral wound (SW)*—SW was developed in the 1970s to make PF modules more compact and less expensive. It provides large specific surface areas at low costs but is suitable only for clear feed water with a low amount of suspended solids due to the narrow feed channels filled with mesh spacers. Typically, spacers with 28 mil (0.71 mm) or 34 mil (0.86 mm) thickness are used, but the actual channel size for feed water is smaller than that. As shown in Figure 2.4, a spacer is placed in the feed channel to secure the spaces in between two membrane sheets and to create turbulence on the membrane surface. The permeate travels around the module through the permeate channel until it is collected in the center pipe. Multiple membrane leaves are used to reduce the pressure loss in the permeate channel by reducing the permeate's traveling distance. This module configuration has been adopted primarily for tight membranes such as NF, RO, and some UF, where dissolved ions and macromolecules are the primary target.
- *Tubular (TB)*—Tubular membranes have been used under crossflow mode with feed water recirculation through the lumen. Module structures are simple because the tubular membrane itself is sturdy enough to retain the pressure in most cases. In some tubular membrane systems, membrane tubes with thin walls are inserted in porous metal pipes (Figure 2.5). Because only the inner membrane tubes are replaced when necessary, maintenance costs can be lower for these modules. Due to the lack of obstructions in the feed channel and the simple flow lines, tubular membranes are compatible with very high amount of suspended solids. However, tubular membranes are mainly for high-value applications with specific needs due to the high capital and operating costs. It is less commonly used for low-value applications compared with hollow fibers, for example, surface water filtration,

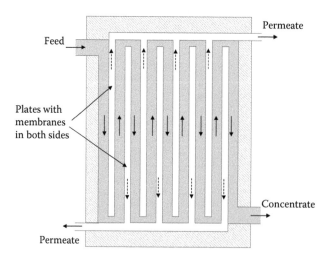

FIGURE 2.3 Schematic of a plate and frame module with multiple membrane plates (or panels) stacked inside a pressure vessel.

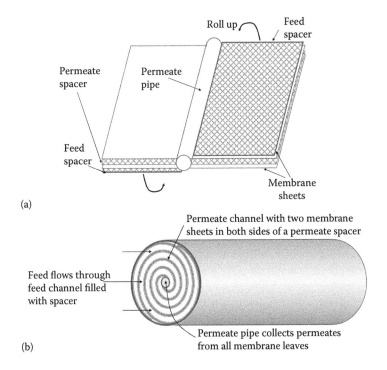

FIGURE 2.4 Schematic of a spiral wound module with two membrane leaves: (a) schematic of an unrolled SW element and (b) SW module. The sides and corners of the unrolled element are sealed with glue before rolling up.

FIGURE 2.5 Tubular membranes operated under positive pressure. (Courtesy of PCI Membranes, a Xylem company.)

municipal MBR, and pretreatment of RO. Recently, the economics of tubular membranes for MBR have been drastically improved by a new system configuration and a new operating method. In Pentair's Airlift™ process shown in Figure 2.6, air bubbles are introduced into the bottom of the vertically mounted modules using multitubes while mixed liquor is circulated through the membrane lumen by a low-pressure pump (Pentair Inc. 2013). Operating costs are much lower for this system compared with that of the traditional tubular membrane systems because high crossflow velocity at high pressure is not required.

• *Hollow fibers in a pressure vessel*—Membranes in a hollow fiber configuration can provide the highest packing density out of all membrane types. In a HF module containing hollow fiber bundles, various operational methods exist. Commonly, feed water is fed to the shell side of the hollow fiber while permeate is obtained from the lumen; however, if feed water does not contain large solids, it can be fed to the lumen. In most commercial modules used for water filtration, feed water is fed to the module at a near constant rate. The transmembrane pressure (TMP) is allowed to increase with gradual cake layer formation on the

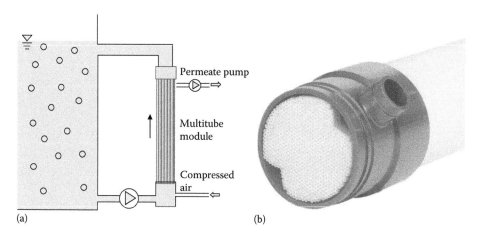

(a)   (b)

FIGURE 2.6   Process diagram of Airlift™ process and the multitube module (33 V model): (a) process diagram and (b) multitube module. (Courtesy of X-Flow, a Pentair company.)

membrane. This module is mostly operated under the dead-end mode without a continuous discharge of concentrate with exceptions. Periodic backwashing is performed to remove the cake layer with an optional air scouring on the shell side. This module is commonly used for surface water treatment, tertiary filtration of secondary effluent, product recovery, etc., where total suspended solid (TSS) in the feed water is not excessive to plug the module during dead-end filtration. Figure 2.7 shows hollow fiber membranes and a module that runs under pressure at crossflow mode.

- *Submerged membrane*—Hollow fiber and flat sheet membranes can be directly submerged into feed water without pressure vessels while permeate is obtained by vacuum. Instead of driving water through the modules to generate shear stress on the membrane surface, air bubbles are sparged underneath the membrane module to scour the membrane surface. Capital costs are the lowest among the membrane modules used for water and wastewater treatment due to the lack of pressure vessels and complex module structures. Operating costs are also low due to the high efficiency of the membrane scouring mechanism and the low vacuum pressure (mostly less than 0.3 bar). A vast majority of the MBR (>99% based on the treatment

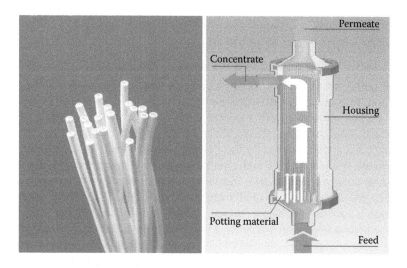

FIGURE 2.7   Microza® hollow fiber membrane and module. (Courtesy of Asahi Kasei Corp.)

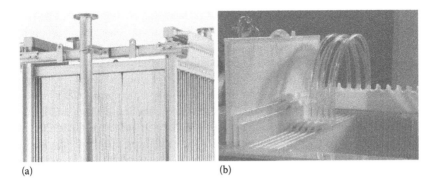

(a)   (b)

FIGURE 2.8 Submerged membrane modules: (a) hollow fiber (Courtesy of Mitsubishi Rayon Co., Ltd.) and (b) flat sheet. (Courtesy of Kubota Membrane USA Corporation.)

capacity) and a significant portion of the surface water filtration adopt submerged membranes. Figure 2.8 shows the two most popular membrane types used as submerged membranes.

• *Disc membrane with mechanical agitation*—Membranes can be rotated to generate mechanical shear stress on the surface and enhance membrane scouring. Alternatively, shearing paddles can be installed in between two membranes and rotated to generate shear stress on the membrane surface. The complexity of the module structure increases capital costs but it allows for lower operating costs due to the lack of a large amount of water moving under pressure as found in PF-type and TB-type membranes. FMX™ (Filmax) was developed by BKT Inc. for water that is difficult to filter, especially water with a high amount of solids and sticky macromolecules. The module structure is similar to conventional PF modules with disc membrane plates, but there are rotating vortex generators in between two membrane plates, as shown in Figure 2.9. The FMX™ system is efficient in filtering water that cannot be filtered by conventional membrane systems, for example, digestate and livestock wastewater, fruit juice, leachate, and frac/flowback water. MF, UF, NF, and RO membranes can be equipped on both sides of the discs depending on the filtration purposes. The commercial

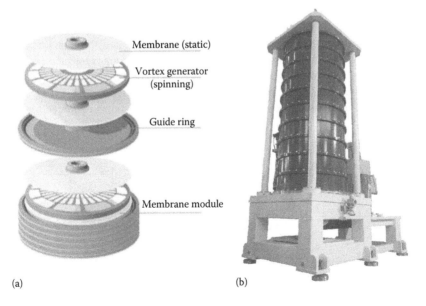

(a)   (b)

FIGURE 2.9 FMX-S plate and frame module with mechanical agitation: (a) module structure and (b) FMX-S class system. (Courtesy of BKT Inc. http://www.bkt21.com.)

FMX-S class systems come with a membrane surface area of 16 to 95 m². The BioBooster module by Grundfos BioBooster A/S of Denmark also adopts the same design principle as FMX™, but the pressure vessels are fabricated as small cylinders like spiral wound RO modules and hence it can be mounted horizontally in a skid. In this module, impeller discs rotate in between the aeration discs and ceramic UF membrane discs. By supplying air through the aeration discs, shear stress is generated on the membrane while oxygen is dissolved. Each 20-foot cylindrical vessel holds 16 sets of the filtration unit and is compatible with up to 50 g/L mixed liquor suspended solids (MLSS) and 5 bar (Judd 2011; Bentzen et al. 2012). Alternatively, membrane discs can be rotated at an eccentric position inside the cylinder without impeller discs at 140 to 200 rpm (Madsen and Bin 2013).

2.1.3 Sidestream MBR versus Submerged MBR

2.1.3.1 Sidestream MBR

MBR had relied mostly on tubular membranes and some plate and frame membranes until submerged membranes were commercialized in the early 1990s. Because mixed liquor is circulated through the membrane systems placed outside of the aeration tank, as shown in Figure 2.10a, it is called a sidestream membrane process. High liquid velocity on the membrane surface, such as 2 to 5 m/s, is required to control membrane fouling. The sidestream MBR provides reliable performance with less concern over membrane plugging or clogging due to the large channel size (e.g., 12–25 mm), compared with submerged MBR. In addition, individual membranes are easy to access for maintenance and can be cleaned with a small volume of cleaning solution.

However, the necessity of building a membrane system compatible with high pressure (3–6 bar) makes the capital costs high. In addition, the energy costs to circulate liquid to be filtered is prohibitively high for any low-value application such as municipal water and wastewater treatment because only 1% to 5% of the feed pumped to the membrane system is recovered as permeate in typical conditions and thereby 20 to 100 times more feed water should be pressurized/circulated than the permeate obtained. The specific energy costs of crossflow filtration are discussed in Section 2.4, but it is higher than 4 kWh/m³ permeate. Despite the high capital and power costs, sidestream MBR can be competitive for small systems, perhaps with less than 100 m³/day capacity, due to the relatively easy operation and maintenance.

The recently developed sidestream MBR systems called Airlift™ rely on a different operating concept. As discussed in Section 2.4.2.2, mixed liquor is circulated through vertically mounted tubular membranes at very low crossflow velocity while scouring air is injected into the inlet side of the tubular membranes. In this system, pressure loss through the membrane module is kept low due to the low water velocity. Permeate is obtained under vacuum. The specific energy consumption of this new system is dramatically lower than those of traditional sidestream membrane systems.

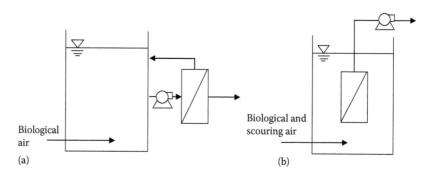

(a)   (b)

FIGURE 2.10 (a) MBR with sidestream membrane and (b) MBR with submerged membrane. Schematics of the MBR with submerged membranes or sidestream crossflow membranes.

2.1.3.2 Submerged MBR

The submerged hollow fiber membrane configuration shown in Figure 2.10b was originally invented to treat wastewater containing radioactive substances in nuclear power plants in Japan (Tajima and Yamamoto 1986). Later, the same concept was adopted by a different research group to separate activated sludge to take advantage of lower capital and operating costs compared with crossflow membrane filtration under pressure (Yamamoto et al. 1989). Due to the modest suction pressure (or TMP) at 5 to 30 kPa and the low flux, cake layer compaction is not as severe as that in the sidestream MBR. As a result, the permeability of the submerged membrane is maintained significantly higher than that of sidestream MBR (100–500 LMH/bar vs. 7–30 LMH/bar) despite of the similar clean water permeabilities, for example, 500 to 2000 LMH/bar for both.

Air scouring is performed in submerged membranes to mitigate membrane fouling. Because air scouring is inherently more efficient than fast crossflow in membrane fouling control, submerged membranes are more energy efficient than the sidestream membranes running under high pressure. It has been also discovered that continuous air scouring can be replaced with intermittent aeration (Côté et al. 2001). The intermittent aeration resulted in reduced specific energy consumption by approximately 30% for hollow fiber membranes when it is compared with continuous aeration. As a result, the submerged membranes, especially with a hollow fiber configuration, need approximately 1/40 to 1/100 of the energy required by tubular membranes (0.1–0.2 vs. 4–10 kWh/m^3).

As summarized in Table 2.2, the predominance of submerged MBR is undisputable as more than 99% of the total membrane surface area installed in Europe in the period of 2003 to 2005 used submerged membranes (Lesjean and Huisjes 2008). This trend is believed unchanged or strengthened for all regions of the world since then throughout 2010s.

TABLE 2.2
General Comparison of the MBRs with Submerged Membrane and Sidestream Membrane

Item	Unit	MBR (Submerged)	MBR (Sidestream)
Typical membrane configuration	–	Hollow fiber (HF)	Tubular (TB)
		Flat sheet (FS)	Plate and frame (PF)
Mode of operation		Crossflow	Crossflow
Operating pressure	kPa	5–30 (vacuum)	300–600
Sustainable flux	LMH (m/day)	15–35 (0.36–0.84)	50–100 (1.2–2.4)
Permeability[a]	LMH/kPa	0.5–5	0.07–0.3
Recycle ratio	m^3 feed/m^3 permeate	–	25–75
Superficial velocity	m/s	0.2–0.3[b]	2–6
SAD$_p$	m^3 air/m^3 permeate	7–30	–
SED[c]	kWh/m^3 permeate	0.1–0.5	4–12
Membrane cost[d]	\$/m^2	<50	>1000
Capital cost		Low	High
Operating cost		Low	High
Cleaning	–	Hard	Easy
Odor/VOC emission potential	–	High	Low
Packing density		Low	High
Market share[e]	–	99%	1%

[a] Permeability in operating condition.
[b] When gas superficial velocity is 0.02 to 0.04 m/s in FS (Yamanoi and Kageyama 2010).
[c] Specific energy demand including energy for permeate suction, but excluding biological aeration.
[d] Including module/frame/housing, if applicable, but not including installation costs.
[e] Membrane surface area based in municipal and industrial (Lesjean and Huisjes 2008).

2.1.4 Integrated versus Separated Membrane Tank

There are two different process configurations in submerged MBR depending on the existence of separate membrane tanks. In the early days, membranes were submerged directly in the aeration tank (Figure 2.11a). However, it was quickly realized that this configuration made membrane maintenance difficult because the entire aeration tank must be drained or membrane cassettes must be hoisted or craned out from the aeration tank before membrane cleaning, air diffuser cleaning, membrane replacement, and so on. As a result, separate membrane tanks started to be installed to the aeration tank (Figure 2.11b). In this configuration, all the mixed liquor in the small membrane tank can be transferred to the larger aeration tank whenever necessary for maintenance works.

Together with the added advantages in maintenance, separate membrane tanks act as a second stage aeration tank. Hence, the system behaves like a plug-flow reactor to some extent, and the chances of leaking untreated contaminants decrease. For example, the NH_4-N that was not oxidized in an aeration tank has a second chance to be nitrified in a membrane tank. Simultaneously, a separate membrane tank also has disadvantages when it is compared with an integrated tank: (1) larger

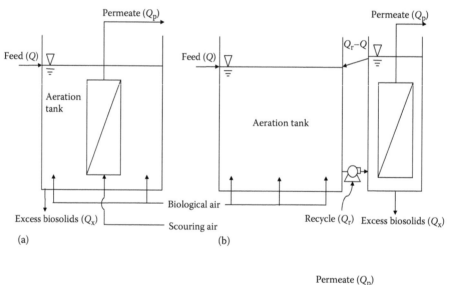

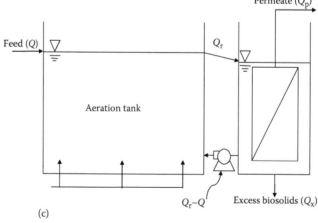

FIGURE 2.11 Configurations of submerged MBR: (a) integrated membrane tank, (b) separated membrane tank type 1, and (c) separated membrane tank type 2.

footprint, (2) higher capital costs, and (3) higher power costs due to the necessity for mixed liquor recirculation and the necessity for more biological aeration in the aeration tank.

The separate membrane tank design again has evolved to the design shown Figure 2.11c, where mixed liquor is pumped from the membrane tank to the aeration tank. This design provides a few advantages: (1) pumping energy savings by reducing the flow to be pumped, (2) potential capital cost savings by using recycle pumps with smaller capacity, and (3) (potentially) reduced membrane fouling by not sending the fine particles generated by the circulation pump directly to the membrane tank, as will be discussed in detail later in this section.

In a separate membrane tank design, mixed liquor is concentrated in the membrane tank depending on the ratio of recycle flow (Q_r) to feed flow (Q). The MLSS in the membrane tank must be controlled at a proper range to avoid the risk of clogging the membrane. The steady-state MLSS in the membrane tank can be calculated using Equation 2.2 based on mass balance around the membrane tank.

$$\text{Biosolids input} = \text{Biosolids output} + \text{Biosolids removal}$$

$$X_0 Q_r = X_m (Q_r - Q) + X_m Q_x \tag{2.1}$$

$$X_m = \frac{X_0 Q_r}{Q_r - Q + Q_x} \tag{2.2}$$

where
  Q feed flow rate (m^3/min)
  Q_r mixed liquor recycle flow rate (m^3/min)
  Q_x excess sludge removal (m^3/min)
  X_0 MLSS in aeration tank (g/L)
  X_m MLSS in membrane tank (g/L)

Equation 2.2 can be further modified to the following equation.

$$\frac{X_m}{X_0} = \frac{1}{1 - Q/Q_r + Q_x/Q_r} \approx \frac{1}{1 - Q/Q_r} \tag{2.3}$$

Equation 2.2 can be rearranged to Equation 2.3, where Q_x can be neglected because it is smaller than 1/100 of Q in most situations. Therefore, the relative ratio of the MLSS in the two tanks becomes a function of the inverse ratio of the mixed liquor recycle flow rate to the feed flow rate, that is, Q/Q_r. If the mixed liquor recycle ratio (Q_r/Q) is 2.5, the concentration factor (X_m/X_0) in the membrane tank will be 1.67, which means that there is 67% higher MLSS in the membrane tank than in the aeration tank. Because concentration factor does not decrease quickly once the recycle ratio exceeds 4, as shown in Figure 2.12, it is hard to justify a higher recycle ratio than that unless there are specific needs. Therefore, the recycle ratio is typically controlled at 3 to 5 to control X_m/X_0 at between 1.5 and 1.25.

As mentioned previously, one disadvantage of the separate membrane tank configuration is the higher energy consumption compared with the integrated tank configuration. In the integrated membrane tank configuration, the oxygen dissolved by the membrane scouring air can quickly be transferred to the surrounding areas and used by microorganisms to treat incoming chemical oxygen demand (COD). Due to the quick dissipation of oxygen, the DO around the membrane module stays relatively low and the driving force for oxygen dissolution is maintained high. On the other

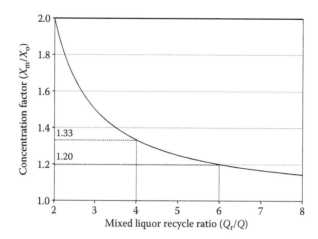

FIGURE 2.12 Relation between mixed liquor recycle ratio and concentration factor in membrane tank.

hand, in the separate membrane tank configuration, dissolved oxygen from the scouring air is con-sumed only for endogenous respiration in the small membrane tank. Some of the dissolved oxygen is transported to the aeration tank by the mixed liquor recycled, but it is not as fast as the convective oxygen transportation occurring in the integrated membrane tanks. Overall, the contribution of scouring air to oxygen dissolution is smaller in a separate membrane tank configuration due to the low driving force and thereby more fine bubbles should be supplied in an aeration tank. Based on a field survey, biological air took only 15% of the total air supplied in a MBR with integrated mem-brane tank designs (Fenu et al. 2010) whereas it took 59% of the total air separate membrane tank designs (Williams et al. 2008). It is also noticeable that the demands of biological air are affected by the membrane cassettes placement in the tank even in the same integrated membrane tank con-figuration. The oxygen dissolution from the scouring air can be best used by spacing membrane cas-settes evenly in integrated membrane tank configuration, although the operation and maintenance might be more challenging.

2.2 HOLLOW FIBER MEMBRANES

2.2.1 Outside-In versus Inside-Out Filtration Mode

2.2.1.1 Outside-In Mode

The biggest advantage of outside-in filtration is the capability of handling water with very high amount of suspended solids. All the commercial hollow fiber modules marketed for MBR run at outside-in modes. These modules are directly submerged in mixed liquor without pressure vessel and permeate is collected from the lumen side by vacuum. Air scouring is performed either continu-ously or intermittently to reduce solids accumulation on membrane surface. MemPulse™ (Evoqua), Microza MUNC-620A (Asahi Kasei), ZW500™ (GE Water), etc., are the examples (Figure 2.13).

Some outside-in modules used to treat the water with a low amount of suspended solids (<50 mg/L) run under dead-end mode without air scouring. In this filtration mode, all the feed water passes through membranes leaving solids on membrane surface. A large amount of solids can be stored in the spaces among membrane fibers before the module fails to produce design flow rate at the maximum pressure allowed. Backwashing is performed periodically to remove the accumulated solids whenever TMP exceeds the threshold or cycle time ends. Most hollow fiber modules for a low amount of suspended solids are enclosed in a pressure vessel and run under posi-tive pressure at dead-end filtration mode with occasional backwashing. Microza UNA-620A (Asahi Kasei), ZW1500 (GE Water), etc., are examples. Meanwhile, submerged hollow fiber modules are

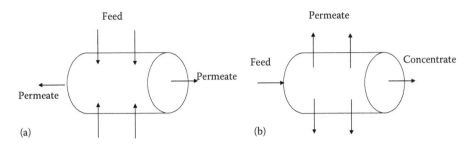

FIGURE 2.13 (a) Outside-in mode and (b) inside-out mode. Outside-in and inside-out filtration modes in hollow fiber membrane.

also used to treat low TSS water. Permeate is obtained by vacuum through submerged hollow fiber membranes and air scouring is performed only during the backwashing periods in order to enhance the cleaning efficiency. Hollow fiber membranes are either horizontally mounted (ZW1000 of GE) or vertically mounted (CMF-s of Memcor).

2.2.1.2 Inside-Out Mode

Hollow fiber membranes with an internal skin layer can run in the inside-out mode. This mode is good to maintain uniform hydrodynamics in the lumen, but it is practically not possible to generate turbulence in the lumen due to the small water channels. In fact, Reynolds number cannot exceed the threshold of 2100 under the practically possible operating conditions applicable for hollow fiber membranes. Another drawback is that only water with a very low amount of suspended solids can be filtered because the fiber lumen is prone to being plugged up by the particles. Once a fiber is plugged, concentrate from other fibers can flow back to the outlet of the plugged fiber, as illustrated in Figure 2.14, but the lack of a through-flow makes the plugged fiber fill with solids.

It is often required to connect membrane modules serially to obtain the desired concentration factor, where the concentrate of one module is fed to the next module just like in RO. One limitation of the hollow fiber modules running under inside-out mode is that it can hardly be connected serially due to the excessive longitudinal pressure loss through the small hollow fiber lumen. If multiple modules are connected serially, feed pressure in the first module becomes excessively high to overcome the pressure loss, thereby the flux of the first module becomes too high. The high flux in turn causes quick membrane fouling whereas the other modules suffer from insufficient feed flow. To solve this system design problem, X-FlowXIGA™ modules have feed water bypass channels through which a portion of the feed water bypasses to the next module without going through the membrane lumen. Because the longitudinal pressure loss is kept low by the bypass channel, the feed pressure to each module can be maintained reasonably constant and multiple modules can be serially connected in a pressure vessel just like in RO.

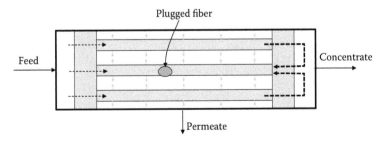

FIGURE 2.14 Hollow fiber plugging mechanism, where plugged fiber is gradually filled up with particles transported by the feed and concentrate from other unplugged fibers.

2.2.2 INTERNAL PRESSURE LOSS

2.2.2.1 Overview

In submerged hollow fiber membrane, internal (or lumen) pressure loss is one of the dominant factors affecting the performance of the membrane. As illustrated in Figure 2.8, permeate flow velocity in the lumen gradually increases toward the exit with additional permeate joining the flow. At the same time, the rising permeate velocity in the lumen causes increasingly larger pressure decline toward the exit. As a result, the suction pressure (or TMP) is the highest in the fiber exit and in turn the corresponding flux is the highest in the same place (Chang and Fane 2001; Chang et al. 2006a). Meanwhile, the static pressure in the lumen does not affect the filtration because it is always cancelled out by the static pressure in the shell side.

The TMP and the flux gradients along the hollow fiber can cause excessively high flux near the fiber exit whereas the average flux of the membrane remains low. Due to the high flux, membrane fouling starts in the fiber exit and as a consequence flux declines in the same area. The lost flux should be compensated by the increased flux in the adjacent area upstream under constant flux modes. Consequently, vacuum pressure (or TMP) must increase to transport more permeate from deeper spaces in the fiber.

2.2.2.2 Theory

In an infinitesimal block of the hollow fiber shown in Figure 2.15, the product of the cross-sectional area of the membrane lumen $\left(\pi D_i^2/4\right)$ and the flow velocity gain (dv) should be equal to the incremental permeate flow rate gain calculated by the product of the surface area ($\pi D_0 dx$) and the flux (J) in the same block as expressed in Equation 2.4. This equation can be rearranged to Equation 2.5.

$$\frac{\pi}{4}D_i^2 dv = \pi D_0 J dx \tag{2.4}$$

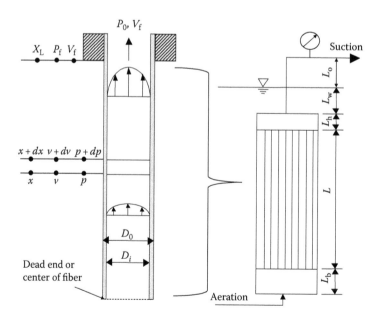

FIGURE 2.15 Internal pressure loss in hollow fiber, where X_L = effective membrane length (m), P_f = suction pressure in the potting entrance (Pa), V_f = flow velocity in fiber exit (m/s), P_0 = suction pressure in the fiber exit (Pa), L_h = potting depth (m), and L_o = elevation of pressure gauge from the water surface (m). (From Yoon, S.-H. et al., *J Membr Sci.* 234(1–2):147–156, 2004.)

$$\frac{dv}{dx} = \frac{4D_0 J}{D_i^2}$$
(2.5)

where
- D_i internal diameter of fiber (m)
- D_0 outer diameter of fiber (m)
- v liquid velocity in fiber lumen (m/s)
- J flux (m/s)
- x distance from the dead-end or the fiber center (m)

Pressure drop in the lumen side can be described by Equation 2.6 based on the Hagen–Poiseuille equation assuming no-slip on the lumen surface.

$$\frac{dp}{dx} = -\frac{32\mu v}{D_i^2}$$
(2.6)

where
- p pressure in fiber lumen (kg/m/s^2 or Pa)
- μ viscosity of permeate (kg/m/s)

Flux should be proportional to TMP for clean water filtration (Equation 2.7). Because R_c and R_f are zero in clean water filtration, the permeability constant, k, corresponds to $1/\mu R_m$ in the resistance in series model (Equation 1.4). Here, p_0 is the pressure outside the hollow fiber, but can be assumed zero because it is always offset by the static pressure in the lumen, thereby the internal pressure, p, is the same as effective TMP. The minus sign before p indicates that it is the vacuum pressure.

$$J = k(p_0 - p)$$
(2.7)

Equations 2.5, 2.6, and 2.7 can be solved simultaneously to obtain internal pressure and flux profiles. The average flux (or the apparent flux measured based on permeate flow rate and membrane surface area) of the hollow fiber is calculated using the following equation:

$$\bar{J} = \frac{\int_0^L J\, dx}{L}$$
(2.8)

The TMP measured by a pressure gauge in Figure 2.15 includes the effect of the pressure gauge elevation from the water surface. The TMP reading can be calculated by Equation 2.9 after adding the static water pressure in the pipeline between the water surface and the pressure gauge. The pressure loss in the permeate pipe is assumed negligible.

$$\Delta P_{T,reading} = \Delta P_T + 9.80 L_o$$
(2.9)

where
- $\Delta P_{T,reading}$ TMP reading by pressure gauge (kPa)
- ΔP_T effective TMP in hollow fiber module exit (kPa)
- L_o pressure gauge elevation from water surface (m)

If the permeate valve is located downstream of the pressure gauge and is closed during the pause mode, the water column between the pressure gauge and the water level of the membrane tank causes a vacuum pressure. In practice, the effective TMP (ΔP_T) is estimated by subtracting the static suction pressure measured during the pause mode from the TMP measured during the filtration mode ($\Delta P_{T,reading}$).

2.2.2.3 Internal Pressure and Flux Profile as a Function of Average Flux

The accuracy of the mathematical model was experimentally verified using a single hollow fiber membrane in clean water (Yoon et al. 2008). GE's hollow fiber membrane with a composite dual layer structure was used to filter clean water in the experiment. Details are found in the reference, but the effective inner diameter was estimated at 0.84 mm based on the analyses of pressure and flux data. As shown in Figure 2.16, the experimentally measured pressures match very well with the curves obtained from the mathematical modeling discussed in the previous section for many different average fluxes.

One consequence of internal pressure loss in hollow fibers is that the measured TMP is always higher than the effective average TMP. The overestimated TMP causes lower membrane permeability compared with the true membrane permeability. In the above experiment, the true permeability of the hollow fiber membrane used was estimated at 1.66×10^{-9} m/s/Pa or 598 LMH/bar based on the TMP gradient along the fiber. On the other hand, the apparent permeability based on TMP reading in permeate exit was much lower at 8.73×10^{-10} m/s/Pa or 314 LMH/bar.

As average flux increases, the magnitude of the imbalance between the highest and the lowest fluxes also increases in the same fiber. According to Equation 2.7, flux is proportional to the TMP for clean membranes so that the flux imbalance can be linearly proportional to the TMP imbalance. As shown in Figure 2.16, TMP imbalance is small when the average flux is 1.82 LMH, but it increases dramatically as flux increases. The increasingly uneven TMP (or flux) distribution at a high average flux can cause accelerated membrane fouling near the fiber exit. Once the membrane fouling is initiated in the fiber exit, it tends to propagate upstream until the filtration finds a new balance between TMP and flux distributions. The extent of the effective TMP and the flux gradients are dependent on the fiber dimension including fiber diameter and length. Therefore, the optimum flux that does not incur excessive TMP/flux imbalance can vary depending on the fiber dimension.

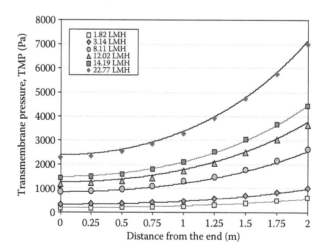

FIGURE 2.16 Comparison of experimental and theoretical data in clean water conditions for the hollow fiber membrane with 0.84 mm inner diameter, where markers and solid lines represent experimental data and theoretically calculated data, respectively. (From Yoon, S.-H. et al., *J Membr Sci.* 310:7–12, 2008.)

2.2.2.4 Road Map of Flux Profile in Hollow Fibers

2.2.2.4.1 Crossflow Filtration

In theory, if the highest local flux in a hollow fiber stays below the critical flux, membrane fouling would not occur at all and filtration could be performed indefinitely without causing a TMP increase. In this case, the only negative effect of internal pressure drop in the hollow fiber membrane would be the increased energy consumption to compensate for the internal pressure loss.

In practical situations, however, no true critical flux exists as discussed in Section 1.2.6. Small particles or macromolecules deposit on the membrane surface at any flux whereas the deposition rate increases exponentially as flux increases. Critical flux can be defined as the maximum flux that does not cause a fast TMP increase in practice although the criterion of judging the fast TMP increase is a bit subjective. In submerged MBR running under a constant flux mode, hollow fiber membranes start to foul from the fiber exit, where the flux is highest, and it propagates upstream. Figure 2.17 illustrates a road map of hollow fiber membrane fouling in submerged membrane modules, where filtration is performed with a continuous or an intermittent aeration. The progress of membrane fouling can be split into the following four stages:

1. *Stage I: Clean membrane*
   When membranes are clean at the very beginning of the filtration cycle, flux is the highest in the fiber exit and the lowest in the dead end or in the middle of the fiber, if the both ends of the fiber are open. This stage lasts for a very short period until initial membrane surface coating is completed by a mixture of colloidal particles and macromolecules.
2. *Stage II: Cake layer formation, line "a"*
   Cake layer formation starts on the exit side of the hollow fiber, where flux is the highest. Flux declines in the fiber exit to the level of the critical flux set by the hydraulic condition on the membrane surface and the characteristics of mixed liquor. The lowered flux corresponds to the flat area in curve "a". Simultaneously, the flux in the upstream increases to compensate the flux loss. TMP also increases to compensate the loss of average membrane permeability. Figure 2.18 shows the time curve of TMP in hollow fiber membrane filtration at three different yeast contents in the feed. At a constant flux (30 LMH), TMP increases until it reaches a steady state. In this steady state, flux profile in the hollow fiber corresponds to curve "a" in Figure 2.17.

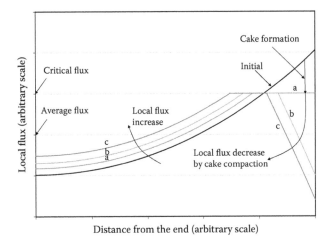

FIGURE 2.17   Conceptual road map of hollow fiber membrane fouling in submerged membrane.

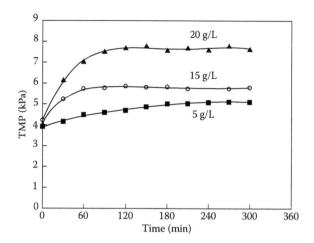

FIGURE 2.18 TMP—time curve for filtration of yeast solution of different concentration. A bundle of poly-propylene membrane with 0.65 mm OD, 0.39 mm ID, 51 cm length and 0.2 μm pore size were used at 30 LMH. Total number of fiber was 118 and scouring air flow was 11 L/min. (From Chang, S. et al., *AIChE J.* 48:2203–2212, 2002b.)

3. *Stage III: Cake layer compaction and its propagation, line "b" and "c"*

As filtration continues, the cake layer formed near the fiber exit becomes more compact due to the rearrangement of the deposits over time and the continuous filling up of void spaces by small particles. Cake layer compaction also plays a role starting from the bottom of the cake layer, as discussed in Section 1.2.3. Local flux near the fiber exit can decrease below even the critical flux as a consequence. The local flux in the upstream must increase to compensate the flux loss in the downstream. TMP also increases to compensate the membrane permeability loss. Hypothetically, if cake layer compaction does not occur at all, internal pressure loss does not accelerate membrane fouling significantly. It is because once the flux profile is set at curve "a," the cake layer resistance increases only through additional deposition, which is a very slow process under a subcritical flux. As a result, the flux profile can be stabilized at curve "a" for a much longer period under the hypothetical conditions.

4. *Stage IV: Rapid propagation of cake layer compaction*

As the cake layer compaction propagates toward the upstream of the hollow fiber, the available clean membrane area that can be used to compensate the flux loss in the downstream gradually decreases. As a result, the incremental flux increase in the upstream becomes larger as flux declines in the downstream. This fast-rising flux in the upstream accelerates cake layer compaction in the area. Simultaneously, TMP increases more rapidly than in stage III to compensate the rapid average permeability loss and to transport more permeate through a longer distance from the upstream. Finally, TMP reaches a threshold and the membrane needs to be cleaned.

2.2.2.4.2 Dead-End Filtration

The membranes used to filter the water with a low amount of suspended solids are often operated under dead-end filtration mode. If hollow fiber modules are used in dead-end filtration mode, they are typically operated under a constant flux mode without air scouring or liquid crossflow. Under these conditions, all the particles originally contained in the permeate deposit on the membrane surface, forming a cake layer. Although critical flux does not exist under the dead-end filtration mode, the membrane fouling sequences are nearly identical to those in the crossflow filtration mode.

1. *Stage I: Clean membrane*
   Because all the particles originally contained in the filtered water deposit on membrane surface, this stage lasts only an infinitesimal period at the beginning of the filtration.

2. *Stage II: Cake layer formation*
   As illustrated in Figure 2.19, the initially clean membrane starts to be fouled by deposits as soon as filtration starts. In dead-end filtration mode, strong particle back-transport does not exist due to the lack of the crossflow essential for shear-induced migration and inertial lift. Although particle back-diffusion from the concentration polarization exists, it is too weak to generate a meaningful back-transport velocity for colloidal particles. As a result, the entire fiber is subject to fouling simultaneously, but the highest flux in the fiber exit causes the quickest membrane fouling in the region. Unlike in crossflow mode, flat areas do not exist in the flux profile due to the lack of a meaningfully high critical flux. The flux profile becomes flatter as the filtration proceeds due to the rising flux in the upstream and the declining flux in the downstream, as shown by curve "a."

3. *Stage III: Cake layer compaction and its propagation*
   Cake layer compaction starts from the fiber exit whereas the upstream flux continues to increase to compensate the flux loss in the downstream. As a result, the flux profile continues to become flatter as shown by curves "b" and "c." Flux reversal can occur between the upstream and the downstream depending on the extent of cake compaction in the downstream as shown by curve "d." Meanwhile, feed pressure (or TMP) increases to compensate the overall permeability loss at a constant feed flow rate. In practice, feed flow can decrease somewhat at the end of the filtration cycle due to the limitation of the feed pump maintaining a constant flow at a high head pressure.

4. *Stage IV: Rapid propagation of cake layer compaction*
   As cake layer compaction propagates toward the upstream, the area of the membrane surface without a dense cake layer decreases. As a result, the same extent of flux loss in the downstream causes increasingly greater flux hike in the upstream toward the end of the filtration cycle. The accelerating flux hike in the upstream again expedites cake layer compaction. This leads to a quicker depletion of the available membrane surface area for filtration. Once the filtration enters this vicious cycle, TMP increases exponentially under a constant flux mode.

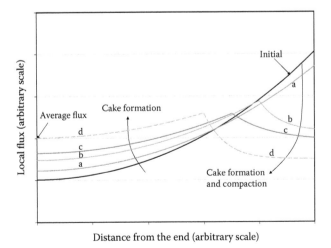

FIGURE 2.19   Conceptual road map of hollow fiber membrane fouling in dead-end filtration mode.

2.2.2.5 Effect of Internal Pressure Drop in Crossflow and Dead-End Modes

In crossflow filtration performed under positive pressure, a certain amount of internal (or longitudinal) pressure loss is inevitable in the feed channel. To overcome the pressure loss and drive feed water through the system, feed pressure should be maintained at a high level. As a result, feed pressure is much higher than necessary to obtain the target flux and hence the flux near the feed inlet tends to be excessively high even if the average flux is low. Membrane fouling can immediately start to occur near the feed entrance with the supracritical flux. The loss of flux is made up by the increased flux in the downstream, which causes a propagation of membrane fouling to the downstream. In addition, the cake layer undergoes a compaction process depending on the magnitude of the pressure exerting on it. The feed pressure required is typically one or two orders of magnitude higher than the minimum pressure required to obtain the target flux in tubular and plate and frame modules. In crossflow filtration, the internal pressure loss plays a crucial role not only in energy efficiency, but also in membrane fouling. To minimize membrane fouling, the internal pressure loss must be kept as low as possible within the intrinsic limitations imposed by the dynamic pressure loss under crossflow conditions.

On the other hand, when dead-end filtration is performed using hollow fiber membrane bundles enclosed in a pressure vessel in outside-in filtration mode, the effect of internal pressure loss on membrane fouling is less significant because the critical flux does not play a role in membrane fouling. As illustrated in Figure 2.20a, flux is the highest near the permeate exit when membranes are clean at the beginning of the filtration cycle. Because all particles contained in the filtered water deposit on membrane surface, cake layer formation is more intense near the permeate exit in the beginning (Figure 2.20b), but the intense particle deposition moves to the upstream toward the end of the filtration cycle. As the average membrane permeability declines, TMP gradually increases. At some point in time, the cake layer near the permeate exit starts to collapse and becomes more compact due to the high TMP in the area (Figure 2.20c). Cake layer growth and compaction continues to spread toward the upstream until the filtration stops when the TMP reaches the threshold (Figure 2.20d). With thinner fibers, the unbalanced filtration along the fiber and the flux loss by cake layer compaction become more severe. However, the effect of cake layer compaction on overall flux performance may not be excessive due to the short filtration cycle at 5 to 30 min in typical conditions. In addition, the benefit of having larger surface area in the same space by using smaller fibers can compensate more than the loss of flux performance. As a consequence, the hollow fiber membranes

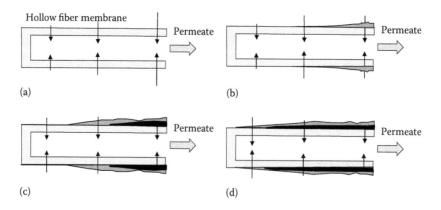

FIGURE 2.20 Progress of membrane fouling in dead-end filtration of hollow fiber membrane module under positive pressure. The length of the arrow indicates the flux in each location. (a) Clean membrane, (b) initial cake formation, (c) spread of cake compaction, and (d) end of filtration cycle.

operated in dead-end modes have much smaller diameters than those used for MBR, for example, Asahi UNA™, Evoqua CMF-S™, and GE ZW1500™.

2.2.2.6 Commercial Modules

As MBR technology develops and its popularity increases, the number of membrane manufacturers providing membrane modules suitable for MBR is growing. There are at least 50 companies around the globe with revenue from the membrane modules suitable for MBR (Wang et al. 2014c). It was often suspected that those companies would be consolidated to a smaller number due to the lack of sufficient revenue and profit to sustain the business in a commoditized market. However, the consolidation has not happened in a massive scale partly due to the decreasing costs of product development and manufacturing. Some of the notable companies are listed in Table 2.3 with their product specifications.

The most dominant technology is submerged hollow fiber membranes with well over half of the market share followed by submerged flat sheet membranes. In terms of membrane surface area installed, submerged hollow fiber and flat sheet membranes take over 99% of the market share whereas crossflow tubular membrane took less than 1% in 2003 to 2005 (Lesjean and Huisjes 2008). Since then, the newly introduced tubular membrane–based Airlift™ technology with a vacuum suction has chipped off some of the market share from submerged membrane technology, but the use of the traditional tubular membranes running under positive pressure is still rare for MBR due to high capital and operating costs.

TABLE 2.3
Commercial Membranes Used in MBR

Supplier	Membrane (Configuration, Material)	Pore Size (μm)	Panel Depth (FS), ID (MT/ST), or ID/OD (HF) (mm)	Panel Height or Fiber Length (m)	Clean Water Permeability (m/s/Pa)	Commercial Name
Asahi Kasei	HF, PVDF	0.1	0.7/1.3	2.0		Microza®
Brightwater	FS, PES	0.08	9			Membright®
Econity	HF, HDPE	0.4	0.41/0.65	0.2	$>2 \times 10^{-8a}$	CF series
Evoqua	HF, PVDF	0.04	0.8/1.3	1.6		MemPulse™
GE Water	HF, PVDF	0.04	0.9/1.8	1.0	1.7×10^{-9b}	ZeeWeed™
Huber	FS, PES	0.04	7			VRM®
Koch	HF, PES	0.05	1.2/2.6	2.0		Puron™
KO Red	FS, PES	0.2	16	1.2		Neofil®
Kubota	FS, cPVC	0.2	8	1.02/1.56	$>10^{-8a}$	Kubota
Mitsubishi Rayon	HF, PVDF	0.4	1.1/2.8	1.0	$>10^{-8a}$	Sterapore™
Norit	MT, PVDF	0.04	5.2 or 8	3.0		Airlift™
Origin Water	HF, PVDF	0.1, 0.3	2.0 (BSY), 2.4 (RF)	2		
Pure Envitech	FS, cPVC	0.4	7			Envis®
Sumitomo	HF, PTFE	0.1	0.8/1.3	2.0	2.5×10^{-9c}	Poreflon™
Toray	FS, PVDF	0.08	7			Membray™

Note: All data from MBR Network (2006) and Pearce, G. *Filtr. Sep.* 45(4):23–25, 2008, unless indicated.

[a] Failed to obtain an exact value due to the very high permeability.

[b] From Yoon, S.-H. et al., *J. Membrane Sci.* 310:7–12, 2008.

[c] Estimated from the data on a website (Sumitomo Inc. 2011. Poreflon™ module. Available at http://www.sei-sfp.co.jp /english/products/poreflon-module_1.html [accessed January 2, 2011]).

2.3 FLAT SHEET MODULES

The dimension of the flat sheet submerged membrane panels developed by Kubota has changed a few times since the 1990s, but the basic structure remains the same. Two flat sheet membranes sit on both sides of a corrugated board with small, evenly spaced holes on the surface. Membrane permeate passes the small holes on the board before it enters into the internal spaces of the corrugated board. After passing the narrow channels, it exits through the permeate outlet located at the top of the panel. The edges of the rectangular membrane unit are sealed with plastic frames that also provide the mechanical strength crucial to prevent the module from bending in two-phase turbulences. The thickness of the membrane panel is approximately 5 to 6 mm, which is essential not only to keep internal pressure loss low but also to provide enough mechanical strength in the turbulence. The panels are stacked to the horizontal direction in a cassette (or submerged membrane unit, SMU) with 6 mm spaces between two panels, as shown in Figure 2.8. In this type of membrane panel, periodic maintenance backwashing using diluted chlorine is less compelling for flat sheet membranes than for hollow fiber membranes because the large permeate channels inside the panel have little chance of being plugged up by the growing biofilm. For recovery cleaning, backwashing can be performed at very low pressure, for example, 0.3 to 0.6 m H_2O, using a chlorine solution (1000–10,000 mg/L) after pausing the aeration. The two membrane sheets bulge from both sides of the support plates by the backflow and meets with other membrane sheets in the middle of the upflow channels. As shown in Figure 2.21, the height of the cassette has been grown taller gradually in the last two decades to use scouring air more efficiently and to reduce the footprint. The initially one-story structure (ES type) built with 0.8 m^2 panels (Type 510) grew to a two-story structure (EK type) to double the membrane area per footprint and save the energy for scouring air. Later, EK cassettes grew to EW cassettes with larger panels with a surface area of 1.45 m^2 each (type 515). The specific aeration demands range 0.3 to 0.4 $m^3/m^2/h$ (SAD$_m$) or 11.6 to 16.6 (SAD$_p$) assuming the flux is 25 LMH. Recently, EW cassette has been replaced with RW cassette. Kubota also introduced a membrane cassette called SP400. This cassette includes 40 subcassettes and each subcassette includes 40 0.25-m^2 panels, which makes for a total number of 1,600 membrane panels and a total membrane area of 400 m^2 per SMU. According to the manufacturer's specifications, specific air demand per surface area (SAD$_m$) is 0.30 $m^3/m^2/h$ as discussed in Section 3.2.3.

Pure Envitech was one of the early starters in the late 1990s and has offered submerged flat sheet membranes. Envis® flat sheet membranes are based on a similar concept from Kubota. Pure Envitech introduced a stacked cassette design called SBM®, which is similar to the SP400, earlier

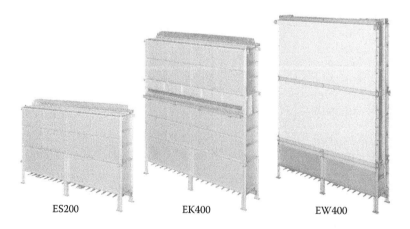

ES200   EK400   EW400

FIGURE 2.21 Various types of submerged membrane unit (SMU) equipped with flat sheet membrane panels: ES200 (0.8 m^2 × 200 ea., 2.0 m high), EK400 (0.8 m^2 × 400 ea., 3.5 m high), and RW400 (1.45 m^2 × 400 ea., 4.3 m high). (Courtesy of Kubota Membrane USA Corporation.)

than Kubota. SBM® cassettes consist of 50 to 230 SBM® blocks with 13 0.3-m³ membrane panels depending on the model.

Microdyn-Nadir's BIO-CEL® membrane and modules are unique from the perspective of membrane panel design. Two flat sheet UF membranes made of polyether sulfone (PES) with 0.04 µm pores are fused on porous nonwoven polyester plates. This module is backwashable at up to 15 kPa (Microdyn-Nadir 2011). Imbalanced filtration may be more significant than in the traditional flat plate panels due to the thin nonwoven polyester support layer (<2 mm) through which the permeate must flow. In addition, the membrane panels sway in the turbulent flow generated by air bubbles due to the flexible nature of the exceptionally thin panels. To reduce the internal pressure drop and control membrane swaying, the membrane cassettes are designed with a permeate collection pipe passing through the center of the membrane panels. The permeate travel distance is minimized whereas the permeate pipe provides an anchoring point for the membrane panels, as shown in Figure 2.22.

Although flat sheet modules can provide robust performance in MBR, they are not suitable for backwashing at high pressure and are often more expensive than hollow fiber membranes due to the high module manufacturing costs per membrane area. To overcome these limitations, fiber sheet membranes were developed by inge (currently BASF) in the late 2000s. In this module, membrane sheets were fused on both sides of porous spacers (supports) as shown in Figure 2.23a. Later, Anaergia Inc. developed a flat sheet submerged membrane called FibrePlate™ by putting two corrugated membrane plates together (Tomescu and Simon 2013). In this module, permeate channels are formed in between the two corrugated plates and the membrane panel appears hollow fiber membranes fused side-by-side, as shown in Figure 2.23b. Modules are placed in the membrane tank to have the permeate channels mounted horizontally. In this configuration, the effective membrane area is approximately 50% larger than a flat plate with the same height and width. Compared with hollow fiber modules, this new membrane configuration does not benefit from the random fiber displacement discussed in Section 3.4.2, but the grooves formed in between the permeate channels may act as turbulent promoters when air–liquid mixture flows transversally over the membrane surface. The actual performances of the backwashable flat plate modules are not well known publicly yet.

1. Fine bubble aeration
2. Activated sludge
3. Filtrate flow inside laminate
4. Effluent
5. Drainage layer
6. Membrane

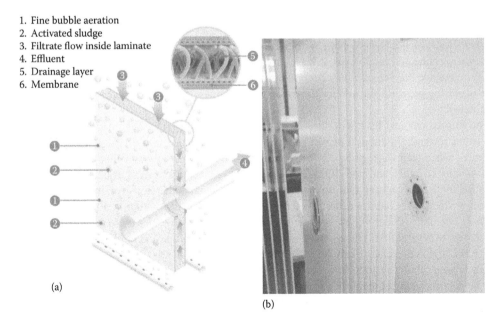

(a)

(b)

FIGURE 2.22 (a) Operating mechanism and (b) cassette (permeate pipe not installed). Operating mechanism and the membrane cassette of BIO-CEL®. (Courtesy of Microdyn-Nadir GmbH.)

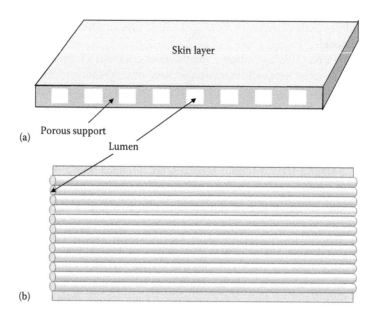

FIGURE 2.23 Backwashable flat sheet submerged membranes: (a) fiber sheet (FiSh) membrane by inge (currently BASF) and (b) FibrePlate™ membrane of Anaergia.

2.4 TUBULAR MEMBRANE

2.4.1 TUBULAR MEMBRANES WITH SINGLE-PHASE FLOW

2.4.1.1 Optimization Issue

A high crossflow velocity is necessary for tubular membranes to control membrane fouling. The high crossflow velocity in turn causes high pressure loss in the lumen. By serially connecting tubular modules, more permeate can be obtained from the pressurized feed and hence the energy might be more efficiently used. However, the longer channel causes more pressure loss under a high crossflow velocity, which is synonymous with higher feed pressure. Therefore, the efficiency gain from the serially connected modules is at least partially cancelled out by the increased feed pressure. It is also noticeable that the number of modules that can be connected serially is limited by the maximum operating pressure allowed for the tubular membranes, that is, 4 to 6 bar for typical polymeric membranes.

The limited channel length in turn limits the water recovery (Q_{perm}/Q_{feed}) from one pass of the feed water through the tubular membrane channel. For example, only 0.15% of feed water can be recovered when feed water proceeds 1 m through the membrane lumen, if flux is 100 LMH, crossflow velocity is 3 m/s, and tube ID is 2.54 cm. If 5% recovery is desired, the channel needs to be extended to 33 m, assuming that flux is uniform along the channel. However, it is unlikely that the crossflow velocity of 3 m/s is achievable in a 33-m channel at the maximum allowed feed pressure of 6 bar. Therefore, obtaining 5% recovery is not readily possible under these conditions unless the flux increases several fold. The typical design flux of tubular membrane in MBR is known to be less than 100 LMH and is more likely 60 to 80 LMH (Larrea et al. 2014).

In a typical tubular system shown in Figure 2.24 feed water is distributed to multiple tubes in the inlet header and merged again in the outlet headers. When the flow splits and merges, a turbulent fluid condition is generated and a significant portion of energy is lost, especially at a high flow rate. Simultaneously, energy is also lost while the velocity head converts back and forth to the pressure head due to the changing flow velocity when the cross-sectional area of the flow channel changes. Therefore, the total length of the membrane channel hardly exceeds 10 to 15 m in practice, whereas

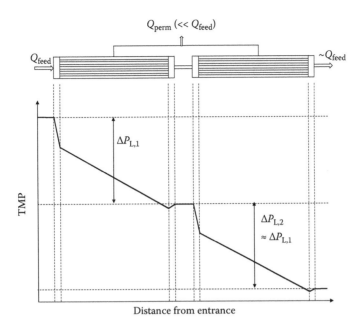

FIGURE 2.24 Longitudinal pressure profile in tubular membrane system with multitube modules in MBR.

TMP in the inlet is controlled at a reasonable level (4–6 bar). It is noteworthy that a high TMP is required to keep the flow moving at a desired velocity through the extended flow channels rather than to obtain the desired flux.

Reducing dynamic pressure loss is the key to improving the efficiency of tubular membranes. It allows either lower feed pressure at the same channel length or longer channel at the same feed pressure. Although substantial pressure is lost in the connecting areas between the two modules, longer membrane tubes help reduce the number of connecting points. The pressure drop caused by the splitting and merging flows in the feed and concentrate headers of a module can be also reduced by using single tube modules with couplers having identical inner diameters with the membrane tubes. In this case, feed water splits and merges only one time each before and after the multiple single channels. For instance, some of Koch ABCOR® tubular modules are 3 m long and consist of a single tube with an ID of 2.54 cm. All the tubes are connected with either straight couplings or U-tube couplings when flow return is required. The pressure loss is minimized by keeping the channel diameter constant at 2.54 cm from the entrance to the exit. If larger surface areas are required, more channels are added in parallel. As a result, the channels share only the entrance and exit in the entire streamline. Due to the constant channel size throughout the channels, longitudinal pressure drop can be kept lower than that with multitube modules.

Figure 2.24 shows the pressure profile in tubular membrane system, where pressure loss occurs not only in the membrane channel, but also in the headers of the module entrance and exit. These pressure losses are mostly uncontrollable because they occur whenever the fluid moves depending on the fluid viscosity, velocity, channel dimensions, etc. In the figure, feed pressure decreases sharply when the feed splits among multiple tubes due to the conversion of the pressure head to the velocity head as explained by Bernoulli's equation. Pressure continuously decreases as the feed water proceeds through the tubes. Pressure may increase slightly when the flows merge in the concentrate header again, where the velocity head converts back to the pressure head. However, it decreases again when the feed water (or concentration of the first module) splits in the second module. These patterns repeat until the feed water passes all the modules and exits the system.

The total pressure loss through the tubular membrane system (or feed pressure) can be expressed as Equation 2.10, where total pressure drop in the channel, $\sum_{i=0}^{n} \Delta P_{L,i}$, is added with the concentrate pressure of the last module, P_{out}, or the backpressure. Pressure loss in the pipeline connecting modules is assumed zero because it can be readily kept low by using properly sized pipes, smoothly curved U-tube connectors, and so on.

$$P_{in} = \sum_{i=0}^{n} \Delta P_{L,i} + P_{out} \qquad (2.10)$$

In the above equation, P_{out} is much smaller than $\sum_{i=0}^{n} \Delta P_{L,i}$ because outlet pressure is kept low, typically at 0.5 bar or below, whereas pressure loss in membrane tubes can be 2 to 6 bars depending on the number of tubes connected. Therefore, inlet pressure (or feed pressure), P_{in} is nearly proportional to the number of modules connected (n). Meanwhile, the energy required to pressurize water is approximately proportional to the feed pressure at a constant flow rate and the amount of permeate obtained is also roughly proportional to the number of modules serially connected. Therefore, connecting tubular membranes serially does not allow large energy savings unlike the popular notion.

Although energy savings are not substantial, the longer membrane channel allows high water recovery from one pass of the feed water. The high recovery results in less frequent passage of the feed water through the feed pump while producing the same amount of permeate. The less frequent exposure of feed water to the high shear stress exerted by the feed pump can be beneficial, if the solids in the feed are prone to breaking down and if the preservation of the solids is important.

2.4.1.2 Optimization of System Design and Operation

The flux of crossflow MF/UF is generally not sensitive to TMP because the system is running under a mass transfer–controlled region, where incremental TMP increase is countered by cake layer growth and compaction that soon brings the flux back to the original level, as discussed in Section 1.2.7.1. In addition, it is not sensitive to membrane permeability either because the cake layer dominates the total filtration resistance rather than the membrane itself, as discussed in Section 1.2.8. Under these circumstances, the most effective way of improving flux is by controlling cake layer depth by raising crossflow velocity. However, tradeoffs exists due to the high pressure loss at high crossflow velocity. The high pressure loss in turn requires higher feed pressures, but if feed pressure is already near the upper limit allowed by the manufacturer, membrane channel length needs to be reduced in order to not exceed the limit. The shorter channel length negatively affects the water recovery and the energy efficiency. In addition, the high crossflow velocity itself is negative on the water recovery and the energy efficiency by reducing the residence time of the water in the membrane system. As a result, increasing crossflow velocity does not indefinitely help improve the overall system efficiency.

The major design parameters of tubular membrane systems, for example, crossflow velocity, feed pressure, and channel length, are closely tied together. As discussed in the previous section, changing the system design in favor of one parameter causes drawbacks in other parameters and tends not to result in a huge gain in the overall system performance due to its self-limiting nature. There are no obvious optimization methods with regard to system efficiency, but simply following the basic design guideline while avoiding an extreme deviation from it is the practice. The inherently low efficiency hinders the tubular membrane from proliferation in the mass market.

To mitigate the cake layer compaction in the early stage of the filtration, permeate flow rate is controlled at a constant level in the BioPulse system (Berghof GmbH 2014). In this system, up to seven tubular modules with hundreds PVDF tubes are connected serially and crossflow velocity is maintained relatively low, that is, 1 to 1.5 m/s. While permeate flow is regulated at a constant rate, TMP is tracked to monitor the extent of membrane fouling. Under these conditions, permeate backpressure decreases as the membrane fouling proceeds and the TMP calculated by subtracting permeate backpressure from the feed pressure increases. If the TMP increases by more than 10% to

15% from the initial, crossflow velocity is increased automatically to 2 to 3 m/s until the initial TMP is recovered. Because the flux is kept low from the beginning of the filtration cycle, cake layer compaction is minimized and hence the filtration cycle is prolonged. If TMP recovery is not sufficient, backwash is performed with or without chemicals. In MBR, 4 m modules with a 10-in. diameter containing 8 mm tubes are used in a sidestream configuration. Because the pressure loss along one module is only 0.1 bar at 1 m/s excluding the pressure loss in the module headers, feed pressure can be kept lower than other tubular membrane systems running without permeate flow control.

2.4.2 Tubular Membranes with Two-Phase Flow

2.4.2.1 Horizontally Mounted Tubular Membrane at High Crossflow Velocity

In this process, compressed gas is injected into the inlet side of the multitube module whereas liquid circulation is performed as shown in Figure 2.25. When liquid velocity is high, for example, more than 1 m/s, the dynamics of the gas bubble is dominated by turbulence rather than gravity. Under these conditions, the system performance is not significantly affected whether the modules are mounted vertically or horizontally. The two-phase flow causes two conflicting effects from the perspective of system performance. The bubbles injected into the tube can increase the flux by generating turbulence, but they also increase pressure loss and reduce the allowable membrane channel length, as discussed in Section 2.4.1.1. Overall, the flux can be increased, but the maximum possible water recovery from the one pass of feed water through the system does not necessarily increase due to the shorter channel length.

If bubbles are injected into an existing system at a constant flow rate, feed pressure increases due to the increased pressure loss by the two-phase turbulence. The better membrane scouring caused by the turbulence improves flux under these conditions, but the gas injection does not necessarily guarantee the improved specific energy consumption because the higher permeate flow is obtained at higher TMP. Therefore, the gas injection rate and the liquid velocity should be optimized to minimize the total energy consumption. With two-phase flow, the optimum liquid flow rate seems to be below 1.0 m/s with same or reduced feed pressure. If flux is maintained similar to that in single-phase flow, the low feed flow rate may allow significant energy savings. In Linde's AXENIS™ process, pure oxygen is injected into the module used to filter the mixed liquor of the activated sludge process. The injected oxygen not only scours the membrane surface but also dissolves in the mixed liquor and is then used for biological oxidation processes.

2.4.2.2 Vertically Mounted Tubular Membrane at Low Crossflow Velocity

If tubular membranes are mounted vertically and gas bubbles are injected from the bottom of the modules, the gas bubbles injected not only generate turbulence but also circulates the liquid through the membrane system by the "airlift pump effect." Imasaka et al. (1989, 1993) came up with this idea during the Aqua-Renaissance Project sponsored by the Japanese government in the late 1980s. This

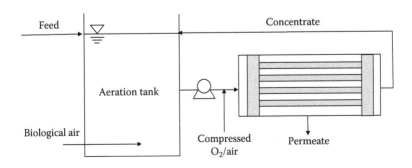

FIGURE 2.25 Schematic diagram of membrane bioreactor running with multitube module.

process aimed energy savings in anaerobic MBR equipped with tubular ceramic membranes by eliminating the high pressure circulation pumps. As shown in Figure 2.26a, anaerobic digester was placed approximately 10 m higher than the membrane to use the static pressure (~100 kPa or 1 bar) as a driving force for filtration. Compressed N_2, H_2, or the recycled headspace gas of the digester was injected to the inlet of the membrane module to circulate the anaerobic broth.

Because the liquid circulation relied on the airlift pump effect without a circulation pump, cross-flow velocity was variable depending on the gas flow rate. The linear velocity of the liquid in the ceramic membrane (ID = 3.8 mm, length = 0.5 m) ranged between 0.27 and 2.7 m/s when gas flow rate varied between 0.12 and 5.12 L/min. At an MLSS of 5000 mg/L and a static pressure of 100 kPa, flux increased from approximately 25 LMH to approximately 75 LMH whereas gas flow increased from 0.12 to 1.51 L/min. Overall, the specific energy consumption for filtration was estimated at 1.78 kWh/m³, which is much less than the observed value from the conventional system with single-phase flow, for example, more than 4 kWh/m³. Although specific energy savings were realized by using the airlift pump effect, the operational flexibility was limited in this process because water circulation and gas/air scouring are coupled together. As a result, this system could not be optimized further by varying the gas flow and liquid flow independently.

Later, Cui et al. (1997) demonstrated that air pressure could be used to provide the TMP for the filtration by throttling the gas/liquid outlet instead of placing the feed container at a high position. A study was performed to compare the energy efficiency of the airlift tubular membrane system running under backpressure with that of single-phase system. By using a 1.2-m-long tubular membrane with 12.7 mm diameter, the superficial liquid velocity of 0.03 to 0.14 m/s was obtained at a superficial air velocity of up to 0.012 m/s. They found that the two-phase airlift system had approximately 30% higher fluxes than the single-phase crossflow system without the need for a circulation pump. However, this system configuration was still limited by the coupled gas and liquid velocities.

The liquid velocity was decoupled from gas velocity by using a liquid circulation pump, as shown in Figure 2.26b, in the Airlift™ system of Pentair (formerly Norit). In this system, vertically mounted modules with hundreds of small tubes (5.2 mm inner diameter) and the crossflow water velocity is controlled at a desired level by circulation pump. Compressed air is injected to the bottom header of the membrane modules independently from the liquid circulation. As a result, the system can be optimized to obtain the maximum overall efficiencies combining the best air and liquid

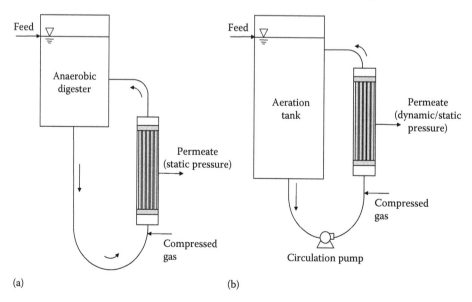

FIGURE 2.26 Schematic diagrams of membrane bioreactors running with multitube module using airlift effect: (a) no circulation pump and no permeate pump and (b) with circulation pump and permeation pump.

flow rates (Futselaar et al. 2007). Although pressure loss occurs when the two-phase flow splits among the small tubes and travels through the tubes in the module, it is at least partially compensated by the static pressure difference in between the aeration tank and the membrane tubes. In fact, membrane tubes are partially filled with air bubbles and hence the static pressure in the membrane system is lower than that in the aeration tank. In a laboratory test performed using 2-m-long tubes with 9.9 mm inner diameter, air injection into the bottom of vertically mounted tubes reduced the total pressure loss of flowing liquid at 2 to 10 cm/s velocity (Ratkovich et al. 2011). As a result of the lowered pressure loss with bubbling, the gas pressure required to bubble the system was also kept low. The low liquid and air pressure eventually allow the high energy efficiency of the process.

In X-Flow's Airlift™ MBR, the pressure differential of the circulation pump can be controlled to less than 0.5 bar at a low liquid crossflow velocity of less than 0.5 m/s in a short channel length (~3 m). The low differential head pressure results in significant energy savings relative to the conventional crossflow tubular membrane processes, where the head pressure ranges from 3 to 6 bar. Vacuum pressure can be used to obtain permeate in this process, but positive pressure is also available by throttling the concentrate flow. Although the design flux of Airlift™ in MBR is known at 30 to 50 LMH, two known municipal systems are designed at 31 LMH and 34 LMH, respectively (Larrea et al. 2014). The specific energy consumption for the filtration measured from the field is reported at less than 0.5 kWh/m³ (STOWA 2009).

Oppositely to Airlift™, mixed liquor is circulated from the top to the bottom in Berghof BioAirDS MBR although overall process configurations are similar (Berghof GmbH 2014). Compressed air is injected into the top of the module containing hundreds of 8-mm tubes. Liquid velocity is controlled at 0.2 to 0.5 m/s and permeate is obtained by vacuum at a TMP of 0.1 to 0.5 bar. Flux of 30 to 55 LMH and a specific energy consumption of 0.4 to 1.0 kWh/m³ are claimed for municipal MBR. If this process is compared to the process with the opposite flow direction, extra energy is required to circulate the liquid because air–liquid flow moves against the higher static pressure in the aeration tank. In addition, air bubbles move slower than liquid in this configuration because the flow moves against the buoyancy. This potentially compromises the air scouring effect to some extent. However, the air bubbles discharged at the bottom of aeration tank contribute to the oxygen dissolution and thereby a portion of the excess circulation energy spent can be recovered by saving the biological aeration.

Tubular membrane processes are classified in Table 2.4. The traditional processes with single-phase feed flow require the largest specific energy due to the necessity of the fast liquid circulation. But the modified processes with two-phase feed flow require much less specific energy because they are operated at low liquid velocity without an excessive cake layer compaction issue.

TABLE 2.4
Classification of Tubular Membrane Process

	Single-Phase (Traditional)	Two-Phase High Flux	Two-Phase Low Flux
Liquid velocity	3–6 m/s	1–3 m/s	<0.5 m/s
Gas/air injection	No	Yes	Yes
Typical flux in MBR	100–150 LMH	N/A	30–50 LMH
Permeate drawing	Positive pressure	Positive pressure	Vacuum
Configuration	Horizontal or vertical	Horizontal or vertical	Vertical
Typical channel length	6–15 m	~10 m	~3 m
Feed pressure	2–6 bar	2–6 bar	<0.5 bar
Specific energy consumption	>4 kWh/m³	N/A	<0.5 kWh/m³
Example	Duraflow DF series KMS ABCOR™	Linde AXENIX™	Pentair Airlift™

2.5 SUBMERGED ROTATING DISC MODULES

In the submerged rotating disc module, flat sheet membranes attached to both sides of the discs that rotates in mixed liquor generating shear stress on the membrane surface. During the Aqua-Renaissance 90 project in Japan in the late 1980s, the prototype rotating disc modules equipped with 0.1 μm ceramic membrane plates were tested in an anaerobic digester broth. The rotating disc module was encapsulated in a pressure vessel and permeate was obtained by pressurizing the vessel. Because the primary moving object was not water, the energy cost of developing shear stress on the membrane surface was low. However, the benefit of the energy savings was offset by the high capital costs of the module with complex structures.

More recently, a similar configuration was commercialized as VRM® (Vacuum Rotation Membrane) by Huber SE, with modifications, as shown in Figure 2.27 (Huber 2013). In this new configuration, ceramic membranes are replaced with less expensive polymeric membranes based on PES chemistry with 0.038 μm pores or a MWCO of 150 kDa. Permeate is obtained by vacuum and collected from one side of the rotation axis in the center of the submerged disc in an open tank. Air scouring is performed through the air pipe running through the hollow rotation axis while the disc assembly is rotating at a low speed at 0.9 to 2.5 rpm (Komesli et al. 2007; Melcer et al. 2009). Specific air demand per membrane area (SAD_m) at typical conditions is approximately 0.25 m³/m²/h. SAD_m is substantially lower than other flat sheet membranes but comparable to state-of-the-art hollow fiber membranes. However, the actual specific energy costs for the scouring air can be less than the SAD_m suggests because the air is introduced through the rotation axis against low static pressure.

Mechanical agitation provided by disc rotation is not a major antifouling mechanism in this module unlike the original rotating disc module tested decades ago. The maximum linear velocity on the edge of the disc is only 0.16 m/s if discs with 3 m diameter rotate at 1 rpm. Moreover, when the disc assembly is rotating, the mixed liquor in between the discs tends to move with the disc and thus the relative liquid velocity to the membrane becomes lower. Therefore, aeration seems to be a major membrane scouring mechanism whereas the disc spinning synergistically assists it. Another aspect of this design is that only less than half of the membrane surface located within a certain angle above the axis is air scoured in any given moment because air is supplied from the rotation axis. If a quarter of the disc area is within the aeration zone in each moment, the aeration frequency will be

FIGURE 2.27   VRM® membrane unit from Huber SE: (1) scouring air, (2) permeate, (3) rotating hollow shaft, (4) membrane segment with stacked membrane discs, and (5) permeate collecting pipe. (Copyright of HUBER Technology, http://www.huber-technology.com.)

15 s ON and 45 s OFF at a disc rotation speed of 1 rpm. Meanwhile, intermittent suction cycles can be adjusted to accommodate the membrane fouling potential in a given condition.

2.6 FILTRATION OF WATER WITH A LOW AMOUNT OF SUSPENDED SOLIDS

2.6.1 PRESSURE FILTRATION

The performance of submerged membranes does not improve too much by lowering MLSS if other operational parameters, such as DO, pH, SRT, F/M ratio, and others, are in a reasonable range. This is partly because the major membrane foulants are not the over micron-sized particles that take the vast majority of MLSS, but the submicron-sized fine particles and macromolecules called SMP, which originate from dead microorganisms and extracellular biopolymers. Because the fine particle concentrations do not decrease proportionally to the MLSS, low MLSS does not guarantee low membrane fouling. In addition, the membrane scouring effect by microbial flocs decreases as MLSS decreases due to the scarce large floc population. For example, although secondary effluent has very low MLSS, for example, less than 50 mg/L, membrane fouling rate is not necessarily low in the filtration process. The fine particles and macromolecules contained in such water can easily adsorb on the surface without being interrupted by the large flocs and their back-transport velocity is not affected by air scouring as much as large flocs do.

Due to the low efficacy of air scouring at very low MLSS, operating membranes without scouring air can be more economical than operating them with scouring air for a slight gain in flux, although membrane fouling can occur quicker. Under these conditions, all the fine particles contained in feed water deposit on the membrane surface, but the cake layer can be removed by periodic backwashing. In fact, this dead-end filtration is commonly practiced in the filtration of secondary effluent, surface water filtration, RO pretreatment, etc., where TSS in feed water are low at approximately 10 to 50 mg/L or below. Hollow fiber membrane bundles packed in a pressure vessel are typically used in outside-in filtration mode. Feed water is fed to the shell side of the membrane at a constant flow rate in a pressure vessel-type module (Figure 2.28). Flux is set at 22 to 45 LMH for secondary effluent recycling (Côté et al. 2004) and at 40 to 80 LMH for surface water filtration. Because the

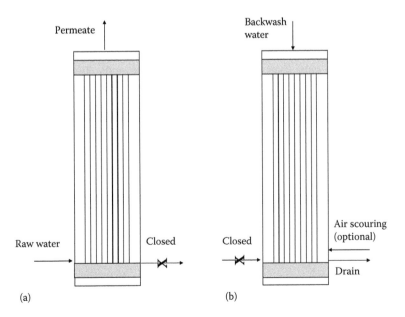

FIGURE 2.28   Dead-end filtration with a hollow fiber membrane in a pressure vessel: (a) filtration mode and (b) backwash mode.

concentrate outlet is closed, all feed water fed to the module is produced as permeate whereas all the particles contained in the feed water deposit on the membrane. Once the TMP reaches a preset threshold, backwash cycle starts with an optional air scouring to enhance cake layer detachment.

The threshold TMP can be set at 100 to 300 kPa (or 1–3 bar), but the lower the threshold TMP is, the easier the membrane cleaning is in general due to the less severe cake layer compaction. Filtration time and backwash time are widely variable depending on the feed water quality, but filtration time ranges of 5 to 30 min and backwash times of 15 to 60 s are typical. The backwash flow rate is set at 100% to 200% of the permeate flow rate to enhance cake layer removal. To enhance the cake layer removal, air scouring can be performed simultaneously by injecting air into the shell side of hollow fibers depending on the module. The commercial membranes falling into this category are Siemens MEMCOR® XP and CP, Asahi's Microza® UNA and UNS, GE ZeeWeed®1500, and others.

2.6.2 VACUUM FILTRATION

Submerged modules can also be used for water with low TSS. Permeate is obtained by vacuum under the outside-in filtration mode, as shown in Figure 2.29. Depending on the manufacturer, hollow fibers can be mounted either vertically (MEMCOR® XS and CS, Evoqua) or horizontally (ZeeWeed®1000, GE). Concentrate can be purged continuously or intermittently to maintain the solids concentration within the maximum limit while filtration is being performed. In some cases, feed water is concentrated in the membrane tank without being purged until a certain concentration factor is reached.

Flux is typically set at 20 to 45 LMH for the secondary effluent filtration and at 30 to 60 LMH for surface water filtration depending on feed water quality. No air scouring is performed during filtration in most commercial processes with some exceptions because the membrane performance increase is not sufficient to justify the cost of performing aeration when suspended solids concentrations are low. Once vacuum pressure reaches a threshold (30–80 kPa) or the cycle time ends (10–30 min), filtration stops and backwash cycle starts using the stored permeate. Air scouring can be performed simultaneously with backwashing to enhance cake layer detachment from the membrane.

2.6.3 EFFECT OF FLUX ON SYSTEM PERFORMANCE

The weight of the cake layer formed on membrane surface is proportional to the cumulative filtrate volume in dead-end filtration mode because all the particles in the feed water deposit on the membrane. However, the total filtrate volume obtainable in one filtration cycle is affected by the flux because the extent of the cake layer compaction is dependent on the flux, as discussed in Section 1.2.3.1. At low flux, more filtrate can be obtained before TMP reaches a threshold because cake

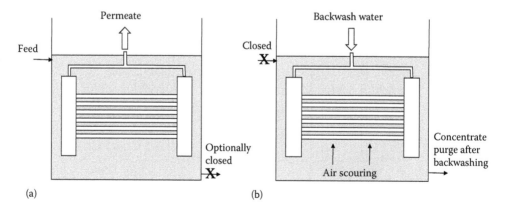

(a)   (b)

FIGURE 2.29 Dead-end filtration with submerged membrane, where vacuum is used to obtain permeate: (a) filtration mode and (b) backwash mode.

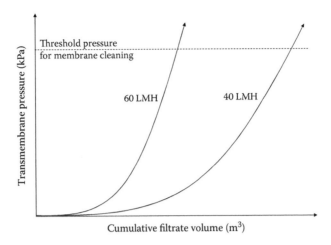

FIGURE 2.30 Effect of flux on the filtration duration and cumulative filtrate volume.

layers are not as compact as at high flux. Figure 2.30 is a conceptual graph showing the effect of flux on the duration of a filtration cycle.

The effect of flux on overall economics of the process can be determined by the average net flux over the period of filtration and cleaning cycles considering the volume of filtrate used for backwashing, chemical costs for backwashing, etc. Although low average flux is beneficial in minimizing the membrane-cleaning frequency and the operating costs, it results in high capital costs to prepare a large membrane surface area. Therefore, the optimum design flux is decided to maximize the overall costs while securing a stable operability without excessively frequent membrane cleaning. In a pilot study (Parameshwaran et al. 2001), secondary effluent from a sequencing batch reactor process was filtered using MF membrane at dead-end modes. TMP reached 20 kPa in 100 min at 30 LMH, but the same TMP was reached in 5 min at 90 LMH. As a result, the filtrate volume obtained at 30 LMH in one filtration cycle was sixfold more than that at 90 LMH. Considering capital and operating costs at various fluxes, it was concluded that 60 LMH was the optimum flux that allows minimum total costs.

2.6.4 PATHOGEN REMOVAL

2.6.4.1 Log Removal Value

The pathogen removal efficiency is commonly expressed as log removal value (LRV) in logarithmic scale. For example, if the treatment process removes 90% of the target pathogen, LRV is 1, and if it removes 99%, LRV is 2, and so on. LRV is calculated using Equation 2.11 by using either the pathogen counts in feed ($C_{p,feed}$) and effluent ($C_{p,eff}$) or the removal efficiency (R) of the target pathogen.

$$\text{LRV} = -\log_{10}\frac{C_{p,\text{eff}}}{C_{p,\text{feed}}} = -\log_{10}(1-R) \qquad (2.11)$$

In the above equation, it is noticeable that the LRV does not count on the concentration polarization effect in the membrane system, which makes the effective pathogen concentration on membrane surface higher than that in the bulk. In addition, the pathogens in feed water are gradually concentrated while the feed water proceeds to the exit. Therefore, the actual pathogen concentration membranes see is higher than $C_{p,feed}$. Considering this concentration effect, the effective LRV is somewhat higher than the apparent LRV, that is, by 0.2 to 0.4 units, depending on the recovery and the filtration condition in spiral wound RO modules. Under the current surface water treatment rule (SWTR) set by the U.S. Environmental Protection Agency, *Cryptosporidium*, *Giardia*, and viruses must be removed or inactivated more than 2 log, 3 log, and 4 log, respectively, in the drinking water treatment process (USEPA 2001a).

MF and UF are capable of removing viruses at high efficiencies, but it is difficult to obtain log removal credits from the government. In fact, MF does not provide an absolute physical barrier for the virus due to the large pore sizes. UF may provide absolute barriers for viruses, but the pressure-based integrity test methods are not capable of detecting the breaches as small as the smallest viruses due to the necessity of the high air pressure most membranes are not compatible with, as discussed in Section 2.7.2.1. The lack of proper integrity test methods is a hurdle in obtaining LRV credits in UF. In theory, RO is an ultimate barrier for the virus that allows very high LRV, but it is difficult to obtain more than 2 LRV credits in general due to the lack of accurate integrity test methods.

In the United States, state agencies consider a variety of factors when granting removal credits, including the demonstration of treatment efficiency, total removal/inactivation requirements, experience with the technology, and the approach toward multiple barrier treatment (USEPA 2001a). Although the actual LRV can be much higher, states rarely grant log removal credit in excess of the federal requirement. Most states grant 2.5 to 3.0 log removal credits for *Giardia* for MF and UF. Only seven states awarded 2 to 4 log for *Cryptosporidium*. Only a few states grant a virus removal credit for MF, but it is only up to 0.5 log due to the larger pore size than the smallest viruses and the lack of the integrity test methods that can detect virus-sized breaches. If the membranes are combined with coagulation, flocculation, chlorination, or clarification, some log removal credits can be given for viruses. Only seven states have awarded virus removal credit for UF up to 4 log.

2.6.4.2 Virus Removal

Viruses are submicron-sized creatures that consist of DNA or RNA enclosed inside a coat of protein. Due to the absence of ribosomes essential to produce proteins (or enzymes), viruses are parasitic to much larger bacteria in which viruses multiply utilizing the bacterial proteins. As a result, viruses are removed by MF/UF at very high efficiencies along with bacteria despite the small sizes (20–300 nm). In addition, the cake layer formed on membrane surfaces act as a dynamic membrane that rejects free viruses and hence the virus removal efficiency is not a strong function of membrane pore size. The following are some of the experimental observations found in the literature:

- When phage T4 (0.11 μm) was used as a model particle, a membrane with a pore size of 0.22 μm resulted in 5.7 log removal initially, but it increased to 8.0 log removal as cake layer matures (Lv et al. 2006).
- The removal efficiency of naturally occurring MS-2 coliphage could be only weakly correlated with membrane pore sizes as shown in Figure 2.31 and Table 2.5 (Hirani et al. 2010).
- When MS-2 coliphage with 0.03 μm diameter was spiked to an MBR treating municipal wastewater, pore size effect became more apparent (Hirani et al. 2010). The microfiltration membrane with 0.1 μm pore size showed 98.0% removal (or 1.7 log), but the membranes with 0.03 μm pore size showed 99.996% removal (or 4.4 log). Perhaps the spiked foreign viruses did not have enough time to be embedded in microbial floc before they were filtered by the membrane. However, the difference was still only approximately 2% although it could be critical with respect to sanitary and regulatory purposes.

The virus rejection efficiency varies depending on the age of the membrane, perhaps due to the different levels of reversible and irreversible membrane fouling in membrane pores and surfaces (Tazi-Pain et al. 2006). An example is shown in Figure 2.32, where 2-month-old membranes have higher MS2 phage removal rates compared with new membranes. It is also noticeable that the initial removal efficiency is somewhat better than the later stage removal efficiency. It is perhaps because

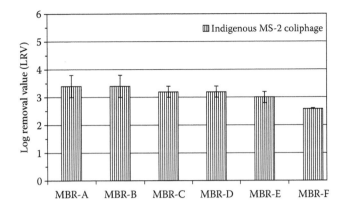

FIGURE 2.31 Indigenous MS-2 coliphage removal by the MBR systems evaluated in the study. Nominal pore sizes are (A) 0.05 μm, (B) 0.04 μm, (C) 0.08 μm, (D) 0.03 μm, (E) 0.1 μm, and (F) 0.1 μm. (From Hirani, Z.M. et al., *Water Res.* 44(8):2431–2440, 2010.)

TABLE 2.5
Effect of Pore Size on Permeate Quality

	Units	Kubota	US Filter	Zenon	Mitsubishi
Commercial designation		Type 510	MemJet B10R	ZW 500d	Sterapore HF
Configuration		Flat sheet	Hollow fiber	Hollow fiber	Hollow fiber
Material		PE	PVDF	PVDF	PE
Pore size		0.4	0.08	0.04	0.4
Effluent turbidity[a]	NTU	0.08 ± 0.02	0.04 ± 0.02	0.06 ± 0.02	0.07 ± 0.02
Effluent COD[a]	mg/L	18.4 ± 9.6	20.5 ± 13.4	17.3 ± 8.6	23.2 ± 5.3
Effluent total coliform[a]	MPN/100 mL	13 ± 69	386 ± 674	807 ± 1314	7 ± 7
Effluent total coliphage[a]	PFU/100 mL	10 ± 24	13 ± 13	1 ± 1	13 ± 22

Source: Reproduced from DeCarolis, J.F., and Adham, S., *Water Environ Res.* 79(13):2536–2550, 2007.
[a] Average ± standard deviation.

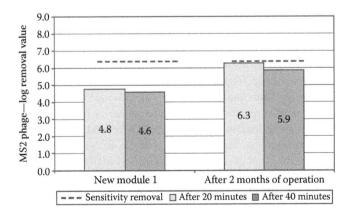

FIGURE 2.32 Virus removal efficiency of new and old membranes at different time frames in a filtration cycle. (From Tazi-Pain, A. et al., *Desalination.* 199:310–311, 2006.)

viruses accumulate in the cake layer during the filtration cycle, but the exact cause is not clear. Similar observations were made when the virus removal performances of removable and unremovable cake layers were compared using MS-2 and ϕX174 in drinking water treatment. The unremovable dense cake layer improved log removal by as much as 2.5 log, but removable loose cake layer improved log removal by only 0.5 log (El Hadidy et al. 2014).

MF and UF can remove viruses at very high efficiencies, but there are barriers to obtaining regulatory approval as a means of removing viruses (USEPA 2005). It is primarily due to the inability to detect breaches as small as the enterovirus, for example, approximately 0.02 μm. To detect such small breaches, air pressure should be increased to at least 3 to 5 bar, but most MF and UF used for bulk water treatment are not compatible with such high pressures. For wastewater recycling projects that may involve indirect human contact, obtaining regulatory approval is often harder for MF than for UF due to the concerns over the lack of an absolute barrier for viruses.

2.6.4.3   Coliform Removal

Fecal coliforms are the bacteria that live in the digestive tract of warm-blooded animals and are excreted through the feces. Fecal coliforms generally do not cause disease for humans and animals, but their population density is used as an indicator of the potential presence of pathogens (UNC 2014). Fecal coliforms are bacteria that are much larger than viruses, and can be easily filtered by 0.45 μm filters even if they exist as single cells. Therefore, the data summarized in Table 2.5 does not show any correlation between the fecal coliform removal efficiency and pore size. According to one study (Nishimori et al. 2010), 5 log coliform removal was achieved with 0.4 μm flat sheet submerged membranes for more than 10 years in two full-scale municipal MBR plants in the United Kingdom.

Despite the very high rejection by membrane, the microbial colonies grown on the lumen side of membranes and the internal surfaces of permeate pipes can be major sources of the microorganisms detected in permeate. Periodic chemical backwashing of the membranes using NaOCl and the sanitation of permeate pipes are necessary if the microbial contamination of permeate are to be controlled tightly. For disinfected tertiary effluent, California Title 22 regulations require a minimum chlorine exposure of 450 mg-min/L or a 5 log virus inactivation in addition to the effluent total coliform concentration not exceeding a 7-day median of 2.2 MPN/100 mL (California EPA 2013).

In the meantime, membrane pore size has little influence on effluent turbidity and COD because the particle removal efficiencies are hardly affected by the pore size. Although very small particles including viruses can be rejected better by the membranes with small pores, the concentrations of these particles are not high enough to noticeably change the bulk water quality parameters in MBR.

2.7   MEMBRANE INTEGRITY MONITORING

2.7.1   Overview

Integrity monitoring has been a critical part of the membrane process in surface and ground water filtration to produce potable water preventing potential biohazards from protozoan parasites resistant to free chlorine, for example, *Giardia* (6–18 μm) and *Cryptosporidium* (3–7 μm). On the contrary, maintaining rigorous membrane integrity is generally not a priority in MBR and tertiary filtration because the effluent is not for direct human contact or consumption in most cases. It is also because the effluent can be readily disinfected by UV or chlorine (or both) to meet regulatory requirements before being discharged or reused. As a result, integrity monitoring beyond turbidity or particle counting (or both) is not commonly performed in MBR. However, integrity monitoring has received more attention recently as the indirect or direct reuse of wastewater for potable purposes is becoming a promising option especially in the arid states in the southwest of the US, as will be discussed in Section 7.3.

All commercial MF and UF membranes commonly used for water treatment have small enough pores (0.02–0.45 μm) to remove bacteria and protozoan parasites as long as membranes are intact without defects. They are also capable of removing a vast majority of viruses typically at an efficiency of more than 99% because the majority of viruses are embedded in bacteria and the dynamic membranes (or cake layer) formed on membrane surfaces are capable of rejecting most individual viruses.

Membrane integrity test methods can be divided into direct methods and indirect methods, as summarized in Table 2.6. The existing direct integrity test methods include air pressure decay test (PDT), diffusive air flow monitoring (DAM), sonic test (ST), bubble point (or visual bubble) test (BPT), etc., where PDT is most commonly used as an off-line test method. Although the PDT cannot locate leak points, the ST and BPT can (Crozes et al. 2002). Membrane integrity can also be checked indirectly by tracking turbidity, particle counting, microbial counting, and the like.

The detection ranges of various integrity test methods are compared in Figure 2.33. The log removals in x axis are based on the rejection efficiencies of protozoa such as *Giardia* and *Cryptosporidium* at the time of detection. For example, if a method can detect the integrity breach when the *Cryptosporidium* removal is 99% or less, the sensitivity of the method is 2 LRV (= − log(1− 0.99)) for *Cryptosporidium*. As shown in the figure, direct methods such as PDT and diffusive air flow (DAF) are much more sensitive than indirect methods such as turbidity and particle counting (Johnson 1998). The extent of leakage is not only affected by the number of compromised membrane fibers but is also affected by the location of the cut or damage. If damage occurs far from the permeate exit in hollow fiber membrane, less water passes through the breaches due to the internal pressure loss in the fiber compared with the case with damages near the permeate exit (Mi et al. 2005).

Most submerged membranes popularly used in MBR are not suitable for integrity tests targeting viruses and bacteria leaks. Submerged composite hollow fiber membranes with two-layered structures (laminated membrane on braids) may be capable of holding up to 50 to 70 kPa of air pressure, but it is only sufficient for detecting breaches for protozoan parasites. Submerged flat

TABLE 2.6
Direct and Indirect Membrane Integrity Monitoring Methods

	Direct Methods	Indirect Methods
Definition	Directly monitoring integrity breaches on membrane	Monitoring water quality
Advantages/disadvantages	Mostly simple and accurate	Continuous monitoring with low accuracy (turbidity/particle monitoring)
	Typically *in situ* measurement	
	Not for continuous monitoring	Intermittent monitoring with high accuracy (bacterial/viral count)
	Filtration must be paused during the test	
Examples	PDT	Turbidity monitoring
	VDT	Particle count monitoring
	DAM	Particle index monitoring
	BPT	Bacterial/viral challenge test
	ST	Nanoparticle challenge test
		Fluorescent dye challenge test

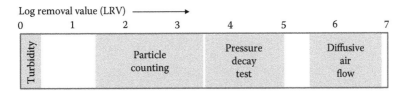

FIGURE 2.33 Relative sensitivity of various monitoring methods in detecting protozoa. (Data from Johnson, W.T., *Filtration & Separation*, 35(1):26–29, 1998.)

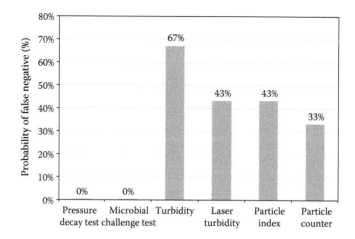

FIGURE 2.34 Comparison of the accuracy of some integrity test methods. False negative means the case where leaks are conceived intact. (Data from Liu, C., and Wachinski, A., Integrity testing for low-pressure membrane systems. *Proceedings of NEWWA Spring Joint Regional Conference & Exhibition.* April 1, 2009. Worcester, Massachusetts.)

sheet membranes in theory can hold air pressure, but it is barely sufficient for determining the leak potential of even protozoan parasites. If submerged flat sheet membranes can be taken out from the tank, vacuum decay tests (VDT) can be performed at a higher vacuum pressure, for example, 50 to 80 kPa, but it is procedurally challenging. Single-layer submerged hollow fiber membranes have an advantage in integrity tests because high air pressure (200–300 kPa or higher) can be applied to the lumen for PDT or DAM tests targeting viruses *in situ*.

Integrity tests fail to detect integrity breaches at times and produce false-negative results. The accuracies of various integrity test methods are compared in Figure 2.34. Although microbial challenge tests (indirect method) are accurate by definition, a direct integrity test method, PDT, can closely match it except for viruses. On the contrary, all other indirect methods such as turbidity and particle indexing/counting are not only insensitive, but also have significant chances of false-negative readings at a probability of 33% to 67% due to the fouling of optical cells, the interferences of air bubbles, and so on.

2.7.2 Direct Method

2.7.2.1 Bubble Point Test

The purpose of this test is to determine the size of the largest pore (or defect) on the membrane. The site of the leak can also be located under certain circumstances. As shown in Figure 2.35, if wet hydrophilic membrane is exposed to air, the water filled in the pore refuses to evacuate from the pore due to capillary suction pressure. As air pressure increases, the water filled in the largest pore/defect is pushed out first once the air pressure exceeds the capillary suction pressure. The lowest pressure that opens up the largest pore is called the bubble point. The size of the largest pore (or defect) is estimated using Equation 2.12 (Côté et al. 2002). Here, the contact angle is defined as shown in Figure 1.29.

$$d = \frac{4B\gamma\cos\theta}{\text{BP}} \tag{2.12}$$

where

BP   bubble point of defect (Pa)

B capillary constant (0–1)

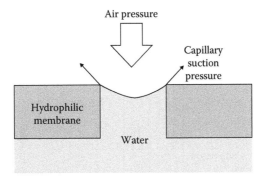

FIGURE 2.35 The role of capillary suction in hydrophilic membrane, where the water inside pore resists against air pressure due to the capillary suction pressure.

  d pore/defect diameter (m)
  γ surface tension at the air–liquid interface (0.0728 N/m at 20°C)
  θ contact angle (degree)

Figure 2.36 shows the relation between bubble point and pore size for three different contact angles at 20°C. Capillary constant was assumed to be 0.5 to reflect the nonideal pore shapes and the foulants adhering near the pore entrance in most practical conditions. According to the graph, at least 10 to 50 kPa is required to ensure the membrane integrity sufficient to reject protozoan parasites such as *Cryptosporidium* and *Giardia* whereas at least 100 to 200 kPa is required for bacterial pathogens. Therefore, all the vacuum-based integrity methods are suitable for ensuring the removal of protozoan parasites. To detect integrity breaches as small as enteroviruses, high pressures such as 300 to 500 kPa or higher are required depending on contact angle and capillary constant. Most, UF membranes used for water and wastewater treatment are not suitable for integrity tests targeting viruses.

BPT is performed by gradually pressurizing either the membrane lumen or shell using compressed air while monitoring bubbles or air flow on the other side, as illustrated in Figure 2.37. This method is best suited for hollow fiber membranes contained in pressure vessels, but it is also applicable for submerged hollow fiber membranes. If the modules are submerged in clean water

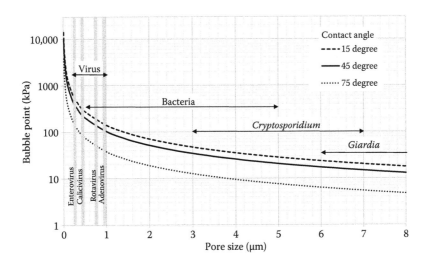

FIGURE 2.36 Bubble point as functions of pore size and contact angle and the size ranges of various microorganisms at 20°C. Capillary constant was assumed 0.5.

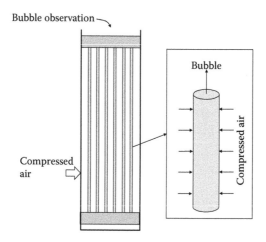

FIGURE 2.37 BPT of a wet hollow fiber module to estimate the largest pore (or defect) size.

and the membranes are visible through the water, the leak points can be identified. BPT is not suitable for most submerged flat sheet modules, where membranes sit on porous support plates without being fused or glued. However, some flat sheet modules with fused membranes on support plates are compatible with BPT although the maximum allowed air pressure tends to be barely sufficient for ensuring protozoan parasite rejection.

Integrity breaches can be fixed by locating the leaky fiber of some hollow fiber modules, for example, X-FlowXIGA (Farahbakhsh et al. 2004). After dislodging modules from the pressure vessel, the modules are placed vertically and the water in the shell side is drained. A small amount of water is poured on the potting cross-section of the module and compressed air is supplied to the shell. Leaky fibers can be identified from the bubbles emerging from the potting cross-section. After plugging all leaky fibers, the same procedure is repeated for the other side of the module.

2.7.2.2 Pressure Decay Test

PDT, which is also called *pressure holding test*, is one of the most commonly used test methodologies in the field. This method can be performed *in situ* at below the bubble point by injecting compressed air into the feed pipe. The water filled in the feed side of the module is drained before the test. Due to capillary suction pressure (Figure 2.35), air does not pass the wet membrane as long as pressure is below the bubble point. After isolating the membrane from the pressure source, air pressure is monitored for the predetermined duration (typically 5–20 min). For intact membranes, only very slow air pressure declines are observed by losing some air to the other side of the membrane by diffusion. As the number of defective fibers increases, more air is lost through the defects and pressure declines faster. CUNO Minicheck from 3M is a commercially available handheld device used for PDT. It can be programmed with up to 19 different test protocols with test parameters specific to the membrane system being diagnosed.

Figure 2.38 shows experimental pressure decay rate (PDR) for one membrane module with 20,000 polysulfone fibers. The baseline PDR without broken fibers is 0.16 psi/min (or 1.1 kPa/min) in the figure. The corrected PDR is obtained by subtracting the baseline PDR from the measured PDR. For example, the corrected PDRs are 0.08 psi/min and 0.24 psi/min when 1 and 3 fibers broken, respectively. The baseline PDR measured from a small-scale system is not necessarily applicable for larger scale systems because the ratio of the void volume and the membrane surface area are different. Therefore, baseline PDR is determined when membranes are new and intact. Later the measured PDR are compared against the baseline. The integrity of the new membrane system should be cleared by individual testings before measuring the baseline PDR. If the measured PDR of the used membrane exceeds the preset threshold, compromised modules need to be located and fixed/replaced.

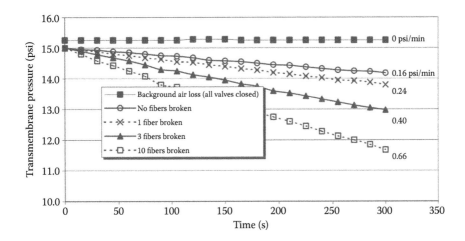

FIGURE 2.38 Pilot-scale PDT result with 20,000 polysulfone hollow fiber membranes in a pressure vessel (114 m²) from Polymem. (From Hugaboom, D.A., and Sethi, S., An evaluation of MF/UF membrane integrity monitoring strategies as proposed in LT2. *AMTA Conference*, Washington, DC, USA. 2004.)

The log reduction value of the hollow fiber modules with pressure vessel can be estimated using Equation 2.13. More details on the equation derivation are found in the ASTMD6908 protocol (ASTM 2010).

$$\mathrm{LRV}_e = \log_{10}\left[\frac{Q_{\mathrm{filt}} P_{\mathrm{atm}}}{CF \cdot PDR \cdot V} f_1 f_2 \right] \tag{2.13}$$

where

$$f_1 = \frac{\mu_{\mathrm{water}}}{\mu_{\mathrm{air}}}, f_2 = \frac{P_{u,\mathrm{test}}^2 - P_{d,\mathrm{test}}^2}{2 P_{\mathrm{atm}} \mathrm{TMP}}$$

Q_{filt} filtrate flow rate (m³/s)
$P_{u,\mathrm{test}}$ average pressure in high pressure side (kPa absolute)
$P_{d,\mathrm{test}}$ average pressure in low pressure side (kPa absolute)
P_{atm} atmospheric pressure (kPa absolute)
CF   concentration factor in module (1 for dead-end mode)
PDR pressure decay rate (kPa/s), can be replaced with VDR for VDT
TMP transmembrane pressure during filtration (kPa)
V volume pressurized during test (m³)
μ viscosity (kg/m/s)
LRV_e estimated log reduction value

For PDT, $P_{u,\mathrm{test}}$ is the average air pressure in module and $P_{d,\mathrm{test}}$ is atmospheric pressure. For VDT, $P_{d,\mathrm{test}}$ is the atmospheric pressure and $P_{d,\mathrm{test}}$ is the average vacuum pressure.

The advantages of the PDT in submerged and pressurized membranes are as follows (USEPA 2005):

- Ability to meet the resolution criterion of 3 μm under most conditions
- Ability to detect integrity breaches on the order of single fiber breaks and small holes in the lumen wall of a hollow fiber depending on test parameters and system-specific conditions
- Applicable for most MF and UF systems
- High degree of automation

- Widespread use by utilities and acceptance by states
- Simultaneous use as a diagnostic test to isolate a compromised module in a membrane unit in some cases

The limitations of the PDT are as follows:

- Inability to continuously monitor integrity without stopping filtration
- Calculation of LRV requires the volume of pressurized air in the system, but it is difficult to measure with high accuracy
- Potential to yield false-positive results, if the membranes are not fully wetted (which may occur with newly installed and hydrophobic membranes that are difficult to wet, or when the test is applied immediately after a backwash process that can fill pores with air)
- Difficult to apply to the membranes that are oriented horizontally as a result of potential draining and air venting problems

2.7.2.3 Diffusive Air Flow Test

DAF test is basically the same as PDT, but it measures air flow rate going into the system to compensate for the air loss under constant air pressure. Alternatively, if the leaked air from hollow fiber membranes pushes out the water filled in the shell, displaced water volume can be measured in a period of 5 to 20 min. The DAF test is more accurate than PDT in general (Trimboli et al. 2001), especially when displaced water volume is measured because measuring volume is more accurate than measuring pressure in general especially when large void volume buffers the pressure change. Because DAF is affected by temperature, temperature correction is required.

2.7.2.4 Vacuum Decay Test

Instead of using compressed air, vacuum can be applied to either the lumen or the shell after draining the water. This method is called VDT and is basically the same as PDT except for the source of the pressure differential. However, this method is not capable of performing integrity tests targeting bacteria because, in practice, the maximum pressure differential is only 0.8 to 0.9 bar, as will be discussed in Section 2.7.2.4. VDT is applicable for hollow fiber modules enclosed in pressure vessels, submerged hollow fiber membranes, and spiral wound modules, but care must be taken to not let the membrane surface dry. Procedures are as follows (USEPA 2005):

1. Drain the water from one side of the membrane. Shell side liquid is drained for hollow fiber modules whereas permeate side liquid is drained in spiral wound modules in general.
2. Apply vacuum to the drained side of the membrane. A vacuum of 20 to 26 in. Hg (or 0.7–0.9 bar) is applied. To comply with the "long-term 2 enhanced surface water treatment rule" (LT2E SWTR), the applied vacuum must be sufficient to meet the resolution criterion of 3 μm that is sufficient to reject protozoan parasites. If membranes are exposed to the air during the test, care must be taken to not dry the membrane to prevent pores from drying out.
3. Isolate the vacuum source and monitor the vacuum decay for a designated period. If there are no leaks in the membrane, then the vacuum should decay marginally over the duration of the test. Typically, vacuum pressure is monitored over a period of 5 to 10 min. The measured vacuum decay rate should be compared with the pre-established upper control limits (UCL).

VDT is applicable for spiral wound membranes or other systems that cannot be pressurized on the filtrate side of the membrane. However, it is not widely used in full-scale systems because it can barely meet the resolution criterion of 3 μm under most conditions due to the low vacuum pressure available.

2.7.3   INDIRECT METHOD

2.7.3.1   Turbidity Monitoring

Turbidity is employed in many MBR plants because it is the simplest and perhaps the most economical method to detect leaks. However, the sensitivity of the instrument is not enough to detect membrane breaches equivalent to the size of protozoan parasites in most applications. In one survey, filtered water turbidity never exceeded 0.15 NTU as measured by a single sensor light or laser-based instrument regardless of the membrane integrity tested. In a pilot study, whereas LRV declined by 3.5 units, no discernible response was observed using the turbidity meter. Therefore, the detection limit of the single sensor–based turbidity is considered very low at 0.0 to 0.3 LRV equivalent (Johnson 1998). A multisensor turbidimeter was somewhat more sensitive that it was able to detect integrity breaches when the LRV of *Bacillus subtilis* decreased from 5.3 to 2.8 under the given experimental conditions (Hugaboom and Sethi 2004).

2.7.3.2   Particle Counting

Particle counting is a real-time test method that does not require pausing membrane filtration unlike the direct methods. It is based on dynamic light obscuration and is considered the most accurate among all indirect integrity monitoring methods devised for porous membranes (Farahbakhsh et al. 2003). It can provide either particle counts or an index of water quality between 0 and 9999, depending on how the optically collected data are processed (Adham et al. 1995). The ability of particle counters to detect breaches on membranes varies depending on the condition, but pilot studies have shown that counting particles 2 μm or above is effective for monitoring parasite leaks. Particle counters are being employed in many MBR and surface water filtration plants despite of its high capital costs; however, they do not necessarily satisfy the expectations on accuracy and reliability. The probabilities of false-negative readings using these methods are reportedly significant, for example, 33% to 43%, as shown in Figure 2.34.

According to one full-scale study, particle counters that can detect particles down to 0.5 to 0.05 μm can provide discernible responses to the low levels of compromised fibers (Hugaboom and Sethi 2004). This study demonstrated that highly sensitive particle counters could detect a single fiber breach in 980,000 fibers in a full-scale membrane rack when feed water turbidity was 0.06 to 0.16 NTU in surface water treatment. In practice, interpretation of the particle counting chart is not always straightforward due to the noise and fluctuation of the readings. According to a chart shown by Liu and Wachinski (2009), it is difficult to judge whether the instantaneous surges indicate integrity breaches or normal fluctuations in a short period. An integrity breach can be judged more accurately after monitoring the trend for a longer period, for example, 2 to 3 h, unless there are catastrophic changes. Drifting of the baseline can be an issue in practical situations. The baseline curve showed (graph is not shown here) a slight sign of increase although there was no integrity breach. The mildly increasing trend continues with the one cut of the hollow fiber. Although the particle count moved up when one fiber was cut, the particle count did not seem to be responsive when more than 0.0006% of fibers were cut. It is not always straightforward to tell when the leak starts and how severe the integrity breach is. As a result, substantial false-negative readings are commonly experienced in field conditions.

2.7.3.3   Microbial Challenge Test

Microbial challenge tests are the ultimate integrity test by definition. All other pressure-based methods are not applicable for most MF/UF membranes to ensure virus removal, but microbial challenge test provides the most direct information on it. For bacterial challenge tests, *Brevundimonas diminata* can be used at a dosage of 10^7 cfu/mL in feed. MS-2 coliphage and *B. atrophaeus* can be used as surrogate species for viruses and parasites, respectively, at a concentration sufficient to detect the leaked virus in the permeate at the target LRV. However, it is challenging to practice in full-scale plants due to the limitations associated with a large number of samples, the long delay time until the test results could be obtained, and the manpower required. It is, however, essential to validate other integrity test methods and to obtain the regulatory approval of the membrane system. In one study, fluorescent dyes such as fluorescein-5-isothiocyanate was tagged on MS-2 phages to make virus detection easier (Gitis et al. 2006). In this method, membrane integrity can be tested in real time by measuring fluorescence in permeate using an in-line fluorometer. However, the maximum LRV traceable is less than 2 at neutral pH in realistic conditions because a large quantity of tagged viruses are required in feed water to make the fluorescence signal strong enough in the permeate at the target LRV.

2.7.3.4   Nanoparticle Challenge Tests

Gold, silver, copper, or polystyrene nanoparticles with a diameter of 1 to 100 nm can be used as surrogates for bacteria or viruses. To detect the trace level of such particles in the permeate, fluorescent dyes can be tagged. Because such colloidal nanoparticles tend to aggregate and form large flocs at neutral pH due to the low zeta potential, special techniques may be required to properly store and handle the nanoparticles. By modifying the surface properties of nanoparticles with respect to surface charge and zeta potential, such problems can be mitigated and the particles behave more similarly to the microorganisms they mimic. Gold nanoparticles can also be detected by anodic stripping voltammetry (ASV) down to 1 ppb (Gitis et al. 2006). In a study performed with gold nanoparticles, up to 4.5 LRV could be obtained when cellulose ester membranes with a MWCO of 20 kDa were used. When nanoparticles tagged with fluorescent dye (FluoSpheres® of Life Technologies) were used to detect leaks in RO, the rejection efficiency of a model virus (MS2 phage) agreed reasonably well with the rejection efficiency of nanoparticles in a compromised membrane with a pinhole, for example, 99.93% to 99.97% virus removal versus 99.90% to 99.91% nanoparticle removal (Mi et al. 2004).

Although the particles tagged with fluorescent dye are technically feasible concept, it is challenging to secure high LRV due to the large amount of nanoparticles required to make the fluorescent signal strong enough in permeate. It is because dyes are only a small portion of the larger nanoparticles. For instance, if fluorescein is detectable at 0.0001 mg/L whereas the LRV target is 4 and the mass ratio of dye/nanoparticles is 500, the nanoparticles tagged with fluorescein are required at 500 mg/L (= $0.0001/10^{-4} \times 500$). Although this method allows for in-line integrity tests without interrupting the system operation, the high costs are a significant hurdle that needs to be overcome for the technology to be commercialized.

2.8   MEMBRANE SUPPLIERS

The number of membrane suppliers for MBR has been growing rapidly since 2000s, especially in East Asia, and it is now believed that there are at least 50 suppliers globally as of 2013. The large number of suppliers for the market size is the main reason for the decreasing membrane prices. The thinning profit margin is the basis of predicting the industry consolidation to a smaller number of players, but it has not yet occurred at a massive scale. It is partly due to the decreasing costs of developing membrane products and the manufacturing as the technology becomes fully mature. MBR employ either submerged membranes or sidestream membranes, but submerged membranes take a vast majority of the MBR market globally regardless of the region. Perhaps a roughly equal number of hollow fiber and flat sheet membrane suppliers exist, but hollow fiber membranes take much more than a half of the market share in terms of treated water volume whereas flat sheet membranes take the rest of the market. Some of the membrane manufacturers are summarized in Table 2.7, but Kubota, Evoqua (formerly Siemens), Asahi Kasei, and GE are the major suppliers playing globally. PVDF, PE, PP, PES, PTFE, PAN, and others have been used as membrane materials, but a growing number of membranes are based on PVDF in the MBR market.

TABLE 2.7
Summary of Commercial MBR Membrane Module Products

Submerged		Sidestream
Flat Sheet	**Hollow Fiber**	**Multitube/Multichannel**
A3—MaxFlow (DE)	Asahi Kase—Microza™ (JP)	Berghof—HyPerm AE, HyPerflux (DE)
Agfa—VITO (BE)	Beijing Origin Water Co. (CN)	Duraflow—DF series (US)
Alfa Laval—Hollow Sheet (SE)	Canpure—Canfil (CN)	MEMOS—Membrane Modules Systems GmbH—MEMCROSS (DE)
Anaergia-FibrePlate™ (CA)	Ecologix—EcoFlon™, EcoFil™ (CN)	Novasep—Kerasep® (FR)
Brightwater—MEMBRIGHT® (IRL)	ENE Co., Ltd.—SuperMAK (KR)	Orelis Environment—PLEIADER®, KLEANSEPR® (FR)
Colloide—SubSnake (NIR)	Econity—KSMBR (KR)	Pentair (formerly Norit X-Flow) (US)
Ecologix—EcoPlate™, EcoSepro™ (CN)	Evoqua—MemPulse™ (US)	
Huber—VRM®, ClearBox®, Biomem (DE)	GE—ZeeWeed® (US)	
Hyflux—Petaflex (SG)	Kolon Industry—Cleanfil® (KR)	
Inge—FiSh (DE)	MEMOS Membranes Modules Systems GmbH—MEMSUB (DE)	
IWHR (DE)	Memstar Technol. Ltd.—SMM (SG)	
KOReD—Neofil® (KR)	Mitsubish Rayon—SUR™, SADF™ (JP)	
Kubota—ES/EK (JP)	Mohua Technology—iMEM-25 (CN)	
LG-Hitachi Water Sol.—G-MBR (KR)	Philos Co. Ltd (KR)	
Pure Envitech Co., Ltd.—ENVIS (KR)	SENUO Filtration Technol—SENUOFIL (CN)	
Suzhou Vina Filter Co.—VINAP (CN)	Sumitomo Electric Industries—POREFLON™ (JP)	
Toray—MEMBRAY® TMR (JP)	Superstring MBR Technol.—SuperUF (CN)	
Weise Water Systems GmbH—MicroClear® (DE)	Suzhou Vina Filter Co.—F08 (CN)	
	Zena SRO—P5 (CZ)	

Source: Modified from Santos, A. and Judd, S.J., *Sep. Sci. Technol.* 45:850–857, 2010a.

Note: AT, Austria; BE, Belgium; CA, Canada; CN, China; CZ, Czech Republic; DE, Germany; DK, Denmark; FR, France; IRL, Ireland; JP, Japan; KR, Korea; NIR, Northern Ireland; NL, Netherlands; SE, Sweden; SG, Singapore; SP, Spain; US, United States.

3 Submerged Membrane Process

3.1 OVERVIEW

The permeability decline associated with foulant deposition on membrane surface is a major hurdle in membrane filtration. In traditional cross-flow filtration performed under positive pressure, high cross-flow velocity is an essential requirement to control the membrane fouling, but it causes excessive feed pressure that, in turn, causes excessive cake layer compaction and low energy efficiency. In addition, the frequent recirculation of the feed water through high-pressure pump imposes negative effects on particles due to the high shear stress caused by circulation pumps (Brockmann and Seyfried 1996; Tardieu et al. 1998). As a result, such traditional filtration methods are primarily used as part of manufacturing processes that created sufficient additional values to cover the costs or as wastewater treatment processes that did not have better alternatives.

As introduced in Section 2.1.3.2, the first submerged membrane filtration concept was invented by Tajima and Yamamoto in 1986 for wastewater treatment in nuclear power plants (Tajima and Yamamoto 1986). Later, the same concept was adopted by different researchers for MBR (Yamamoto et al. 1989). In those original configurations, the U-shaped polyethylene-based hollow fiber bundles with 0.1 μm pores were submerged in a container filled with wastewater. While sparging air bubbles underneath the membrane bundle, permeate was obtained using a vacuum pump from the top header that holds both ends of the U-shaped fiber. The center of the U-shaped bundle located in the lowest point was fixed in the bottom of the tank from which aeration was performed to scour the membrane. Membrane fouling was observed to be very slow at low flux settings (e.g., <10 LMH).

From the economics stand point, submerged membrane filtration is superior to traditional cross-flow filtration operated under positive pressure. The capital costs of submerged systems are much lower because it does not require housed membrane modules and the associated plumbing compatible with high pressures. Low-cost membrane modules allow the use of much larger surface areas in MBR. The larger surface areas in turn allow much lower flux than in traditional crossflow filtration, thus the membrane fouling rate dramatically decreases. In addition, air scouring is inherently more energy efficient than circulating mixed liquor through membrane systems in controlling membrane fouling. For instance, submerged hollow fiber membranes need approximately 1/40 to 1/100 of the energy required by tubular membranes (0.1–0.2 vs. 4–10 kWh/m^3). Due to the high efficiency, air sparging is used not only for submerged membranes but also for some hollow fiber membranes housed in pressure vessels (Bérubé et al. 2006).

Other benefits of submerged membrane process is that air scouring produces less shear stress than circulation pumps used in traditional crossflow filtration and hence disrupts biological floc less. The disadvantages of submerged membrane process include the necessity for tighter pretreatment to prevent membrane fiber clogging (hollow fiber) and channel blocking (flat sheet), more volatile organic carbon emission when treating some industrial wastewaters, and so on, compared with traditional crossflow filtration.

Today, more than 99% of MBRs around the globe rely on submerged membranes based on the treated water volume and this trend is expected to continue in the foreseeable future. Therefore, understanding how submerged membranes work is crucial in understanding MBR technology. In this chapter, details on submerged membranes will be discussed including the principle operating methods, air scouring mechanisms, module design factors, etc.

3.2 OPERATION OF SUBMERGED MEMBRANE

3.2.1 PERMEATE DRAWING

3.2.1.1 Intermittent Permeate Drawing

Intermittent permeate drawing under vacuum is performed in virtually all MBR plants using sub-merged membranes to improve the net (or average) flux. It consists of repeated on-and-off cycles according to predetermined cycle times. During the off time, the cake layer becomes loose without having the transversal permeate flow through it. Hence, a portion of the membrane foulants can be detached from the surface by the two-phase turbulence generated by air scouring. Although no permeate is obtained during the off time, net flux improves by the much improved flux during the on time in a vast majority of cases.

There are no universal rules in setting the cycle time, but the off time typically takes 10% to 20% of the total cycle time depending on membrane fouling potential in the given condition. Although 9 min on and 1 min off seems most common, any combination of 8 to 11 min on and 1 to 2 min off is also commonly adopted in the field. Full-scale plants have the capability of modifying their cycle times depending on the membrane fouling rate and wastewater flow rate.

Centrifugal pumps have been widely used to draw permeate primarily due to its low cost. The sharp decline of flow rate at high transmembrane pressure (TMP) should be factored in when centrifugal pumps are used as permeate pumps. In addition, priming the pump when the system starts should be properly planned out. This pump may also require an additional device to stop the forward or backward flow syphoning during the off time depending on the elevation of permeate discharge pipe. Due to the vacuum pressure in permeate pipes, a portion of the dissolved gas in the permeate is desorbed and forms air bubbles, thereby the pump impellers are exposed to the risk of cavity damage. Especially in large-scale systems, a large number of bubbles are formed in the pipe-line and the chance of pump damage may increase. To prevent cavity damage, the permeate can be passed through bubble separation chambers in which bubbles rise to the headspace and are removed by vacuum pumps. Instead of centrifugal pumps, positive displacement pumps such as lobe pumps are also commonly used. Despite the higher pump costs, the overall costs for permeate drawing can be reduced because they are self-priming, reversible for membrane backwashing, constant in flow at varying vacuum pressure, and resistant to cavity damage.

3.2.1.2 Gravity-Driven Permeate Drawing

Gravity filtration has occasionally been adapted when permeate could be naturally transferred to lower levels without an additional pumping, such as in mountain slopes, as shown in Figure 3.1. One obvious benefit of gravity filtration is the lack of permeate pump that needs electric power. In addition, the TMP cannot exceed the static pressure, $P_{S1} + P_{S2}$, in any case. The low maximum TMP eliminates the possibility of excessive cake layer compaction that hampers the membrane cleaning efficiency. However, the lack of capability to increase TMP above the static pressure can also be disadvantageous when flux should be increased during peak flow conditions beyond the level required during normal flow. If permeate must be pumped to a remote location or if it needs to be pumped up to be discharged to a river or lake, gravity filtration no longer allows net energy savings. Therefore, gravity filtration can be justified when additional pumping is not necessary after drawing the permeate.

The maximum driving force available can be calculated using Equation 3.1. The pressure loss in the pipeline, P_L, can be controlled to a negligible level compared with other pressure losses by properly sizing the pipes. If the outlet for the permeate discharge pipe is placed in the same level with the pressure gauge, P_{S2} can also be neglected. Flow rate is controlled by using a flow control valve and its opening is feedback-controlled by the readings from the permeate flow meter in the downstream. If the permeate valve is completely closed, P_G is equivalent to P_{S1} in Equation 3.1 and both P_{S2} and P_L are effectively zero due to the isolation from the system and the lack of flow, respectively, thereby the TMP becomes zero according to the equation. If the permeate valve is partially open,

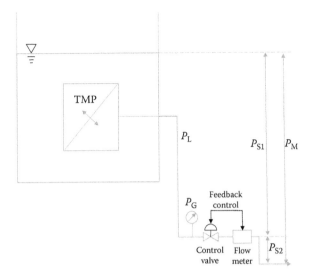

FIGURE 3.1 Conceptual diagram of gravity filtration.

P_{S2} comes into play as the two water bodies before and after the valve are hydraulically connected. As the valve opening increases, flow rate increases and P_G decreases, which leads to increasing TMP according to the equation. Assuming large enough pipes are used to draw permeate, P_L can be neglected in practice. If the flow control valve is undersized, it exerts a substantial P_L at the full opening and the TMP cannot be increased to the full potential set by the static pressure.

$$TMP = P_{S1} + P_{S2} - P_G - P_L \tag{3.1}$$

where

P_G	gauge pressure (kPa)
P_L	pressure loss in permeate and discharge pipes (kPa)
P_{S1}	static pressure at the valve (kPa)
P_{S2}	suction pressure in discharge pipe (kPa)
P_M	maximum pressure available (kPa)
TMP	transmembrane pressure (kPa)

3.2.2 INTERMITTENT AERATION

3.2.2.1 Intermittent Aeration with Defined Cycle

In submerged membrane processes, most of the energy is used for membrane scouring. Even if state-of-the-art hollow fiber membranes are used, aeration costs take at least 30% to 40% of the total plant energy consumption, as shown in Figure 6.43. Therefore, saving energy in membrane scouring is the most important part of the total plant energy optimization to make the process more economical. In the late 1990s, it was discovered that aeration could be performed intermittently without causing permanent membrane fouling (Rabie et al. 2001; Guibert et al. 2002). The cake layer built during the no-air cycle seemed to be removed sufficiently in the subsequent aeration cycle. Based on pilot tests, the effectiveness of intermittent aeration was demonstrated. In Figure 3.2, membrane fouling rates at different air cycle times were plotted against flux. When average airflow rates were maintained at 0.36 m³/m²/h, membrane fouling rates decreased with intermittent aeration except when aeration cycles were very short at 5 s–5 s (5 s on and 5 s off) under the experimental conditions used. This result

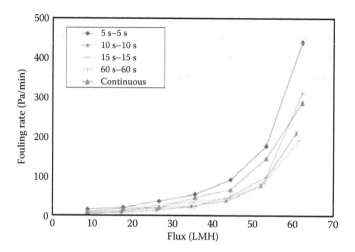

FIGURE 3.2 Effect of various intermittent aeration cycles on membrane fouling rate when GE ZW500a module with 46.5 m² surface area was used to filter 1.5 g/L bentonite solution at an aeration rate of 0.36 m³/m²/h. (From Guibert, D. et al., *Desalination* 148:395–400, 2002.)

suggests that overall aeration rate can be reduced with intermittent aeration without increasing membrane fouling. Currently, GE's ZW500® membranes are relying on intermittent aerations, mostly at 10 s–10 s cycle, with a potential to further reduce the aeration by using a 10 s–30 s cycle when membrane fouling potential is low (Buer and Cumin 2010). When intermittent aeration is compared with continuous aeration for the same instantaneous airflow rate, intermittent aeration causes slightly more fouling because membrane fouling occurs when air is off. Therefore, slightly stronger instantaneous aeration is required during the on time to maintain the same antifouling efficacies overall. As a result, when on and off cycles are equally timed, less than 50% of net aeration savings can be obtained without affecting the membrane performance by intermittent aeration. Based on the shear stress analysis, the net air savings seem to be approximately 30%, which will be discussed later.

Intermittent aeration can be performed in two different ways in a full-scale plant: (1) alternating on/off cycles among cassettes and (2) alternating on/off cycles between two groups of diffusers in each cassette. In the former method, all the diffusers in one cassette are either on or off simultaneously, which provides strong upflow during the on time in half of the cassettes. In the latter method, half of the diffusers in a cassette are on at any given time. The upflow generated from half of the diffuser scours not only the membranes above them but the adjacent membranes through the wake effect. However, the upflow can be slower during aeration compared with the other method.

Fulton and Bérubé (2010) measured shear stress on membrane surfaces with and without intermittent aeration. Three commercial-scale GE ZW500c® modules were used in clean water at a gross airflow rate of 0.17 m³/m²/h. Intermittent aeration was performed either by alternating the aeration among diffusers (alternating) or turning all diffusers on/off simultaneously (pulsing). The shear stress on the membrane surface was directly measured by electrochemical sensors embedded in five different vertical columns parallel to the fiber bundles (c), four different levels (d), and three different vertical columns vertical to the fiber bundles (e) in the cassette, as shown in Figure 3.3. Two different aeration cycle times at 3 s–3 s (fast) and 6 s–6 s (slow), respectively, and continuous aeration were tested.

As summarized in Table 3.1, average shear stress remained almost identical for all intermittent aeration modes regardless of the cycle time and the method of intermittent aeration. Only the average airflow rate seems to be important with regard to shear rate in the intermittent aeration mode. Meanwhile, the average surface shear stress with continuous aeration was measured to be only approximately 40% greater than those of intermittent aeration modes despite the 100% greater air consumption. If the same average shear stress is desired in intermittent aeration with 50% on time,

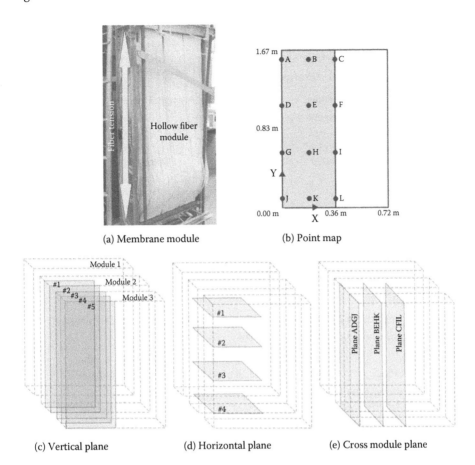

(a) Membrane module

(b) Point map

(c) Vertical plane

(d) Horizontal plane

(e) Cross module plane

FIGURE 3.3 Membrane system used to characterize surface shear stresses induced by sparging. (a) Cassette frame with three membrane modules (only outer front module is visible); (b) locations where shear forces were measured on each vertical plane (rows ABC, DEF, GHI, and JKL at "y" of 1.57, 1.08, 0.59, and 0.10 m, respectively, and columns ADGJ, BEHK, and CFIJ at "x" of 0, 0.18, and 0.36 m, respectively); (c–e) shaded vertical, horizontal, and probe planes, respectively, in quarter of interest within the three-membrane module cassette. Three ZW500c modules (GE) were used at an airflow of 0.17 m³/m²/h. (From Fulton, B., and Bérubé, P.R. Optimal module configuration and sparging scenario for a submerged hollow fiber membrane system. *Proceedings of WEF Membrane Applications Conference*. June 6–9, 2010, Anaheim, California.)

instantaneous (or gross) airflow rate must be increased somewhat. These observations suggest that the intermittent aeration can save air demand by somewhat less than 50%, perhaps by approximately 30%, although it is difficult to draw a definitive value out of such systems behaving nonlinearly.

Other remarkable finding from the same experiment was that aeration pattern greatly affected the distribution of surface shear stresses over the membrane cassette. The median shear stresses in the hypothetical planes defined in Figure 3.3c through 3.3e are summarized in Table 3.2. In continuous aeration mode, shear stress tends to be much greater in the top of the module than in the bottom. At the same time, it is substantially greater in the core of the cassette and in the middle of the vertical height than in the areas exposed externally. All of these suggest that air bubbles tend to form a channel (or plume) in the center of the cassette through which bubbles move up fast. The maldistribution of shear stress causes inefficient air use. Membrane fouling can start at the low shear area and propagate to the adjacent area because the lost flux in the fouled area must be compensated by the higher flux in the adjacent areas at a constant flux mode. Meanwhile, the average shear rate is lower with the intermittent aeration in all planes compared with those with the continuous aeration, but the shear stress was much more evenly distributed due to the lack of the air channels through

TABLE 3.1

Average Surface Shear Stress at 90% Confidence Level at Different Aeration Modes

Sparging Conditions	Average Shear Stress (Pa)
Slow alternating (3 s–3 s)	1.02 ± 0.02
Slow pulse (6 s–6 s)	0.96 ± 0.02
Fast alternating (3 s–3 s)	1.05 ± 0.06
Fast pulse (6 s–6 s)	0.96 ± 0.05
Continuous	1.41 ± 0.18

Source: Fulton, B., and Bérubé, P.R. Optimal module configuration and sparging scenario for a submerged hollow fiber membrane system. *Proceedings of WEF Membrane Applications Conference.* June 6–9, 2010. Anaheim, California.

TABLE 3.2

Comparison of the Average Median Shear Stresses at Different Planes in a Membrane Cassette

Location	Continuous Aeration Median Shear (kg/m/s²)	Slow Pulse Aeration Median Shear (kg/m/s²)
Horizontal plane 1	2.24	0.61
Horizontal plane 2	1.20	0.74
Horizontal plane 3	0.94	0.68
Horizontal plane 4	0.80	0.66
Plane ADGJ	0.63	0.52
Plane BEHK	1.20	0.73
Plane CFIL	1.66	0.83
Vertical plane 1	1.39	0.73
Vertical plane 2	1.67	0.47
Vertical plane 3	1.13	0.67
Vertical plane 4	1.43	0.78
Vertical plane 5	0.83	0.71

Source: Data from Fulton, B., and Bérubé, P.R. Optimal module configuration and sparging scenario for a submerged hollow fiber membrane system. *Proceedings of WEF Membrane Applications Conference.* June 6–9, 2010. Anaheim, California.

which bubbles are aspirated. Therefore, air can be more efficiently used at the intermittent aeration mode by reducing the area with low shear rates in submerged hollow fiber membrane processes.

3.2.2.2 Intermittent Aeration with Undefined Cycle

The long-term reliability of air cycle valves has been a significant issue in practice in intermittent aeration with defined cycle times. In fact, the air cycling valves must change their position approximately 3 million times a year at a 10 s–10 s cycle. With larger plants, the chances of having mechanical issues increases proportionally due to the increased number of air cycling valves. To improve the mechanical reliability simultaneously with the air scouring effect, a system called MemPulse™ was introduced by Evoqua (formerly Siemens) in 2008. In this system, air chambers are installed underneath the vertically mounted hollow fiber module and compressed air is supplied to the air chamber at a constant rate. Once the air chamber is filled up with compressed air, air slugs are quickly discharged from the chamber for a few seconds scouring the hollow fibers vigorously. The frequency of air slugs is dependent on the size of the air chamber and the compressed airflow rate, but the bubbling time is only a quarter or less of the total cycle time. Although intermittent aeration with undefined cycle times allows less freedom for air cycling time control at a normal operating condition, it was claimed that the new method could save 20% to 30% of compressed air compared with intermittent aeration with defined cycle times. Recently in 2005, it was announced that LEAPmbr was adopted for a wastewater plant upgrade in Stockholm, Sweden, with a peak flow of 864,000 m³/d.

MemPulse™ is based on the intermittent airlift pump effect that was described by Kondo (1998). The operation of the airlift pump consists of four steps, as shown in Figure 3.4: (a) the air chamber is filled up by continuous airflow; (b) once the air level reaches down the entrance to the vertical airlift channel, air bubbles start to leak; and (c) the rising air bubbles bring up water with them through the airlift channel in the center. Because the rising gas/liquid mixture pulls more air from the chamber according to Bernoulli's theorem, "step c" accelerates by itself and generates a large number of air slugs in a short period. Finally, the air slugs stop when air is depleted in the air chamber in "step d."

More recently in 2011, GE Power & Water introduced LEAPmbr, which uses similar air cycling methods as MemPulse™. Intermittent aeration with undefined cycling times reportedly save up to 30% of air and 50% of capital for aeration-related equipment when it is compared with intermittent aeration with defined cycle times (GE Water 2013). Specific aeration demand per permeate volume (SAD_p) is aimed at approximately 6 m³/m³. No sufficient empirical data published by third-party experimental data exists in public domains (Barillon et al. 2013).

According to a pending patent (Cumin et al. 2011), the air chamber of LEAPmbr appears to have a slightly different structure as shown in Figure 3.5. The vertical airlift channel does not have an opening at the bottom of the chamber. As a consequence, when air slugs rise, no additional liquid other than the liquid already filled in the airlift channel moves upward during the air purging step (Figure 3.5c in the figure). No experimental data comparing the efficacies of the two different air chambers (Mempulse™ and LEAPmbr) are available in the public domain.

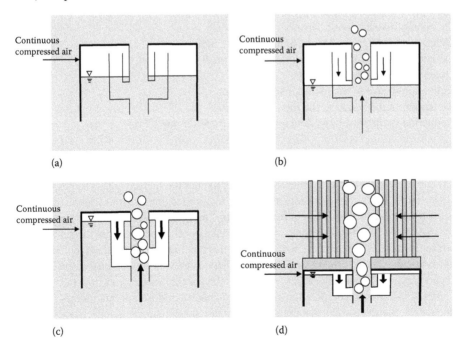

(a)   (b)

(c)   (d)

FIGURE 3.4 Mechanism of intermittent airlift pump with a continuous airflow: (a) air filling, (b) air purging (start), (c) air purging (end), and (d) overview.

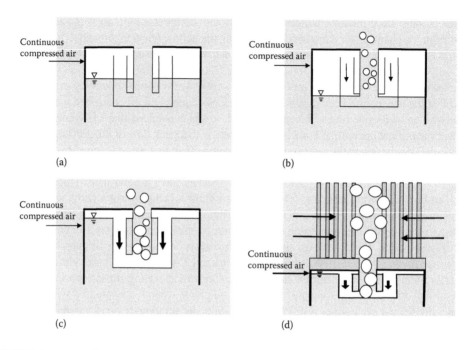

FIGURE 3.5 Mechanism of intermittent airlift pump with continuous airflow: (a) air filling, (b) air purging (start), (c) air purging (end), and (d) overview.

In one ongoing project in the city of Riverside, California, as of 2012, total net savings of $800,000 with power savings of $74,538/year was projected in a proposed MBR with 98,400 m³/day (or 26 MGD) by adapting LEAPmbr instead of the defined air cycling. The savings were largely from the following items (Ciccotelli 2012).

- Deletion of two air scour blowers
- Deletion of 1420 ft. (or 433 m) of air scouring piping
- Deletion of 112 cassette air scour piping connections
- Deletion of 24 air scour valves
- Reduction in size of the main air scour header
- Reduction in the height of the piping supports
- Reduction in the traveling crane and canopy by 5.5 ft. (or 1.7 m)

3.2.3 Specific Air Demand and Specific Energy Demand

The SAD can be based on either permeate volume or membrane surface area. The permeate volume–based SAD, that is, SAD_p, represents the air volume required to obtain a unit volume of permeate. For example, if 10 m³ of scouring air is required to obtain 1 m³ of permeate, SAD_p becomes 10. It must be noted that SAD_p is only an approximate measure of energy efficiency because the actual energy consumption for aeration is also a function of the head pressure that the air must overcome. If two different membrane cassettes have the same SAD_p, the taller cassette will be less energy efficient due to the higher head pressure. Meanwhile, SAD_m represents the specific scouring airflow rate per membrane surface area. It is useful to compare the aeration efficiencies of different membranes modules, but it is less indicative of energy efficiency than SAD_p because the actual energy efficiency is also affected by flux and membrane packing density not considered in SAD_m. Therefore, SAD_m is primarily used as a system design factor when calculating air blower capacities from membrane surface area.

SAD_p tends to be lower with hollow fiber membranes than flat sheet membranes due to the higher packing density as discussed in Section 3.4.5. For a given membrane system, SAD_p varies by influent flow rate because scouring aeration rate is not proportionally controlled to the flow. In general, SAD_p increases as influent flow rate decreases because a minimum airflow is necessary to maintain the air scouring effect.

- It has been reported that SAD_p can be as low as 7.3 with intermittent aeration in well-controlled systems if influent flow rate is kept at design flow and all operating parameters are in favorable ranges. In field conditions, such low SAD_p values are not commonly observed. Instead, much larger SAD_p are observed in the field than projected based on pilot tests. The low SAD_p may be obtained only when hydraulic flow rates are at or above the design flow allowing high fluxes. If the hydraulic flow rate (or flux) is below the design level, SAD_p increases because air scouring rate does not proportionally decrease. SAD_p in the field have been observed mostly from 14 to 30 except in the peak flow conditions that last only short period of time. This trend is apparent in Figure 3.6, where SAD_p is above 10 to 20 except the peak flux conditions.
- The low SAD_p is obtainable only in well maintained pilot systems where every operating parameter is tightly controlled to maintain good mixed liquor quality. Full-scale MBRs inevitably suffers from varying degrees of nonideal operating conditions in terms of hydraulic and organic loading (or food to microorganisms [F/M] ratio), dissolved oxygen (DO) level, mixed liquor suspended solids (MLSS), mixing, nutritional balance, temperature, etc.

Flat sheet membranes generally need more scouring air than hollow fiber membranes primarily due to their lower packing density. According to a survey (Section 3.4.5.2), the packing density of flat sheet membranes is approximately half that of hollow fiber membranes. As a result, scouring aeration takes a greater portion of the total aeration in flat sheet membranes. According to a field survey performed by Barillon et al. (2013), aeration energy demands are equal or less for membrane scouring than for biological aeration in hollow fiber membranes, but energy demands for membrane scouring are nearly double the energy demands for biological aeration in flat sheet membranes.

Specific energy demand (SED) is an ultimate measure of the energy costs. SED can be estimated not only for membrane scouring, but also for biological aeration, mixed liquor pumping, permeate drawing, etc. SED can be also calculated for the overall process by adding all SED for unit processes. It is typically estimated as kilowatt hour per cubic meter of wastewater treated. When the energy efficiencies of membrane scouring are compared, SED tends to be proportional to SAD_p in a given system, but only a weak proportionality exists between two different membrane systems due to the different head pressures required for aeration.

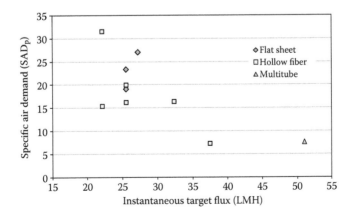

FIGURE 3.6 SAD_p as a function of net flux in various MBR plants. (Data from DeCarolis, J.F. et al. *Water Environ Res.* 79(13):2536–2550, 2007.)

TABLE 3.3
The Energy Consumption of Municipal MBR

	Capacity (PE)	Membrane	Year	SED (kWh/m³)
Rödingen (Germany)	3000	HF/GE	2001	2.03
Markranstädt (Germany)	12,000	HF/GE	2001–2003	1.8–1.5
Knautnaundorf (Germany)	900	FS/Huber	2002–2003	1.3–2
Seelscheid (Germany)	11,000	FS/Kubota	2004–2005	0.9–1.7
Monheim (Germany)	9700	HF/GE	2003–2005	~1
Brescia (Italy)	46,000	HF/GE	2003–2005	0.85
Nordkanal (Germany)	80,000	HF/GE	2004–2005	0.9
			2005	1.00
Varsseveld (Netherlands)	23,150	HF/GE	2006	0.88
			2007	0.83

Source: Giesen, A. et al., Lessons learnt in facility design, tendering and operation of MBR's for industrial and municipal wastewater treatment. *Proceedings of the Water Institute of South Africa (WISA) Biennial Conference and Exhibition.* May 18–22, 2008. Sun City, South Africa; Courtesy of http://www.royalhaskoningdhv.com.

Note: Population equivalent (PE) varies country by country but it is mostly ~60 g BOD_5/day/person in European countries, which is equivalent to 0.3 m³ wastewater/day/person with a BOD of 200 mg/L.

The lowest field SAD_p for membrane scouring claimed in full-scale MBR appears at less than 10 m³/m³, which was obtained in UluPandan, Singapore, when the influent is at a daily average design flow rate (Judd 2011; Chen et al. 2012a). The SED for membrane system excluding sludge circulation is also low at less than 0.1 kWh/m³ permeate with an intermittent aeration cycle at 10 s–30 s mode (10 s on and 30 s off) with GE's ZW500d cassettes. The more recently deployed GE LEAPmbr uses irregular air slugs generated by the air chamber installed underneath the membrane module, as discussed in Section 3.2.2.2. This system aims at SAD_p values of approximately 6 similarly to the MemPulse™ system of Evoqua. However, the efficacy of the air slug without a fixed cycle time is not yet supported by abundant field data (Barillon et al. 2013).

The total SED including all direct energy use such as bio/scouring aeration, pumping, mixing, etc., has been known to be 0.8 to 2.0 kWh/m³ in municipal MBRs depending on the membranes used, the capacity of the plant, and the level of optimization (Giesen et al. 2008; Krzeminski et al. 2012). The SED can be further reduced by optimizing membrane and biological systems with better flow rate modulation, but the practical minimum with the current technologies is considered to be approximately 0.6 kWh/m³. The SEDs of MBR plants including biological air, mixed liquor circulation, mixing, etc., are summarized in Tables 3.3 and 3.4.

3.2.4 REMOVAL EFFICIENCY

3.2.4.1 Effluent Quality

MBR effluent does not contain any measurable total suspended solids (TSS) in theory because the pore sizes of most membranes are smaller than the nominal pore size of the filter (0.45 μm) used for the TSS test. In addition, much smaller particles than the pore size are also removed by the dynamic membrane formed on membrane surfaces, as discussed in Section 1.2.4. With respect to organic contents, MBR effluent contains less soluble organics than conventional activated sludge (CAS) effluent because it runs at much lower F/M ratio. At the food stringent condition, slowly degrading organics are treated better and less soluble microbial products (SMP) are formed. Moreover, some macromolecules are rejected along with colloidal particles by the cake layer formed on the

TABLE 3.4
Commercial Membranes Used in MBR

Manufacturer	Configuration (Model)	Flux, Net (m/day)	Area (m²)	Air, Net (m³/h)	SAD$_m$ (m³ air/m² membrane/h)	SAD$_p$ (m³ air/m³ permeate)	Remark	Reference
GE Water	HF(ZW500d)	0.6	98	18	0.18	7.3	Manufacturer's specifications	Côté et al. 2004a
Kubota	FS(SP400)	0.8	400	120	0.30	9.0		
Kubota	FS(EW400)	0.6	500	192	0.38	15		
Asahi	HF(MUNC-620A)	0.6	25	6	0.24	9.6		
Econity	HF(4005CF)	0.4	1000	150	0.15	9.0		Herold 2011
Koch Puron					0.14–0.53			
Kubota	FS(EK200)	0.6	160	96	0.60	24	Pilot test	Adham et al. 2004
Siemens	HF(B10R)	0.58	37	13.4	0.36	16	Pilot test	Adham et al. 2004
Mitsubishi	HF(SADF)	0.72	29	13.7	0.47	16	Pilot test	EUROMBRA 2006
A3 Water Solutions	HF(SADF)	0.6	140	28	0.2	8	Pilot test	Grélot 2010
GE Water	HF(ZW500a)	0.6	5280	2300	0.44	17	Rödingen, Germany	Brepols 2011
GE Water	HF(ZW500c)	0.6	4088	1577	0.39	15	Key Colony, FL	
		0.73	67	25	0.37	13		Adham et al. 2004
		0.9	67	18	0.27	7.2	Pilot test	
		0.48	?	?	0.26	13	Pilot test	Côté et al. 2004b
		?	?	?	?	15	Varsseveld, Netherlands	
		0.6	84,480	34,000	0.40	16	Nordkanal, Germany	Brepols 2011
	HF(ZW500d)	0.6	12,100	4000	0.33	13	Giessen, Germany	

Note: Design parameters were used for the calculation. Actual SAD can increase during dry season because flux decreases more sharply than aeration rate.

membrane. As a result, when treating the same wastewater, MBR effluent quality is always better than CAS effluent not only in terms of suspended solids but also in terms of soluble organics.

The other aspect of the MBR is that the effluent quality is much more stable than those of CAS. Even if the CAS runs at a reasonable operating condition, effluent quality fluctuates mainly due to the suspended solids that do not settle in the secondary clarifier. Sludge bulking, organic or hydraulic overloading, foaming, toxic components in feed, etc., are considered the culprits of sludge bulking. The fine organic particles contained in the effluent contribute to not only TSS but also biological oxygen demand (BOD). On the other hand, the effluent quality of MBR is more stable because all the fine particles are completely removed by the membrane regardless of the biological conditions.

MBR effluent is also distinguished from CAS effluent by its high COD/BOD ratio. Because MBR removes slowly degradable BOD better than CAS at a food stringent condition, MBR effluent contains very little biodegradable materials. Non-biodegradable soluble materials, which are detected only as COD, are not degraded in both MBR and CAS. As a result, MBR effluent contains disproportionally low BOD comparing to COD and this results in high COD/BOD ratio. For example, typical effluent COD is 10 to 20 mg/L in municipal MBR whereas typical BOD is only 1 to 2 mg/L or below the detection limit. MBR is also advantageous over CAS for nitrification due to its long solids retention time (SRT), which is essential for enriching sufficient nitrifier population in the mixed liquor. According to Hirani et al. (2013), 90% of MBR produce effluent with equal or better quality than the water quality shown in the following list, which is based on a survey of 38 MBR across the United States.

- NH_4-N < 0.4 mg/L-N
- TOC < 8.1 mg/L
- Turbidity < 0.7 NTU
- Total coliform bacterial count < 100 CFU/100 mL
- MS-2 bacteriophage < 20 PFU/100 mL

3.2.4.2   Heavy Metals Removal

Heavy metals such as Co, Ni, Mn, Cr, Cu, Pb, and Zn contained in wastewater may have acute toxicity for microorganisms in biological systems, especially when the concentrations sharply surge above the tolerable levels. The concentration thresholds of inhibition vary depending on the environmental factors and the existence of other heavy metal ions. Typically, oxygen uptake rate (OUR) reduction, increasing SMP levels, increasing membrane fouling, etc., have been observed when certain heavy metal ions in wastewater surge for short periods (Katsou et al. 2011; Amiri et al. 2010). Despite the known short-term effect, it is harder to know the long-term effect of heavy metals on biology, but it can be postulated that the long-term negative effect is not as much as those observed in short-term tests. In fact, the MBR treating oil field wastewaters containing substantial amounts of heavy metals run under similar operating conditions as municipal MBR, as discussed in Section 6.12. Moreover, some microorganisms are specifically effective for removing certain metal ions. It was observed that Hg/Cu, Cd/As/ Co, and Cd/Cu were effectively removed by *Bacillus* sp., *Pseudomonas* sp., and *Staphylococcus* sp., respectively, at 30% to 60% efficiencies (Nanda et al. 2010.)

MBR removes a portion of the trace heavy metals from wastewater through bioassimilation, ion exchange, and filtering metal oxides (Santos and Judd 2010b). When wastewater containing varying concentrations of Cu(II), Pb(II), Ni(II), and Zn(II) at 3 to 15 mg/L each was treated in a laboratory-scale MBR at 15 days SRT and 5.77 g/L MLSS, removal efficiencies were observed at 80%, 90%, 50%, and 77%, respectively (Katsou et al. 2011). OUR was reduced substantially with the spikes of metal ions, but approximately half of the loss was recovered by adding vermiculite at 5 g/L to the aeration tank. It was postulated that the lowered soluble metal ion concentrations by the adsorbent (vermiculite) was the reason for the recovery. In the same experiment, vermiculite improved the heavy metal removal efficiency by 88% (Cu), 98% (Pb), 60% (Ni), and 85% (Zn), respectively. In another study, removal efficiencies of Cr(III), Zn(II), and Pb(II) by MBR and CAS were observed (Moslehi et al. 2008). For feed water containing chromium and zinc at less than 50 mg/L, MBR removed 95% and 76% of chromium and zinc, respectively. The substantially higher lead removal efficiency of MBR compared with CAS, that is, 65% versus 44%, suggested that the lead adsorbed on suspended solids or colloidal lead oxides suspended in mixed liquor were filtered by the membrane. In a comparison study of MBR and CAS in two different municipal wastewater treatment plants, MBR demonstrated higher removal efficiencies than CAS for Al, Fe, Zn, Cu, Hg, and Cr whereas Pb and Ni removal efficiencies turned out oppositely (Carletti et al. 2008). Heavy metal removal efficiencies in a municipal MBR were observed at 92% (Hg), 90% (Cu), 74% (Pb), 72% (Ni), 72% (Cr), 68% (Zn), 89% (Fe), and 89% (Al).

3.2.5 Membrane Cleaning

3.2.5.1 Backwashing

Periodic backwash, which is also called backflush, can be performed to dislodge a portion of the cake layer attached on the membrane surface by reversing the permeate flow. This method is primarily for hollow fiber membranes and is used to remove cake layer more actively than simply pausing the permeate drawing. A portion of permeate is flowed backward through the membrane lumen for a short period, for example, 5 to 30 s, every 10 to 15 min in MBR, but the duration and the cycle time can be adjusted depending on membrane fouling potential. This method is effective for improving net flux in many situations, but the system becomes more complex because automatic valve and pump controls are required. A portion of the permeate obtained is used as backwash water and as a result the net permeate flow declines.

In MBR, when periodic membrane backwash is compared with the basic on/off cycling of the permeate drawing (or intermittent suction), the advantages of periodic backwashing are often found to be not great enough to justify the partial loss of permeate. In fact, when filtration is performed with air scouring, sticky and dense cake layer is formed mostly by the macromolecules called SMP or EPS rather than by floc particles. In this case, backwashing is not as effective as it is for removing the cake layers formed by large particles due to the nature of the cake layer. As a result of the low efficacy, periodic backwashing is performed less commonly today than decades ago for MBR, although it is still a standard procedure for the hollow fiber membranes used for secondary effluent and surface water treatment, where all the particles in feed water deposit on the membrane in the absence of air scouring.

3.2.5.2 Maintenance Cleaning

For hollow fiber membranes, maintenance cleaning or cleaning in place is performed periodically to recover the membrane performance. It is essentially backwashing with a diluted chlorine (or bleach) solution and has largely replaced conventional backwashing in MBR. The frequency of maintenance cleaning varies widely depending on the membrane fouling rate, for example, once every few hours to once every week. Typically, 100 to 600 mg/L of chlorine solution is backflowed through the permeate line at 15 to 20 LMH for 10 to 20 min while scouring air is paused. The total volume of cleaning solution required is 3 to 6 L/m² each time maintenance cleaning is performed.

The primary purpose of this procedure is cleaning the membrane surface by oxidizing the deposit, but the maintenance cleaning is also crucial to sanitize and clean the hollow fiber lumen. Without periodic chemical backwashing, the narrow hollow fiber lumen can be partially or fully plugged up by microbial colonies and the macromolecules passing through the membrane. This leads to high-pressure losses in the lumen, which in turn causes a maldistribution of the permeate flow among membrane fibers. Eventually, it can lead to the loss of effective membrane surface area and cause unbalanced filtration among fibers. This, in turn, can accelerate membrane fouling, as discussed in Section 3.4.7. The effect of lumen plugging by microorganisms can be similar to the effect of lumen plugging by air bubbles, as shown in Figure 3.28.

3.2.5.3 Recovery Cleaning

Recovery cleaning or off-line cleaning is performed when maintenance cleaning is no longer effective to improve the membrane permeability substantially. This laborious process is designed to be performed every 6 to 12 months, but the actual frequency varies widely depending on the operating conditions. The permeability thresholds that trigger membrane cleaning also vary depending on membrane manufacturer and the specific operating environment, but recovery cleaning is considered when permeability decreases to less than 0.8 to 1.0 LMH/kPa or TMP increases to more than 20 to 25 kPa. It must be noted that permeability recovery becomes increasingly harder as the recovery cleaning is postponed until TMP increases excessively, for example, beyond 30 kPa. In fact, the cake layer becomes more compact as TMP increases, as discussed in Section 1.2.3. Therefore, recovery cleaning should be performed as soon as either permeability hits the minimum threshold or when TMP hits the maximum threshold recommended by the manufacturer.

In MBR with submerged hollow fiber membranes in a combined aeration and membrane tank (Figure 2.11a), membrane modules/cassettes are hoisted out from the aeration tank and cleaned with a water jet either right above the tank or in another designated cleaning tank. Then, the modules/cassettes are submerged in the cleaning tank filled with 1000 to 5000 mg NaOCl/L for 1 to 4 h or longer depending on the membrane manufacturer's guidelines. If membranes are submerged in a separate membrane tank, as shown in Figure 2.11b, modules/cassettes can be cleaned without being removed from the tank. In this case, mixed liquor is transferred to aeration tanks followed by water jetting the tank walls and the membranes to remove large debris. If fibrous materials or other large debris are caught on the membrane bundles or cassettes, they must be removed manually before water jetting. After draining the contaminated water, cleaning solution is filled up and membranes are immersed for a designated period.

Because chlorine tolerance is not only affected by the base material of membrane but also by additives and the manufacturing method, it is difficult to generalize chlorine tolerance based on the base membrane material. The effect of chlorine on membrane longevity is discussed in detail in Section 6.10.1. If the chlorinated organic compounds generated during cleaning are a concern, hydrogen peroxides (H_2O_2) can be used at 0.1% to 2% at surrounding or elevated temperatures following the manufacturer's guideline. Other oxidants such as peracetic acids (CH_3COOOH), ozone, sodium perborate, etc., have been used in laboratory-scale studies with varying extents of success, but these are not practical in large-scale systems due to the high costs associated with the chemical itself and the additional equipment required (Wang et al. 2014c).

If scales such as $CaCO_3$, $Fe(OH)_3$, $CaSO_4$, etc., are a concern, the membrane modules can be submerged in diluted oxalic or citric acids after NaOCl treatment. The recommended or allowed dosages of oxalic and citric acids vary depending on the manufactures, but it is generally 0.5% to 2%. These organic acids can provide sufficiently low pH to dissolves scales (e.g., 1–3), yet are safe to the membranes used in MBR. Scales are dissolved in acidic solution due to the high solubility in low pH. In addition, such organic acids form metal oxalates or metal citrate complexes scavenging the soluble metal ions in the liquid, thereby scale dissolution is expedited by the lowered free metal ion concentrations. Simultaneously, the high buffer capacity of organic acids provides relatively stable pH conditions during the course of the scale-dissolving process. Oxalic acids have much smaller molecular weight and lower pK_a than citric acids and hence they are more efficient in reducing cleaning solution pH at the same dosage as summarized in Table 3.5 and Figure 3.7. The unit prices of the two organic acids are comparable, but less oxalic acids are required to prepare cleaning solutions. As a result, oxalic acids are considered more cost-effective in general. The dosage of citric acids is higher than oxalic acids, but it is safer for the membranes sensitive to pH providing greater buffer capacity to dissolve scales.

The less expensive strong acids such as H_2SO_4 and HCl can also be used after being diluted, but extra care must be taken to avoid accidental overdoses that can damage membranes at low pH of less than 1 to 2. Complete mixing of acids with dilution water is essential to avoid membrane damage by low pH in local areas. Other disadvantage of inorganic acids is that the pH easily increases by

TABLE 3.5
Comparison of Oxalic Acid and Citric Acid as Cleaning Solution

		Oxalic Acid	Citric Acid
Formula		$C_2H_2O_4$	$C_6H_8O_7$
MW (anhydrous)		90.03 g/mol	192.12 g/mol
Solubility (25°C)		14.3 g/100 mL	164 g/100 mL
pK_{a1}		1.23	3.14
pK_{a2}		4.19	4.75
pK_{a3}		N/A	6.39
pH	1 g/L	2.02	2.79
	5 g/L	1.46	2.40
	10 g/L	1.25	2.24
	20 g/L	1.05	2.08

the organic and inorganic contaminants attached on membranes due to the low buffer capacity. It is somewhat controversial, but using NaOCl before acids tends to be more effective than the other way around because NaOCl removes most organic foulants and exposes scale particles to acids in the subsequent acid soaking. Spent cleaning solution can be sent to the equalization tank to be mixed with raw wastewater and treated biologically. The chlorine shock, organic loading shock, and pH drop by the spent acid solution must be carefully reviewed to determine the allowable mixing ratio of the spent cleaning solutions and the raw wastewater especially in small plants.

EDTA can also be used to remove scales from the membrane, but the environmental effects must be carefully reviewed in conjunction with the regulations set by the local government. EDTA sequesters multivalent ions from water by strongly binding with them, for example, Ca^{2+} and Mg^{2+}. The reduced free multivalent ion concentration in the water induces scale dissolution by equilibrium reaction. If EDTA is not allowed by the regulation or is not available, sodium triphosphate (STPP) or diethylenetriaminepentaacetic acid (DTPA) can be used as alternative chealants. For persistent organic foulants, cleaning can be also performed at high pH (12–14) using NaOH within the limit imposed by membrane material. At high pH, fat and oil are hydrolyzed to smaller fatty acids and dissolved. Proteins and carbohydrates also obtain negative charges by losing proton at high pH and tend to dissolve in the water phase. In some cases, NaOH and NaOCl may have a synergistic effect when they are used

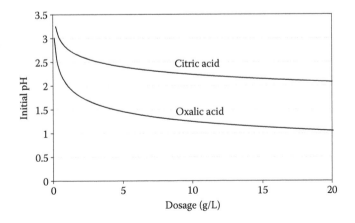

FIGURE 3.7 Initial pH of citric and oxalic based cleaning solutions as a function of dosage.

simultaneously. It can be postulated that persistent organics are swelled at high pH due to the increasing negative charges on the organic molecules whereas easier access to the internal space is secured for oxidants. Alkaline conditions are known to be more effective in dissolving proteins than in dissolving carbohydrates. In some industrial MBR, surfactants can also be used with or without a combination with oxidants and alkalines at 20°C to 35°C to remove persistent organic compounds that might have originated from the manufacturing process. Screening tests using various surfactants and chemical combinations might be required to identify proper surfactants and formulations for target foulants.

Recently, the use of reducing agents such as ascorbic acids and sodium dithionite ($Na_2SO_2O_4$) has been proposed when organic acids are not effective in dissolving ferric salts, although the practicality of the method is still up for further analysis (Zhang et al. 2015). Those reducing agents can increase the solubility of iron by reducing ferric ions (Fe^{3+}) to ferrous ions (Fe^{2+}). In this study, ferric chloride was fed to the anoxic tank at a Me/P ratio of 2 or 4 (mole/mole). The fouled membrane was cleaned sequentially using NaOCl (500 mg/L) and citric acid (0.1 mM or 1.92 g/L with a pH adjustment to 4.0 using acetate buffer), but the cleaning was not effective in removing ferric salts from the membrane surface. However, the membrane cleaning efficiency was improved significantly by replacing the citric acid with ascorbic acid (10–20 mM, 24–48 h) or sodium dithionite (300 mM, 12–24 h).

Maintenance and recovery cleanings can also be performed in flat sheet membrane systems in a similar manner. Because membrane sheets simply sit on rigid porous supports that act as a permeate channel, membrane sheets swell and meet in the center of the upflow channel during the backwash. The net head pressure of the cleaning solution must remain less than 0.3 m to 0.6 m of H_2O (or 3–6 kPa), depending on the membrane panel, to prevent membrane damage. To guarantee low backwash pressure, the cleaning solution tank is raised 0.3–0.6 m above the water level in membrane tank so that the cleaning solution is fed to the membrane system by gravity (or static pressure) without an injection pump. Because most flat sheet submerged modules have big enough lumen spaces in the support panel, microbial plugging of the lumen is much less likely than in hollow fiber membranes. Therefore, maintenance cleaning with chlorine is less compelling than for hollow fiber membranes.

The exposure of membrane to free chlorine can cause changes in membrane porosity, loss of tensile strength, changes in surface property and contact angle, etc. The increase of porosity and permeability is mainly caused by the gradual loss of the additives mixed with the base polymer, as discussed in Section 1.5.3, for example, PEG, PVP, PVA, PMMA, sulfonated polystyrene, and various alcohols and surfactants. Such changes can also reduce the hydrophilicity of the membrane surface and make the membrane surface rougher. Meanwhile, the chain scission of the polymer can lead to the formation of phenyl sulfonates, which increase the electronegativity of the surface, especially in polyethersulfone (PES) membranes (Arkhangelsky et al. 2007). For GE membranes (ZW500), a maximum cumulative chlorine exposure is known to be 500,000 ppm/h assuming that exposure concentration and exposure time are equally influential to membrane oxidation (Fenu et al. 2012). However, according to a recent study, chlorine damage turned out to be more influenced by exposure time rather than exposure concentration when chlorine concentration was higher than 3600 mg/L as NaOCl (Abdullah and Bérubé 2013). For PES membranes, chlorine tolerance is known to be 250,000 ppm/h (Pilutti and Nemeth 2003).

3.2.6 DIFFUSER CLEANING AND MAINTENANCE

All diffusers are destined to be fouled eventually by organic and inorganic foulants. The major cause of diffuser fouling is known to be the dried sludge inside and outside the diffuser pores in activated sludge process (Kim and Boyle 1993; Hung and Boyle 2001). Sludge drying is particularly fast in air diffusers because the air passing the diffuser is hot and dry due to the adiabatic compression occurring in the blower, which raises the temperature and lowers the relative humidity. In general, sludge hardly intrudes into the diffuser during aeration but it can intrude when airflow stops during system maintenance. Sludge intrusion occurs more prevalently in coarse bubble diffusers than in fine bubble diffusers due to the large pores that do not have automatic shutoff mechanisms. Sludge intrusion also occurs in fine bubble diffusers gradually over a number of aeration cycles. If fact, when aeration stops, sludge can wet the pore entrance and is dried by the hot air when aeration resumes. The hydrophilic surface of the dried sludge helps additional sludge intrusion into the pores in the next pause cycle. As the aeration cycle repeats, sludge intrudes gradually deeper into the fine pores and eventually plug up the pores. To minimize sludge intrusion, some fine pore diffusers are made of thermoplastic or elastomeric sheets (plasticized PVC, EPDM rubber, neoprene rubber, etc.). In these diffusers, the fine pores on the diffuser surface are closed at pause mode, not allowing sludge to reach the internal spaces, but the pores become larger to generate bubbles when the diffuser surface bulges because of the internal air pressure during aeration modes. In many commercial fine-bubble diffusers, internal check valves (or nonreturn valves) are also used inside the diffuser to prevent backflow, as shown in Figure 3.8. In this diffuser, the elastic cap enclosing the internal air nozzle bulges and passes air only when the air pressure exceeds the static pressure.

Scaling can also occur simultaneously with organic fouling in the external surface of the diffuser by calcium carbonate, ferric hydroxides, aluminum hydroxides, calcium sulfate, etc., depending on the wastewater source and the chemicals used for pretreatment. Although the scaling potential is not too high in the bulk liquid in most municipal MBR, scales can form on diffuser surfaces due to the lower solubility of salts at high temperature, for example, calcium carbonate and calcium sulfate (Hoang et al. 2007; Ketrane et al. 2009). The internal fouling of diffuser and air pipe can also occur by the deposition of airborne particles carried by the compressed air if the air filters do not function properly.

Diffuser fouling causes the increase of resistance for airflow and hence increases the backpressure. It can also increase the average bubble size due to the larger specific airflow rate through the remaining open pores in the diffuser. The reduced airflow and the larger bubble sizes in turn compromise DO concentration in the bioreactor. Localized diffuser fouling tends to occur at different rates depending on the location of the diffusers in the aeration tank. This can ultimately cause uneven air distribution in the tank and disrupt mixing patterns. Oftentimes, the unevenly distributed air is visually observed from the bioreactor surface. In extreme cases, excessively large bubbles (or slugs) can be observed on the water's surface. Various methods exist to maintain and clean diffusers as below.

In situ *chemical cleaning*: To remove calcium carbonate scaling from diffuser disc, formic acid can be injected to the air pipe through a nozzle in the discharge side of the blower while blowers are under operation (Bretscher 2014). Material compatibilities of formic

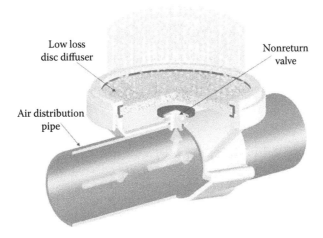

Low loss disc diffuser

Nonreturn valve

Air distribution pipe

FIGURE 3.8 The low-loss KKI-type disc diffuser attached to the air distribution pipe by a saddle-mount design. (Courtesy of Sulzer.)

acid with pipes, valves, joints, etc., must be reviewed before the procedure. The injected formic acid is either vaporized or atomized to the mist in the hot and dry air moving fast in the air pipe. The formic acid transported to the diffuser dissolves primarily calcium carbonates and magnesium carbonates and dislodges volatile dry matters. However, formic acids are less effective in removing ferric and phosphoric compounds. The exact amount required for cleaning depends on the extent of the fouling and the alkalinity of the mixed liquor, but up to 100 mL of 85% to 98% formic acid can be injected per diffuser within 1 h without decreasing pH excessively in typical condition. Chemical cleaning can be performed twice a year, but the frequency is highly dependent on the system's design, operating condition, wastewater source, etc. In one study (Stevens 2008), disc-type fine bubble diffusers were used in a municipal wastewater treatment plant. The diffusers consisted of double membranes and a check valve to minimize sludge backflow and to plug the internal air nozzles during the pause mode. However, initially low backpressure at around 50 kPa gradually increased and reached 65 kPa over time, whereas DO in the aeration tank decreased noticeably. Diffusers were cleaned by injecting 85% formic acid into the air pipe at a dosage of 15 mL/disc. As a result, backpressure decreased to 50 kPa though the extent of organic foulant removal was not known.

Mechanical cleaning: If diffuser fouling is caused by the sludge intruding into the diffuser assembly, mechanical cleaning can be the most effective mean. The bioreactor should be drained and cleaned before the diffusers are disassembled. Vacuum tankers may be required to remove the accumulated sludge and grit from the bioreactor floor. Both sides of the discs are cleaned with soft brushes or water jet after dislodging them from the diffuser assembly. Additionally, diffuser pipes can also be flushed using water to remove the deposit. The flushing water can be acidified to improve the cleaning efficiency. The mechanically cleaned discs can be cleaned further by acid soaking as shown below. Compromised discs, o-rings, and other parts need to be replaced.

Acid soaking: If the *in situ* chemical cleaning is not sufficient, acid soaking can be performed *ex situ* after disassembling the diffusers from the air pipes. It is a time-consuming and costly process because bioreactors need to be drained before the procedure. Diluted sulfuric acids are the least expensive option and does not produce BOD that needs to be treated biologically after the cleaning procedures, but it requires a great deal of care on safety. Formic, citric, or oxalic acids can also be used instead of sulfuric acid.

3.3 FUNDAMENTALS OF AIR SCOURING IN SUBMERGED MEMBRANE

3.3.1 Membrane Scouring Mechanism

Air sparging induces shear stress at the membrane surface as rising air bubbles travel near the membrane. The shear stress enhances the particle backtransport away from the membrane surface by enhancing the mass transfer in the concentration polarization layer. There are two common mechanisms responsible for the high surface shear stress in both submerged flat sheet and hollow fiber membranes:

- *Falling film effect*—If rising bubbles in water are seen from the perspective of water movement, it can be described that the water placed above the bubble continuously falls down to the bottom of the bubble by gravity. Therefore, when bubbles rise near the membrane surface, the thin water film between the rising bubble and the membrane moves down to fill the gap created by the rising bubble. This is the so-called falling film effect. Localized high liquid velocities in the falling film could increase the mass transfer at a membrane surface by up to two orders of magnitude depending on the size of the air bubbles, when it is compared with nonsparging conditions (Ghosh and Cui 1999; Judd et al. 2001).

• *Wake effect*—When the falling water around the rising bubble reaches the bottom of the bubble, it swirls around the corner causing eddies. The eddies that occur in the wake of a rising air bubble can generate a short period of very high shear stress twice as large as that induced by the localized high liquid velocities in the falling film depending on the air bubble size (Bérubé and Lei 2006; Bérubé et al. 2006).

The shear stress in local membrane area can be measured using electrochemical shear sensors as described in Cabassud et al. (2003), Bérubé and Lei (2006), and Bérubé et al. (2006, 2008). Figure 3.9 shows an experimentally measured shear stress profile using electrochemical shear sensors as a function of bubble position on the membrane surface. In this method, the shear stress is proportional to the voltage reading from the sensor. When a rising ellipsoidal bubble is approaching the sensor marked by a white dotted line, the initially stable shear signal starts to increase (frames 1 and 2). When the bubble passes the sensor (frame 3), shear stress increases sharply due to the falling film effect. Right after the bubble passes the sensor (frames 4, 5, and 6), shear stress reaches its maximum due to the wake effect of the rising bubble. The patterns of shear stress profile are variable depending on the bubble shape, but the general trend remains the same (Chan et al. 2007).

In submerged flat sheet membranes, falling film and wake effects are two of the major mechanisms of antifouling because all the bubbles must pass through the narrow channels in between two panels in contact with the membranes in its proximity. On the contrary, in submerged hollow fiber membranes, bubbles tend not to rise along the membrane surface, but rise through the spaces among fiber bundles, where the larger spaces allow less hydraulic resistance for the bubble's rise. This is more apparent in full-scale hollow fiber membranes, where bubbles have greater chances to escape from the fiber bundle while rising especially in the upper part of the fiber. Therefore, the

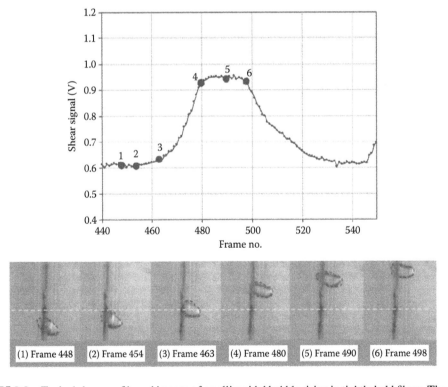

FIGURE 3.9 Typical shear profile and images of an ellipsoidal bubble rising in tightly held fibers. The translucent vertical streaks seen in the pictures are the eight fibers in the bundles. The white horizontal dashed line across the six images shows the position of the shear probe on the test fiber. Gas sparging rate = 0.5 mL/min; diffuser nozzle size = 2 mm. (From Chan, C.C.V. et al., *J Membr Sci.* 297:104–120, 2007.)

significance of falling film and wake effects are not as great as in submerged flat sheet membranes. In addition, the antifouling mechanisms can be somewhat different in the lower and the upper parts of the same fiber.

It has been known that not only the magnitude of the average upflow velocity (or shear stress) on membrane surface but also the variability of the upflow velocity (Ueda et al. 1997; Chan et al. 2011). The variability of flow velocity can be quantified by various measures in practice, for example, the average magnitude of the time derivative of upflow velocity, and the standard deviation of the multiple instantaneous upflow velocities measured in different moments. In one study, nine hollow fibers were used to filter 1 g/L of bentonite (Yeo et al. 2006). It was observed that the membrane fouling rate quantified by the final TMP tended to be lower when the standard deviation of upflow velocity was high, as shown in Figure 3.10. This suggests that using variable airflow rates at a constant average airflow rate can be an effective approach to enhance membrane scouring without increasing energy costs. In fact, the intermittent aeration performed widely in submerged hollow fiber membrane processes is an extreme form of airflow rate variation that generates high standard deviation in upflow velocity.

Specifically for hollow fiber membranes, the following mechanisms are responsible for the high shear forces on membrane surfaces.

- *Random fiber displacement (or sway)*—If a portion of the hollow fiber is suddenly pulled by a random turbulent motion of air or liquid, the rest of the fiber is also pulled toward that same direction. Even if a certain portion of the fiber is not directly exposed to the turbulence, it is still affected by this random motion generated in other parts of the same fiber. As the fiber length increases, not only the chance of the fiber being pulled increases but also the strength of the pulling increases because the longer fibers are exposed to the turbulence with larger surface areas. In fact, the pulling force (N) is a product of the membrane area (m²) exposed to the turbulence and the shear stress (N/m²) on it. Therefore, long hollow fibers are particularly benefited by this antifouling mechanism that can at least partially compensate the performance loss from the internal pressure loss discussed in Section 2.2.2.
- *Physical contacts among fibers*—In turbulent water, hollow fibers can physically hit each other, generating high shear forces on the membrane surface. The magnitude of the surface shear forces resulting from lateral velocity is dependent on a number of complex parameters relating to the configuration of the submerged membrane system (e.g., fiber length, looseness, density, and flexibility) and the air sparging practices.

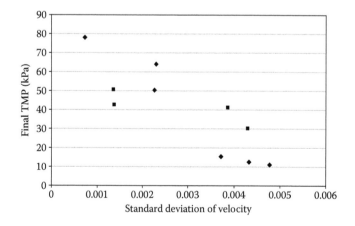

FIGURE 3.10 Effect of standard deviation of upflow velocity on final TMP. Particle image velocimetry was used to monitor flow velocity around hollow fiber membranes. (From Yeo, A.P.S. et al., *J Membr Sci.* 280:969–982, 2006.)

- *Lateral liquid migration*—Lateral fiber movement in turbulent conditions can generate lateral liquid migration in and out of the fiber bundle. It facilitates the transfer of concentrated sludge in the fiber bundle out to the bulk (Cui et al. 2003). It is crucial in preventing hollow fiber bundles from being clogged by the excess sludge accumulated in them.

3.3.2 Effect of Airflow on Membrane Fouling

As expected intuitively, membrane fouling rates can be mitigated by increasing scouring airflow rate. Air bubbles not only contact directly with membranes but also increase bubble-induced water flow on the membrane surface. In addition, rising bubbles increase random fiber movements causing the acceleration and deceleration of fibers in the liquid. When fibers turn the direction abruptly to the opposite side in the turbulent flow, the shear stress exerted on the membrane surface greatly increases and the foulants on the membrane surface have a chance to be dislodged. However, it has been known that air scouring can increase the shear stress only to a certain extent and as a consequence sustainable flux cannot be increased infinitely by increasing the air scouring rate in a given condition. The existence of the maximum sustainable flux in turn implies that the incremental benefit of increasing airflow rate diminishes as airflow rate increases. Conceptually, the optimum combination of aeration rate and flux exists, which allows the minimum lifetime costs of MBR.

In a submerged membrane system, upflow increases as airflow increases. When horizontally mounted hollow fiber membranes made of polyethylene (Mitsubishi Rayon, Japan) were used in MBR, upflow velocity quickly reached its maximum at 40 cm/s at an airflow rate of 0.4 m^3/min as shown in Figure 3.11a. Upflow velocity plateaued at or above 0.4 m^3/min, but the standard deviation of the upflow velocity continued to increase until the airflow reached 0.75 m^3/min. Both upflow velocity and its standard deviation decreased when airflow rate increased from 0.75 m^3/min to 0.95 m^3/min and this suggested that excessive airflow might be negative on membrane scouring (Ueda et al. 1997). In the filtration experiment (Figure 3.11b), as airflow rate increased, the TMP required to obtain constant flux decreased. It is noteworthy that the TMP at 15 LMH significantly decreased when airflow rate was increased from 0.3 to 0.5 m^3/min, although there was little difference in upflow velocity. It can be attributed to the high standard deviation in upflow velocity at 0.5 m^3/min (Figure 3.11a), where the standard deviation indicates the extent of the acceleration and declaration of the upflow. The existence of the maximum upflow velocity was later confirmed in many other studies, although the reported maximum upflow velocity varies widely between 0.4 and

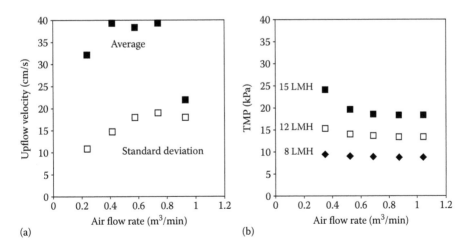

(a) (b)

FIGURE 3.11 Effect of airflow rate on (a) upflow velocity and (b) TMP in MBR. Horizontally mounted hollow fiber membranes (Mitsubishi Rayon Co., Japan) were used. TMP was measured 8 min after the start of suction on the 280th day. (From Ueda, T. et al., *Water Res.* 31(3):489–494, 1997.)

0.8 m/s depending on the experimental condition used (Liu et al. 2000; Cui et al. 2003; Bérubé and Lei 2006; Yeo et al. 2007).

When a bentonite solution (0.65 g/L) was filtered using vertically mounted hollow fiber membranes with 1 mm inner diameter (ID), the membrane fouling rate measured by TMP rising rate decreased rapidly as superficial air velocity increased, as shown in Figure 3.12. However, once the superficial velocity reached 0.8 mm/s, only a marginal amount of membrane fouling reduction was observed (Xia et al. 2013). This observation is in line with the observation of the existence of maximum upflow velocity as discussed in the previous paragraphs.

As superficial velocity on the membrane surface increases, the total filtration resistance calculated by Equation 1.4 tends to decrease to a certain extent. According to Chang (2011), bubbling can reduce mainly reversible fouling resistances caused by loosely bound particular matters (Figure 3.13). The irreversible fouling resistance caused by the adsorption of small molecules such

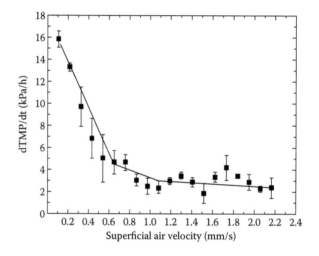

FIGURE 3.12 Fouling rate of vertically mounted PVDF hollow fiber membranes (Memstar) in 0.65 g/L bentonite solution, where superficial air velocity is measured by dividing airflow rate (mm³/s) by the cross-sectional area of tank (mm²). (From Xia, L. et al., *Water Res.* 47(11):3762–3772, 2013.)

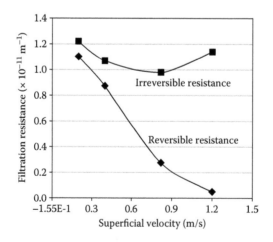

FIGURE 3.13 Effect of superficial velocity on reversible and irreversible resistance of hollow fiber membranes installed in a plate and frame module with liquid and bubble circulation. (From Chang, S., *Desalination.* 283:31–39, 2011.)

as SMPs or extracellular polymeric substances (EPS) is not only greater than reversible fouling resistance but is also not significantly affected by the superficial velocity. It can be attributed to the very low backtransport velocity of small cell debris and biopolymers that deposit on membrane even at high shear rates. The strong binding tendencies of such species can make them firmly attached on the surface. At very high superficial velocity, little reversible resistance exists due to the thinning cake layer depth, as shown in Figure 3.13. The lack of sufficient cake layer allows direct contacts of SMP and EPS with membrane, and this might be the cause of the slight increase of irreversible fouling resistance at high superficial velocity. The increasing irreversible resistance also suggests the possibility of negative consequences in membrane performance when scouring airflow rates are excessive.

3.3.3 Effect of Bubble Size on Flux

3.3.3.1 Overview
In early studies in the 1990s, diffusers with 6 to 8 mm pores were considered the optimum to control membrane fouling in submerged membranes instead of fine bubble diffusers with less than 3 mm pore sizes regardless of the membrane configuration. The better membrane performance with such coarse bubble diffusers was attributed to the faster bubble rising as bubbles experience less drag forces for their buoyancy. Starting from the horizontally mounted submerged hollow fiber membranes (Sterapore™) of Mitsubishi Rayon Co. of Japan in the early 1990s, coarse bubble diffusers had been predominantly used in MBR until counterevidences were observed in 2000s. Due to the complexity of the hydrodynamics in submerged membrane processes and the lack of standardized testing protocols, publicly available data sets are often conflicting. However, many of the later studies have revealed that small bubbles were more effective than large bubbles in certain conditions. This implies that the greater number of small shear events on membrane is superior to the infrequent strong shear events at a constant airflow rate. The conflicting observations were made mostly with flat sheet modules (Sofia et al. 2004) and hollow fiber modules with short fiber lengths, for example, less than 0.5 m (Fane et al. 2005). In the meantime, the optimum bubble size for commercial-scale hollow fiber membranes seems to be consistently larger than those for flat sheet membranes. The different optimum bubble sizes can be attributed to the different membrane scouring mechanisms of the two different membrane configurations.

As will be discussed in the following sections, small bubbles tend to be more efficient than large bubbles for flat sheet membranes, tightly held hollow fiber membranes, and short hollow fiber membranes (<500 mm) held loosely. For these types of membranes, the direct contact of bubbles with the membrane surface and the upflow generated by rising bubbles are the major antifouling mechanisms. On the other hand, large bubbles seem to be more effective for commercial-scale modules with loosely held hollow fibers, where random fiber displacements are the major antifouling mechanism. Most of all, the optimum bubble size must be experimentally determined for each specific membrane module configuration under the condition in which modules will be used. It is because the dynamics of two phase flow and the complex interaction between environmental factors and membrane fouling is not fully understood.

3.3.3.2 Flat Sheet
In flat sheet submerged membranes, membrane panels are stacked against each other with 5 to 7 mm spaces in between the panels. All the bubbles released from the diffusers located underneath the membrane panels rise through the in-between spaces while physically contacting the membrane surface. Large bubbles are likely more effective in membrane scouring due to the strong falling film and wake effects, but a large amount of bubbles are required to cover the entire membrane surface. In contrast, small bubbles generate weaker falling film and wake effects, but they can cover

the entire membrane surface at a lower airflow rate. The optimum bubble size must be determined considering the membrane scouring effect and the cost of maintaining the airflow rate.

In one experiment, flat sheet membrane panels in a cassette with well-defined upflow and down-flow channels were used in an aeration tank filled with mixed liquor. The membrane fouling propensity measured by the TMP rising rate was much lower with fine bubbles than with coarse bubbles at identical airflow rates, as shown in Figures 3.14 and 3.15 (Sofia et al. 2004). It was also observed that the average upflow was faster with fine bubbles in the experimental conditions used in the study.

To elucidate the effect of bubble size and channel size on shear stress on membrane surface, mathematical simulation studies have been performed based on computational fluid dynamics (CFD; Ndinisa et al. 2006a; Prieske et al. 2010; Böhm et al. 2012). It has been revealed that the shear stress tends to increase as channel size decreases. The shear stress also increases as bubble size increases, but it only marginally increases once the bubble size reaches the size of the channel. The relations among shear stress, bubble size, channel size (or spacing between two membrane panels), and bubble rising velocity are plotted in Figure 3.16. The maximum shear stress ranges widely depending on the existence of the downcomer, which improves upflow velocity by facilitating the downflow, for example, 0.7 Pa without downcomers (Ndinisa et al. 2006a) and 4 Pa with downcomers (Prieske et al. 2010).

Because the upflow channel sizes are typically approximately 6 mm for flat sheet membranes and the actual bubbles released from the diffuser are larger than the diffuser pore size, diffusers with pores smaller than 6 mm are optimum for flat sheet membranes. In fact, Kubota initially used diffusers with 10 mm pores but changed them to those with 4 mm pores (Ndinisa et al. 2006a). Recently, the nozzle size for the new module (SP400) was reduced further to less than 3 mm,

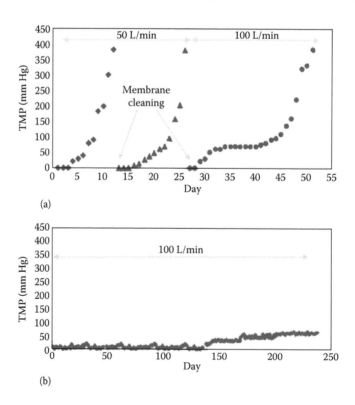

(a)

(b)

FIGURE 3.14 Effect of diffuser pore size on membrane fouling: (a) two coarse bubble diffusers with six 2-mm holes and (b) two fine bubble diffusers with an unknown number of 0.5-mm holes. For both graphs: MLSS = 8 g/L; water temperature = 30°C; gross flux = 16.9 LMH (or 0.48 m/day); and suction cycle = 8 min on/2 min off. (From Sofia, A. et al., *Desalination* 160:67–74, 2004.)

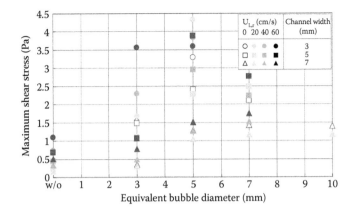

FIGURE 3.15 CFD determination of maximum wall shear stress exerted by differently sized bubbles rising at terminal rising velocity in channels of different widths. (From Prieske, H. et al., *Desalin Water Treat.* 18:270–276, 2010.)

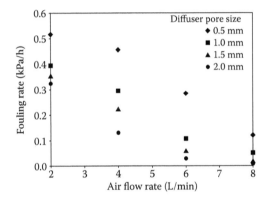

FIGURE 3.16 Membrane fouling rate determined after 2 h of yeast filtration. One A4 sized module (Kubota/ Yuasa, Japan) with 0.1 m² was installed in an acrylic plastic case. The space between membrane and the case wall was kept at 7 mm and MLSS was 6 g/L. (From Ndinisa, N.V. et al., *Separ Sci Technol*, 41, 1383–1409, 2006.)

whereas the height of the cassette was increased to 2600 mm to improve air utilization efficiency. It was claimed that a larger number of fine bubbles not only scour membranes more efficiently but are also distributed more evenly among the upflow channels. As a result of all these efforts, the design SAD$_m$ has decreased to 0.3 m³/m²/h, which falls into the high end of the hollow fiber modules', for example, 0.1 to 0.3 m³/m²/h (De Wilde et al. 2008).

On the contrary, some literature suggests that large bubbles are more effective for membrane scouring than small bubbles in flat sheet submerged membrane systems. Figure 3.16 shows one example where a 6 g/L yeast suspension was filtered while air was supplied through the diffusers with 0.5 to 2.0 mm pores. In the range of airflow rate tested, the larger the diffuser pore size was, the lower the membrane fouling rate was under the experimental condition.

The conflicting bubble size effects are caused by the different specific airflow rates used in different experiments. If the airflow rates used in the experiment were grossly high, even large bubbles could cover the entire membrane surface. Then, the strong shearing effect of the large bubble dominates the antifouling mechanism rather than the coverage effect of the small bubbles on the membrane surface. In commercial submerged flat sheet membrane processes, scouring airflows are kept at a low level to obtain greater energy efficiencies. If large bubbles are used, they cannot cover the

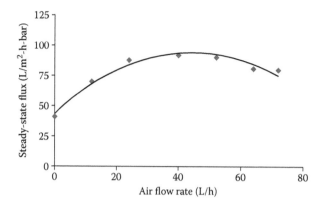

FIGURE 3.17 Influence of airflow rate on steady-state flux. BIO-CEL® flat sheet membranes (Microdyn-Nadir) were used to filter 10 g/L yeast solution. (From Qaisrani, T.M., and Samhaber, W.M., *Desalination* 266:154–161, 2011.)

entire membrane surface at the condition. As a consequence, small bubbles would be superior to large bubbles in controlling membrane fouling by covering larger membrane surface areas.

The consequence of bubble dynamics is often unpredictable due to its complex nature. When flat sheet membranes (BIO-CEL®, Microdyn-Nadir) were used to filter 10 g/L yeast solution, steady-state flux increased as airflow rate was raised when the airflows were low at less than 40 L/h. However, the steady-state flux started to decrease as airflow rate increased to more than 40 L/h, as shown in Figure 3.17 (Qaisrani and Samhaber 2011). The initial flux increase can be attributed to the increasing number of bubbles at high airflow rates. It is not clear what caused the flux decline after the peak, but bubbles were considered too large at very high airflow rates that they started to hinder the liquid to reach the membrane surface. The smaller floc/particle size at high shearing conditions may also contribute to the flux reversal. Another potential explanation is that the lack of cake layer at high aeration rates allows direct access of small particles to the membrane, which may cause pore plugging.

The specifications of some commercial flat sheet membranes are summarized in Table 3.6. Depending on the manufacturer, diffusers with various pore sizes are being used.

TABLE 3.6
Specifications of Commercially Available Flat Sheet Modules

Manufacturer/Model	Membrane Spacing (mm)	Panel Height (mm)	SAD$_m$ (m^3/m^2/h)	Superficial Gas Velocity (m/s)	Aeration
A3 Watersolutions/M70	7	1050	0.31	0.025	Fine
Brightwater Eng./Membright	9	950	1.28	0.076	Coarse
ColloideEngineering/Sub snake	10	1000	0.5	0.028	Fine
PureEnviTech/SBM® 8S20L (multideck)	8	195 (Panel) 3260 (Cassette)	0.30	0.032	Fine
Kubota/510 ES (single deck)	7	1000	0.75	0.047	Coarse
Kubota/SP400 (multideck)	7	2600 (Cassette)	0.30	0.032	Medium
MicrodynNadir/Bio-Cel®	8	1200	0.3–0.8	0.0125–0.033	Fine
Toray/TRM140-100S	6	1608	0.3	0.037	–

Source: Modified from Prieske, H. et al., *Desalin Water Treat.* 18:270–276, 2010.

3.3.3.3 Hollow Fiber

The bubble size effect on hollow fiber membrane performance varies depending on the fiber length and the looseness (or tightness) in the module, among many other factors. When a single hollow fiber with 1 mm outer diameter (OD) and 400 mm length was installed tightly in a square channel (20 mm × 20 mm) and superficial liquid velocity was controlled by a circulation pump, slower membrane fouling was observed with small bubbles than with large bubbles at the same airflow rate (Yeo et al. 2007). As shown in Figure 3.18a, membrane fouling rate was the lowest with small bubbles at a low superficial liquid velocity (16 mm/s). When the liquid velocity was increased to 160 mm/s, however, the bubble size effect became less apparent, as can be seen in Figure 3.18b.

Similar bubble size effects have been observed in many other studies. When shear stress on multiple 120 mm hollow fiber membranes was measured, the high number of weak shear events generated by small bubbles resulted in more favorable hydrodynamic conditions for fouling reduction (Chan et al. 2007). When tightly held hollow fiber membranes with an 820 mm length were used, TMP increased slower with fine bubbles under identical filtration conditions (Martinelli et al. 2010). Fine bubbles were more effective in the ultrafiltration of river water using 110-mm-long membrane fibers (Tian et al. 2010).

In contrast, fine bubbles do not tend to be more advantageous over coarse bubbles if long hollow fibers are held loosely in a module. When hollow fibers with 0.65 mm OD and 500 mm length were held loosely at 1% looseness (the ratio of excess fiber length to the distance between two headers), no significant differences were observed between 0.5 and 1.0 mm bubbles with regard to membrane fouling (Wicaksana et al. 2006). In commercial membranes with approximately 2000 mm length held with 2.5% looseness, coarse bubbles were superior to fine bubbles.

The exact causes of conflicting bubble size effects are not known, but it seems linked to the varying bubble dynamics depending on the fiber length and the fiber tightness/looseness in the module. At least two plausible mechanisms exist, which can cause the conflicting bubble size effects. First, membrane scouring by direct contact with bubbles becomes less significant as fiber length increases. This is because bubbles tend to escape from the hollow fiber bundle as they rise due to the greater hydraulic resistance inside the bundle. Although small bubbles are more effective in covering the larger membrane surface area, this benefit becomes less significant as fiber length increases. Second, random fiber displacement becomes a more significant antifouling mechanism as the length of the hollow fiber increases only in loosely bound hollow fiber membranes, as will be discussed in Section 3.3.1. As a matter of fact, in short hollow fiber membranes (110–400 mm),

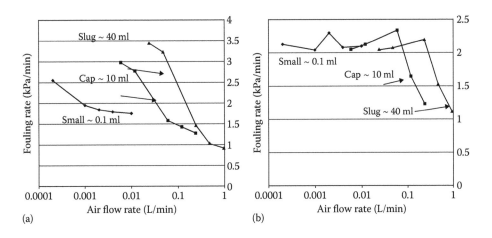

FIGURE 3.18 Membrane fouling rate measured by TMP increasing rate at different linear liquid velocity: (a) 16 mm/s and (b) 160 mm/s. A single hollow fiber membrane with 1 mm OD and 400 mm length was used to filter 2 g/L bentonite. (From Yeo, A.P.S. et al., *J Membr Sci.* 304:125–137, 2007.)

small bubbles are more effective than large bubbles regardless of the tightness of the membrane fiber at a constant airflow. The small bubbles are also effective for the longer fibers held tightly in a module without moving laterally because they cover a larger membrane surface area causing more shear events. However, large bubbles are more effective for modules with long hollow fibers held loosely because they cause stronger irregular lateral fiber movements that transmit to the entire fiber. All the commercial-scale modules with loosely held fibers with a length of approximately 2 m use large bubbles, for example, SADF™, Microza®, and ZW500d™.

3.4   OPTIMIZATION OF SUBMERGED HOLLOW FIBER MODULE

3.4.1   EFFECT OF FIBER LOOSENESS

It has been claimed that 0.1% to 5% redundancy (or looseness) in fiber length relative to the distance between the two headers greatly reduces membrane fouling (Henshaw et al. 1998). This is because the hollow fibers with a little slack can make random lateral movements (or sway) in two-phase turbulent flow. The random movement generates strong shear forces on the membrane surface especially when the fibers change the direction abruptly. The amplitude of the fiber movement increases as the looseness increases, but excessive looseness (e.g., 5%–10%) can lead to membrane breakages at the turbulence especially in long fibers (Cui et al. 2003). Figure 3.19 shows the effect of fiber looseness on membrane fouling rate. The TMP rising rate significantly decreases when fiber looseness increases from 0% to 1%. Further increase of looseness to 4% only marginally decreases membrane fouling under the experimental condition used.

The frequency and the amplitude of the random fiber movements are the crucial factors affecting the antifouling efficacy of air scouring and they are positively correlated with airflow. However, both of them have their maximum limits set by the hollow fibers' characteristics and the module configuration. When hollow fiber movements were monitored using a video camera, the frequency of the fiber movant did not increase further despite the rising airflow rate once it reached the maximum. In addition, the maximum fiber amplitude was also limited depending on fiber length and looseness (Wicaksana et al. 2006). These observations suggest that aeration is effective in generating shear stress on the membrane surface only to a certain extent depending on fiber looseness, length, and flexibility. It is noteworthy that this notion is in line with the old observations on the existence of the maximum sustainable flux achievable by increasing airflow, as shown in Figure 3.11 (Ueda et al. 1997).

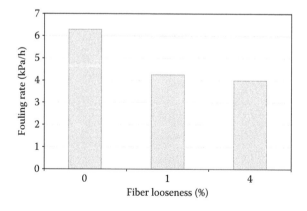

FIGURE 3.19   Effect of fiber looseness on membrane fouling rate. Fiber length = 50 cm; OD = 0.65 mm; ID = 0.39 mm; pore size = 0.2 μm; polypropylene, flux = 30 LMH, 5 g/L yeast; scouring airflow = 1.5 L/min; nozzle size = 1 mm. (Data from Wicaksana, F. et al., *J Membr Sci*. 271, 186–195, 2006.)

Figure 3.20a shows the experimental results obtained with tightly or loosely held fibers in two-phase flow (Bérubé and Lei 2006). The figure also includes baseline data obtained with tightly held fibers scoured by one-phase flow without air. In this experiment, reservoir water was filtered at a vacuum pressure of 28 kPa using a single hollow fiber (ZW500 of GE). Flux was the lowest with the tightly held fiber used in single phase flow without aeration. The steady-state flux increased significantly with aeration at an identical water flow rate. The highest flux was obtained with the membrane held loosely with a 2.4% looseness (420 mm fiber was held in between the two headers with 410 mm space). A similar observation was made in an independent study (Chang and Fane 2002). Figure 3.20b compares two TMP profiles of tightly held and loosely held fibers operated at 30 LMH. No TMP increase was observed for 300 min with 5% looseness, but a large TMP increase was observed from 4 to 50 kPa when hollow fibers were held tightly. The minimum fiber length required to take advantage of the fiber looseness is dependent on the experimental conditions including airflow rate, fiber diameter, fiber flexibility, airflow rate, viscosity of the medium, etc., but it seems that short fibers with a length of less than 200 to 400 mm receives little benefit from the loosely held fibers.

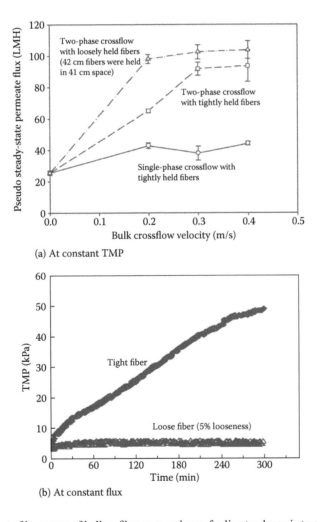

(a) At constant TMP

(b) At constant flux

FIGURE 3.20 Effect of looseness of hollow fiber on membrane fouling tendency in two-phase flow: (a) measured by pseudo steady-state permeate flux, where GE's ZW500 fibers were used after cutting fibers at 420 mm length and (b) measured by TMP. (a: From Bérubé, P.R., and Lei, E., *J Membr Sci.* 271(1–2):29–37, 2006; b: From Chang, S., and Fane, A.G., *J Chem Technol Biotechnol* 77:1030–1038, 2002.)

3.4.2 Effect of Fiber Flexibility

Random fiber displacement is one of the most crucial mechanisms that controls hollow fiber membrane fouling as discussed above, but the fiber flexibility determines the magnitude of the random fiber displacement under the given aeration conditions. Intrinsic fiber flexibility is decided by the properties of the polymeric materials used to build the fiber, but membranes can acquire additional flexibility by proper combinations of fiber structure, length, diameter, wall thickness, looseness, manufacturing methods, chemical additives, etc.

For a given membrane fiber with a fixed fiber looseness, fiber length determines the maximum fiber amplitude. The longer the fibers are, the larger the maximum fiber amplitudes are. Wicaksana et al. (2006) observed the amplitude of the fibers with three different lengths using a video camera. All three fibers were held loosely at 1% looseness and aeration rate was fixed at 2 L/min. As expected, the longest fiber with a 90 cm length had the largest amplitude out of the three fibers, as shown in Figure 3.21. In addition, as fiber length increases, the actual amplitude more closely reached the maximum theoretical amplitude at a lower aeration rate. In fact, the ratio of actual to theoretical maximum increased from approximately 45% to approximately 70% to 85% as fiber length increased from 50 cm to 70 to 90 cm at the same airflow. This observation suggests that long fibers are more advantageous than short fibers in maximizing random fiber displacement effect at a low airflow rate.

The hollow fiber membranes with small IDs undergo a large internal pressure decrease that affects the membrane performance negatively, as discussed in Section 2.2.2., but thin fibers are subject to more vigorous random lateral movements compared with their thicker counterparts, thereby the disadvantages from the internal pressure loss can be at least partially offset. In one experiment (Chang and Fane 2002), two hollow fibers with 0.65 and 2.7 mm OD were compared under identical conditions. The smaller fiber with 0.65 mm OD fouled slower than the larger fiber with 2.7 mm OD despite the larger internal pressure loss as shown in Figure 3.22. This suggested that the random fiber movement effect was dominant over the internal pressure loss effect under the experimental conditions. In a more elaborate experiment (Wicaksana et al. 2006; Fane et al. 2005), three 50-cm polypropylene hollow fibers with OD of 0.65, 1.0, and 2.7 mm, respectively, were tested. As fiber diameter increased from 0.65 to 2.7 mm, fiber amplitude decreased sharply from 3 to 1 cm at an identical aeration rate of 2 L/min (Figure 3.23). As a consequence, the membrane fouling rate measured by the TMP rising rate was reciprocally proportional to the fiber diameter. In another experiment, when 50 and 70 cm fibers with 0.65 mm fiber OD and 1% looseness were used to filter 5 g/L yeast at 30 LMH, 70 cm fibers fouled faster at 1.5 L/min airflow rate

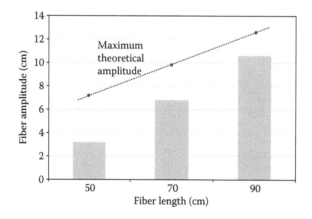

FIGURE 3.21 The variation of fiber amplitude with fiber length: 0.65 mm fiber OD, 1% looseness, 2 L/min airflow rate, and 1 mm diffuser. (Data from Wicaksana, F. et al., *J Membr Sci.* 271, 186–195, 2006.)

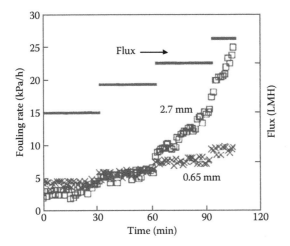

FIGURE 3.22 Effect of fiber diameter on membrane fouling. Scouring airflow = 7.3 L/min; medium = 5 g yeast/L. (From Chang, S., and Fane, A.G., *J Chem Technol Biotechnol.* 77:1030–1038, 2002.)

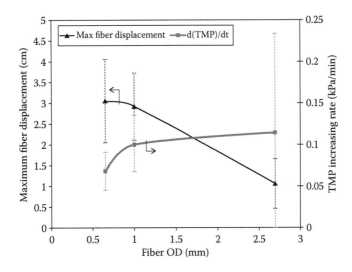

FIGURE 3.23 Fiber displacement and TMP increase (dTMP/dt) versus fiber OD. Fiber length = 50 cm; tightness = 99%; MLSS = 5 g/L yeast; flux = 30 LMH. (From Fane, A.G. et al., *Desalination* 185(1–3):159–165, 2005.)

(Wicaksana et al. 2006). This suggests that the performance gain by the enhanced fiber movement failed to completely offset the performance loss by the internal pressure loss under the experimental conditions used.

As discussed previously, the net effect of the hollow fiber diameter on membrane fouling can be either positive or negative, depending on the experimental conditions, such as aeration rate, fiber length, diameter, and feed water characteristics. This is because the magnitudes of the positive and negative effects are case-specific and it might turn out oppositely in slightly different experimental setups. Therefore, the effect of the size of the hollow fibers can only be found out empirically by performing the filtration at the condition the membranes are targeting. Due to the complex nature of the hydrodynamics in such two-phase flow, theories are useful only to perceive the factors that potentially affect the membrane performance, and thereby need to be closely monitored.

3.4.3   EFFECT OF INTERNAL PRESSURE LOSS

3.4.3.1   Membrane Fouling Induced by Internal Pressure Loss

It has been well known that the internal pressure loss can cause excessively high flux near the permeate exit whereas the average flux of the membrane remains low. It causes quick cake layer formation and flux decline near the permeate exit. Under constant flux mode, upstream flux must increase to maintain the constant average flux, which in turn causes higher fouling potential in the upstream. As the fouling spreads, the lost flux in the downstream must be compensated by the gradually shrinking clean membrane surface in the upstream. The increasing flux in the clean membrane surface leads to accelerated TMP increase. Therefore, internal pressure loss in the hollow fiber must be kept as low as possible.

The effect of pressure loss on imbalanced filtration and membrane fouling has been experimentally proven in hollow fiber membranes. Lee et al. (2008) analyzed the porosities of the cake layer formed in different positions of a 15 cm hollow fiber membrane. After cutting the used membrane to six pieces (2.5 cm each), the average porosities of the cake layer were measured for each piece of membrane using confocal laser scanning microscopy (CLSM). As shown in Figure 3.24, the cake layer porosity is the lowest near the permeate exit and tends to increase toward the upstream. It is noticeable that such imbalanced filtration occurred even in a short membrane with only a 15-cm length operated at a moderate flux, that is, 30 LMH. If a longer membrane fiber is used, the gradient of the cake layer porosity would be more dramatic. In the figure, cake porosity ranges were unrealistically high at 60% to 75%, but it occurs in CLSM because the threshold that distinguishes void spaces from solids is set somewhat arbitrarily by the operator.

Figure 3.25 shows the profile of a cake layer weight in five different segments of a hollow fiber membrane with both ends open (Li et al. 2014). Membrane samples were taken after filtering 5 g/L of yeast suspension with air sparging for 6 h. In the figure, TD1 to TD5 represent five equally spaced segments of a hollow fiber from one end of a membrane to the other end. It is apparent that more foulants deposit at both ends of the hollow fiber than in the middle due the high local flux in the permeate exits.

3.4.3.2   Advantages of Long and Thin Hollow Fibers

Long and thin hollow fibers are generally disadvantageous with respect to internal pressure loss, but they provide substantial benefits that can at least partially offset the disadvantages. The performance

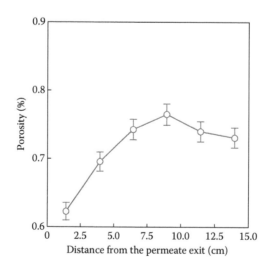

FIGURE 3.24   Cake layer porosity profile in hollow fiber membrane used in a laboratory-scale MBR. (From Lee, C.H. et al., *Desalination* 231: 115–123, 2008.)

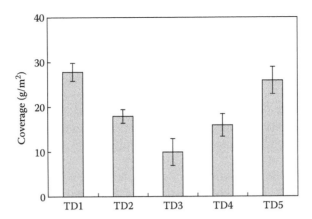

FIGURE 3.25 Gravimetric analysis of the membrane samples obtained from the corresponding areas detected by TD1–TD5 after 360 min of fouling operation, where TD1–TD5 are the five pieces of hollow fibers with identical lengths from one open end to the other open end. (From Li, X. et al., *J Membr Sci.* 451:226–233, 2014.)

of longer and thinner fibers can be better than shorter/thicker counterparts especially when air scouring is weak.

- First, as shown in Figure 3.21, the maximum theoretical amplitude of the fiber is proportional to the fiber length at a given looseness. In addition, the actual fiber amplitude can easily approach the maximum theoretical fiber amplitude at a low airflow when fibers are long and thin. Because the random lateral fiber motion is the principal antifouling mechanism of submerged hollow fiber membranes, the larger fiber amplitude can significantly offset the drawbacks from the internal pressure loss.
- Second, the frequency of random fiber movement is positively correlated with the fiber length. If one segment of a long hollow fiber is suddenly pulled by the turbulence, the pulling motion can spread throughout the fiber. Long fibers are exposed more to such sudden movements than short fibers because they have more exposure to the turbulence.
- Third, scouring air can be used more efficiently with long fibers. Although more energy is required to generate the same scouring airflow at higher static pressure, power consumption does not increase linearly with the head pressure in most blowers. Therefore, specific airflow rate per footprint can be raised with long fibers without increasing the energy consumption. Overall, higher energy efficiency can be obtained with long fibers submerged in a deep membrane tank.
- Fourth, module costs can be reduced because more membrane area is available per permeate header. In general, potting hollow fibers in permeate headers is a major source of cost in module production rather than the hollow fiber spinning because the fiber potting is not fully automated. Therefore, module costs per membrane area tend to decrease as fibers grow longer.

3.4.3.3 A New Direction to Improve Hollow Fiber Membrane Performance

There are a number of factors affecting membrane performance in submerged MBR, but permeability (or porosity), surface properties, internal pressure loss, mobility of membrane fiber, etc., can be some of the factors related to membrane properties. Of those, membrane permeability (or porosity) is rarely a major factor determining membrane performance in porous membrane processes, as discussed in Section 1.2.8. Instead, cake layer resistance takes most of the total filtration resistance in a vast majority of conditions and controlling cake layer formation is the key to obtaining good membrane performance. Internal pressure loss is another major factor affecting membrane performance in submerged hollow fiber membrane processes as discussed in Section 3.4.3. Meanwhile, there is no strong evidence that membrane surface properties are a critical factor when it comes to the membrane materials popularly used for MBR, as discussed in Section 1.4. As a result, instead of focusing on improving membrane permeability or surface properties, more efforts have been made to reduce internal pressure loss and to make membranes more flexible. For a given membrane material, tradeoff exists because thicker fibers suffer less from internal pressure loss, but the antifouling mechanism by random fiber displacement weakens. On the other hand, thinner fibers are advantageous in terms of random fiber displacement but they suffer from greater internal pressure loss and are more prone to being cut.

A new attempt was made using more advanced membrane materials and fabrication techniques to reduce the internal pressure drop without losing fiber flexibility and strength (Suk et al. 2012). In this study, the ID increased from 0.8 to 1.0 mm while keeping the OD the same at 2.1 mm compared with popular commercial submerged membranes for MBR. As a result, the wall thickness of the membrane, including the braid and membrane layer, decreased from 0.65 to 0.55 mm. This thinner membrane wall allowed greater flexibility and larger fiber amplitude at the same airflow rate, yet was strong enough to be fabricated as modules with 2-m-long fibers without hampering mechanical strength. In a pilot test, the new hollow fiber membranes with larger inner fiber diameters showed more stable TMP profiles than the benchmarked commercial hollow fiber membranes at an identical flux.

3.4.4 HOLLOW FIBER DIMENSION OPTIMIZATION

A number of factors are involved in hollow fiber module optimization. The goals of hollow fiber optimization include the following:

- Maximization of random fiber movement in two-phase turbulent flow
- Minimization of internal pressure drop
- Fabrication of mechanically stable module in two-phase turbulent flow

The challenge is that those three goals are conflicting each other in many ways. For example, random fiber movement can be maximized by reducing fiber diameter and increasing fiber length, but these directly cause higher internal pressure drops as well as fragile fibers in the turbulence. By increasing fiber diameter, lower internal pressure loss can be achieved whereas membrane modules become mechanically more durable. However, fiber flexibility would be compromised and hence the fiber amplitude decreases. The low fiber amplitude leads to quicker membrane fouling. Therefore, it is inevitable to compromise the three goals to some extent to come up with the optimum module design.

Depending on the materials, structure, and the spinning process condition, etc., the optimum combination of ID, OD, and length can vary. If the membrane material has a high tensile strength, hollow fibers can be made with a thin wall by reducing the OD without compromising the ID. These thin hollow fibers benefit from the random fiber movement in the air–liquid flow. The thin fibers also allow high packing density that in turn allows savings of scouring air, although fiber clogging is another factor to be considered. Alternatively, material with a high tensile strength can be used to make hollow fibers longer while maintaining the same diameter. Although longer fibers suffer more from the internal pressure loss, they benefit from the enhanced random fiber movement that can compensate for the drawback at least partially. In contrast, if the membrane's tensile strength is low, hollow fibers with larger diameters are the only plausible way. Although the thick fibers do not move as much as its smaller counterparts, reduced internal pressure drop can help reduce the cake layer formation near the permeate exit.

The tensile strength of the membrane can be greatly enhanced by laminating the membrane layer on woven or nonwoven hollow braids. The membrane layer can be either chemically integrated or physically laminated on the braid. Alternatively, the membrane layer can be firmly fixed on the braid by letting the membrane polymers smear into the spaces among microfibers in woven braids forming anchors. Depending on the laminating condition, the maximum allowable backwash pressure varies. The enhanced tensile strength allows larger and longer fibers, which in turn allows low internal pressure loss. Although the intrinsic fiber flexibility is inevitably compromised to some degree due to the dual layer structure, the drawback can be at least partially compensated by the improved fiber flexibility of long fibers. The important properties of hollow fiber membrane and their effect on membrane performance are summarized in Table 3.7.

Ideal membrane materials should be highly crystalline having a high tensile strength. It should also have a low glass transition temperature (T_g) to be highly flexible at the typical MBR operating condition (e.g., 5°C–40°C). High tensile strength allows thin walls and long fibers simultaneously. Thin walls also allow larger ID without increasing OD, which mitigates internal pressure loss and enhances random fiber displacement. Long fibers allow lower module costs and lower aeration costs. Membranes made of polymers with low T_g tend to be flexible at room temperature and maximize the antifouling actions in membrane tank.

There are two commercial hollow fiber membranes that rely on two opposite design concepts. As summarized in Table 3.8, GE's ZW500d® membranes have a larger ID and a shorter effective length than the Asahi Microza® module, for example, 0.9 versus 0.7 mm for ID and 1000 versus 2000 mm. Due to the smaller ID and the longer fibers, internal pressure drop is much larger for Microza® than that for ZW500d. Both membranes are known to be made of polyvinylidene difluoride (PVDF), but Microza® membranes are made using a thermally induced phase separation (TIPS) method being considered as a method to produce membranes with high tensile strength. ZW500d membranes are known to be made using a nonsolvent induced phase separation (NIPS) method and the membrane layer is laminated on woven hollow braid supports to enhance mechanical stability. If only the internal pressure drop is considered, sustainable flux should be much lower for Microza® than for ZW500d because the severe TMP gradient in the fiber accelerates membrane fouling. However, the average design and operating fluxes of the two membranes have been reported to be comparable to each other for typical municipal MBR.

TABLE 3.7
Important Properties of Hollow Fiber Membrane and Their Effect on Membrane Performance

Property	Benefit	Drawback
Large ID	• Reduced internal pressure drop allows long fibers	• Compromised packing density
Thin membrane wall	• Increased fiber flexibility • Increased random movement • Increased packing density	• Reduced fiber strength
Extended fiber length	• Reduced aeration energy • Reduced footprint • Increased random fiber movement • Increased scouring effect • Reduced specific module costs	• Increased fiber breakage • Increased internal pressure drop that induces membrane fouling • Reduced handling property
High flexibility	• Increased random fiber movement • Reduced membrane fouling	• Fiber abrasions from random collision among fibers
Dual layer structure	• High tensile strength allows long fibers • Reduced fiber breakage	• Compromised random fiber movement • Compromised lumen diameter

TABLE 3.8

Comparison of Two Commercial Hollow Fiber Membranes

	GE	Asahi	Remark
Model	ZW500d®	Microza® MUNC-620A	
Membrane structure	Double layer	Single layer	
Spinning method	NIPS	TIPS	
Material	PVDF/Polyester	PVDF	
ID/OD (mm)	0.9/1.9	0.7/1.2	
Pore size (um)	0.04	0.1	
Height (mm)	2000	2000	Membranes only
Effective suction length (mm)	1000	2000	Two-sided suction for ZW500®
Module dimension (mm)	2198 mm (H) × 855 mm (L) × 49 mm (D)	150 mm diameter × 2000 mm length	
Area per module (m²)	31.6 or 34.4	25	

This unexpected effect of thin hollow fibers can be attributed to the enhanced random fiber movement and the larger fiber amplitudes. The fiber ID/OD reported in the literature for the two membranes are 0.9/1.9 mm and 0.7/1.2 mm, respectively. Therefore, the cross-sectional area of the Microza membrane based on OD is less than half of the ZW500d membrane. In addition, the membrane wall thickness is only half (0.025 mm versus 0.05 mm). Therefore, Microza® membranes are subject to more vigorous random fiber movement. It is considered that the enhanced fiber amplitude/movement by the small fibers and the higher packing density compensate the disadvantages from the internal pressure loss.

3.4.5 PACKING DENSITY

3.4.5.1 Effect of Packing Density on Productivity

Packing density is one of the determining factors of membrane productivity. Various definitions of packing density exist depending on module configuration and the purpose. Packing density affects the energy efficiency, space or footprint required, costs for cleaning chemicals, transportation and installation costs, etc., of the membrane system.

For submerged hollow fiber and flat sheet membrane modules, the packing density can be defined based on either the footprint or the three-dimensional space required to install the modules. The specific total membrane surface area in a membrane tank is more relevant packing density in comparing the compactness of various membrane configurations than individual modules' packing density. For the hollow fiber membranes held in a pressure vessel, specific membrane area per housing volume is relevant to compare the approximate cleaning solution volume required, but the membrane area per cross-sectional area of housing or the membrane area per footprint of the membrane rack is more relevant to represent the relative size of the system.

It has been known that excess packing density causes high solids flux toward the membrane especially when the flux is high and hence the solid flux toward the membrane bundle is excessive. This solids flux issue is much more significant in hollow fiber modules than in flat sheet membranes due to the high packing density and the hindrance of the fiber bundles. At optimal conditions, particle concentration in the fiber bundle is balanced at a slightly higher level than in the bulk as a result of the enhanced transversal (or lateral) mass transfer by the two-phase turbulence. However, the transversal flow velocity decreases as packing density increases, which can cause an excessive accumulation of solids in the fiber bundle. If the MLSS in the fiber bundle reaches a critical level, the slow transversal flow and the high MLSS can lead to fiber clogging. Although solids accumulation can be controlled by increasing the scouring airflow to some extent, the increased energy costs and operational risks may not be able to justify the productivity gains from the compact module. These interconnectivities suggest that the optimum membrane packing density is tied up not only to the fiber characteristics but also to the operating parameters that affect the overall economics of the module, for example, MLSS of the feed water, flux, aeration rate, and footprint. Due to the complexities of the dynamics around membrane modules, the optimum fiber packing density can be decided only empirically under the condition the module is targeted for with respect to MLSS, airflow rate, temperature, F/M ratio, etc.

3.4.5.2 Packing Density of Commercial Modules

The specific surface area can be defined either by the filtration area per membrane tank volume (m^2/m^3) or by the filtration area per membrane tank footprint (m^2/m^2). Because the footprint is more of a concern in typical situations, footprint-based specific surface area is commonly used to compare the packing densities of commercial membranes.

Large specific surface not only allows small tank size/footprint, but also allows savings of aeration power costs and cleaning chemicals, etc. (Chang 2011). Hollow fiber membranes are more compact than flat sheet membranes in general. The average specific surface areas based on footprint are reported at 292 m^2/m^2 and 118 m^2/m^2 for commercial hollow fiber and flat sheet membranes, respectively (Santos et al. 2011). These high packing densities allow hydraulic retention times to be somewhat less than 1 h for hollow fiber membranes and approximately 2.0 h for flat sheet membranes in municipal plants. In one publicly available proposal in 2012, hollow fiber technology offered 263 m^2/m^2, where up to 480 ZW500d modules with 34.4 m^2 are packed in 10 cassettes placed in a tank with a dimension of 22.9 $m^L \times 2.7$ $m^W \times 4$ m^H. Flat sheet membranes offer around 100 m^2/m^2, where 4800 Kubota flat sheet membranes with 1.25 m^2 are packed in 12 cassettes (aka submerged membrane unit) placed in a tank that is 14 m long and 4.5 m wide (ACWA Ltd. 2007).

Many flat sheet submerged membrane modules are stacked upward to improve the specific surface area per footprint. As a result, many flat sheet cassettes are taller than comparable hollow fiber cassettes. However, the maximum stacking height is limited by the inefficient membrane scouring caused by bubble coalescence, airflow channeling, as well as the challenges in cassette handling, installation, and maintenance. Historically, for hollow fiber membranes, developing compact modules was considered one of the focus areas to appeal to the market in the 1990s, but it has evolved to less compact modules in general optimizing the aeration costs, fouling risks, and the module costs. Despite the less compact modules, the packing density per footprint in the membrane tank has been maintained at a similar or somewhat greater level by increasing the fiber length and optimizing the spaces among modules, cassettes, and walls.

Figure 3.26 shows a commonly available hollow fiber membrane cassette in MBR. Hollow fibers are potted in relatively narrow bottom header, and the spaces between two modules are reserved equal or larger than the bundle depth. As a result, the thin curtain fiber bundles are subject to the enhanced mass transfer in and out of them. Due to the loosely held fibers with a slack (or extra length), membrane fibers sway laterally and are more equally spaced in the middle of the module height. Because of the large spaces in between modules, the total cross-sectional area fibers take in the footprint of a cassette is only a small portion of the footprint. On the other hand, high-fiber packing densities are applied for the hollow fiber modules used for low TSS water. In these modules, the amount of suspended solids in the feed is low, and hence large spaces are not required among fibers. The stored solids among fibers can be removed by performing periodic backwashing with an optional air scouring.

3.4.6 Effect of Fiber Location

In a hollow fiber module, direct air scouring effect, upflow velocity, particle concentrations, etc. vary depending on the location in the module. Therefore, shear stress on the membrane surface is not uniform, but varies depending on the location in a fiber and the location of the fiber in a module. For instance, even in a very short hollow fiber module with a length of 120 mm, some bubbles

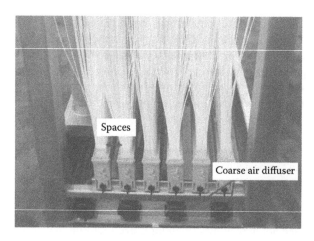

FIGURE 3.26   Hollow fiber modules installed in a cassette.

escape from the fiber bundle although an air nozzle was installed in the center of the module. This means that only a small number of bubbles come into direct contact with the top portion of the fibers and this tendency becomes more prevalent as fiber length increases in full-scale commercial modules (Chan et al. 2007). Liquid upflow velocities are also quite different inside and outside the fiber bundle due to the resistances the bubbles experience when rising. In addition, the particle concentration inside the fiber bundle is higher than in the bulk because mixed liquor is concentrated as a result of permeate drawing through the membrane. Overall, the fibers located near the center of the membrane bundle are exposed to higher membrane fouling potential than other fibers.

To investigate the location effect on membrane fouling in a hollow fiber module, nine hollow fiber membranes were tightly mounted through a perforated plate as shown in Figure 3.27 (Yeo and Fane 2005; Yeo et al. 2006). The hollow fibers were connected to a vacuum chamber with transparent wall using separate tubings. The flux of each individual fiber was measured by counting the permeate drops after video taping the vacuum chamber. As a feed, 0.5 to 2 g/L of bentonite was used and the average particle size was 5 µm. The highest flux was found from the fibers potted in the four corners (#1) whereas the lowest flux was from the fiber in the center (#3). It is apparent that the bubbles and liquid flow passing near the fiber in the corner experience the least resistance because these fibers are neighboring with only two other fibers and one additional fiber in the diagonal direction. Therefore, the membranes in the four corners are exposed to the fastest flow movement that causes the least membrane fouling. On the contrary, the upflow velocity and the turbulence around the fiber in the center (#3 in the figure) is the lowest due to the hydrodynamic resistances imposed by the four neighboring fibers in the immediate location (#2) and the four additional fibers in the diagonal direction (#1). In addition, due to the concentration effect in the center of the fiber bundle, the bentonite concentration at the center fiber is higher than in the bulk although the concentration

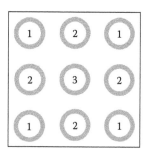

FIGURE 3.27   Position of nine fibers potted through a perforated plate.

effect is not supposed to be high for this nine-fiber system. In a full-scale module with a large membrane potting area, membrane performance can be much more severely affected by the hydraulic hindrance effect and the concentration effect imposed by the large fiber bundles.

3.4.7 VERTICAL MOUNT VERSUS HORIZONTAL MOUNT

Membrane performance can be affected by the orientation of the hollow fiber membrane. Many different orientations such as vertical, horizontal, slanted, radially slanted around a center like an umbrella, U-shaped or reverse U-shaped, etc. have been tested. However, only vertical and horizontal configurations are considered practical because they are easy to build while providing high packing densities. Slanted modules have been studied, but no particular advantages were found in terms of membrane performance and packing density (Sridang et al. 2005).

In a laboratory study, to simulate the vertically and horizontally mounted fibers used in submerged membrane system, hollow fibers were mounted either parallel or transversal to the flow direction in a channel of plate and frame module. When yeast suspension (5 g/L) was filtered at a constant TMP mode, the fibers mounted parallel to the flow showed higher steady-state flux (Chang and Fane 2000). Similar observations were made in other experiments (Fane 2002; Chang et al. 2002a,b). It was postulated that the rising bubbles near the membrane fiber generated strong falling film and wake effects that scoured the membrane surface efficiently. On the other hand, the top half of the horizontally mounted hollow fiber is hidden from the upflow and is supposedly vulnerable to the fouling under the experimental condition. Perhaps the membrane scouring effect by eddies in the hidden half of the hollow fiber is not as strong as the falling film and the wake effect occurring in vertically mounted hollow fibers. However, the horizontally mounted hollow fibers can be advantageous when the packing density is high. Under this condition, the physical contacts among fibers can generate a strong membrane scouring effect exceeding the falling film and the wake effect. Because the hollow fiber modules with a high packing density is only suitable for water with a low amount of suspended solids, horizontally mounted hollow fiber modules are primarily used for tertiary filtration or raw water filtration.

The membrane fouling is a consequence of very complex interaction among many factors. It is nearly not possible to predict the performance of larger scale modules with a sufficient accuracy based on the observations from small scale experiments. As a result, despite the theories and the experiences from small scale tests, the actual field tests often ends up with surprises. For instance, small bubbles are optimum in small scale experiment, where bubbles tend to rise along the membrane fibers, but large bubbles are more optimum in large-scale modules, where bubbles tend to rise outside the fiber bundle without having too many of direct contacts with the membrane. Because the antifouling mechanisms in a laboratory-scale module and a full scale module can be substantially different, the most part of module optimization process requires full scale modules to be tested in realistic conditions.

One potential disadvantage with horizontally mounted membranes is lumen blockage by bubbles, which can interrupt permeate transportation partially or fully. In fact, the excess dissolved gas in permeate can escape and form bubbles in the fiber lumen due to the lower gas solubility at negative pressure. Some of those bubbles can stick on the hydrophobic domains on the lumen surface in the hollow fiber (Figure 3.28). The presence of an air bubble induced by "dry point" or "unwetted point" within the hollow fiber can cause a hydraulic resistance that interferes the permeate flow through the lumen (Chang et al. 2008). As more fibers are exposed to the lumen blocking by air bubbles, other intact fibers without lumen blocking must produce more permeate to compensate for the permeate loss. Although the attachment of bubbles in the membrane lumen can occur in any membrane configuration, it can be more significant in horizontally mounted membranes because buoyancy does not help discharge the bubble from the lumen.

Another disadvantage of horizontally mounted hollow fiber membrane is the vulnerability to fiber breakages due to the stress given on both ends of the fibers, if aeration is performed during the filtration cycle. Although the rising two-phase flow pushes the horizontally mounted fibers upward,

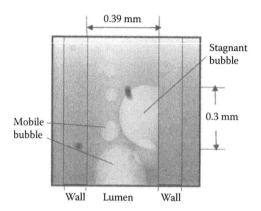

FIGURE 3.28 Images of stagnant and mobile bubbles detected in hollow fiber lumen by x-ray micro-imaging. (From Chang, S. et al., *J Membr Sci.* 308, 107–114, 2008.)

the eddies formed behind the fiber also pull the fibers to the same direction. As all the drag forces transfer to both ends of the fiber, the repeated stress increases the material fatigue, which can eventually lead to fiber breakage. To minimize the chances of fiber being cut by fatigue, the maximum allowable fiber length is limited to varying degrees depending on the tensile strength of the fiber. As a consequence, the length of horizontally mounted hollow fibers is shorter than that of vertically mounted hollow fibers in general. The relatively short membrane length in the horizontally mounted hollow fibers acts as a pressure for the specific module costs by increasing the number of headers per membrane area.

Horizontally mounted submerged hollow fiber modules were first commercialized in the 1990s as SteraporeSUR™ by Mitsubishi Rayon Co. of Japan targeting MBR market. The polyethylene membrane fibers were known to be surface-treated to hydrophilic, and ID and OD were 0.27 and 0.41 μm, respectively. Although two-sided suction was performed, the small fiber diameter and relatively long fiber length (~900 mm) in the original version only allowed a low sustainable flux (e.g., <12 LMH), despite the high scouring airflow rate at approximately 100 m^3/m^2/h based on the footprint. The membrane length was later reduced to 400 to 600 mm in the newer version to reduce the internal pressure loss while improving mechanical stability. A similar horizontal membrane concept was adopted by GE (formerly Zenon Environmental Inc.) in its ZW1000® module in 2003 targeting water with a low amount of suspended solids, for example, surface water, secondary wastewater from clarifier, and RO pretreatment. The packing densities of these modules are much higher than those of MBR modules due to the lower TSS in the operating condition. Air scouring is performed only during the backwash cycles, which allows energy cost savings while reducing the chances of fiber damage.

3.4.8 HEADER DESIGN

3.4.8.1 Bottom Header

In submerged hollow fiber modules, the rising flow generates a swirling zone immediately above the bottom header, as shown in Figure 3.29a. The swirling liquid develops local downflow that in turn develops dead zones. This is more apparent when scouring air is supplied from the external space of the bottom header. The very limited fiber amplitudes in the bottom header fail to disrupt the dead zone that not only promotes membrane fouling but also makes the area vulnerable to ragging or clogging by unfiltered fibrous materials in the pretreatment process. A similar event could also occur in the top header where the upflow meets with the dead end.

To mitigate the dead zone issue, the depth of the fiber bundle has been decreased gradually over the last decade. In some modules, air nozzles are embedded in the middle of the bottom header,

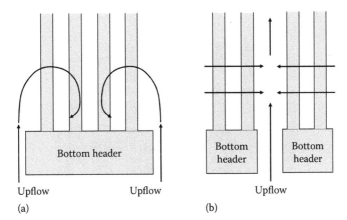

FIGURE 3.29 (a) External air nozzle and (b) internal air nozzle. Effect of bottom header structure on flow pattern.

which holds a curtain fiber bundle, but it increases the complexity of the module configuration by a great deal. In well-established commercial systems, curtain fiber modules are stacked horizontally with spaces among them. Air–liquid mixtures rise through the spaces between the modules, as can be seen in Figure 3.27. If prescreening of raw wastewater functions properly, the accumulation of solids in the bottom header is not detrimental not only because it occurs in a small portion of the large module but also because it can hardly propagate upward due to the vigorous lateral fiber movement.

Unlike in curtain fiber bundle modules, module depth (or diameter) cannot be reduced too much in round-shaped modules due to economical constraints. If the module diameter is too small, too many modules with a small surface area are required. The low membrane surface area per module increases the specific cost of membrane surface area. Therefore, the depth of the membrane module tends to be greater in round modules, thus the dead zone problem tends to be more significant in round-shaped modules. To avoid this drawback, scouring air can be introduced through the nozzle in the center of the fiber bundles (Figure 3.29b). The bubbles rising through the center of the module generate lateral in-flows from outside to inside the bundle mitigating the dead zone issue. One potential drawback of this method is that, if large debris exists in the medium as a consequence of prescreening failure, the lateral flow coming from the bulk into the center of the module can exacerbate fiber clogging. The packing density of the cylindrical modules per footprint tends to be slightly lower than that of curtain fiber bundle modules due to the unused spaces around the module (Santos et al. 2011).

3.4.8.2 Open Top Header

It has been known that the top headers of vertically mounted hollow fiber modules are vulnerable to sludging especially when prescreening fails. Figure 3.30 shows a hollow fiber module compromised by sludging when prescreening failed in a municipal MBR plant. This is largely a preventable problem by maintaining prescreens properly, but once it occurs, debris needs to be picked out manually.

The top header can be removed to eliminate the sludging issue in the place. In this open top header design, the ends of the fiber are sealed individually and the permeate is obtained from the bottom header (Johnson et al. 1998). This module design was originally devised for gas transfer membranes used for oxygen transfer in mixed liquor, but the concept was adopted for vertically mounted hollow fiber modules by Puron, a division of Koch Membrane Systems. The top ends of the hollow fibers are sealed individually with caps and the fibers are stood inside a rack moving freely in the boundary set by the rack. PES-based cast solutions are applied on braided supports and are allowed to penetrate into the spaces among the strings that consist of braids. The firmly attached membrane layer on the

FIGURE 3.30   Sludging in the top header when prescreening fails in municipal MBR.

braid allows membrane backwashing at a high pressure. Although the sludging potential in the bottom header remains the same as with other hollow fiber modules, this module is inherently free of the sludging in the top header. Despite the significant benefits, open top header designs have a few potential drawbacks compared with the curtain fiber bundle designs with top and bottom headers.

First, because the permeate obtained from the top portion of the membrane must travel to the bottom due to the lack of the top header. If the internal pressure drop is excessive, internal pressure drop can cause accelerated membrane fouling. Second, the open top header modules require larger and more rigid fibers compared with the curtain fiber modules to let the membranes stand without anchoring points in the top. The large fiber diameter is also a requirement to mitigate the large internal pressure loss caused by fibers with one opening in the bottom. Consequently, the fibers used in this module are not as flexible as those in curtain fiber bundle modules. The low fiber flexibility limits the random fiber movement in two-phase flow and can affect the membrane performance negatively as discussed in Section 3.4.2. Third, if gas bubbles are formed in the lumen, it is harder to remove because bubbles must travel against gravity (or buoyancy). In fact, the vacuum pressure in the hollow fiber lumen is responsible for the gas bubble formation as discussed in Section 3.4.7. If bubbles are captured by hydrophobic domains of the lumen surface and obstruct the permeate flow, the flux from the membrane area above the bubble will be compromised. Consequently, the flux from the other membrane fibers should increase at a constant flux mode. The bubbles in the lumen can be removed by flushing the fiber at a higher than regular flux for 0.5 to 1.0 min periodically, but the high flux can trigger accelerated membrane fouling to some extent. The same phenomenon also can occur in modules with two headers, but it is less significant because bubbles move upward at least in the top half of the fibers.

3.4.9   Connections among the Factors Affecting Module Optimization

Due to the complex connections among the factors affecting membrane performance, any finding in a specific condition cannot be deemed applicable to other conditions. For instance, higher steady-state flux might be achievable with the fibers with smaller diameters in one condition, but it can turn out to be the opposite in other conditions depending on the length of the fiber used, airflow rate, MLSS of the mixed liquor, etc. In other instances, large bubbles might be more beneficial in one condition, but it can turn out to be the opposite in another condition. Therefore, all the module design parameters should be considered holistically under the conditions that the module is intended to be used in. This also suggests that there are no membrane modules optimum for all conditions.

The following is a list of factors affecting membrane performance that need to be carefully considered in analyzing empirical data:

- Fiber dimension: ID, OD, and length
- Fiber property: permeability, pore size distribution, and flexibility
- Module structure: depth and width of fiber bundle, air nozzle position, etc.
- Aeration method: bubble size, airflow rate, bubbling frequency, etc.
- Mixed liquor property: MLSS range, SRT, viscosity, rheological property, etc.

Some complex interactions among the parameters listed above are found in the literature. For instance, fiber amplitudes are a crucial factor affecting the antifouling action of hollow fibers, but they significantly vary depending on the medium's viscosity. Again, the medium viscosity varies depending on both biological condition and MLSS. It was observed that the effect of fiber looseness on fiber amplitudes turned out to be dependent on medium viscosity, as can be seen in Figure 3.31. In a low-viscosity medium (water, 1 cP), fiber amplitudes continue to increase when fiber looseness increased from 0% to 5%. However, in a high-viscosity medium (water–glycerol mixture, 3.2 cP), fiber amplitude reached a plateau at a much lower amplitude and at a looseness of 2% (Wicaksana et al. 2006).

Contradictory observations are often made from seemingly identical experiments perhaps due to the subtle differences in experimental condition overlooked or not revealed. As discussed in Section 3.3.3.2, the experimental results shown in Figures 3.14 and 3.16 were all obtained with one 0.1 m^2 flat sheet membrane manufactured by the same company, Yuasa Co. (Japan), although the module used for Figure 3.14 was fabricated by Kubota and the module used in Figure 3.16 was by Yuasa Co. Despite the very similar experimental conditions, the former study found that small bubbles were superior to large bubbles, whereas the latter study found the opposite. The causes are not clear, but they can be attributed to the differences in specific airflow rates per surface area, mixed liquor properties, diffuser specifications, flow channel dimension, etc. If the specific

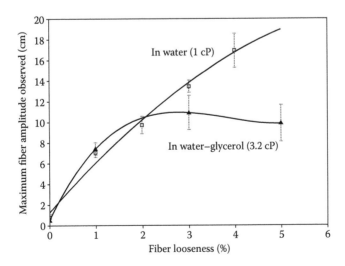

FIGURE 3.31 Effect of fiber looseness and medium viscosity on fiber amplitude. Fiber length = 70 cm; OD = 0.65 mm; ID = 0.39 mm; pore size = 0.2 μm, polypropylene; scouring airflow = 2 L/min; nozzle size = 1 mm. (From Wicaksana, F. et al., *J Membr Sci.* 271, 186–195, 2006.)

airflow rate is high enough, both small and large bubbles can cover the entire membrane surface area. Because large bubbles have stronger falling film and wake effects, stronger scouring effects can be realized. On the contrary, if the specific airflow rate is low, only small bubbles can cover the entire membrane surface at a constant airflow rate, thereby small bubbles can turn out to be superior.

Contradictory observations also exist in hollow fiber membranes. In one study, small bubbles (5 mm) were more effective than large bubbles (20 mm) in laboratory-scale experiments performed with short membranes installed in a confined spaces of a plate and frame module (Fane 2005). However, the larger bubbles generated from the diffusers with 6 and 8 mm pores are more effective than the bubbles generated from the diffusers with less than 3 mm pores in commercial hollow fiber modules, where actual bubble sizes are much larger than the diffuser pore size. The exact causes are not clear, but perhaps the difference came from the fact that more air bubbles come into direct contact with the membrane surface in the small-scale module than in the full-scale module with 2-m-long fibers. In addition to the complexities in the dynamics associated with the mixed liquor, bubble, and membrane fibers, mixed liquor property is an additional factor affecting membrane performance in MBR. In fact, the biology-related factors are notoriously difficult to control because they are affected by uncontrollable parameters such as wastewater flow rate, strength, composition, etc.

3.5 OPTIMIZATION OF SUBMERGED FLAT SHEET MEMBRANE

3.5.1 OVERVIEW

Substantially less literature exists for the optimization of flat sheet membranes than for hollow fiber membranes. The literature dealing with the fundamentals of air scouring are even rarer for flat sheet membranes. The relatively simple hydrodynamics in addition to the smaller market share might fail to draw enough attention from the academe. The data available in the public domain today has largely come from commercial membrane suppliers, as a result, much of the know-how is not published (Ndinisa et al. 2006a,b).

The scarcity of research is also affected by the fact that the membrane panel design parameters are determined mainly by the necessity of making membrane panels sturdy enough to resist lateral vibrations in two-phase turbulence. For instance, most flat sheet submerged membrane panels have 5 to 7 mm depth to secure a sufficient physically sturdiness in two-phase flow with some exceptions. The panel dimension decided to secure the physical strength automatically provides large enough channels to transport permeate without a significant pressure loss in between two membranes. The width and height are again decided to allow sufficient physical sturdiness and to secure easy handling/installation properties. In the meantime, 5 to 7 mm spaces between two membrane panels are required at a minimum primarily for running the system without upflow channel plugging at high MLSS (e.g., 6–20 g/L).

3.5.2 DESIGN PARAMETERS

Panel depth—The depth of the membrane panel is decided mainly to secure enough mechanical strength to resist against bending by the lateral vibrations in the turbulent flow. Most commercial products adopt a 5 to 7 mm depth, which is also sufficient to secure the permeate channel without causing internal pressure loss. Typically, corrugated plates with internal permeate channels are used as a supporting structure. The membrane fouling induced by the internal pressure loss is not a significant issue in most submerged flat sheet membranes due to the much larger permeate channels. If the panel thickness can be reduced substantially by using new frame materials compatible with the turbulence, membrane packing density can increase.

Panel width and height—Wider and taller panels are advantageous in securing larger surface areas in a cassette and in minimizing specific manufacturing costs. However, as panels become wider, it is exposed to stronger lateral vibrations and become prone to bending in the turbulence. In general, if panels become wider, the thickness of the panel must increase to prevent panels from bending in the turbulence for a given framing material and structure, which may in turn compromise the packing density. As a result, there is an optimum panel width that maximizes the packing density without compromising mechanical strength depending on the frame material and panel structure. Meanwhile, the height of the membrane panel is decided to secure handling and installation properties. In addition, each panels' height must be coordinated within the maximum height of the multistack cassette.

Spaces between two panels—Securing sufficient spaces in between two panels is important to prevent upflow channels from being plugged by large debris not removed in pretreatment or by thickened sludge due to the lack of sufficient flow movement. However, the upflow channel size is somewhat compromised for the sake of obtaining a reasonable packing density. Although small channels allow a high packing density and a high energy efficiency, 5 to 7 mm spaces are reserved between two panels in most commercial systems.

Space for flow development—In a membrane cassette equipped with multiple flat sheet membranes, air bubbles and upflow must be distributed evenly among the channels. It has been known that a certain space is required in between diffusers and the lower end of flat sheet membrane modules to allow two-phase flow to be fully developed and distributed among the channels. The minimum length required is not only affected by the length and the width of the cassette cross-section but is also affected by the existence of walls surrounding the flow development region of the cassette (see Figure 3.34). Bubble sizes are also a factor determining the minimum length required for flow development. The commercial flat sheet membrane modules typically use 30 to 60 cm of spaces underneath the membrane module.

Bubble size and airflow rate—As discussed in Section 3.3.3.2, bubble size is an important factor affecting the operating costs as well as the membrane scouring efficiencies. The optimum bubble size that allows high-energy efficiency tends to be smaller for flat sheet membranes than for hollow fiber membranes. This is because flat sheet membranes are scoured directly by rising bubbles through the narrow channels among membrane panels. At a given airflow rate, a large number of small bubbles can cover a larger surface area than a small number of large bubbles at the same airflow rate.

Multideck cassette height—The height of the membrane panels directly affect the power consumption for membrane scouring. In general, energy efficiency increases as panel height increases because the injected air bubbles scour larger membrane surface areas before escaping from the mixed liquor. Moreover, energy consumption for air compression increases less than proportionally to the head pressure. Thus, panels can be stacked to double or triple decks in a cassette to improve the overall energy efficiency. Typically, double-deck cassettes are used for midscale to large-scale projects. The total frame heights are between 3.50 m (EK model with Type 510 panel) and 4.29 m (RW model, double-deck with Type 515 panel) for Kubota's systems.

3.5.3 Upflow Pattern and Its Effect on Fouling

The ideal flow pattern in a submerged flat sheet membrane is a uniform upflow induced by evenly distributed air bubbles covering the entire membrane surface. However, it has been observed that the actual upflow is far from uniform. In fact, air bubbles are concentrated on both sides of the rectangular channel (or column). Swirling and meandering flows can cause even local downflows on the membrane surface. Nonetheless, bubbles cannot escape from the upflow channel and the upflow pattern in the flat sheet membrane is still deemed more uniform than in hollow fiber membranes, where bubbles escape from the fiber bundle while rising. The existence of defined upflow channels gives advantages to the flat sheet membranes when they handle high levels of suspended solids (Lebegue et al. 2008).

The two-phase upflow pattern was simulated using CFD for the condition, in which a flat sheet submerged membrane panel with 0.1 m² is installed in a rectangular channel with 7 mm spaces from the wall on both sides of the panel (Ndinisa 2006). The results showed that the rising bubbles move toward the edges of the channel as soon as they enter into the channel at a flow rate of 2 to 8 L/min. The results from the CFD also confirmed experimentally that the bubbles divide into the two main air plumes formed near the two edges of the column. The degree of flow meandering was small at a low airflow rate of 2 L/min and was much larger at a high airflow of 8 L/min. The flow meandering became more significant as bubble size grew.

Contrary to the fast-moving two-phase flow near the edges of the channel, the slowly moving upflow in the center can reverse its direction and become a downflow depending on the airflow rate. In addition, the narrow flow zones right on the edges can also move down to compensate for the fast upflow through the two air plumes. Figure 3.32 shows the distribution of liquid only velocity across the horizontal line in the membrane panel (Ndinisa 2006). When airflow is low (e.g., 2 L/min), the maximum downflow velocity can be as much as the upflow velocity, that is, −15 cm/s versus 20 cm/s. However, the liquid flowing downward contains not many bubbles that hinder the liquid flow and is not as effective as the two-phase upflow in mitigating membrane fouling. Overall, the shear stress acting on the membrane surface is much larger in the two air plumes near both edges of the channel than in the center or right on the edges.

Similar observations were made empirically in another independent study (Zhang et al. 2009). When shear stress was measured directly with electrochemical sensors, two to four times stronger shear stresses were measured near the edges of the membrane panel than in the center. Figure 3.33a illustrates the shear stress distribution on a flat sheet submerged membrane with aeration. It has also been observed that the least amount of cake layer was formed near both edges of a membrane panel as shown in Figure 3.33b. It is also noticeable in the CFD study that the average shear stress on the membrane surface increased only by approximately 50% from 0.07 to 0.10 Pa when airflow rate increased from 2 to 8 L/min (Ndinisa 2006). This is in line with the numerous prior observations made in other studies discussed in Section 3.3.2, in which the incremental gains in the scouring effect become marginal as airflow rate increases.

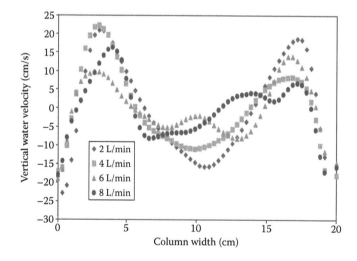

FIGURE 3.32 Vertical water velocity profile distribution at 30 cm from the bottom. (From Ndinisa, N.V., Experimental and CFD Simulation Investigations Into Fouling Reduction by Gas–Liquid Two-Phase Flow for Submerged Flat Sheet Membranes. PhD dissertation, The University of New South Wales, Sydney, Australia, 2006.)

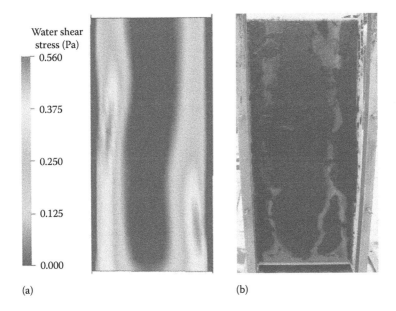

Water shear
stress (Pa)
- 0.560
- 0.375
- 0.250
- 0.125
- 0.000

(a)   (b)

FIGURE 3.33 Comparison of CFD flow pattern and actual cake layer formation: (a) flow pattern by CFD and (b) cake layer formed on used membrane with a modified aspect ratio. (a: From Ndinisa, N.V., Experimental and CFD Simulation Investigations Into Fouling Reduction by Gas–Liquid Two-Phase Flow for Submerged Flat Sheet Membranes. PhD dissertation, The University of New South Wales, Sydney, Australia, 2006; b: From Geilvoet, S.P. The Delft Filtration Characterisation Method Assessement. PhD dissertation, Delft University of Technol, The Netherlands, 2010.)

The poor bubble distribution can be prevented by inserting vertical baffles in the upflow channel. In a laboratory-scale study, multiple thin baffles were vertically placed with 1 cm spaces in between the membrane and the wall (Ndinisa 2006). The baffles divided the large channel into multiple subchannels, preventing bubbles from crossing the vertical line and forming large air plumes. As a result, bubbles could not gather to form large plumes and the shear stress could be evenly distributed. It was found that the same sustainable flux was obtained with the baffles at half of the airflow rate. However, such baffles could exacerbate the clogging issue under field conditions, in which unfiltered fibrous debris or microbial flocs could plug the entrance of the narrowed upflow subchannels formed by the baffles.

3.5.4 MODIFIED MODULE DESIGN

In submerged hollow fiber membrane processes, specific air demand per permeate volume (SAD_p) has been substantially reduced mainly by implementing intermittent aeration since the early 2000s. In the meantime, SAD_p for submerged flat sheet membranes has been reduced mainly by stacking up membrane panels and improving scouring efficiency by reducing bubble sizes (see Section 3.3.3.2); however, SAD_p can be further reduced by improving the flow dynamics around the membrane cassette.

It has been pointed out that the abrupt flow direction change from the down comer to the riser causes a significant frictional loss and slows down the upflow, as shown in Figure 3.34a (Prieske et al. 2010). A smoother draft tube edge was introduced to achieve lower bend loss and thus higher circulation velocities (Figure 3.34b). Because the aerator located at the entrance of the draft tube can interrupt the flow, it can be moved to the bottom of the membrane tank. Consequently, a much more homogenous bubble distribution across the whole cassette can be achieved with smaller chances of channel clogging. With this configuration, either higher shear forces can be achieved at the same aeration rate or scouring air flow can be reduced while achieving a same shear stress. As shown in Figure 3.35, 30% to 50% faster liquid circulation was observed with the modified cassette design.

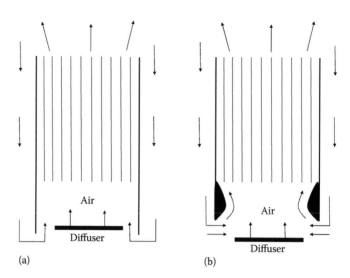

FIGURE 3.34 (a) Conventional design and (b) improved design. Enhanced upflow velocity by positioning diffusers outside the frame and adding baffles. (Reproduced from Prieske, H. et al., *Desalin Water Treat.* 18:270–276, 2010.)

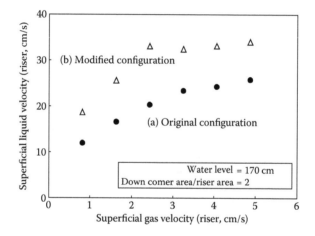

FIGURE 3.35 Effect of enhanced upflow velocity by positioning diffusers outside the frame and adding baffles—comparison of superficial liquid velocities of conventional and modified designs. (From Prieske, H. et al., *Desalin Water Treat.* 18:270–276, 2010.)

Spacers that promote turbulence can be inserted in between the membrane channels. In one study, spacers with helical shape were inserted vertically in the upflow channel (Melin et al. 2011). While air and water flowed through a helical water path, lateral mixing was promoted in the channel and the concentration polarization weakened. It was observed that the helical spacer reduced TMP increasing rates greatly at 35 LMH by creating additional turbulences in the channel without increasing aeration rate. However, the helical spacers can make the system vulnerable for plugging, especially when large debris and fibrous materials exist in mixed liquor in MBR.

4 Activated Sludge Process

4.1 OVERVIEW

In the activated sludge process, organic contaminants are oxidized by microorganisms under aerobic condition. It is called the *activated sludge process* because the sludge produced is an activated biomass. Since its invention in the early 20th century, many variations have been introduced to enhance its treatment efficiency while reducing capital and operating costs or to accommodate unconventional needs. The original activated sludge process based on the suspended culture with air bubbling is also called the *conventional activated sludge* (CAS) process to distinguish it from the variations such as those based on fixed or floating biofilm carriers, membranes, and unconventional biological reactor configurations. Membrane bioreactor (MBR) has been evolved from CAS by replacing the gravity-based clarifier (or the settling tank) with the membrane separation system.

Figure 4.1 shows a typical CAS process. Raw wastewater filtered by a coarse screen is stored in the equalization tank before it is fed to the primary clarifier. Flow equalization is a crucial step to secure an efficient use of the entire system without causing hydraulic or organic overloads. Equalization tanks are sized to buffer the peak flow, allowing an aeration tank size reduction. One method used to size the equalization tank is discussed in Section 6.7. Raw wastewater is pumped to the primary clarifier to remove easily removable suspended solids. By removing insoluble biological oxygen demand (BOD) in the form of suspended solids, organic loading to the aeration basin can be substantially reduced. The sequence of equalization tank and primary tank can be switched if the hydraulic fluctuation is insignificant and the hydraulic overloading to the primary clarifier is unlikely. The settled wastewater is fed to the aeration basin, where the mixed culture microorganisms are suspended by air bubbles under aerobic conditions. Various microorganisms cooperate to oxidize biodegradable organics and nitrogen. Depending on the residence time of the microorganisms, approximately 30% to 60% of the carbons in the biodegradable organics are assimilated to live microorganisms while the rest of them are oxidized to CO_2. Organic and inorganic nitrogen are also oxidized to nitrate at various efficiencies by nitrifiers depending on the process condition. A portion of the nitrogen and phosphorus contained in raw wastewater are assimilated to microorganisms and removed from the water with the excess sludge. The removal of nitrogen and phosphorus can be enhanced by adding additional tanks with no aeration and recirculating the mixed liquor through them properly, as will be discussed in Sections 4.4 and 4.5. Then the mixed liquor is sent to the secondary clarifier and settled to separate sludge (or biosolids) from the effluent. The efficiency of clarification is dependent on the extent of floc formation occurring naturally by floc-forming bacteria in the mixed liquor. If the floc formation is not satisfactory, coagulants and/or flocculants may be added before the secondary clarifier. The effluent may be required to be posttreated for disinfection, trace organic removal, color removal, recalcitrant chemical oxygen demand (COD) removal, etc., before being discharged to the nature as discussed in Section 4.7. The settled solids are called *sludge* or *biosolids*, which mainly consist of live microorganisms and their derivatives. Typically, most of the settled sludge is sent back to the aeration basin as a seed sludge to treat fresh wastewater, but a portion of it can be discharged as excess sludge. This sludge is also called *excess biosolids* to distinguish it from the primary sludge. Excess sludge is posttreated to reduce the moisture content before it is disposed as discussed in Section 6.13.3.

There are many variations of CAS. In the sequencing batch reactor (SBR), wastewater filling, aeration, settling, and scanting occur sequentially and repeatedly in one tank by controlling valves and

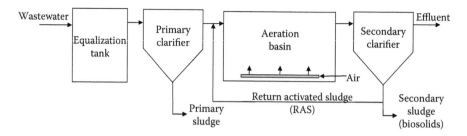

FIGURE 4.1 Typical CAS process.

blowers properly. A portion of the settled sludge is used as seed sludge for the next cycle. Trickling filter uses fixed media contained in a column over which wastewater is dripped slowly. While the wastewater flows down, the biofilm on the fixed media breaks down the organics in it. Oxygen is supplied through the void spaces among the media usually by natural convection, but sometimes air is blown through the column using fans. Rocks/gravel, slag, polyurethane foam, ceramic, or plastic media are used as fixed media. In a process called *integrated fixed-film activated sludge* (IFAS), floating media are added to the aeration tank to enrich the microorganisms, especially when microbial washout is a concern. In this process, floating media are filtered out before the mixed liquor is sent to the secondary clarifier. A portion of the settled biosolids are recycled back to the aeration tank to enrich mixed liquor suspended solids (MLSS) in IFAS; however, if the settled biosolids in the subsequent clarifier are not recycled back to the aeration tank, the process is called *moving bed biofilm reactor* (MBBR). In the contact stabilization process, raw wastewater is contacted using the activated sludge for a short period to allow microorganisms absorb BOD. The BOD-bearing sludge is separated in a clarifier and sent to an aeration tank to oxidize the BOD absorbed on the sludge. After settling the aerated mixed liquor in other clarifier, the settled sludge is returned to the contact tank. This process allows the fast treatment of BOD using compact biological tanks, but the effluent quality is not as good as that of CAS. In the oxidation ditch process, the mixed liquor is circulated around a donut-shaped circular reactor without walls dividing sections by paddles. Depending on the location of the air nozzles and the wastewater injection point, aerobic, anoxic, and anaerobic zones are formed sequentially. The sequential exposure of the mixed liquor to the different zones enhances nitrogen and phosphorus removal. The mixed liquor is settled in a clarifier, and a portion of the biosolids settled in the clarifier may be returned back to the oxidation ditch.

The CAS process is a simple and economical process widely used for organic wastewater treatment, but a few limitations exist as follows:

- The secondary clarifiers can remove biological flocs that settle easily, but it is not effective in removing fine particles such as cell debris and single cells. The final effluent with fine particles has a limited value because it can be repurposed only for insensitive applications. The varying effluent quality depending on biological condition is another limiting factor when the effluent is discharged to the nature or is repurposed. Maintaining constant biological condition is nearly impossible in most field conditions because of the variability of hydraulic and organic loading rates. Sludge bulking and foaming are the common issues affecting effluent quality, as discussed in Section 4.3.2.
- CAS processes are designed and operated to secure good sludge settling properties in the secondary clarifier. In fact, bulky aeration tanks are required to obtain the sludge that settles quickly in the secondary clarifier, although a much smaller aeration tank is sufficient to obtain satisfactory BOD removal. It has been known that a low MLSS, for example, 1 to 4 g/L, at a low specific organic loading rate measured by the food-to-microorganism (F/M) ratio, is essential to obtain the sludge with good settling properties.

- The biological treatment process occurring in the aeration basin is tied up with the clarification performance of the secondary clarifier. Therefore, the aeration tank cannot be designed to obtain the highest efficiency of its own.

The above-mentioned limitations can be solved by replacing the secondary clarifier with a membrane filtration system. By having a physical barrier capable of separating solids from water regardless of the characteristics of solids, the biological process can be optimized to obtain the maximum efficiency without having a concern over the subsequent clarification process. The footprint requirement of the secondary clarifier can be much reduced by using a membrane system. The size of the aeration tank also can be much reduced by raising MLSS. Moreover, the space required for the separation process can be eliminated completely, if submerged membranes are used instead of side stream membrane systems.

Those conceptual benefits of MBR have been found largely true, but it has turned out that there are significant interactions between the biological process and the membrane separation process. In fact, MLSS cannot be raised too much in order not to hamper membrane performance as discussed in Section 5.2.3. The specific organic loading rate (or the F/M ratio) of MBR should be maintained lower than that of CAS to obtain the mixed liquor that can be filtered by the membrane without causing excessive membrane fouling. As a result, despite the three- to fourfold higher MLSS, the total aeration tank volume can be reduced by only 25% to 50% in a typical municipal MBR. Primary clarifiers are often replaced by fine screens with a pore size of 0.5 to 3 mm, and the secondary clarifiers are not required in MBR. Overall, up to 50% to 70% of footprint reduction is readily possible. With respect to excess sludge production, MBR produces less sludge than CAS because of the greater solids retention time (SRT), but the actual sludge production is not noticeably less than that in CAS in many municipal MBR sites because no solids are lost through the effluent.

4.2   WASTEWATER MICROBIOLOGY

4.2.1   Microorganisms

4.2.1.1   Virus

Viruses are the simplest and the smallest life-forms that can self-reproduce. They are basically encapsulated DNA or RNA mostly by proteins alone, but sometimes they are enclosed by additional lipid envelopes that surround the proteins. They are unable to reproduce independently without a host cell because of the lack of complex metabolic systems required for self-reproduction, thereby they are parasitic. Viruses are divided by animal virus, plant virus, and bacteriophages, depending on the host preference. There are 31 families of DNA viruses and 47 families of RNA viruses, depending on the nucleic acid type/strandedness and the presence or absence of envelop. Viruses do not participate in the biological degradation of organic materials, but some pathogenic viruses, for example, influenza viruses, noroviruses, and rotaviruses, draw attention because of the chance to survive through the wastewater treatment processes. In MBR, virus removal efficiency is very high (>99%) despite their small sizes (0.02–0.3 μm) because a vast majority of them reside in much larger bacteria that can be easily removed by the dynamic membrane formed on porous membrane even if the bacteria exist as single particles, as discussed in Section 2.6.4.1.

4.2.1.2   Prokaryote

All prokaryotes are single-cell organisms having cell walls except a genus of bacteria called *mycoplasma*. Cell walls made of peptidoglycan are not only responsible for the shape of the microorganisms but also provide the protection from the osmotic damage caused by the salt concentration difference in between the cytoplasm and the external water. The peptidoglycan layer can be stained by crystal violet stain according to the Gram test. If the thin peptidoglycan layer is sandwiched between inner and outer cell membranes made of lipopolysaccharides (or endotoxins), crystal violet stains are not retained on the cell wall, and the bacteria are called *Gram negative*. Prokaryotes do not have a nucleus membrane, and cell constituents are dispersed in the cytoplasm. They are usually larger than 0.15 µm but smaller than 2 µm. Ribosomes are suspended in cytoplasmic matrix, and chromosomes are in circular shapes. Prokaryotes reproduce asexually, usually by binary fission. Bacteria and archaea fall into this category.

As summarized in Figure 4.2, prokaryotes are typically classified based on their morphology, although the morphological similarity does not necessarily mean they are identical in genetic level. Prokaryotes take a vast majority of the cell mass in the activated sludge and are responsible for the most part of BOD treatment, nitrification, and excess phosphorus removal. As summarized in Table 4.1, the microorganisms responsible for BOD removal, for example, cocci and bacilli, grow fast; thereby, BOD removal is rarely problematic in the activated sludge process. By contrast, nitrifiers require 1 to 2 days to replicate even in laboratory condition. As a result, retaining a sufficient amount of nitrifier in the mixed liquor is crucial for the successful removal of nitrogen.

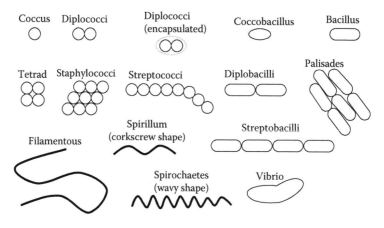

FIGURE 4.2 Morphological classification of prokaryotes.

TABLE 4.1
Microorganisms Participating in the Activated Sludge Process

	Prokaryote (Bacteria)	Eukaryote		
Role in the activated sludge process	BOD removal, nitrification, denitrification, and luxury uptake of phosphorus	BOD and trace organic removal (fungi), grazing small microorganisms, and clear supernatant (higher life-forms)		
Population in the activated sludge	Vast majority	Rare		
Cell wall	Yes	No		
DNA	One circular molecule	Multiple linear chromosomes		
Nucleus membrane	No	Yes		
Division	Binary fission	Mitosis		
Typical size (with exceptions)	0.15–2 µm	5–10 µm (fungi), 50–300 µm (protozoa), 100–500 µm (rotifers), and >500 µm (nematodes)		
Growth rate	Fast	Slow	Slow	Slow
Replication time (in test tube)	20–30 min (*Bacillus* sp.) 1–2 days (nitrifiers) 10–30 days (methanogens)	–	3–10 days	

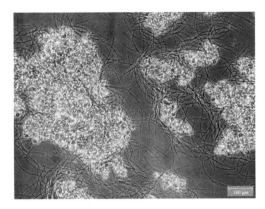

FIGURE 4.3 Abundant filamentous bacteria at a low F/M ratio, for example, less than 0.05 g COD/g MLSS/day. This mixed liquor had a fairly low membrane fouling potential in a pilot study using synthetic feeds.

Denitrification is rarely a problem in the activated sludge process as long as sufficient amounts of electron donors (or BOD) are present because the bacteria responsible for denitrification are largely overlapped with those for BOD removal.

Filamentous bacteria play a crucial role in providing a structural stability to the microbial floc and are essential for building the mixed liquor with a good settling property. However, if their population is excessive, it hampers floc formation and deteriorates sludge settling. The cause of filamentous bulking is not always clear, but it typically occurs when there is a deficiency of oxygen/micronutrients, a scarcity of BOD at low organic loading, the presence of toxic chemicals, etc. Abundant filamentous bacteria are often observed in MBR mixed liquor, as shown in Figure 4.3, primarily because of the low F/M ratio.

Some prokaryotes, especially the rod-shaped ones, have one, two, or more flagella on the external surface, which provide mobility to the cell. Flagella are the long and thin structure composed of a protein, flagellin, arranged in long fibers. Spirochetes can move by rotating the axial filaments that extend through the cell wall. Many bacteria are enclosed in a layer of thick and gummy polysaccharides and proteins called *glycocalyx*. The glycocalyx serves as a reservoir for nutrients and protects the organism from the environmental changes. If the glycocalyx is tightly bound on the cell, it is called a *capsule*, but if it is a loosely bound, it is called a *slime layer*. A portion of the glycocalyx is detected as an extracellular polymeric substance (EPS) in biopolymer quantification, as discussed in Section 5.2.2.1.

The bacteria that belong to the *Bacillus* and *Clostridium* genera are able to form spores with highly resistant structures to the external environment. Spores are primarily formed when the environment is not favorable for the growth. Spores can survive through the harsh environment that normal bacteria cannot deal with, for example, high temperature, long dry time, starvation, and toxic chemicals.

4.2.1.3 Eukaryote

Eukaryotes have a nuclear membrane that encapsulates multiple linear DNA molecules associated with histones and other proteins from which ribosomes are produced and exported to the cytoplasm. Eukaryotes are usually larger than 2 μm and multiplied by mitosis. Most eukaryotes have membrane-bound mitochondria that play a crucial role in energy metabolism by producing most adenosine 5′-triphosphate (ATP) cells need. The direct contribution of eukaryotes to BOD removal is minimal, but they graze suspended bacteria and cell debris and clarify the supernatant. All higher life-forms such as fungi, protozoa, rotifers, worms, plants, and animals fall into this category. The fungi found in the activated sludge appear similar to filamentous bacteria, but they are distinctive by the thick filament structure. Protozoa are unicellular and are typically motile. Amoeba, flagellates such as *Giardia lamblia*, *Cryptosporidium*, and ciliates are some examples of protozoa. Rotifers and nematodes are animals with motility, and the large species are visible to the naked eye (up to 2.5 mm).

Eukaryotes grow much slower than prokaryotes in general, perhaps because of the small surface area-to-volume ratio that can limit the rate of absorbing nutrients. Therefore, eukaryotes are prone to be washed out at low SRT, for example, 4 to 10 days, in CAS, and their population is difficult to be recovered once wiped out. MBR is not only running at a long SRT, for example, 12 to 30 days, but also retaining all microorganisms in the system. Thus, MBR have much more higher life-forms in the mixed liquor than CAS.

4.2.2 Bacterial Composition and Nutrient Requirement

Water takes approximately 85% of biosolid weight. Therefore, it is very hard to obtain lower moisture contents than 85% in sludge cake by dewatering the biosolids without performing any sort of drying process. In practice, lowering moisture content becomes exponentially harder once it reaches 90% by using one of the dewatering methods discussed in Section 6.13.3.

The elemental compositions of microorganisms somewhat vary depending on the species, but C, H, O, and N takes approximately 50%, 8%, 20%, and 14% of the cell mass, respectively, in the activated sludge. The composition can be written approximated as $C_5H_7O_2N$. From the following chemical equation, the theoretical oxygen demand is calculated at 1.98 g O_2/g VSS (= 7 × 32/113), including nitrification but excluding the oxygen credit from denitrification:

$$C_5H_7O_2N + 7O_2 \rightarrow 5CO_2 + 3H_2O + HNO_3$$

$$MW\ 113 \quad 7 \times 32 \quad 5 \times 44 \quad 3 \times 18 \quad 63$$

The COD equivalence of the cell mass can be calculated based on the chemical equation below. The theoretical specific COD of the cell mass is calculated at 1.42 g COD/g VSS (= 5 × 32/113). Because the hexavalent chromium used for COD measurement does not oxidize nitrogen, the measured COD can be considered equivalent to the oxygen required for oxidizing carbon only.

$$C_5H_7O_2N + 5O_2 \rightarrow 5CO_2 + 2H_2O + NH_3$$

$$MW\ 113 \quad 5 \times 32 \quad 5 \times 44 \quad 2 \times 18 \quad 17$$

In the activated sludge process, VSS takes approximately 80% of the biosolids while the rest of the biosolids are considered ash. Therefore, COD/MLSS becomes around 1.14 g COD/g MLSS, assuming MLVSS/MLSS is 0.80. In MBR, MLVSS/MLSS tends to be lower than that in CAS because of the longer SRT; hence, COD/MLSS can range from 1.0 to 1.1 g COD/g MLSS.

A more detailed composition of bacteria is summarized in Table 4.2, along with the source of each element and its function. Nitrogen and phosphorus take a significant portion of the dry cell mass, which are often a limiting nutrient in the activated sludge process. In general, the ratio among BOD, TKN, and TP is recommended as 100:5:1 to grow healthy ecosystem. If TKN and TP are not sufficient, bacteria can still remove BOD by converting it to exocellular polysaccharides and store them among microorganisms in the microbial floc. The loosely bound polysaccharides called *slime* can be easily released to the water phase and foul membrane in MBR or deteriorate effluent quality in CAS by interfering sludge settling. Nutrient deficiency can also cause abundant filamentous bacteria such as *Thiothrix*, type 021N, and *Nostocoida limicola* (Jenkins et al. 2004).

Trace elements such as manganese, zinc, cobalt, copper, and molybdenum also play a crucial role in microbial metabolism by being a cofactor of enzyme, which function as active sites in the catalytic reaction. In a series of studies, additional calcium (4.2 mg/L), cobalt (0.05 mg/L), copper

TABLE 4.2
The Source and Function of Major Elements of Bacterial Cells

Element	% Dry Weight	Source	Function
Carbon	50	Organic compounds or CO_2	Main constituent of cellular materials
Oxygen	20	H_2O, organic compounds, CO_2, and O_2	Constituent of cell material and cell water, O_2 is electron acceptor in aerobic respiration
Nitrogen	14	NH_3, NO_3^-, organic compounds, and N_2	Constituent of amino acids, nucleic acids nucleotides, and coenzymes
Hydrogen	8	H_2O, organic compounds, and H_2	Main constituent of organic compounds and cell water
Phosphorus	3	Inorganic phosphates $\left(PO_4^{3-}\right)$	Constituent of nucleic acids, nucleotides, phospholipids, LPS, and teichoic acid
Sulfur	1	SO_4^{2-}, H_2S, S^0, organic sulfur	Constituent of cysteine, methionine, glutathione, and several coenzymes
Potassium	1	Potassium salts	Main cellular inorganic cation and cofactor for certain enzymes
Magnesium	0.5	Magnesium salts	Inorganic cellular cation and cofactor for certain enzymatic reactions
Calcium	0.5	Calcium salts	Inorganic cellular cation, cofactor for certain enzymes, and a component of endospores
Iron	0.2	Iron salts	Component of cytochromes and certain nonheme iron proteins and a cofactor for some enzymatic reactions
Trace	<1	Mn, Zn, Co, Cu, and Mo	Cofactor for some enzymatic reactions

Source: Todar, K. Online textbook of bacteriology. Available at http://textbookofbacteriology.net/nutgro.html, 2005.

(0.5 mg/L), iron (4.2 mg/L), or magnesium (4.2 mg/L) improved COD removal by 7% to 12% when the initial removal efficiency was 70% in treating paper mill wastewater in laboratory-scale studies (Barnett et al. 2012). However, additional molybdenum (0.05 mg/L) or zinc (0.5 mg/L) reduced COD removal slightly by 2%. In other studies, individual or combinations of metal ions were tested using wastewater from fine chemical manufacturing (Burgess et al. 1999, 2000). It was found that additional manganese (1 mg/L) increased both respiration rate and removal efficiencies. However, cobalt (1 mg/L) and zinc (1 mg/L) showed an exact opposite effect, reducing the organic removal efficiencies by 5% and 40% to 50%, respectively.

Despite anecdotal evidence, the effects of trace metals on biological activities are fuzzy at best. It is perhaps due to the complex interactions among the factors affecting microbial health and treatment efficiencies, which can turn out differently depending on the environment. The known demand of trace metals by microorganisms is only a very small amount for the cofactors that constitute the active site of enzymes. Trace metals also possibly play a role in forming the three-dimensional structure of biomolecules by intercalating among different segments of the biomolecules. Trace metals are often found at elevated levels in healthy biomass, but that does not directly prove the necessity of the excess trace metals for metabolic purposes. As discussed in Section 3.2.4.2, some microorganisms have a tendency to accumulate specific metal ions, thereby playing an important role in removing trace metals from wastewater in the activated sludge process. The vitamins that belong to the B group have drawn attention as metabolic stimuli, for example, thiamine, riboflavin, niacin, pantothenic acid, pyridoxine, biotin, and folic acid, but the empirical results are fuzzy just like for trace metals. As a result, there are no known commercial agents with supplemental trace metals and/or vitamins recognized as the products universally applicable to enhance the activated sludge process.

4.2.3 KINETICS

Microbial growth typically follows Monod equation as shown in Equation 4.1, where μ (per second) is a specific growth rate, S (mg/L) is a substrate concentration, μ_{max} is a maximum specific growth rate when substrate availability is not a limiting factor for the growth, and K_S (mg/L) is a half reaction constant, that is,

$$\mu = \mu_{max} \frac{S}{K_S + S} \tag{4.1}$$

The relationship between the normalized substrate concentration and the normalized specific growth rate is shown in Figure 4.4. According to this graph, the specific growth rate rises rapidly as the substrate concentration increases when substrates are scarce (low concentration). However, the specific growth rate rises slowly when substrates are abundant (high concentration). As a consequence, the specific growth rate only approaches the asymptotic line set by the maximum specific growth rate. It is noticeable that the specific growth rate becomes half of the maximum growth rate μ_{max} when the substrate concentration S is equal to half of the reaction constant K_S. If S is equal to $4K_S$, the growth rate is $0.8\mu_{max}$, but it increases only marginally to $0.9\mu_{max}$ even if S increases to $9K_S$.

One implication of this equation is that the growth of microorganisms is self-correcting as long as microbial health remains in a reasonable range. For example, if specific microbial activity or the maximum specific growth rate (μ_{max}) is lowered by the environmental factor, the substrate concentration can increase as a result of the lowered reaction rate. At this condition, although the specific activity per microorganism is lowered, the high substrate concentration boosts the microbial growth rate (μ). As a consequence, the substrate concentration may not be affected too much by the lowered specific growth rate in the long run. In the nitrification process, if water temperature decreases, the nitrification rate also decreases. As a result, the ammonium concentration in water increases, but the abundant ammonium can promote nitrifier growth. In the long run, the nitrifier population will increase using the excess ammonium in water, and the nitrification efficiency can be recovered.

Enzymatic reactions generally follow the Michaelis–Menten equation drawn from the enzyme–substrate reaction mechanism (Atkins and de Paula 2009). The outline of this equation is basically identical to the Monod equation obtained experimentally to describe microbial growth. The enzymatic reaction can be written as the following chemical reaction, where E indicates enzyme,

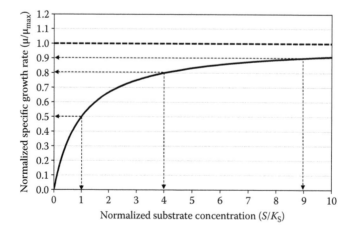

FIGURE 4.4 Abundant filamentous bacteria at a low F/M ratio, for example, less than 0.05 g COD/g MLSS/day. This mixed liquor had a fairly low membrane fouling potential in a pilot study using synthetic feeds.

S indicates substrate, ES is the enzyme bound with substrate, and k is the reaction constant for each subreaction:

$$E + S \quad \overset{k_a}{\underset{k_a'}{\underset{\leftarrow}{\rightarrow}}} \quad ES \overset{k_b}{\rightarrow} P + S$$

At a steady state, the bound enzyme concentration [ES] is constant. Therefore, the production rate of ES written as $k_a [E][S]$ should be equal to the decay rate written as $k_a'[ES] + k_b[ES]$ or $(k_a' + k_b)[ES]$. From this relation, [ES] can be written as Equation 4.2:

$$[ES] = \frac{k_a[E][S]}{k_a' + k_b} \tag{4.2}$$

The initial total enzyme concentration, $[E]_0$, is the sum of $[E]$ and $[ES]$ (or $[E] + [ES] = [E]_0$). Equation 4.2 can be rewritten as Equation 4.3 and again be rearranged to Equation 4.4, as follows:

$$[ES] = \frac{k_a([E]_0 - [ES])[S]}{k_a' + k_b} \tag{4.3}$$

$$[ES] = \frac{k_a[E]_0[S]}{k_a' + k_b + k_a[S]} \tag{4.4}$$

The overall reaction rate, d[P]/dt, can be written as Equation 4.5 by multiplying the bound enzyme concentration with the corresponding reaction constant. By inserting Equation 4.4 to Equation 4.5 and then rearranging it, Equation 4.6 is obtained, shown as follows:

$$\frac{d[P]}{dt} = k_b[ES] \tag{4.5}$$

$$\frac{d[P]}{dt} = \frac{k_a k_b[E]_0[S]}{k_a' + k_b + k_a[S]} = \frac{k_b[E]_0[S]}{(k_a' + k_b)/k_a + [S]} \tag{4.6}$$

Equation 4.6 can be rewritten as Equation 4.7, which is called the *Michaelis–Menten equation*, by defining the constant, $(k_a' + k_b)/k_a$, as a half reaction constant, K_S. In this equation, $k_b[E]_0$ corresponds to the maximum reaction rate because $[E]_0$ is the maximum value [ES] can reach in Equation 4.5. If the substrate concentration S is the same as K_S, the reaction rate becomes half of the maximum reaction rate $k_b[E]_0$:

$$\frac{d[P]}{dt} = k_b[E]_0 \frac{S}{K_S + S} \tag{4.7}$$

In the activated sludge model (ASM) of the International Water Association (IWA), the modules similar to Equation 4.7 are used to describe the reaction rates (Henze et al. 1986; Henze et al. 2000). For instance, the concept of a half reaction constant is used to describe the anoxic growth of heterotrophs, where heterotrophs grow by moving electrons from the electron donor, that is, organic matters, to the combined oxygen in nitrate. Therefore, the reaction rates increase as COD and nitrate concentration increases. On the contrary, the reaction is hampered by the molecular oxygen dissolved in the mixed liquor because microorganisms prefer to use them. This anoxic growth can be described as Equation 4.8. In this equation, $X_{B,H}$ (mg/L) is the heterotrophs, μ_H (/s) is specific growth rate of heterotrophs, η_g is a correction factor for anoxic growth, S_S (mg/L) is soluble COD in mixed liquor, K_S (mg/L) is a half reaction constant for soluble COD consumption, $K_{O,H}$ (mg/L) is a half inhibition constant for oxygen consumption, S_O (mg/L) is a dissolved oxygen, K_{NO} (mg/L) is a half reaction constant for nitrate, and S_{NO} (mg/L) is a nitrate concentration. The terms for COD (S_S) and nitrate (S_{NO}) are in the same form found in Equation 4.7. However, the term for dissolved oxygen (S_O) is written slightly differently to express the inhibition effect, wherein the higher the S_O, the smaller the term:

$$\frac{dX_{B,H}}{dt} = \mu_H \eta_g \frac{S_S}{K_S + S_S} \frac{K_{O,H}}{K_{O,H} + S_O} \frac{S_{NO}}{K_{NO} + S_{NO}} X_{B,H} \tag{4.8}$$

4.2.4 Distinguished Microbial Properties of MBR from CAS

The microbial aspect of MBR has been one of the prime interests of the academic society partly because it is a critical factor affecting membrane performance and partly because the complex nature of the microbial community provides intriguing subjects to investigate. Numerous research papers have been published on this topic in conjunction with the membrane performance.

Microbial species found in the MBR mixed liquor are basically the same as those in the CAS process, but the microbial population as a whole has some distinguished characteristics because of the low F/M ratio (or long SRT) and the complete retention of non-floc-forming microorganisms.

- The average floc sizes of MBR are generally smaller than those of CAS. One convincing explanation is that MBR does not lose non-floc-forming bacteria; hence, they can grow without being washed out. On the contrary, non-floc-forming bacteria can be easily lost through the effluent in CAS, whereas floc-forming bacteria are selected and proliferate in the system. In extreme case, flocs can be grown to large granules with up to approximately 4 mm by intentionally washing out non-floc-forming bacteria, for example, the Nereda® process (Giesen et al. 2013).
- Filamentous microorganisms tend to be more abundant in MBR. The cause is not completely clear, but it is popularly believed that filamentous microorganisms outcompete floc formers at food-scarce conditions in MBR due to the large specific surface area. In some cases, filamentous population can be an order of magnitude higher in MBR than that in CAS (Merlo et al. 2004).
- The *Zoogloea* population tends to be low in MBR perhaps because of the scarce food at the low F/M ratio. However, once the *Zoogloea* population surges by insufficient nutrients or toxic chemicals, the exocellular polysaccharides released to the liquid phase expedite membrane fouling.
- Higher life-forms grow much slower than bacteria, and hence a long sludge residence time is necessary to enrich them. Because of the long SRT, abundant levels of higher life-forms exist in MBR sludge in general, for example, rotifer, nematode, amoeba, and ciliate. The higher life-forms graze on fine cell debris and bacteria. It is not clear how much the grazing effect influence the membrane fouling, but abundant higher life-form population is somewhat correlated with low membrane fouling rates. In one study, when sessile ciliates and free-swimming ciliates populations increased, the contents of particles smaller than 10 µm, especially the particles around 1 µm, significantly decreased (Luxmy et al. 2000).

The high filamentous population itself does not necessarily cause accelerated membrane fouling. In fact, the high SRT and the low F/M ratio tend to increase filamentous population in MBR, but the membrane fouling potential decreases as the SRT increases or the F/M ratio decreases in general. By contrast, if the filamentous bloom occurs by the external factors that stresses microorganisms, for example, low dissolved oxygen (DO), toxic feed components, and nutrient deficiency, some extent of viscous bulking and foaming occurs simultaneously. Under such stress conditions, microbial product concentrations increase in the mixed liquor and the membrane fouling rate also tends to increase. In most laboratory and pilot studies, filamentous blooms are induced by imposing the stress conditions listed previously. As a consequence, the sludge property becomes unfavorable for membrane, and membrane fouling is found to be strongly linked to the filamentous population. On the contrary, the relationship between filamentous population and membrane fouling rate is not straightforward in a full-scale MBR running under pseudo-steady state.

4.3 OPERATIONAL ISSUES

4.3.1 FOAMING

Most foams owe their existence to the presence of the surfactants concentrated at the form surface, as shown in Figure 4.5. Surfactants not only reduce the surface energy/tension associated with surfaces but also stabilize the thin films and prevent them from rupture by tightly aligning side by side on the foam surface. In aqueous foam, surfactant molecules have both hydrophobic and hydrophilic moieties. The hydrophobic tail sticks out to the air, and the hydrophilic head is dissolved in water. Losing the water through evaporation and drainage, foams are inherently unstable and eventually disappear. If foam particles form a group, numerous borders are formed among the individual foam droplets. Liquid mainly exists in the plateau borders shared by multiple foam particles. Correspondingly, an individual polyhedral cell has its sharp edges and corners rounded off. Water is drained through the plateau borders downward by the gravity while evaporation occurs on the film. If enough water is lost and the film thickness reaches a minimum threshold, it will rupture.

The molecules and cell debris, which are more hydrophobic than other constituents of the mixed liquor, tend to move to the air–water interface and stabilize the foam. Some filamentous bacteria having hydrophobic surfaces also contribute to the foam stabilization by constructing a physical structure that supports the foam particles and interfere with the water drainage. *Nocardia*,

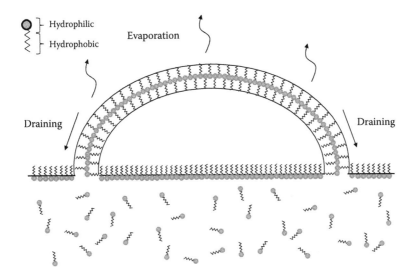

FIGURE 4.5 Mechanism of foam formation.

Microthrix parvicella, type 1863, etc., are commonly known as *foam stabilizers*. *M. parvicella* tend to be more frequently observed in winter when water temperature is low, whereas *Nocardia* are more common in summer. The hydrophobic debris in the plateau borders in the foam layer interfere with the water drainage and stabilize the foam layer. Under this condition, defoaming occurs mainly by the slow evaporation of water from the foam film.

Excessive foaming is one of the most common problems in CAS. It occurs regardless of the aeration methods, but it is more pronounced when CAS is oxygenated by diffused aeration rather than mechanical mixing. If excess foam overflows from the aeration basin, it contaminates the surrounding areas and the associated mechanical equipment causing hygienic issues. In addition, foaming can also cause a safety issue by making the floor slippery. In MBR, excessive foaming often indicates a high membrane fouling potential because foaming predominantly occurs when SMP levels are high in the mixed liquor (Luxmy and Yamamoto 2003). Filamentous bacteria such as *Nocardia* and *M. parvicella* having hydrophobic surfaces are commonly associated with a severe foaming. *Nocardia* are mostly contained within floc particles, whereas the long and thin *M. parvicella* extrudes from the floc particles. However, the free *Nocardia* and *M. parvicella* suspended in liquid phase can be accumulated in the foam layer and stabilized. Filamentous microorganisms are known to grow faster than floc-forming microorganisms under the low organic loading condition used by MBR. Moreover, they are easy to be enriched because they are not discharged with effluent at all in MBR.

Although foaming is positively correlated with membrane fouling in general, studies have found that the relationship between foaming and membrane fouling is more complicated than it seems (Cosenza et al. 2013). Depending on the operating condition, when foaming occurs, materials with hydrophobic nature are accumulated on the mixed liquor surface, and their concentration in liquid phase declines. As a result, the membrane fouling rate can decline when severe foaming occurs. Detailed mechanisms are discussed in Section 6.5.2.7.

Meanwhile, antifoams consist of more hydrophobic molecules and/or particles than the naturally occurring foam-causing materials, for example, petroleum hydrocarbons, silicone oils, long-chain alcohols, vegetable oils, and long-chain fatty acids. If antifoam is added, the hydrophobic constituents move to the foam layer and spread rapidly throughout the foam film. Once hydrophobic antifoam particles are interposed onto the foam film, the contact angle at the three-phase boundary among water, air, and antifoam particles increases, as shown in Figure 4.6. Consequently, water film thickness near the antifoam particle decreases, and it eventually causes foam rupture by breaking the force balance around the foam particle. Meanwhile, water-soluble antifoams consist of surfactant-like molecules with a higher hydrophobicity than the foam-causing materials. If they

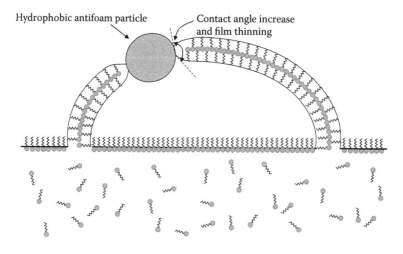

FIGURE 4.6 Mechanism of foam collapse by hydrophobic antifoam particle.

are sprayed directly to the foam layer, the artificial hydrophobic molecules quickly displace the surfactant molecules. These liquid antifoams eliminate foam particles by changing the local surface tension and by breaking the force balance around the foam.

Traditionally, emulsified silicon oils, hydrocarbons, polymers, glycerins, polyglycols, etc., have been used in CAS. However, the antifoams containing silicon oils, hydrocarbons, and polymers are not preferred in MBR because of the concerns over the direct membrane fouling by the hydrophobic molecules. Despite the negative notion on certain antifoams, no sufficient empirical data collected from full-scale MBR under realistic condition exists. The lack of extensive empirical data is partly because it is difficult to justify risking the membranes to the unknown effect from oil-based antifoams in a full-scale MBR. It is also because applying the suspected antifoam may result in losing the membrane manufacturer's warranty. The antifoams approved by one of the major membrane manufactures are summarized in Table 5.1. The effects of foaming and defoaming on oxygen transfer efficiency (OTE) and membrane fouling will be discussed in Section 6.5.2.7.

4.3.2 Sludge Bulking

Sludge bulking in CAS is defined as the condition, where the separation of solids from water is poor in the secondary clarifier because of the poor sludge settling and/or the poor sludge compaction properties. Severe sludge bulking can cause failure of the CAS process by raising suspended solids contents in the effluent. Sludge bulking can be classified to various categories: (1) filamentous bulking, (2) viscous bulking, and (3) pin floc. Filamentous bulking occurs when filamentous bacteria grow excessively, interfering floc formation and sludge settling. Any abnormal operating condition can cause filamentous bulking, as summarized in Table 4.3. On the basis of historical data obtained from the CAS treating various wastewaters from different sources, some indicator filamentous bacteria have been identified (Jenkins et al. 2004). Types 0041, 0675, 1851, and 0803 tend to outcompete at low F/M condition. *Thiothrix* I and II, *N. limicola*, and types 021N, 0092, 0914, 0581, 0961, and 0411 tend to outcompete in septic condition. The high population of *Thiothrix* I, II, and type 021N is often linked to nitrogen deficiency, whereas *N. limicola* III, *Haliscomenobacter hydrossis*, and *Sphaerotilus natans* are linked to phosphorus deficiency. *Nocardia* spp., *M. parvicella*, and

TABLE 4.3
Factors Affecting Sludge Bulking and Clarification

Factors	Effect
Organic loading	Organic loading fluctuates as wastewater BOD and/or flow rate fluctuates
	Fluctuating organic load may cause low DO that promotes sludge bulking
	Low F/M can cause pin floc and poor sludge settling
Wastewater characteristics	pH/temperature/septicity/nutrient contents
	Toxic components in wastewater can negatively affect floc formers while promoting filamentous bacteria
	Nutrient deficiency can cause filamentous or viscous bulking
System design	Insufficient air supply causes low DO
	Poor mixing causes dead zones with no DO
	Short circuit in aeration tanks causes incomplete treatment
	Short circuit in clarifier causes high TSS in effluent
	Clarifier design (sludge collection and removal)
	Limited return sludge pumping capacity
Others	Insufficient nutrients
	High oil/grease causes low DO
	Excessive aeration can cause dispersed sludge with a poor settling property

type 1863 can be linked to high grease and oil content in wastewater. If type 1701, *H. hydrossis*, and *S. natans* are blooming, low DO can be suspected. Fungi are not filamentous bacteria, but they have filamentous-like morphologies and tend to grow fast at pH lower than 6.

Viscous bulking, which is also called *nonfilamentous bulking*, occurs when excessive polysaccharides are produced by microorganisms. The excessive polysaccharides make the floc bulky and cause poor sludge compaction. Some loosely bound polysaccharides are detached and increase SMP levels in the sludge. In addition, it causes excess flocculant consumption along with poor sludge dewaterability in the sludge thickening process. Excessively high F/M ratio, nutrient deficiency, low DO, and toxic compounds such as chromium and sulfides are the common cause of viscous bulking. Excessive *Zoogloea* populations are often observed under this condition. Meanwhile, pin floc occurs when small and weak flocs are formed and hence activated sludge flocs are easily sheared. These small flocs (<50 µm) do not settle quickly in the secondary clarifier and deteriorates the effluent quality in CAS. It occurs most commonly at excessively low organic loading condition (or low F/M ratio, or long SRT), but it can also occur by toxic components contained in influent.

In CAS, the extent of sludge bulking is quantified by sludge volume index (SVI, mL/g MLSS). To estimate SVI, a mixed liquor sample taken from the aeration tank is settled for 30 min in a graduated cylinder. Then the volume ratio of the settled sludge and the total sludge are measured, for example, $V_{\text{Settled sludge}}$ (L) and V_{Sludge} (L), respectively. In addition, the MLSS of the unsettled sludge (g/L) is measured. Finally, SVI is calculated as shown in Equation 4.9. SVI represents the volume (mL) of the mixed liquor that contains 1 g of MLSS. The lower the SVI, the better the sludge settling property. If the SVI is less than 100 mL/g, the sludge is considered to settle fast and form a dense sludge layer. If SVI is 100 to 200 mL/g, sludge settles slower, but the effluent quality may improve depending on the condition because sludge can sweep fine particles better when it settles. If SVI is more than 250 to 300 mL/g, sludge settles excessively slowly and effluent quality deteriorates. However, SVI is not a relevant parameter to link the sludge property to the process performance in MBR. The SVI of MBR sludge is fairly low even if the sludge does not settle at all. For example, SVI is 100 mL/g by definition for the sludge with an MLSS of 10 g/L.

$$\text{SVI} = \frac{V_{\text{Settled sludge}}}{\text{MLSS} \cdot V_{\text{Sludge}}} \times 1000. \tag{4.9}$$

4.3.3 SLUDGE RISE IN THE SECONDARY CLARIFIER IN CAS

Sludge blankets in the secondary clarifier can rise and hamper the solid–liquid separation, if the retention time in the secondary clarifier is too long and/or the nitrate concentration in the mixed liquor is too high. It is because the nitrate is reduced to nitrogen gas by heterotrophic bacteria and trapped by the sludge blanket. The rising sludge blanket can be distinguished from the bulking sludge by the presence of small bubbles on the sludge. The rising sludge problem can be fixed by reducing the SRT in the secondary clarifier, by increasing the return activated sludge (RAS) flow, and by reducing the sludge blanket depth. Alternatively, the nitrification rate can be reduced by increasing the excess sludge removal. At the lowered SRT, the slowly growing nitrifier is partially washed out, and the nitrate formation can be hampered partially. However, this method increases the ammonia nitrogen contents in the effluent.

4.3.4 CHEMICAL TREATMENT

If excessive filamentous microorganism population causes foaming and sludge bulking, it can be controlled chemically by using oxidation agents, for example, NaOCl, H_2O_2, and O_3. Although such oxidants are not selective and all microorganisms are affected, non-floc-forming microorganisms are more prone to oxidation agents than floc formers because of the high surface area-to-volume ratio. In particular, the filamentous bacteria suspended in the mixed liquor or poked out from a large floc are vulnerable. Meanwhile, the slowly growing filamentous bacteria grow slower than fast floc formers; hence, they suffer more from population loss than floc formers. This kinetic control should be performed periodically to suppress the excessive growth of filamentous bacteria. If oxidizing agents are added excessively, they can cause diffused floc and increase turbidity of the effluent. In addition, they can also cause a decline in the performance of the nutrient removal process by damaging the phosphorus-accumulating organisms (PAOs) and/or nitrifiers in enhanced biological phosphorus removal (EBPR) process (Chang et al. 2004). Such microorganisms grow slower than heterotrophic microorganisms involving BOD removal, and it can take a long time to recover the nutrient removal efficiency once their population is wiped out.

Chlorine can be added to the RAS, the secondary clarifier center well, and/or the side stream channel of the aeration tank. Alternatively, it can be sprayed directly to the aeration basin surface. Aqueous bleach (or NaOCl) is commonly used because it does not need a special equipment to handle/inject it other than metering pumps. If chlorine gas is used for large-scale plants, a specially designed chlorinator is necessary to inject it. The target dosage is 2 to 8 g/kg biosolids per day based on the total MLSS retained in the system, but it can widely vary depending on the response of the system. For example, if the total amount of biosolids is 1000 kg in the system, including clarifiers, 2 to 8 kg NaOCl (or 13–53 L of 12.5% bleach with a density of 1.2 g/cm^3) can be added to RAS intermittently throughout the day. To avoid overdose, the daily chlorine dosage is started from low and increased until observing the filamentous population decline. Nitrification and phosphorus removal should be monitored carefully while chlorination program is applied. If NaOCl is aimed to control filamentous foaming, it can be sprayed directly to the foam layer. Surface spraying of diluted chlorine solution (50 mg/L) can be effective for controlling *Nocardia* foam (Jenkins et al. 2004).

Although chlorine is the most widely used oxidant, H_2O_2 can also be used for the same purpose. It is added to RAS at 100 to 200 mg/L in the beginning until filamentous microorganisms are sufficiently controlled. Once the filamentous population falls below the target level or SVI becomes sufficiently low, the dose may be reduced to 25 to 50 mg/L to maintain the condition. Alternatively, ozone can be sparged with air in the aeration tank through diffusers or mechanical mixers (Nagasaki and Nakazawa 1996). The low-level ozone mixed with air, for example, 0.01 to 0.16 wt.%, kills filamentous microorganisms in the aeration tank. Ozone can also be added to a portion of the RAS (Wijnbladh 2007). When ozone was dosed to the 10% of RAS in a municipal WWTP at a dosage of 6.5 g O_3/kg SS, sludge settling property improved within first few weeks. Further, ozone injection at the same dosage for a total of 8 weeks followed by injecting ozone at 4.4 g O_3/kg SS in the next 4 weeks substantially reduced filamentous population and improved sludge settling. After ozonation was stopped, the settling property returned to the original level in 3 weeks. At the dosages used, no effect on nitrification was observed. It was observed that *M. parvicella* was initially wiped out, but it returned after ozone dosage was reduced from the ninth week.

4.3.5 Effect of Foaming, Sludge Bulking, and Pin Floc in MBR

In the activated sludge process, biological foaming starts by the viscous polysaccharides and proteins with hydrophobic moieties secreted by microorganisms. In normal condition, the foam particles collapse relatively quickly by water drainage and evaporation. However, the foam can obtain the necessary stability if water drainage and evaporation are interfered by the high biopolymer concentration in the form and/or the networks of particles and macromolecules. Some filamentous bacteria with hydrophobic surfaces, for example, *Nocardia*, *M. parvicella*, and type 1863, can accumulate in the foam layer and stabilize by interrupting water drainage. It is worthwhile to mention that abundant filamentous bacteria do not necessarily mean severe foaming because they are more likely a stabilizer rather than a starter of foam. In fact, MBR sludges tend to contain abundant levels of filamentous bacteria perhaps due to low organic loading rate (or F/M ratio), but foaming is not necessarily severe. Meanwhile, filamentous bacteria are not a direct cause of membrane fouling either because the back transport velocities of such large over-micron-sized particles are high enough. Instead, the viscous biopolymers with hydrophobic moieties secreted by floc formers are suspected to be the major cause of membrane fouling.

Because biopolymers are the common factors causing both foaming and membrane fouling, membrane fouling rates have a good chance to increase when foaming increases (You and Sue 2009). However, if foaming helps hydrophobic biopolymers scavenged up to the foam layer, the hydrophobic biopolymer level in the mixed liquor can be lowered and membrane fouling rate can decrease (Sharp et al. 2006; Cosenza et al. 2013). Foam layers can be killed by adding antifoams, but the viscous biopolymers returning back to the mixed liquor may not only elevate the membrane fouling rate but also interfere with oxygen transfer efficiency. Therefore, abrupt defoaming should be avoided for stable system operation. Detailed mechanisms of foaming and defoaming are discussed in Section 6.5.2.7.

Filamentous populations are often high in MBR relative to those in CAS perhaps due to the low F/M ratio at which filamentous bacteria with a large surface area-to-volume ratio have advantages in uptaking food at starving conditions (Figure 4.3). High filamentous population is not a direct indicator of high membrane fouling tendency. However, abnormal or fluctuating operating conditions that cause filamentous bloom also tend to stress the floc formers and induce the elevated EPS/SMP level in mixed liquor. In many cases, the rising filamentous population tends to be correlated with membrane fouling as a result (Chang and Lee 1998; Meng et al. 2006). It has been commonly observed that EPS content increases when filamentous bulking occurs in MBR.

The viscous bulking caused by excessive polysaccharides secreted by microorganisms is directly linked to the elevated membrane fouling rates (Ferré et al. 2009). The common causes of viscous bulking are nutrient deficiency, excessively high F/M ratio, low DO, and toxic chemicals such as chromium and sulfides. The elevated level of the *Zoogloea* population is often observed when viscous bulking occurs (Figure 4.7). A portion of the excess slime secreted by microorganisms can be detached from the microorganisms and can increase SMP levels in the mixed liquor. The relations between the SMP levels and the membrane fouling rate are not completely clear, as discussed in Section 5.2.2.2, but the rising SMP concentration in the short period often causes accelerated membrane fouling in a given plant in general.

Pin floc also commonly occurs in MBR due to the low F/M ratio and the complete retention of non-floc-forming bacteria. This results in a much smaller average particle size in the MBR compared with those in the comparable CAS, but the pin flocs (<50 μm) are still big enough to have a sufficient back transport velocity that moves them away from the membrane surface, as discussed in Section 1.2.5. Thus, the MBR sludge with an appearance of pin floc does not necessarily cause high membrane fouling. However, if the average particle size declines with the elevation of SMP/EPS in mixed liquor, membrane fouling can accelerate (Chang et al. 1999). The cause and effect of common abnormal symptoms in CAS and MBR are summarized in Table 4.4.

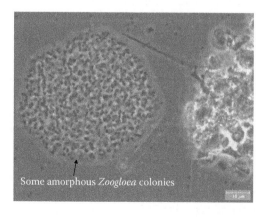

Some amorphous *Zoogloea* colonies

FIGURE 4.7 *Zoogloea* colony filled with slimy polysaccharides in the spaces among cells.

TABLE 4.4

Comparison of the Effect of Abnormal Symptoms on CAS and MBR

	Effect on Process	
Symptom	**CAS**	**MBR**
Filamentous population	• Poor sludge settling • High COD and TSS in effluent • Increases foaming potential • Occurs by nutrient deficiency, low F/M (filamentous) or high F/M (viscous bulking), low DO, toxic chemicals, etc.	• High filamentous population does not necessarily indicate fast membrane fouling unless high SMP levels are accompanied. However, filamentous population and membrane fouling tend to be correlated positively because both occur simultaneously when abnormal condition occurs
Viscous bulking		• Increases the membrane fouling potential due to high SMP
Pin floc	• Occurs at a low F/M ratio or by toxic chemicals • Effluent quality deteriorates	• No direct effect on membrane fouling unless elevated SMP levels are accompanying
Foaming	• Occurs by hydrophobic SMP and stabilized by some filamentous microorganisms and hydrophobic cell debris	• Tends to increase the membrane fouling potential due to the hydrophobic foam-causing materials • If foam-causing materials are accumulating in form layer, membrane fouling may decrease temporarily

4.4 NITROGEN REMOVAL

4.4.1 NITRIFICATION

4.4.1.1 Mechanism

Nitrification is a two-step process, where ammonia nitrogen $\left(NH_4^+\right)$ is oxidized to nitrite $\left(NO_2^-\right)$ by *Nitrosomonas* and nitrite is further oxidized to nitrate $\left(NO_3^-\right)$ by *Nitrobacter* as shown in the chemical equation below. Because nitrate formation from nitrite is far faster than nitrite formation from ammonia, the nitrite concentration is negligible at less than 1 mg/L at steady state in a typical condition, shown as follows:

$$
\begin{array}{ll}
\text{Nitritation by } \textit{Nitrosomonas} \text{ (Slow):} & NH_4^+ + 1.5O_2 \rightarrow NO_2^- + 2H^+ + H_2O \\
\text{Nitratation by } \textit{Nitrobacter} \text{ (Fast):} & NO_2^- + 0.5O_2 \rightarrow NO_3^- \\
\text{Overall reaction:} & NH_4^+ + 2O_2 \rightarrow NO_3^- + 2H^+ + H_2O \\
\quad \text{Molecular weight (Da)} & \underline{14\,(N) \quad 2 \times 32} \\
\quad \text{Alkalinity} & \underline{0 \qquad 0 \qquad 0 \qquad -2 \qquad 0}
\end{array}
$$

$$(4.10)$$

When 1 mol of NH_4^+ is oxidized, 2 mol of acidity is produced or 2 mol of alkalinity is consumed according to the previously mentioned chemical equations. Here, NH_4^+ is considered neutral because it consists of one alkalinity (NH_3) and one acidity (H^+). The specific oxygen requirement for ammonia nitrogen oxidation can be calculated based on the above equations. Assuming no new autotrophic microorganisms are produced during the reaction, 4.57 mg O_2 (= 2 × 32/14) and 7.14 mg alkalinity as $CaCO_3$ (= 2 × 50/14) are consumed to oxidize 1 mg NH_4-N.

In practical situations, however, some nitrogen atoms are used to produce new microorganisms. Because of the partial loss of nitrogen to the new cells, O_2 and alkalinity consumptions based on the treated NH_4-N are slightly lower than the values calculated in above paragraph. Although the exact value slightly varies system by system depending on the autotrophic sludge yields, equivalent oxygen and alkalinity consumptions are often assumed at 4.3 mg O_2/mg N and 6.8 mg $CaCO_3$/mg N considering the nitrogen loss to cell syntheses. The following equations represent nitritation and nitratation reactions in typical nitrification condition (Tchobanoglous et al. 2003):

Nitritation:

$$55NH_4^+ + 76O_2 + 109HCO_3^- \rightarrow C_5H_7O_2N\,(cell) + 54NO_2^- + 57H_2O + 104H_2CO_3 \quad (4.11)$$

Nitratation:

$$400NO_2^- + NH_4^+ + 4H_2CO_3 + HCO_3^- + 1950O_2 \rightarrow C_5H_7O_2N\,(cell) + 3H_2O + 400NO_3^- \quad (4.12)$$

The autotrophic bacteria responsible for nitrification not only grow slower than heterotrophic bacteria but also are more susceptible to the environmental fluctuations such as temperature, pH, and the toxic chemicals contained in wastewater. The high ammonia concentration can inhibit nitrification, but the threshold concentration varies depending on the pH. In general, a high level of ammonia is more toxic at a high pH, for example, above 8, because charge neutral ammonia can penetrate microbial cells easier and disrupt microbial physiology. A high nitrite $\left(NO_2^-\right)$ level in the mixed liquor indicates the problem in nitratation. DO, pH, temperature, and BOD/TKN ratio affect nitrite accumulation. Nitrites are generally toxic to aquatic lives if discharged to the environment.

Nitrification occurs by various autotrophic microorganisms. Although energy is derived from the oxidation reaction of ammonia (NH_4-N), a portion of it is spent to reduce the dissolved CO_2 and to obtain the carbon required to build the cell mass. Because the relatively small energy stored in N-H bonding is used to produce the C–C bonding with high chemical energy, autotrophic cell growth is not as efficient as heterotrophic cell growth. As a result, autotrophic biosolid yields (Y_A) are much lower than those of heterotrophic biosolid yields (Y_H). In the ASM#1 (Henze et al. 1986), Y_A is assumed at 0.24 g COD/g N whereas Y_H is 0.67 g COD/g COD. Meanwhile, nitrogen contents in most wastewaters are typically much lower than carbon contents, for example, the TKN/COD ratio is around 0.10 to 0.15 for municipal wastewater. Because of the low nitrogen contents and the low biosolid yield, autotroph biomass is typically only 2% to 5% of the total amount of biosolids in municipal wastewater treatment. Therefore, autotrophic biosolid yields are often neglected when calculating an approximate excess of biosolid production.

4.4.1.2　Effect of pH and Alkalinity

The optimum pH for nitrification is known to be around 7.5, which is the middle ground of the two optimum pH for *Nitrosomonas* and *Nitrobacter* that grow fastest at pH 7.8 to 8.0 and at pH 7.3 to 7.5, respectively. It is well known that the nitrification rate slows down at a pH below 7.0 and ceases completely at around pH 6.0 in CAS. Therefore, it is recommended to maintain the mixed liquor pH higher than 7.0 while not allowing it below 6.5 in all circumstances. However, it must be noted that the nitrification rate slows primarily when the pH swings in a short period because the microbial community cannot adapt to the new pH quickly. If the mixed liquor is acclimated at a low pH for a sufficient period, nitrification can occur even at a pH below 6.0 at a fairly high rate (Tarre and Green 2004). Although full-scale studies are rare, many laboratory-scale studies have suggested that some of the minor species classified as *Nitrosomonas* and *Nitrobacter* genera grow fast at a low pH and eventually replace the dominant species that grow quickly only above a neutral pH.

If wastewater alkalinity is low, excessive pH fluctuation can occur during the nitrification and the denitrification processes because of alkalinity consumption and production in the process. The pH of the aeration tank can be substantially lower than that in the anoxic tank as a consequence, and the microorganisms are exposed to the two different pH periodically. The fluctuating pH negatively affects the microorganisms involving the nitrogen removal when mixed liquor is circulated through the tanks. Therefore, sufficient alkalinity is crucial for the successful removal of biological nitrogen, especially when TKN loading is high and fluctuating. The low dissolved CO_2 concentration at low alkalinity may limit the growth of autotrophic bacteria by limiting the availability of carbon source, but there is insufficient empirical evidence supporting the notion.

4.4.1.3 Temperature Effect

It is well known that the specific growth rate of nitrifiers decreases as temperature decreases. Nitrifiers can be even washed out under a persistently low water temperature in CAS. The maximum growth rate of the nitrifier decreases less than a half when temperature decreases from 14°C to 6°C in long-term tests (Gujer 2010). Alongside, the maximum specific nitrification rate of the nitrifiers grown at 14°C decreased by 2/3 when temperature decreased to 6°C. By contrast, the maximum specific nitrification rate of nitrifiers grown at 6°C did not increase threefold when the temperature was raised to 14°C, as shown in Figure 4.8.

From the kinetics stand point, the slow nitrifier growth at low temperature does not directly mean a slow nitrification rate at low temperature if biological systems run under a consistently low temperature for an extended period. First, there is a great deal of redundancy in the nitrifier population because of the long SRT in biological nutrient removal (BNR) processes, especially when it is combined with MBR. The high SRT of 12 to 30 days enables the sufficient enrichment of autotrophs in an ordinary condition. Therefore, poor nitrification is not commonly observed in MBR treating municipal wastewater as long as there are no drastic changes in water pH, temperature, TKN loading, etc. Second, if the ammonium concentration increases as a result of a slow nitrification at a low temperature, the nitrifier can grow faster at the high ammonium concentration. This self-correcting mechanism can keep the nitrification efficiency high even at low water temperatures as long as the temperature does not change quickly.

4.4.1.4 Dissolved Oxygen Effect

Nitrification rate is affected by the DO level. A DO of 1 mg/L is considered as a requirement to prevent any inhibition effect caused by the insufficient DO in practice whereas 0.3 mg/L is considered

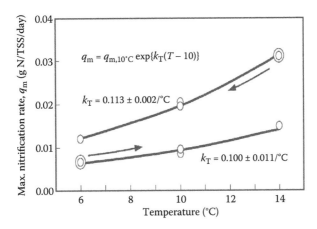

FIGURE 4.8 Short-term effects of temperature on ammonium oxidizing organisms based on batch results with the activated sludge grown at 6°C and 14°C. In the equation, T represents liquid temperature in Celsius. (Reproduced from Gujer, W., *Water Res.*, 44, 1–19, 2010. With permission.)

a minimum (Stenstrom and Poduska 1980; Tchobanoglous et al. 2003). The optimum DO is also dependent on floc sizes because it affects the oxygen diffusion/supply to the nitrifiers inside the floc. If floc particles are large, for example, more than 1 to 2 mm, the excess DO at more than 2 to 3 mg/L helps improve the nitrification rate. The contact time also plays a role in nitrification. For instance, in one full-scale dairy wastewater treatment plant running at very long hydraulic retention time (HRT) (20 days) and SRT (>100 days), nitrification was accomplished nearly 100% despite the negligible DO at below the detection limit, for example, less than 0.1 mg/L. It has been known that ammonia-oxidizing bacteria (AOB) in Nitrosomonas genus, for example, Pseudomonas, Xanthomonadaceae, Rhodococcus, and Sphingomonas, are responsible for low-DO nitrification (Fitzgerald et al. 2015).

4.4.2 Denitrification

4.4.2.1 Effect of Oxidation Reduction Potential

Biological denitrification occurs when molecular oxygen (O_2) is not sufficient for the respiration. Under this condition, the combined oxygen contained in nitrate (NO_3-N) can be used as an alternative oxygen source by heterotrophs. The reduced nitrogen eventually escapes as molecular nitrogen (N_2) from the mixed liquor. The condition in which denitrification occurs is called *anoxic* to distinguish it from the *anaerobic* condition in which no oxygen sources exist whether they are molecular oxygen or combined oxygen.

Oxidation reduction potential (ORP) is used to monitor and control anoxic and anaerobic conditions. Once aeration stops, dissolved molecular oxygen serves as an electron acceptor until it is depleted and ORP decreases to around +50 mV, as illustrated in Figure 4.9. At around +50 mV, nitrate starts to serve as an electron donor until it depletes at the ORP of around −50 mV. If there is no additional nitrate, sulfate becomes an electron donor and ORP decreases further below −50 mV. It must be noted that all the ORP values in the figure are rough guidelines with ±50 to 100 mV error because ORP is not only affected by nitrate and sulfate concentrations but also affected by various factors such as ionic composition/strength, pH, temperature, presence of other oxidants, etc. For example, the triggering ORP for denitrification can vary between −100 and 100 mV, depending on the environment and ORP can decrease below −200 mV in the anoxic tank in the actual process. If the granular sludge with a large diameter is grown in the reactor, nitrification can occur inside the granule even under the presence of molecular oxygen in the liquid phase.

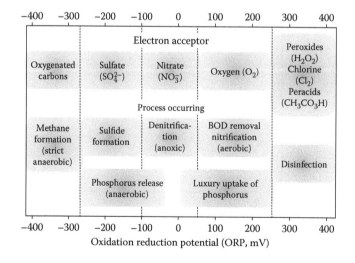

FIGURE 4.9 Relationship between ORP and metabolic processes.

4.4.2.2 Requirement of Readily Biodegradable COD

Denitrification requires electron donors to reduce the combined oxygen in nitrate. If methanol is used, the reaction can be written as Equation 4.13, where 1.905 mg methanol is required to reduce 1 mg NO_3-N (= $5 \times 32/6/14$). If the methanol requirement is converted to COD requirement, it becomes 2.86 mg COD/mg NO_3-N, considering the conversion factor of 1.5 mg COD/mg methanol:

$$6NO_3^- + 5CH_3OH \rightarrow 3N_2 + 5CO_2 + 7H_2O + 6OH^-$$
$$MW\ 3 \times 28\ 5 \times 32$$

$$(4.13)$$

The long HRT in the anoxic tank helps improve denitrification efficiency, but the HRT cannot be prolonged too much because of the adverse effect on aerobic microorganisms. The apparent HRT of the anoxic tank is typically 0.5 to 2 h in municipal wastewater treatment, but the actual can only be half or less due to the internal sludge recycle. Because of the short residence of wastewater in the anoxic tank, only readily biodegradable COD (or BOD) can be used as reductant. Denitrification is a fast reaction performed by majority of the heterotrophs; thereby, it is rarely a rate determining step in nitrogen removal as long as readily biodegradable COD is sufficient. In general, the BOD/TKN ratio in wastewater need to be higher than 3 or more preferably higher than 4 in municipal wastewater treatment to secure a good nitrogen removal in BNR processes. The approximate specific denitrification rates (SDNRs) obtainable with some of the popular carbon sources are summarized in Table 4.5.

Figure 4.10 shows the SDNR as functions of the F/M ratio in the anoxic tank and the percentage of readily biodegradable BOD in the total BOD. For example, if the F/M ratio is 0.5 g BOD/g MLSS/day in the anoxic tank and readily biodegradable BOD takes 30% of the total BOD, SDNR becomes around 0.125 g NO_3-N/g MLSS/day. An example of the anoxic tank sizing is shown in Section 6.15.6, and the following procedures can be used to determine the required anoxic tank volume:

- Assume the HRT of the anoxic tank and calculate the tank volume.
- Calculate the SDNR required to denitrify the nitrate transported to the anoxic tank considering the nitrogen load, MLSS, and anoxic tank volume.
- Calculate the F/M ratio in the anoxic tank.
- Find the point that matches the calculated SDNR and F/M in Figure 4.10.
- If the calculated SDNR required is less than the SDNR found from the curve, the initial HRT assumption is valid. Otherwise, the above sequences are repeated with a higher HRT.

TABLE 4.5
Typical Denitrification Rates for Various Carbon Sources

Carbon Source	SDNR, mg NO_3-N/g VSS/day	Temperature (°C)
Methanol	2.3[a]	10
	5.4–6.8[a]	20
	8.8–13.3[b]	25
Ethanol	0.12–0.90[b]	20
Acetate	3.6[a]	10
	11.7–15.5[a]	20
Wastewater	1.3–4.6[b]	15–27
Endogenous metabolism	0.7–2.0[b]	12–20

[a] From Cherchi et al. *Water Env. Res.* 81(8), 788–799, 2009.
[b] From Tchobanoglous, G. et al., *Wastewater Engineering: Treatment and Reuse.* Boston: McGraw-Hill, 2003.

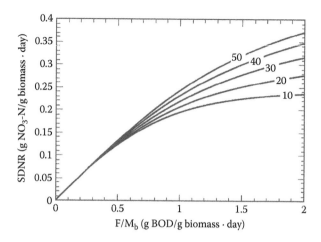

FIGURE 4.10   SDNR as function of the F/M ratio and the % of readily biodegradable BOD in the total BOD. (From Tchobanoglous, G. et al., *Wastewater Engineering: Treatment and Reuse*. Boston: McGraw-Hill, 2003.)

4.4.2.3   Oxygen Credit from Denitrification

When denitrification occurs, a portion of the readily degradable COD in wastewater is treated in the anoxic tank using the combined oxygen contained in the nitrate transported from aerobic tank. Therefore, the oxygen demand in the aeration tank is reduced as much as the amount of COD treated in the anoxic tank.

The theoretical oxygen credit delivered by nitrate is 2.86 g O_2 per gram of NO_3-N removed, as discussed in the previous section. Because 4.57 mg O_2 is consumed when 1 mg NH_4-N is oxidized in theory, the net oxygen consumption during the removal of 1 mg NH_4-N is 1.71 mg O_2 per milligram of NH_4-N removed. In practice, however, the oxygen demand for nitrification and the oxygen credit from denitrification are assumed at 4.3 mg O_2/mg NH_4-N and 2.4 mg O_2/mg NO_3-N, respectively, considering the nitrogen assimilated to the new microbial cells produced as excess sludge. Overall, the net oxygen demand for nitrogen removal becomes around 1.9 mg O_2 per gram of NH_4-N removed.

4.4.2.4   Alkalinity Production from Denitrification

When 1 mol of nitrate is removed, 1 mol of alkalinity (OH^-) is produced, shown as follows:

$$6NO_3^- + 5CH_3OH \rightarrow 3N_2 + 5CO_2 + 7H_2O + 6OH^- \tag{4.14}$$

Considering the 2 mol of alkalinity consumed in the nitrification process as shown in Equation 4.10, net alkalinity consumption is 1 mol per 1 mol of NH_4-N removed. The net 1 mol alkalinity consumption is in fact the disappearance of NH_3 itself, which is counted as alkalinity. The alkalinity production during the denitrification can help prevent the pH from dropping excessively, especially when influent TKN is high and wastewater alkalinity is low. Table 4.6 summarizes oxygen and alkalinity consumption and production during nitrification and denitrification.

4.4.2.5   Simultaneous Nitrification and Denitrification

The steady-state oxygen concentration in the center of the floc can be low because oxygen must diffuse through the microbial floc or biofilm while being consumed biologically. Hence, denitrification can occur at the center of the floc or in the bottom of biofilm attached on biocarriers even at high DO conditions (Kaempfer et al. 2000; Daigger et al. 2007). When the DO profile inside the microbial floc with 3.6 mm diameter was measured using a micro-DO probe, the high DO at 3 mg/L in the

TABLE 4.6
Oxygen and Alkalinity Consumption/Production during Nitrification and Denitrification

	O$_2$ Consumption			Alkalinity Consumption		
	Without Cell Production		With Cell Production	Without Cell Production		With Cell Production
Process	mol O$_2$/mol N	g O$_2$/g N	g O$_2$/g N	eqv. Alk/mol N	g CaCO$_3$/g N	g CaCO$_3$/g N
Nitrification	2.0	4.57	4.3	2.0	7.14	6.8
Denitrification	−1.25	−2.86	−2.4	−1.0	−3.57	−2.9
Overall	0.75	1.71	1.9	1.0	3.57	3.9

external floc surface decreased to zero quickly before the probe penetrated 1 mm into the floc. This strongly indicated that denitrification can occur in the internal space of the floc. However, growing biological floc to such large size is not readily possible in MBR due to the strong turbulence in membrane tanks.

In a commercial process called Nereda®, the granular sludge with up to 4 mm diameter is grown at a high superficial water velocity through the reactor. Under the condition, nonfloc formers are washed out, leaving floc formers suitable for building giant floc particles. Because of the large floc size, the cross section of the floc is stratified as aerobic, anoxic, and anaerobic toward the center of the floc (de Bruin et al. 2006). Similar to the SBR, Nereda relies on repeated fill and draw, aeration, and settling cycles, where anaerobic conditions occur during the fill and draw and the settling cycles. Simultaneous nitrification and denitrification occurs mainly during the aeration cycle. The PAOs growing inside the floc are enriched, while the floc is repeatedly exposed to aerobic and anaerobic conditions. Because of the high MLSS (10–12 g/L), the high food-to-volume ratio (0.8–1.05 g COD/L/day) is possible despite the low F/M ratio of 0.05 to 0.06 g BOD/g MLSS/day (Keller and Giesen 2010). It has been claimed that 60% to 80% of nitrogen removal can be readily achieved.

4.5 PHOSPHORUS REMOVAL

4.5.1 BIOLOGICAL PHOSPHORUS REMOVAL

Phosphorus takes around 1% of the dry cell mass in general. It is an essential element of DNA that stores the genetic information of microorganisms. It is also a constituent of ATP, which plays an important role in energy metabolism. Phosphorus exists in wastewater as orthophosphate $\left(PO_4^{3-}\right)$, polyphosphate, organically bound phosphorus, etc. Around 6 mg of P is removed per gram of COD removed from raw wastewater by the assimilation mechanism in the typical CAS process treating municipal wastewater. It corresponds to approximately 25% to 50% of the total incoming phosphorus to the process (van Haandel and van der Lubbe 2012).

Additional phosphorus can be removed by enriching PAOs in the microbial population. PAO is a group of bacteria that can accumulate phosphorus in the cell mass at the level much higher than ordinary level, that is, up to 38% (van Haandel and van der Lubbe 2012). *Acinetobacter* has been identified as one of the primary PAOs and is a Gram-negative genus belonging to the Gammaproteobacteria. In aerobic conditions, PAOs uptake excess phosphorus and store it as polyphosphate in the cell mass using the energy obtained from the heterotrophic oxidation of organic materials. In anaerobic conditions, where little molecular and combined oxygen is present, PAOs obtain energy from the hydrolysis of the accumulated polyphosphates and use it to uptake volatile fatty acids (VFAs). VFAs are stored in the cell as polyhydroxyalkanoates (PHAs). When PAOs are circulated back to the aeration tank, the stored PHAs are either used to produced energy through an oxidation process or used to produce cell mass. The metabolism of PAOs and the proposed molecular structure of polyphosphate and PHAs are illustrated in Figure 4.11. Polyphosphates are the polymerized phosphate ions of which charges are partially neutralized by cations, for example, Na^+, K^+, Ca^{2+}, and Mg^{2+} (Figure 4.12a). PHAs are the linear polyesters with various alkyl groups (R_1 and R_2) (Figure 4.12b). For instance, if R_1 is methyl group ($-CH_3$) and R_2 is hydrogen (H), it is called *polyhydroxybutyrates*.

In the EBPR process, the mixed liquor is periodically circulated through aerobic, anoxic, and anaerobic conditions. In such cyclic environment, PAOs have survival advantages over non-PAOs because PAOs can sustain its metabolism in the anaerobic condition without molecular or combined oxygen. The soluble phosphate level is the highest in the anaerobic tank and the lowest in aerobic tank in EBPR because of the phosphate release and uptake in each tank. The ORP required in the anaerobic tank to enrich PAOs is −200 to −50 mV, as can be seen in Figure 4.9. However, the actual ORP required can vary significantly by ±50 to 100 mV, depending on pH and water chemistry. The mechanisms of biological phosphorus removal process are illustrated in Figure 4.13.

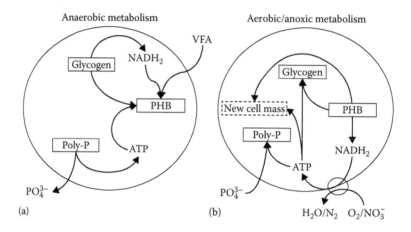

FIGURE 4.11 Proposed metabolism of PAOs under anaerobic and aerobic conditions: (a) poly-P and (b) PHA. (From Smolders, G.J.F. et al., *Biotechnol Bioeng* 43(6):461–470, 1994.)

FIGURE 4.12 Molecular structure of (a) polyphosphate and (b) PHA.

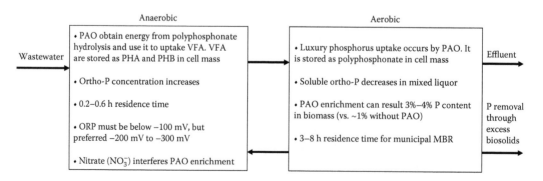

FIGURE 4.13 Summary of biological phosphorus removal mechanism.

4.5.2 Chemical Phosphorus Removal

4.5.2.1 Overview

Phosphorus discharge limit is becoming tighter because it is an essential and limiting nutrient that triggers eutrophication along with nitrogen. However, in a typical aquatic environment, phosphorus is more limiting than nitrogen because the sole source of phosphorus is an external supply through runoff water, wastewater, direct chemical spills, etc., whereas nitrogen can be supplied from the gaseous nitrogen by nitrogen fixing organisms.

Although many different forms of phosphorus species exist in wastewater, orthophosphate $\left(PO_4^{3-}\right)$ is the most common form in most wastewaters. The orthophosphate forms highly insoluble precipitates when it is reacted with inorganic coagulants such as iron, aluminum, calcium, etc. Polyphosphates are the oligomers of orthophosphate and somewhat react with coagulants because of the opposite charges they have. Organic phosphorus mainly exists as phospholipids that form lipid bilayers on the cell membrane, but they do not react with coagulants directly. However, once wastewater undergoes biological degradation process, a vast majority of nonorthophosphates convert to orthophosphates except those bound with refractory organics such as pesticides, herbicides, etc.

As phosphorus discharge limits are becoming tighter, chemical phosphorus removal is practiced more frequently along with the BNR. Unlike its simple appearance, however, chemical phosphorus removal is a complex thermodynamic and kinetic phenomenon of which mechanisms have never been clearly understood. Many mathematical models have been developed to explain the equilibrium of the chemical reactions among metal ions, phosphorus, hydroxyl ions, etc., as functions of temperature, pH, ionic strength, etc. Unfortunately, none of them appear to fit with the observations in the field due to the following reasons:

- Unidentified ionic species and microorganisms in water make the equilibrium far more complicated than those observed in laboratory conditions.
- The kinetic competition between metal hydroxide formation and metal phosphate formation makes all the predictions based on chemical equilibrium (or thermodynamics) invalid. If fact, as explained in Section 4.5.2.3, the inorganic coagulant forms metal hydroxides immediately after contacting water. Then the hydroxyl groups surrounding the metal ions and the complex structure of the inorganic floc become a kinetic barrier for phosphate reacting with the metal ions after diffusing through the metal hydroxide floc.
- Predicting the kinetics of phosphate ion exchange reaction by the metal hydroxides is nearly impossible because many factors affect the kinetics, for example, mixing intensity, inhibition effects from unidentifiable organic and inorganic compounds, pH, temperature, and floc structure.

As a consequence of the inability to predict the kinetics of chemical phosphorus removal, the dosage and the use of inorganic coagulants rely on the empirical data obtained in other plants with a similar hydraulic condition.

4.5.2.2 Chemicals Used

Various aluminum- and iron-based inorganic coagulants can be used to remove phosphorus in biological wastewater treatment as summarized in Table 4.7. Alum and ferric chlorides are the most commonly used coagulants in the MBR process because they can reduce the phosphorus concentration in effluent down to 0.005 to 0.04 mg/L (Takács et al. 2006). Besides the phosphorus removal, aluminum- and iron-based coagulants can improve membrane performance by not only reducing the soluble microbial product (SMP) concentration but also forming larger floc that has less effect on membrane fouling (Lee et al. 2001b). Both aluminum- and iron-based coagulants form very insoluble compounds with orthophosphate at near neutral pH or at slightly acidic pH. Typically, the specific prices of iron-based coagulants are less than that of aluminum-based ones.

TABLE 4.7

Chemicals Used to Remove Phosphorus in Biological Wastewater Treatment

Chemical/Trade Name	Chemical Equation in Solids	Molecular Weight (g/mol)	Equivalent Ratio, g Metal/g P	Precipitates	Appearance as Aqueous Solution
Aluminum sulfate (alum)	$Al_2(SO_4)_3 \cdot 14H_2O$	594.37	0.871	$Al_x(PO_4)_y(OH)_{3x-3y}$	Transparent
Aluminum chloride	$AlCl_3 \cdot 6H_2O$	241.43	0.871		Bright yellow
Sodium aluminate	$NaAlO_2$	81.97	0.871		Transparent
PACl	$Al_2Cl(OH)_5$	174.45	0.871		Transparent
Ferric sulfate	$Fe_2(SO_4)_3 \cdot 5H_2O$	489.96	1.803	$Fe_x(PO_4)_y(OH)_{3x-3y}$	Reddish brown
Ferric chloride	$FeCl_3 \cdot 6H_2O$	270.30	1.803		Reddish brown
Pickle liquor	Fe^{2+} and Fe^{3+}	–	–		Reddish brown
Calcium hydroxide (lime)	$Ca(OH)_2$	74.09	1.941/2.157	$Ca_3(PO_4)_2/Ca_5(OH)(PO_4)_3$	Sold as powder

Most of iron- and aluminum-based coagulants are sold at strong acidic forms to maintain the high solubility of metal ions. If those are sold as solids, they should be dissolved in water before being fed. When metal ions in acidic aqueous coagulant are added to the mixed liquor with a near-neutral pH, they turn to metal hydroxides immediately. The high acidity of coagulant drops the mixed liquor pH to some extent. The nitrification process can be hampered if the pH drop is excessive. Alternatively, sodium aluminate or polyaluminum chloride (PACl) can be used instead of acidic aluminum salts, but they may not be as effective as acidic aluminum salts because aluminum ions are preoccupied by hydroxyl ligands, if solids retention time in the biological system is low, for example, below a few days. Preoccupied hydroxyl ions act as a kinetic barrier for phosphate ions and interfere with the formation of aluminum phosphate, as explained in Section 4.5.2.3. Lime is not commonly used in MBR mainly because it forms insoluble metal phosphate only at a pH of 9 or higher. However, lime can be added to supply alkalinity if the acidic coagulants decrease the pH excessively.

4.5.2.3 Mechanism

Aluminum- and iron-based coagulants are strongly acidic, except sodium aluminate and PACl. Metal ions exist mainly as bare metal ions without ligands in the concentrated stock solution because of the extremely low hydroxyl ion (OH^-) concentration at the strong acidic pH. However, as soon as coagulants are introduced to mixed liquor, metal ions rapidly form insoluble amorphous metal hydroxides, which is the mixture of metal ions with varying numbers of hydroxyl ligands as illustrated in Figure 4.14. The formation of amorphous metal hydroxides occurs instantaneously without high kinetic barriers as soon as metal ions contact with water because hydroxide ions are produced very rapidly from the abundant water molecules surrounding the metal ions.

Fresh metal hydroxides form insoluble flocs with open pore structure that can be easily accessed by phosphate ions. Moreover, fresh hydroxide flocs have reactive surface properties that can exchange hydroxyl group with phosphate ions relatively quickly to form metal–phosphate complexes such as $\equiv MeH_2PO_4$, $\equiv MeHPO_4^-$, and $\equiv MePO_4^{2-}$, where $\equiv Me$ indicates the metal ions on the surface of the amorphous metal hydroxide. Overall, forming strong metal–phosphate bonding is thermodynamically favored instead of weak metal–hydroxide bonding. However, the easily accessible free metal ions deplete as the metal hydroxide flocs are aged. Thus, phosphate ions must travel deeper into the floc to be adsorbed and the phosphate ion exchange reaction slows down.

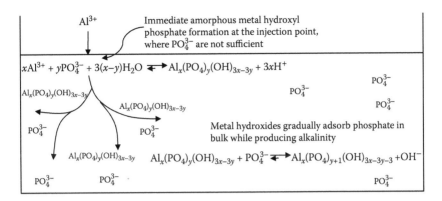

FIGURE 4.14 Conceptual diagram of soluble phosphate removal process by ferric ions in bubble column. The chemical formulae in the diagram do not exactly represent the composition of precipitate.

When ferric ions are added to mixed liquor, hydrous ferric oxides (HFO) that consist of amorphous ferric hydroxide (am-$Fe(OH)_3$(s)), ferric hydrite ($Fe_pO_r(OH)_s \cdot nH_2O$), goethite ($\alpha$-FeOOH), lepidocrocite ($\gamma$-FeOOH), and hematite ($\alpha$-$Fe_2O_3$) are formed (Neethling 2008). The initially reactive ferric hydroxides undergo aging processes and gradually lose their reactivity as hydroxyl groups are exchanged with phosphate ions starting from the easily accessible areas. In fact, a large difference was observed regarding phosphate absorption rate between 3- and 20-min-old metal hydroxides (Smith et al. 2008).

As discussed previously, the formation of metal–phosphate complex is a very slow reaction that may take days or weeks to reach the thermodynamic equilibrium because of the kinetic competition with metal hydroxide formation. Therefore, allowing sufficient contact time with enough mixing energy is crucial to fully use the added metal ions and to obtain a low Me/P ratio. In MBR, the added metal hydroxides to the mixed liquor stay in the system for an extended period because of the long SRT; hence, the contact time is naturally sufficient. Meanwhile, in tertiary phosphorus removal process, metal hydroxide slurry is recirculated from the clarifier to the reaction tank to use the maximum phosphate absorption capacity by increasing the contact time.

4.5.2.4 Factors Affecting Removal Efficiency

- *Mixing intensity*—Vigorous mixing near the injection point may increase the chance of metal ions contacting with phosphate ions before they form metal hydroxides in theory. Moreover, the freshly formed metal hydroxides with reactive surface can contact with more phosphate ions before their activity decreases. As shown in Figure 4.15, the initially identical soluble phosphorus concentration (1 mg/L) decreases faster when the mixing intensity estimated by the G value is high. The effect of the G value tapers off as the G value increases. The phosphorus removal rate only marginally increases when the G value increases from 182 to 425 s^{-1}. According to Mueller et al. (2002), the typical aeration basin of CAS with a depth of 4.6 m has a G value of 80 to 125 s^{-1}. Because MBR requires more intensive aeration because of its greater OUR at lower oxygen transfer efficiency than CAS as discussed in Section 6.3.1, the G value is likely greater than 80 to 125 s^{-1}. Thus, the reaction rate in a typical MBR might be best represented by the curve for 182 s^{-1}, which is close to the curve for the highest G value of 425 s^{-1}. This suggests that the additional mixing energy for the MBR aeration tank would not result in faster phosphorus removal rates by inorganic coagulants.
- *Contact time*—Although metal hydroxide formation is kinetically favored, metal phosphates are favored thermodynamically. Therefore, the hydroxyl ligands surrounding the metal ions in metal hydroxides can be exchanged with phosphate if enough contact time is

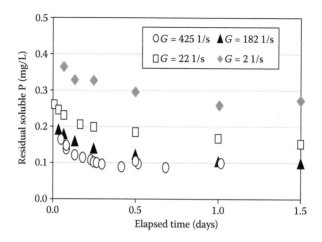

FIGURE 4.15 Phosphorus removal kinetics by ferric chloride at different G values. Initial phosphorus concentration = 1 mg P/L. Initial Fe/P = 3.0 mol/mol. (From Szabó et al. *Water Environ. Res.* 80(5), 407–416, 2008.)

given. As can be seen in Figure 4.15, initially formed metal hydroxides continue to absorb soluble phosphate ions for more than 24 h under the most vigorous mixing condition. The long SRT of MBR is a significant advantage to achieve a low Me/P ratio, allowing extended contact time.

In MBR, contact time effect may overwhelm the mixing intensity effect because strong mixings are performed in every corner of the system by biological and membrane scouring aerations. Injecting coagulant to the location without vigorous mixing may slow down the initial metal phosphate formation, but if contact time is long enough, the metal hydroxides can eventually absorb phosphate ions through ligand exchange reaction. Therefore, with a sufficiently long contact time, the mixing intensity at the chemical injection point is not likely a dominant factor affecting the final phosphorus removal efficiency.

To estimate the coagulant dosage to achieve the target phosphorus concentration in the effluent, jar tests are often performed. However, the results obtained from jar tests are heavily influenced by the kinetics of metal hydroxides formation. A great portion of the metal ions are used to produce metal hydroxides that cannot be converted to metal phosphate in the short duration of the test. As a consequence, the estimated coagulant dosage requirements are prone to be overestimated. In jar tests performed in the laboratory, a high Me/P ratio such as 5 to 10 was required to obtain a phosphorus concentration of 0.1 to 0.2 mg/L (Song et al. 2008), but it has been observed at only 1.5 to 2.0 in the full-scale system, as summarized in Table 4.8. Therefore, when jar tests are performed to estimate coagulant dosage, mixing time effect must be carefully reviewed.

- *Coagulant dosage*—As coagulant dosage increases, soluble phosphorus is removed quicker from the liquid phase because the abundant metal hydroxides provide greater chances of collision with phosphate ions. However, as the phosphate ion becomes scarce, more excess metal hydroxides are required to capture the trace residual phosphate ions. Hence, the soluble phosphate ion concentration does not drop to zero even at high coagulant dosages (Smith et al. 2008).
- *Alkalinity*—Alkalinity plays an important role in nitrification process by stabilizing the pH against the acid produced from the reaction. Conceptually, however, the high alkalinity may hamper chemical phosphorus removal by providing more hydroxide ions that compete with orthophosphates for the same metal ions. At high alkalinity, the pH level near the

TABLE 4.8
Phosphorus Removal Efficiency of Various Municipal MBR

Anaerobic Tank	TP in Effluent (mg/L)	Removal Rate (%)	Metal/P (mole/mole)	Remark
No	0.23	98	0.53	Brepols (2011)
No	0.29	97	0.58	
Yes	0.50	?	1.0 (alum)	Pilot MBR, Hirakata, Japan
	0.10	?	1.0–2.0 (alum)	(Trivedi 2004)
	0.05	?	2.6 (alum)	
Yes	<0.1	>98	1.0–2.5 (alum)	Broad Run, VA, USA (Daigger et al. 2009)
Yes	0.38	94	1.2 (FeCl₃)	Traverse City, MI, USA (Daigger et al. 2009)

dosing point would not decrease much when coagulants are dosed, which expedites the formation of metal hydroxides instead of metal phosphates. There are no sufficient data available with respect to this hypothesis, but the extended contact time provided by the long SRT may compensate any negative effect of the alkalinity in practical MBR conditions.

- *Target phosphorus level*—As target phosphorus level decreases, more reaction time and/or greater coagulant dosages are required because it is increasingly harder for metal ions to randomly collide with phosphate ions. Alternatively, the slow kinetics can be covered by high coagulant dosages, but it causes a high Me/P ratio. The required Me/P ratio increases as the target soluble phosphorus concentration decreases, especially at below 0.1 mg/L.

- *Injection point*—In the BNR process, soluble phosphorus concentrations vary among the biological tanks depending on DO and nitrate levels in the tank. The phosphorus concentration is the highest in the anaerobic tank, where PAOs release orthophosphate, and is the lowest in the aeration/membrane tank, where phosphorus uptake by PAOs takes place. Depending on experimental condition, different observations can be made with respect to the injection point effect.

Conceptually, adding coagulant to anaerobic tanks may enhance the initial metal phosphate formation and leads to an efficient coagulant use because the soluble phosphate concentration is the highest in the tank. According to a pilot test, alum consumption increased four times when dosing point was changed from anaerobic tanks to aerobic tanks while obtaining 0.1 mg/L total phosphorus (TP) in permeate (Trivedi 2004). In other pilot tests, ferrous sulfates were added to the membrane tank to remove phosphorus, but some ferrous ions were discharged before they were oxidized to ferric ions because excess sludge was taken from the membrane tank. By moving the injection point to the anoxic tank, ferrous ions can be better used because they convert to more useful ferric ions. However, no significant changes were observed in terms of effluent quality regardless of the dosing point (Wang et al. 2014b). Overall, an effluent phosphorus level was observed at less than 0.05 mg/L, less than 0.15 mg/L, and less than 0.30 mg/L for 29%, 77%, and 95% of the time, respectively, at a Me/P ratio of 2.6 to 3.0. In other laboratory tests (Johannessen et al. 2006), alum addition to anaerobic tank resulted in lower phosphorous concentration in the effluent, whereas significant quantity of PAO and glycogen accumulating organisms (GAO) were observed.

Meanwhile, many conflicting observations have been reported on the potentially negative effect of adding coagulants to the anaerobic tank. Conceptually, adding a coagulant to anaerobic tanks may suppress PAO growth by not leaving a sufficient amount of free phosphate ions in the subsequent aeration tank, thereby the biological phosphorus removal system can be hampered.

The coagulant dosages should be raised to compensate the lost phosphate absorption capacity of PAOs. Therefore, it is often claimed that coagulants are better to be added to in the downstream of either aeration tanks or membrane tanks (Crawford et al. 2006; Johnson and Daigger 2009; Daigger et al. 2010).

Despite all conflicting observations and theories, the injection point may not be critical in MBR. Because coagulants do not form metal phosphate precipitate immediately after they were added to water, adding a coagulant to any tank does not necessarily mean the depletion of phosphate ions in the tank. In fact, the initially formed metal hydroxides gradually absorb phosphate ions and are circulated among the tanks. In terms of time scale, phosphate exchange (or absorption) reaction by metal hydroxides takes place slowly in a few days to a few weeks, but the complete mixing of the mixed liquors in different tanks takes place in a few hours or no more than a day. Therefore, the added coagulants spread to the entire system before they lose their capability of adsorbing phosphate.

- *Number of injection points*—Conceptually, adding a coagulant through multiple injection points may help expedite capturing phosphate ions from a wider range of reactor spaces before amorphous metal hydroxides are aged. However, just like the injection point effect above, no rigorous empirical data sets supporting this hypothesis exist. The slow phosphate ion exchange reaction may overwhelm this effect in the field condition.
- *Preneutralized coagulants*—Conceptually, partially or fully preneutralized coagulants such as sodium aluminate and PACl would not result in better phosphorus removal than alum and aluminum chlorides. Phosphate ions must replace the existing hydroxyl ligands to be absorbed to the metal hydroxide, which acts as a barrier for the reaction. However, it is also not evident that such preneutralized coagulants are inferior to the other inorganic coagulants because the reaction kinetics is overwhelmed by the contact time effect under the long SRT conditions.

4.5.2.5 Effect of the Me/P Ratio on Removal Efficiency

The Me/P ratio required to remove a certain amount of phosphorus varies widely depending on the experimental condition. When a mixed liquor taken from a municipal plant was treated with alum on a jar tester followed by settling and supernatant filtering with 0.45 micron syringe filters, the Me/P ratio required to achieve 3 mg/L phosphorus in the supernatant was 3.12 (Song et al. 2008). By contrast, in a full-scale MBR, much lower phosphorus concentrations than 3 mg/L are obtained at a much lower Me/P ratio at 0.5 to 2.6 as summarized in Table 4.8. This large discrepancy is caused by the vastly different contact times in the laboratory-scale setup and in the full-scale system. In fact, the mixing (or contact) time in the jar test above was only 5 min whereas the contact time (or SRT) in a full-scale MBR is at least 10 days. According to the table, the lowest phosphorus concentration obtained with an inorganic coagulant in a full-scale MBR is reportedly around 0.05 mg/L at a Me/P ratio of 2.6, but there is a potential to lower it below 0.05 mg/L by adding more excess coagulants (Yoon et al. 2004c).

The mixing intensity at the point of coagulant addition is a conceptually important factor. With intense mixing at the point of coagulant addition, more metal ions can bind with orthophosphate before they form metal hydroxides because the chance of collision between metal ions with orthophosphate ion increases. Although many references emphasize the intense mixing at the point of coagulant addition to achieve high chemical use efficiencies, no direct empirical evidence obtained from realistic MBR conditions is available. Perhaps the notion was originated from tertiary treatment, where the intense mixing at the point of coagulant feeding is crucial for high metal utilization efficiencies due to the relatively short residence time of the metal hydroxides in the process. Likewise, a multipoint injection is conceptually plausible, but the efficacy is not proven in MBR.

4.5.2.6   Toxicity of Inorganic Coagulant

High levels of iron and aluminum ions have been suspected toxic to nitrifiers in biological waste-water treatment, although there is no definitive evidence. The following are the claims found in literature, and all of them are based on laboratory experiments:

- The ferrous ion (Fe^{2+}) concentration higher than 20 mg/L appeared toxic to *Nitrosomonas*, and as a result, the nitrification rate slowed down by 20% (Seyfried 1988).
- When residual ferric ion in the primary effluent was 1.68 mg/L, the nitrification rate in the aeration tank decreased 20% to 34% (Lees et al. 2001).
- With 100 mg/L of PACl, the nitrification rate decreased by 16% while the denitrification rate decreased by 43% (Iversen et al. 2009).
- Ferric ion (Fe^{3+}) was found more detrimental than ferrous ion to the microbial community in the activated sludge process. The pH drops upon the formation of iron hydroxides, the impairment of the floc structure, and the formation of nitrogen oxides and could partially explain the toxicity of the iron. However, the reduced free phosphorus concentration due to the insoluble ferric phosphate formation did not appear a significant factor affecting the microbial activity (Philips et al. 2003).

Despite the above mentioned observations in the laboratory, the theoretical bases of the toxic effect are not well understood. In theory, adding more Al- and Fe-based coagulant only increases the insoluble metal hydroxide contents in the mixed liquor because the solubility of Fe and Al ions is fixed at trace levels at a neutral pH. Therefore, it is not certain how a coagulant dose affects the microbial metabolism without having elevated Al and Fe ion concentrations. Meanwhile, ferric and aluminum salts rarely affect microorganisms negatively in a full-scale MBR. The claims based on long-term full-scale tests are as follows:

- When alum was used at a Me/P ratio of 1 to 2.5 in a 1-year MBR pilot test in Broad Run WWTP in Asheville, Virginia, USA, NH_4-N in the effluent was controlled at 0.03 mg/L on average and total phosphorus was less than 0.05 mg/L (Daigger et al. 2010).
- In the long-term application of $FeCl_3$ in a full-scale plant in Traverse City, Michigan, USA, no inhibition in nitrification has been reported, and 30 to 150 mg/L of $FeCl_3$ was added based on influent flow (Crawford et al. 2006; Daigger et al. 2010).

It is apparent that the negative effect on the microbial community and the performance of the activated sludge has been mostly observed in short-term laboratory tests. In fact, laboratory tests are performed in short term in general, and microorganisms do not have enough time to adapt with the coagulants. Moreover, because of the small reactor volume, strongly acidic coagulants can cause more severe pH fluctuation compared with the full-scale system. On the contrary, in long-term full-scale systems, microorganisms can adapt to the coagulant during the long-term operation while the pH is maintained stably because of the large reactor volume. It is also possible that the species with high tolerances against metal salts replace the other species with low tolerances, if there is any negative impact of coagulant on microorganisms.

In the meantime, the short-term toxicity of aluminum on *M. parvicella* has been well known in CAS (Roels et al. 2002; Nielsen et al. 2005). If foaming or sludge bulking occurs in CAS by *M. parvicella*, PACl, $AlCl_3$, or alum can be added directly to the aeration tank, RAS line, or between the aeration tank and the clarifier at a dosage of 1.5 to 4.5 mg Al/g MLSS/day. However, there is insufficient evidence that aluminum salts are toxic to other filamentous organisms such as *N. limicola* and *Nocardia* spp. The mechanism of the toxic effect is yet to be discovered, but it has been postulated that the coagulation of filamentous microorganisms reduced the accessibility to food while physiology, particularly the lipase production, was partly inhibited. As a result, the uptake of long chain fatty acids can be interfered (Nielsen et al. 2005). Nonetheless, aluminum salts eventually lose their efficacy on the inhibition of filamentous organisms due to the microbial adaptation.

4.6 BNR PROCESS

4.6.1 Constraints in Process Design

The implementation of BNR processes in MBR relies on the same principle used in the CAS. However, there are some unique factors that must be taken into consideration (Crawford et al. 2006; Daigger et al. 2010):

- *Excess DO returns to the anoxic tank*—The mixed liquor in the membrane tank typically contains DO at high concentrations, such as 4 to 8 mg/L (Figure 4.16a). If this mixed liquor is recycled to the anoxic tank directly at a rate of $4Q$ (400% of influent flow rate), effectively 16 to 32 mg/L of oxygen is carried to the anoxic tank based on fresh wastewater volume, which will consume the equivalent amount of readily biodegradable COD in the anoxic tank. Considering the fact that typical municipal wastewater contains only 50 to 100 mg/L of readily biodegradable COD, losing 16 to 32 mg/L to the oxygen transported from the membrane tank is a significant drawback that can hamper the denitrification process that needs a large amount of readily biodegradable COD.
- *Limited flexibility in recycle rate*—In the modified Ludzack–Ettinger (MLE)–type MBR shown in Figure 4.16a, the mixed liquor is concentrated in the aeration/membrane tank due to the loss of permeate. To avoid the accumulation of solids in the aeration/membrane tank, the mixed liquor should be recycled at a certain rate. However, the mixed liquor circulate rate required for solids balance is not necessarily the optimum for the nutrient removal. This constraint can be mostly removed by using the separated membrane tank as shown in Figure 4.16b, where the mixed liquor in membrane tank is recycled to aeration tank and the mixed liquor in aeration tank is recycled to anoxic tank.

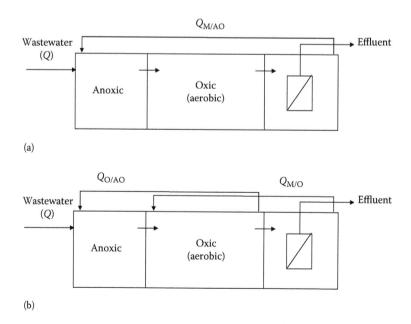

FIGURE 4.16 MLE process: (a) direct recycle to anoxic tank and (b) two-step recycles to anoxic tank.

- *Relatively low MLSS in the anaerobic/anoxic tank*—To mitigate the excessive oxygen carry over to the anoxic tank, a two-step recycle is preferred in MBR, as will be discussed in Section 4.6.2. The mixed liquor with the highest MLSS in the membrane tank is recycled to the aeration tank first, and then the mixed liquor in the aeration tank is again recycled to the anoxic tank. As a result, substantial MLSS gradient is developed among anoxic, aerobic, and membrane tanks with the lowest MLSS in the anoxic tank. The low MLSS in the anoxic tank should be considered when tank sizes are determined.

- *The interference of a coagulant on biological phosphorus removal*—Inorganic coagulants rarely hamper nitrogen removal in a long-term full-scale MBR operation, but biological phosphorus removal can be hampered in the BNR process. The coagulant dose to the biological system reduces the available free phosphorus concentration for PAOs and diminishes their growth rate. If the biological phosphorus removal is hampered, the coagulant required to achieve the target phosphorus concentration will increase. Therefore, if inorganic coagulants are required to meet the target effluent phosphorus, biological interference must be factored in when coagulant dosage is estimated.

- *Mixed liquor short circuit in small reactor*—The biological tanks in MBR are smaller than those in CAS because of the high MLSS. In general, smaller tanks are more prone to short circuit because the time allowed for the liquid to be mixed is shorter. In addition, the effective HRT in the aeration tank of MBR can be much shorter than the apparent HRT due to the high internal recirculation rate through the tanks. For example, effective HRTs are only 1/3 to 1/8 of an apparent HRT in the modified University of Cape Town (UCT) process, as summarized in Table 4.10. Therefore, securing a good mixing is more important in MBR than that in CAS. The sufficient mixing at the point, where the mixed liquor is recycled to, can mitigate short circuit problems.

4.6.2 MLE PROCESS

MLE is the most commonly used BNR process in MBR primarily targeting BOD and nitrogen removal. In the original MLE–MBR process, the mixed liquor is recycled from the membrane tank to the anoxic tank, as shown in Figure 4.16a. However, the excess DO transferred from the membrane tank (4–8 mg/L) to the anoxic tank often hampers the denitrification process by depleting the readily biodegradable COD contained in fresh wastewater. In addition, the flexibility of the process is limited because the mixed liquor recycle rate from the membrane tank to the anoxic tank ($Q_{M/AO}$) has to be set to prevent excessive MLSS accumulation in the aeration/membrane tank rather than to obtain the maximum nitrogen removal efficiency.

To reduce the effect of recycled oxygen, the original configuration was modified, as shown in Figure 4.16b. In this configuration, the mixed liquor in the membrane tank is recycled to the aeration tank, and in turn it is recycled to the anoxic tank. Because the mixed liquor with 1 to 2 mg/L DO is recycled to the anoxic tank instead of 4 to 8 mg/L DO, maintaining low ORP in the anoxic tank is readily possible without consuming too much of readily biodegradable COD. One potential drawback of this two-step recycle is that maintaining high MLSS in the anoxic tank is harder than the traditional arrangement (Figure 4.16a) because the mixed liquor with lower MLSS is transported to the anoxic tank. In the modified MLE with a two-step recycle, $Q_{O/AO}$ can be controlled independently from $Q_{M/O}$ depending on the process goal. The benefit and the drawback of traditional and modified MLE–MBR process are summarized in Table 4.9.

The approximate $Q_{O/AO}$ required to obtain the target nitrogen removal efficiency can be calculated as Equation 4.15. For instance, $Q_{O/AO}$ needs to be approximately $2Q$ if the target nitrogen removal efficiency R is 0.67 according to the following equation:

$$Q_{O/AO} = \frac{R}{1-R}Q \tag{4.15}$$

TABLE 4.9

Comparison of the Advantages of Original and Modified MLE–MBR, Where the Advantage of One System Is the Disadvantage of Another System

Original MLE–MBR	Modified MLE–MBR
• Simpler flow configuration	• Less DO carry over to the anoxic tank
• Less energy for circulation	• Better for the wastewater with low readily biodegradable COD
• Less MLSS dilution in the anoxic tank	• Denitrification can be optimized without being affected by the requirement of mixed liquor recycle to keep MLSS low in the membrane tank
	• Potentially higher denitrification efficiency

The MLE process is not primarily for phosphorus removal, but extra phosphorus removal can occur beyond the level achievable without the anoxic tank depending on the ORP and SRT in the anoxic tank. Inorganic coagulants can be added to the aeration or membrane tank if further phosphorus removal is required.

4.6.3 HYUNDAI ADVANCED NUTRIENT TREATMENT PROCESS

The Hyundai Advanced Nutrient Treatment (HANT) process was developed in the 1990s, targeting both nitrogen and phosphorus removal. It is the simplest BNR process with only one internal recycle line (Yoon et al. 2004c). As shown in Figure 4.17, a deaeration tank is placed after the aeration tank to reduce the DO before it is recycled to the anoxic tank. Anaerobic tanks are placed after the anoxic tank, which is opposite from the MLE process. Because the denitrified mixed liquor proceeds to the anaerobic tank, ORP in the anaerobic tank can be readily kept low enough to encourage PAOs to release phosphate. The other advantage is that the MLSS in the anoxic tank is relatively high because the mixed liquor recycled to this tank is originated from the membrane tank without being diluted. The mixed liquor recycle rate (Q_{RAS}) ranges from $2Q$ to $4Q$. TN and TP in the effluent are known to be 5 to 10 mg/L and 1 to 2 mg/L in municipal wastewater treatment without inorganic coagulant addition.

One drawback of this process is that the recycle rate cannot be independently controlled to achieve the optimum nitrogen removal because it is tied up with the recycle rate required to prevent excessive MLSS accumulation in the membrane tank. For instance, when lower ORP is desired in the anoxic and anaerobic tanks, sludge recycle rate should be reduced to limit the oxygen and nitrate supply to those reactors, but it is not always possible because of the accumulation of solids

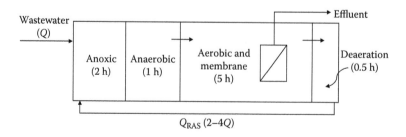

FIGURE 4.17 Process diagram of the HANT process.

in the membrane tank. Therefore, the operational flexibility to counter the varying influent quality is somewhat limited in this process.

4.6.4 MODIFIED UCT PROCESS

The modified UCT process is virtually the same as the Virginia Initiative Plant (VIP) process, but typically UCT processes are designed for longer SRT, 13 to 25 days versus 5 to 10 days for VIP. However, when it comes to MBR, no differences between the two processes exist. This process is also called A^2/O process in MBR because it includes anaerobic, anoxic, and oxic (or aerobic) reactors (Banu et al. 2009). The major benefit of the modified UCT process is that it does not directly return the mixed liquor with nitrates to the anaerobic tank; thereby, the ORP remains low enough to maximize the phosphate release from PAOs in the anaerobic tank.

In the modified UCT process, similar to the MLE process, the mixed liquor in the rear end is recycled to the front end using the three-step mixed liquor recycle, as shown in Figure 4.18, where the mixed liquor in the membrane tank is recycled to the aeration tank, the mixed liquor in the aeration tank is recycled to the anoxic tank, and the mixed liquor in the anoxic tank is recycled to the anaerobic tank. The internal recycles $Q_{M/O}$, $Q_{O/AO}$, and $Q_{AO/AA}$ typically range from $2.5Q$ to $4Q$, from $1Q$ to $2Q$, and from $1Q$ to $2Q$, respectively, depending on the wastewater quality and the goals in nutrient removal. The cascade-type multistage mixed liquor recycle causes diluted MLSS in the anaerobic and anoxic tanks. Hence, the anaerobic and anoxic tank sizes need to be increased to compensate the MLSS dilution effect.

Because of the high recycle rates, effective HRT in the tanks are much lower than the apparent HRT calculated based on the reactor volume and the wastewater flow rate. If median recycle rates are taken ($Q_{M/O} = 3Q$, $Q_{O/AO} = 3Q$, and $Q_{AO/AA} = 1.5Q$), the actual HRTs in anaerobic, anoxic, and aerobic tanks are only 40%, 18%, and 14% of their apparent HRT. For instance, if the apparent HRT in the anoxic tank is 2 h, the effective HRT can only be 0.36 h due to the additional flow caused by the mixed liquor recycle. Under this short HRT, flow short circuit can play a significant role, especially when the mixing power is not sufficient or the reactor configuration is improper. Therefore, extra care must be taken to allow extra mixing, especially in the spot to which the mixed liquor is introduced. Table 4.10 summarizes the recycle ratio of each stream, apparent HRT, and effective HRT.

To size a tank properly, the relative MLSS in each tank should be calculated based on the mass balance around the tank. The four assumptions used in this calculation are as follows:

- MLSS produced by fresh feed water or MLSS decayed in each tank is negligible compared with the preexisting MLSS in the tank.
- MLSS in each tank is in steady state.
- MLSS in raw wastewater is negligible compared with the MLSS in the system.
- Excess biosolid removal from the membrane tank is negligible compared with $Q_{M/O}$.

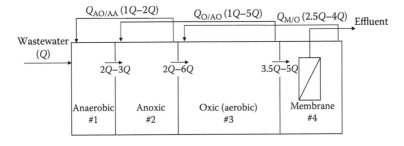

FIGURE 4.18 Process diagram of the UCT process applied for MBR.

TABLE 4.10

Mixed Liquor Recycle Rate and Effective HRT in Each Tank of UCT

Flow Destination →		Anaerobic	Anoxic	Aerobic	Membrane
Flow origin	Feed	$1Q$			
	Anaerobic		$2Q–3Q$ $(3Q)$		
	Anoxic	$1Q–2Q$ $(1.5Q)$		$2Q–6Q$ $(4Q)$	
	Aerobic		$1Q–5Q$ $(3Q)$		$3.5Q–5Q$ $(4Q)$
	Membrane			$2.5Q–4Q$ $(3Q)$	
Apparent HRT (h)		0.5–1.0	1–3	3–5	1–2
Effective HRT based on average flow (h)		0.2–0.4	0.17–0.50	0.43–0.71	0.25–0.5

In the following equations, Q (m³/h) indicates the raw wastewater flow rate and X (g/L) indicates MLSS:

$$\text{Input} = \text{Output}$$

$$\text{Aeration tank } Q_{AO/AA}X_{AO} = (Q_{AO/AA} + Q)X_{AA} \tag{4.16}$$

$$\text{Anoxic tank } Q_{O/AO}X_O + (Q_{AO/AA} + Q)X_{AA} = Q_{AO/AA}X_{AO} + (Q_{O/AO} + Q)X_{AO} \tag{4.17}$$

$$\text{Membrane tank } Q_{M/O}X_M = (Q_{M/O} + Q)X_O \tag{4.18}$$

The solutions of the above equations are as follows:

$$X_{AA} = \frac{Q_{AO/AA}}{Q_{AO/AA} + Q}\frac{Q_{O/AO}}{Q_{O/AO} + Q}\frac{Q_{M/O}}{Q_{M/O} + Q}X_M \tag{4.19}$$

$$X_{AO} = \frac{Q_{O/AO}}{Q_{O/AO} + Q}\frac{Q_{M/O}}{Q_{M/O} + Q}X_M \tag{4.20}$$

$$X_O = \frac{Q_{M/O}}{Q_{M/O} + Q}X_M \tag{4.21}$$

For example, if $Q_{AO/AA}$, $Q_{O/AO}$, and $Q_{M/O}$ are assumed at $2Q$, $2Q$, and $4Q$, respectively, the relative MLSS among the four tanks are $X_{AA}{:}X_{AO}{:}X_O{:}X_M = 0.36{:}0.53{:}0.80{:}1.00$. It is apparent that there are significant MLSS imbalances among the tanks. The low MLSS in the anaerobic and anoxic tanks ultimately affects the nutrient removal efficiency by affecting the exposure time of microorganisms in each different environment.

The effective SRT in each tank can be calculated using the following equation, where Q_X is the mixed liquor removal rate from the membrane tank and X_M is MLSS in the same tank:

$$\text{SRT}_i = \frac{V_i X_i}{Q_X X_M} \tag{4.22}$$

In a pilot test performed with a 20 m^3 reactor, nitrogen and phosphorus removal efficiencies were investigated using municipal wastewater with 34 to 62 mg/L TN and 4.52 to 7.30 mg/L TP. Complete nitrification was obtained in the aeration tank, and NO$_3$-N was less than 1.6 mg/L in the anoxic tank. Nitrogen removal efficiency was 68% to 75% when $Q_{O/AO}$ was set at 3Q, which agreed very well with the theoretical removal efficiency calculated by Equation 4.15. Phosphorus removal efficiency was 74% to 84% with $Q_{AO/AA}$ at 1Q (Banu et al. 2009).

The performance of the UCT process was compared with that of Sammamish BNR (SmBNR) process in laboratory scale (Johannessen et al. 2006). The SmBNR process is identical to the HANT process, but it does not have a deoxygenation tank and the internal recycle rate is set at very high (5Q). In this study, the modified UCT process had a combined aeration tank and membrane tank. The $Q_{M/AO}$ was set at 5Q, which was much higher than the typical (1Q–2Q), whereas the $Q_{AO/AA}$ was set at a typical level, 2Q. The major difference between the two systems is that fresh wastewater is fed to the anaerobic tank in the UCT process, whereas it is fed to the anoxic tank in the SmBNR process. Although COD removal efficiencies were nearly identical in the two processes, UCT was better in P removal, whereas SmBNR was better in nitrogen removal due to the opposite arrangement of the anaerobic and anoxic tanks in the two processes. The initial COD of 747 mg/L was treated to approximately 11 mg/L in both processes. The influent P (20 mg/L) was treated to 6.6 and 11.4 mg/L by UCT and SmBNR, respectively. The influent TN (57.9 mg/L) was treated to 16.4 and 10.4 mg/L by UCT and SmBNR, respectively.

4.6.5　Step Feed Process

It is important to maintain proper SRT in the anaerobic and anoxic tanks to obtain high nutrient removal efficiencies. As discussed in the previous section, one drawback of the modified UCT process is the low MLSS in the anaerobic and anoxic tanks, which makes it necessary to have large anaerobic and anoxic tanks.

A modified BNR with step feed has been devised as shown in Figure 4.19. This configuration allows higher MLSS in the anaerobic and anoxic tanks than the modified UCT. Approximately half of raw wastewater is fed to the first anaerobic tank whereas the other half is fed to the second

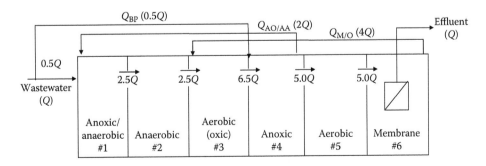

FIGURE 4.19　Process diagram of the BNR process with step feed.

anoxic tank after the aerobic tank. Because the mixed liquor in the membrane tank is recycled to the aerobic tank (no. 3), which is the upstream of the second anoxic tank (no. 4), no pumping is required from aerobic tank to the anoxic tank unlike in the modified UCT process. As a result, the overall pumping cost can be kept lower while obtaining higher nitrogen removal efficiency.

One significant advantage of this process is that the MLSS in the anaerobic and anoxic tanks are maintained higher than that in other competing processes. The relative MLSS in each tank as functions of sludge recycle rates can be calculated as follows:

$$\text{Input} = \text{Output}$$

$$\text{First anaerobic tank } Q_{AO/AA}X_4 = (Q_{AO/AA} + 0.5Q)X_1 \tag{4.23}$$

$$\text{Second anaerobic tank } X_2 = X_1 \tag{4.24}$$

$$\text{First aerobic tank } (Q_{AO/AA} + 0.5Q)X_2 + Q_{M/O}X_6 = (Q_{AO/AA} + 0.5Q + Q_{M/O})X_3 \tag{4.25}$$

$$\text{Anoxic tank } (Q_{AO/AA} + 0.5Q + Q_{M/O})X_3 = Q_{AO/AA}X_4 + (Q + Q_{M/O})X_4 \tag{4.26}$$

$$\text{Membrane tank } (Q + Q_{M/O})X_5 = Q_{M/O}X_6 \tag{4.27}$$

In the above-mentioned equations, Q is the flow rate (m^3/h) and X is the MLSS (g/L). By solving the above equations, the MLSS in each tank are obtained as follows:

$$\text{First anaerobic tank } X_1 = \frac{Q_{AO/AA}}{Q_{AO/AA} + 0.5Q} \frac{Q_{M/O}}{Q + Q_{M/O}} X_6 \tag{4.28}$$

$$\text{Second anaerobic tank } X_2 = X_1 \tag{4.29}$$

$$\text{First aerobic tank } X_3 = \frac{Q + Q_{AO/AA} + Q_{M/O}}{0.5Q + Q_{AO/AA} + Q_{M/O}} \frac{Q_{M/O}}{Q + Q_{M/O}} X_6 \tag{4.30}$$

$$\text{Anoxic tank } X_4 = \frac{Q_{M/O}}{Q + Q_{M/O}} X_6 \tag{4.31}$$

$$\text{Aerobic tank } X_5 = X_4 \tag{4.32}$$

For the recycle rates shown in Figure 4.19, the relative MLSS in the tanks are $X_1:X_2:X_3:X_4:X_5:X_6 = 0.64:0.64:0.86:0.80:0.80:1.00$. It is noticeable that the relative MLSS in the anaerobic and anoxic tanks are substantially higher than those in the modified UCT process. Therefore, same or better BNR efficiency can be achieved with smaller tanks than the modified UCT process.

4.6.6 Case Studies

4.6.6.1 Traverse City, Michigan, USA

The existing VIP process, which was virtually identical to UCT process except its shorter SRT, required a capacity increase by 40% without increasing footprint. Simultaneously, no increase of pollutant discharge to the nearby lake was desired. MBR was chosen for the upgrade project and was commissioned in the summer of 2004 (Crawford et al. 2006; Daigger et al. 2010).

TABLE 4.11
Monthly Average Discharge Limits

Effluent Quality	Discharge Limit Set by State	Voluntary Limit
BOD_5 (mg/L)	25	4
TSS (mg/L)	30	4
NH_4-N (mg/L)	11	1
TP (mg/L)	1	0.5

The plant was designed to treat maximum monthly loads of 9200 kg/day BOD at 32,000 m³/day. The annual average design flow rate was 27,000 m³/day, and the maximum hourly design flow was 64,000 m³/day. As summarized in Table 4.11, tighter discharge limits than what the state government imposed were voluntarily adopted.

A process diagram is shown in Figure 4.20, where four separate membrane tanks are added to the existing biological tanks. Mixed liquor is recirculated between the membrane tanks and the biological tanks using circulation pumps at up to 4 times of the influent flow rate (4Q). The biological tanks consist of anaerobic, anoxic, and oxic (or aerobic) tanks through which influent flows by gravity. The nitrate-containing mixed liquor taken from the aerobic tanks is pumped back to the front end of the anoxic tank at a flow rate of 1Q–2Q for denitrification. Simultaneously, the denitrified mixed liquor in the end of the anoxic tank is pumped back to the front end of the anaerobic tank to induce phosphorus release by PAOs.

The plant consists of two identical treatment trains sharing a 6 mm coarse screen, grit removal system, and primary clarifier with overall 830 m² surface area. The following conditions were added to each train:

- New 2 mm automatic in-channel traveling band screen rated at 38,000 m³/day (screenings are dewatered, compacted, and discharged to a dumpster for landfill disposal).
- Additional aeration capacity of 7500 m³/h by adding new blowers to the existing bioreactors based on the VIP process (6700 m³ each).
- Additional pumps to recycle 28,000 m³/day ($Q_{O/AO}$) from oxic (aerobic) tank to the anoxic tank and to recycle 28,000 m³/day ($Q_{AO/AA}$) from the anoxic tank to the anaerobic tank.

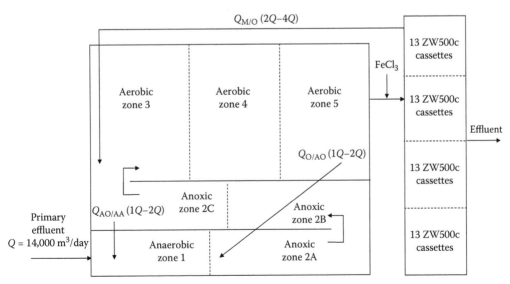

FIGURE 4.20 Process diagram of the MBR in Traverse City, Michigan, USA.

- A new membrane tank with four equally divided cells containing 13 ZW500c membrane cassettes (GE Water) each (the total membrane tank volume is 1700 m³; each membrane cell has a dedicated 10,000 m³/day permeate pump).
- Mixed liquor recycle pumps providing up to 400% recirculation rate from the membrane tank to the oxic tank based on the annual average flow of 13,500 m³/day in each train.
- One 2400 m³/day pump for waste-activated sludge, a membrane backpulse pumping system, a 40 L/min sodium hypochlorite pump, and a 65 L/min citric acid pump along with storage and feed systems for membrane cleaning.

The total membrane surface area in the eight cells dedicated to two trains is 45,760 m² (= 440 m²/cassette × 13 cassettes/cell × 4 cells/train × 2 trains), assuming each ZW500c cassette holds 20 ZW500c modules with 22 m² surface area each. Considering the design flow rates, the annual average net flux is calculated at 25 LMH, the maximum daily flux at 36 LMH, and the maximum hourly flux at 58 LMH, as summarized in Table 4.12.

The membrane system has an air-blowing capacity of 8240 m³/h for membrane scouring using five new blowers. Membrane scouring is performed at a cyclic mode at 10 s on and 10 s off. The specific aeration demand per membrane surface area (SAD_m) is calculated at 0.18 m³ air/m² membrane/h net (= 8240/45,760) or 0.36 m³ air/m² membrane/h gross. The specific aeration demand per permeate volume (SAD_p) is calculated at 7.2 m³ air/m³ permeate at the design flow rate.

The overall performance of the process was checked during the trial performed in November 2004. To enhance phosphorus removal, 38 mg/L of ferric chloride was added before the membrane tank. The actual MBR influent (or primary effluent) and MBR effluent qualities during the trial in November 2004 are also summarized in Table 4.13. All the discharge limits were within the limits set voluntarily, which were much tighter than the limits set by the government.

The process parameters during the trial are summarized in Table 4.14. It is a short-term snapshot of the process performance that can change a bit depending on how the process is adjusted to handle the given hydraulic and organic loads. However, the following observations were derived:

- Solids in the anaerobic and anoxic tanks are quite scarce relative to those in aeration (or oxic) and membrane tanks. SRT in anaerobic, anoxic, and oxic tanks (including membrane) are calculated at 0.18, 1.24, and 6.38 days, respectively. The low SRT in the anaerobic tank is caused by the low mixed liquor recirculation rates ($Q_{O/AO}$ and $Q_{AO/AA}$ are $1Q$) and can be corrected somewhat by increasing the recirculation rates from $1Q$ to $2Q$, but the low SRT in anaerobic and anoxic tanks caused by low MLSS is an intrinsic disadvantage of VIP (or UCT) process.

TABLE 4.12
Design Flow Criteria and Corresponding Net Flux

Design Criteria	Flow Rate m³/day	Net Flux	
		LMH	m/day
Annual average	27,000	25	0.59
Maximum monthly	32,000	29	0.70
Maximum weekly	33,000	30	0.72
Maximum daily	39,000	36	0.85
Maximum hourly	64,000	58	1.40

Source: Data from Crawford, G. et al., Enhanced biological phosphorus removal within membrane bioreactors. *Proceedings of WEFTEC*. Oct. 21–25. Dallas, Texas, 2006.

TABLE 4.13

Traverse City MBR Influent and Effluent Data from the 30-Day Test in November 2004

Parameter	MBR Influent (Primary Effluent)	MBR Effluent (Secondary Effluent)
BOD_5 (mg/L)	179	<2
COD (mg/L)	361	21.7
TSS (mg/L)	104	–
TKN (mg/L)	35	4.9
NH_4-N (mg-N/L)	27	0.5
TP (mg/L)[a]	6.2	0.38

Source: Data from Daigger, G.T. et al., *Water Environ. Res.* 82(9):806–818, 2010.

[a] Ferric chloride dose = 38 mg/L.

TABLE 4.14

Operational Parameters from One of the Two MBR Trains in November 2004 in Traverse City, Michigan, USA

	Anaerobic	Anoxic	Aerobic	Membrane	Total
Incoming flow rate	28,000	42,000	56,000	42,000	
(m³/day)	(2Q)	(3Q)	(4Q)	(3Q)	
MLSS (mg/L)	1582	3061	5839	9044	
Relative MLSS (actual)	0.17	0.34	0.65	1.00	
Relative MLSS	0.17	0.33	0.67	1.00	–
Volume (m³)	335	1173	1843	850	4200
Apparent HRT (h)	0.6	2.0	3.2	1.5	7.0
Effective HRT (h)	0.29	0.67	0.79	0.49	–
Solids in tank (kg)	1582	3061	5839	9044	
Solids distribution (%)	2.3	15.9	47.7	34.1	
SRT (days)	0.18	1.24	3.72	2.66	7.80

Source: Data from Daigger, G.T. et al., *Water Environ. Res.* 82(9):806–818, 2010.

Note: The conditions/assumptions in the previously mentioned calculation are as follows: Influent flow rate (Q) = 14,000 m³/day; $Q_{AN/AA}$ = 14,000 m³/day (1Q); $Q_{O/AN}$ = 14,000 m³/day (1Q); $Q_{M/O}$ = 28,000 m³/day (2Q); Biosolid production = 2893 kg/day; Water temperature = 16.1°C; Ferric chloride dose = 38 mg/L influent; MLVSS/MLSS = 0.71.

- Because $Q_{O/AO}$ ranges from 1Q to 2Q, the maximum TN removal efficiency remains around 70%, assuming raw wastewater carries a sufficient amount of readily biodegradable COD (see Equation 4.15).
- The ratio of MLSS in the aeration tank and the membrane tank ranges from 0.67 to 0.80.

4.6.6.2 Nordkanal, Germany (Brepols 2011)

The Nordkanal wastewater treatment plant, owned and operated by a German public utility institution for water management, Erftverband, discharges effluent to a nearby canal and needed to improve effluent quality to meet the new effluent standard. At the same time, its treatment capacity needed to be increased to relieve the chronic hydraulic overloading issue. The old WWTP was dismantled and a new MBR plant was built 2.5 km away from the original plant. The new MBR plant

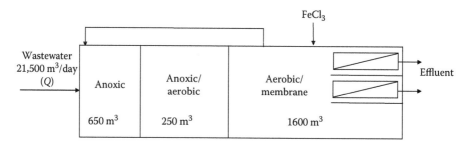

FIGURE 4.21   Process diagram of one of the four parallel MBR systems in Nordkanal, Germany.

needed only 50% of the land that the comparable CAS process would require. Existing rain water storage tank with 2000 m³ capacity in the old facility was converted to a wastewater holding tank. Mechanical screens with 1.0 mm mesh size were used to remove the solids and the fibrous materials carried by raw wastewater, but no primary clarifier was used. When this plant was commissioned in 2004, hollow fiber clogging by fibrous materials was observed because of insufficient prescreening. The screen was replaced with a different type with a self-cleaning function using water jet.

The modified MLE process was adopted, targeting nitrogen removal, whereas chemical phosphorus removal was performed using $FeCl_3$, as shown in Figure 4.21. The mixed liquor in the aerobic/membrane tank was recycled back to the anoxic tank, where nitrate is denitrified consuming readily biodegradable COD carried by raw wastewater. Nitrogen removal efficiency was estimated at 85%. $FeCl_3$ was dosed to the anoxic tank to enhance phosphorus removal. Effluent TP was maintained around 0.5 mg/L, far below the discharge limit of 1.5 mg/L. The molar ratio of Me/P was between 0.22 and 1.30. Because of the inorganic solids generated from $FeCl_3$, the apparent sludge yields were much higher than that without chemical phosphorus removal, that is, 0.55 g solids/g COD at an SRT

TABLE 4.15
Design Parameters of Nordkanal MBR

	Unit	Values	Remark
Membrane type	–	HF modules	4 aeration tanks
		ZW500c (GE Water)	2 trains/aeration tank
Total membrane area	m²	84,480 (total)	24 cassettes/train
		10,560 (per train)	20 modules/cassette
			22 m²/module
Aeration rate	m³/h/trains	4250	Scouring air only
SAD_m	m³/m²/h	0.40	= 4250/10,560 (gross)
SAD_p	m³ air/	33	= 4250 × 24/(24,600/8)
	m³ permeate	18	= 4250 × 24/(45,100/8)
Flow rate	m³/day	24,600 (dry)	Net flux = 12 LMH (0.29 m/day)
	m³/day	45,100 (wet)	Net flux = 22 LMH (0.53 m/day)
Tank volume	m³	2600 (anoxic)	9200 m² total
		1000 (swing)	
		5600 (aerobic/membrane)	
HRT (dry/wet)	h	2.5/1.4 (anoxic)	9.0/4.9 h total
		1.0/0.5 (swing)	
		5.5/3.0 (aerobic/membrane)	

Source: Data from Brepols, C. *Operating Large Scale Membrane Bioreactors for Municipal Wastewater Treatment.* London: IWA Publishing, 2011.

TABLE 4.16

Daily Loading (Quantile $Q_{0.85}$ and Average Values of Sample Data)

Daily Loading	Unit	COD	TN	Ammonia	Nitrate	P_{tot}
Influent ($Q_{0.85}$/avg)	kg/day	9000/6000	810/570	620/450		150/80
Effluent ($Q_{0.85}$/avg)	kg/day	400/230	160/80	4/1	130/60	6/4
Biosolid production	kg/day	4500	265			145
Design influent	kg/day	9600[a]	897			123

Source: Brepols, C. *Operating Large Scale Membrane Bioreactors for Municipal Wastewater Treatment.* London: IWA Publishing, 2011.

[a] BOD = 5250 mg/L.

of approximately 25 days. Process parameters are summarized in Table 4.15. Contaminant loading to the MBR is summarized in Table 4.16.

Fine pore membrane plate diffusers made of silicone were used for biological aeration at an effective diffuser depth of 4.0 m. The standard oxygen transfer rate (SOTR) in clean water was measured at 15.6 g O_2/m³/min, and αSOTR was measured at 8.3 g O_2/m³/min from which the α-factor was calculated at 0.53. The average MLSS and MLVSS in the aerobic/membrane tank were at 12 and 7 g/L, respectively. The F/M ratio averages at 0.05 g COD/g MLSS/day. Intermittent membrane scouring was performed at 10 s on – 10 s off mode by channeling the air to different parts of the train using automatic valves. A periodic permeability swing was observed depending on water temperature. Permeability follows a sine curve, as shown in Figure 4.22, proportional to water temperature.

It was found that the lag time between the surge of nitrogen loading and the surge of nitrogen discharge was significantly shorter than HRT because of the longitudinal mixing in the bioreactors. During dry weather flow, lag time was observed at 2 to 4 h, whereas HRT was longer than 9 h. Likewise, with a wet weather flow, the time lag was 0.5 to 1.5 h and the HRT was 4.7 to 6.0 h.

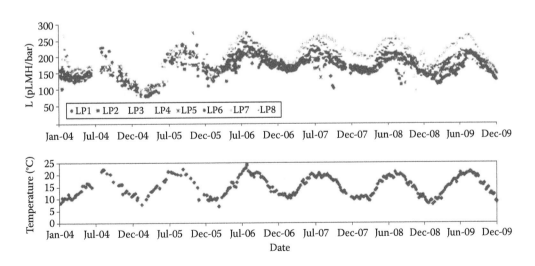

FIGURE 4.22 Weekly average permeability and temperature values at Nordkanal MBR in 2004–2009. (From Brepols, C., *Operating Large Scale Membrane Bioreactors for Municipal Wastewater Treatment.* London: IWA Publishing, 2011.)

4.7 POSTTREATMENT

4.7.1 DISINFECTION

The effluent from CAS and MBR is often disinfected before being discharged to the surface water. The primary purpose of disinfection is to stop pathogens from spreading. Chlorination is the most commonly practiced disinfection method for the secondary effluent. Sodium hypochlorite (NaOCl), chlorine gas (Cl_2), calcium hypochlorite ($Ca(OCl)_2$), and chlorine dioxide (ClO_2) are available in the market. Of these compounds, sodium hypochlorite is the only agent available in a liquid form (12%–15%) and is most commonly used. Choline gas is in the form of liquefied compressed gas, but they are not popularly used due to the safety issues during transportation and application. It also needs injectors specially designed for the purpose. Chlorine dioxide is a yellowish-green gas. It is more effective than other alternatives in disinfection, but it is much costlier than NaOCl due to the necessity of special equipment to generate it from sodium chlorite ($NaClO_2$) onsite. Calcium hypochlorite is mostly available in powder form with more than 70% available chlorine. It should be dissolved in water before use. The calcium ions contained in the chemical tend to form scales in the feeding line/pump by forming calcium carbonate or calcium sulfate. Careful feeding system design and maintenance are required.

Once NaOCl is added to water, it turns to hypochlorite ions (OCl^-), which in turn enter into the equilibrium with hypochlorous acid (HOCl), as shown in the following chemical equations. This charge neutral species is easier to penetrate into microorganisms. Thereby, hypochlorous acid is a more powerful disinfectant than hypochlorite ion. Because the pK_a of hypochlorous acid is 7.5, HOCl and OCl^- concentrations are equal at pH 7.5. HOCl becomes increasingly dominant as the pH decreases further from 7.5 and as a result disinfection becomes faster. On the contrary, HOCl becomes scarce as the pH increases beyond 7.5. For instance, the molar ratio of $[HOCl]/[OCl^-]$ becomes 0.1 at pH 8.5 and 0.01 at pH 9.5. Therefore, longer contact time and/or higher chlorine dosage may be required for adequate disinfection when the effluent pH is high.

$$NaOCl \quad \begin{aligned} &NaOCl \rightleftarrows Na^+ + OCl^- \\ &OCl^- + H_2O \rightleftarrows HOCl + OH^- \qquad pK_a = 7.5 \end{aligned}$$

$$Cl_2 \quad \begin{aligned} &Cl_2 + H_2O \rightleftarrows HOCl + HCl \\ &HOCl \rightleftarrows H + OCl^- \qquad\qquad pK_a = 7.5 \end{aligned}$$

As chlorine is added to the secondary effluent, easily breakable polymeric compounds with double bonds and oxidizable inorganic ions are attacked, for example, biopolymers, residual organics from raw wastewater, Fe^{2+}, Mn^{2+}, and H_2S. After that, free chlorines react with ammonia in the effluent to form chloramines. If the chlorine addition was sufficient, most ammonia convert to nitrogen trichloride (NCl_3). With additional chlorine, NCl_3 eventually breaks down to nitrogen gas (N_2) and nitrogen oxides (NO_2). Chlorine is reduced to chloride and stays in the water. The chlorine dosage sufficient to lead the reaction up to this point is called a *breakpoint dosage*. Any chlorine dose in excess of the breakpoint dosage remains as free chlorine in water and acts as disinfectant. Because chloramines are much weaker biocides than free chlorine, much higher dosages are required to obtain the same log removal values in a given disinfection time. Thus, chloramines are not preferred for the disinfection process that should be completed in a short period, for example, 15 to 60 min for the secondary effluent disinfection. Therefore, excess chlorine must be added to secure a sufficient free chlorine concentration after the breakpoint reaction.

The disinfection kinetics can turn out quite differently at the same residual chlorine concentrations depending on the effluent composition. For example, the compounds with polycyclic rings containing hydroxyl groups and the compounds with sulfur groups react with chlorine and detected as residual free chlorine in titration tests, but those are only weakly biocidal. As a result, required chlorine dosages vary site by site even if gross effluent quality parameters are similar. When the contact time is 30 min and the initial coliform count is 10^4 to 10^6 MPN/100 mL, the required chlorine dosages are 4 to 8, 5 to 15, 6 to 20, and 8 to 30 mg/L to obtain 1000, 200, 23, and less than 2.2 MPN/100 mL, respectively, for the filtered secondary effluent at a neutral pH. A minimum of 15 to 30 min contact time is required, but 1 h is often recommended. Treatment requirement also depends on geographical location. The state of Florida requires total residual chlorine (TRC) higher than 1 mg/L in the effluent discharged and a 15 min contact time at peak hourly flow. The state of Washington requires a TRC higher than 1 mg/L and 30 min contact time. The state of California requires a $C_R t$ higher than 450 mg·min/L in normal time, and the modal contact time is 90 min or longer at peak dry weather flow (USEPA 2012), where the $C_R t$ is a product of chlorine concentration (C_R) and modal contact time (t). Alternatively, disinfection processes that can achieve a 5-log inactivation of MS2 or poliovirus are approved.

The largest concern with chlorination is the production of disinfection by-products (DBPs) such as trihalomethanes (THM) and haloacetic acids, especially in drinking water process, but it is less of concern in the posttreatment of the secondary effluent, where the treated water is not intended for direct human consumption. If dechlorination is required before treated water is discharged, sulfur dioxide (SO_2) gas, sodium sulfite (Na_2SO_3), sodium bisulfite ($NaHSO_3$), and sodium metabisulfite ($Na_2S_2O_5$) can be added before the disinfected effluent is discharged.

Ozone (O_3) is a strong oxidant that has been mainly used for potable water treatment, especially when DBPs are a concern. It requires ozone generator, contactor, off-gas treatment device, etc., and may also need pure oxygen storage tank. Because of the high capital costs and the concern about the safety issues, ozone is not popularly used for the secondary effluent disinfection. Air can be used as an oxygen source, but the energy costs to generate ozone is much higher than those with pure oxygen, for example, 13.2 to 19.8 kWh/kg O_3 versus 6.6 to 13.2 kWh/kg O_3. In general, air based ozone generation is performed primarily in small scale applications.

UV deactivates microorganisms by breaking the bonding of enzymes, DNA, lipids, etc. Many different types of UV lamps are available, but medium pressure lamps with strong and broad spectrum at 200 to 400 nm and low-pressure lamps with a specific wavelength at 254 nm are typically used. High-pressure lamps have much greater outputs at broader spectrum, but low-pressure lamps are more energy efficient, durable, and inexpensive. UV lamps typically have 10,000 h life span until their UV output reaches 70% to 75% of the initial level, but the life span varies depending on the frequency of the on–off cycle (Solsona and Méndez 2003).

The effect of water temperature on the disinfection efficiency is limited, but the energy efficiency of UV lamp may be affected. Because water itself absorbs UV, the UV chambers are designed to have multiple narrow channels with less than 75 mm depth to limit the traveling distance of the UV. The UV transmission efficiency can drop significantly, if effluent turbidity is higher than 5 to 10 NTU or the turbidity is caused by the particles larger than that in the order of 10 μm. In MBR, however, the low effluent turbidity (<0.2 NTU) allows high and stable transmittance, that is, 60% to 70%. UV radiation strength in a device is measured as the power per surface area (μW/cm²), and the dosage is calculated as μWs/cm² (or μJ/cm²) by multiplying the exposure time with the radiation strength. The dosages of UV light required to destroy common bacteria are 6 and 10 mJ/cm², but higher life-forms such as protozoa, rotifers, nematodes, etc., need much higher dosages. The standards dosages are guided mostly at 100 mJ/cm² for the granular media filtration effluent and 80 mJ/cm² for the membrane filtered effluent. The UV exposure time depends on the strength of the UV, but it is typically 10 to 20 s. UV can break large molecules to smaller by-products by breaking weak chemical bonding, but it rarely oxidizes organics to water and carbon dioxides. It generally does not destroy trace organics, but it is effective in destroying some UV-susceptible compounds under the presence of catalysts. UV disinfection is not known to produce bromate and THM commonly produced in ozonation and chlorination processes, respectively, but the genotoxicity of the UV-treated effluent is still under debate (WRF 2011) (Table 4.17).

TABLE 4.17

Estimated Rate of $C_R t$ Values or Various Levels of Inactivation of Dispersed Bacteria, Viruses, and Protozoan Cysts in the Filtered Secondary Effluent at Approximately pH 7 and 20°C

Target	Disinfectant	Unit	Target Log Removal Value			
			1	**2**	**3**	**4**
Bacteria	Chlorine (free)	mg·min/L	0.1–0.2	0.4–0.8	1.5–3	10–12
	Chloramine	mg·min/L	4–6	12–20	30–75	200–250
	Chlorine dioxide	mg·min/L	2–4	8–10	20–30	50–70
	Ozone	mg·min/L		3–4		
	UV radiation	mJ/cm²		30–60	60–80	80–100
Virus	Chlorine (free)	mg·min/L		2.5–3.5	4–5	6–7
	Chloramine	mg·min/L		300–400	500–800	200–1200
	Chlorine dioxide	mg·min/L		2–4	6–12	12–20
	Ozone	mg·min/L		0.3–0.5	0.5–0.9	0.6–1.0
	UV radiation	mJ/cm²		20–30	50–60	70–90
Protozoan cysts	Chlorine (free)	mg·min/L	20–30	35–45	70–80	
	Chloramine	mg·min/L	400–650	700–1000	1100–2000	
	Chlorine dioxide	mg·min/L	7–9	14–16	20–25	
	Ozone	mg·min/L	0.2–0.4	0.5–0.9	0.7–1.4	
	UV radiation	mJ/cm²	5–10	10–15	15–25	

Source: Tchobanoglous, G. et al., *Wastewater Engineering: Treatment and Reuse.* Boston: McGraw-Hill, 2003.

4.7.2 ADVANCED OXIDATION PROCESS FOR TRACE ORGANIC REMOVAL

The oxidation power of UV can be enhanced by combining it with hydrogen peroxide (H_2O_2). The added H_2O_2 works synergistically with UV and improves the oxidation power by producing hydroxyl radicals that are strong oxidants with little selectivity. The hydroxyl radicals destroy molecules by attacking mainly double bonds and scavenging hydrogen atoms from C-H bonds. UV alone can break down macromolecules only to smaller by-products, but UV/H_2O_2 can destroy small molecules such as pharmaceutical and personal care products and endocrine disrupting compounds, as discussed in Section 7.3.3.2. Simultaneously, UV and hydroxyl radicals directly and indirectly kill microorganisms. Most notably, the UV/H_2O_2 process is used to destroy trace organics in the groundwater replenish system (GWRS) for producing potable quality water from municipal wastewater, as discussed in Section 7.3.3.3. The hybrid process of UV and H_2O_2 destroys organics more completely than UV alone, but it still leaves some intermediate products. In addition, unreacted residual H_2O_2 may remain in the treated water. Therefore, it may be required to pass the effluent through the granular activated carbon (GAC) column to further remove intermediate products and residual H_2O_2.

UV can also be combined with ozone. Ozone is more efficient than H_2O_2 in absorbing UV, and this combination can provide a more powerful oxidation environment than UV/H_2O_2. However, the use of ozone involves a burden on off-gas treatment, and it can cause potential safety issues in handling the gas onsite. If the water has poor UV transmission, H_2O_2/ozone can be used by replacing UV with ozone. This process is called the *peroxone process*, where two oxidants synergistically work together, producing hydroxyl radicals rapidly. However, the handling of ozone and the off-gas treatment can still be a hurdle in the peroxone process. As a result, UV/H_2O_2 combinations appear to be more commonly used than UV/ozone or H_2O_2/ozone especially in large scale applications.

In one study, UV/H_2O_2 and ozone/H_2O_2 systems were compared (Xie et al. 2013). The UV/H_2O_2 system was equipped with a medium-pressure UV reactor (four lamps with 3 kW each at 60% power output) with H_2O_2 dosages of 0 to 10 mg/L. The ozone/H_2O_2 system was dosed with ozone and H_2O_2 at 0 to 4 mg/L and 0 to 1.5 mg/L, respectively. 2-MIB, geosmin, diclofenac, bisphenol-A, 17β-estradiol (E2), and other organic contaminants were spiked at 150 to 1000 ng/L for each of the target compounds. In the UV/H_2O_2 system, removal efficiencies of the compounds tended to increase as H_2O_2 dosage increased, but declofenac removal efficiency was high even without H_2O_2. The removal efficiencies were 63% to 83% at the H_2O_2 dosage of 10 mg/L. Meanwhile, the ozone/H_2O_2 system showed 90% to 99% removal efficiencies for 2-MIB, geosmin, bisphenol-A, and E2 at 10 min retention time.

The oxidation reaction of UV can be enhanced by using photocatalysts (TiO_2). In this process, very reactive hydroxyl radicals and electrons with very short lifetime are produced on the catalyst surface and attack organic molecules. The photocatalysts are typically coated on the surface of photoreactors to avoid the costly catalyst recovery process. However, the contaminations of the catalysts by inorganic scales and organic deposits are always a concern. There are few commercial products based on this principle in the market.

Fenton reagents, for example, Fe^{2+} and H_2O_2, can also be used to destroy trace organics. In the following chemical equations, ferrous ions (Fe^{2+}) are oxidized by H_2O_2 when hydroxyl radicals are produced, but some of them are reduced back to ferrous ions:

$$Fe^{2+} + H_2O_2 \rightarrow Fe^{3+} + HO\bullet + OH^-$$

$$Fe^{3+} + H_2O_2 \rightarrow Fe^{2+} + HOO\bullet + H^+$$

Ferrous ion is used as a catalyst, but a large amount is required to push this reaction forward because ferric ions (Fe^{3+}) form insoluble ferric hydroxides and precipitate as sludge.

4.7.3 ADSORPTION

4.7.3.1 Principle

Activated carbon has a large number of small pores that constitute a specific surface area larger than 500 m^2/g. It can be made of soft wood, nutshells, charcoals, lignite, etc., by heating them at a high temperature under the presence of oxygen, steam, CO_2, etc. Activated carbons with a grain size larger than 0.3 mm (or 50 mesh) are typically called *granular activated carbon* (GAC), and smaller ones are called *powdered activated carbon* (PAC).

The adsorption of solutes on activated carbon surface occurs only when it causes a decrease in Gibb's free energy (*G*, J/mol). In Equation 4.33, adsorption always causes negative entropy changes (Δ*S*, J/mol/K) because of the reduction of the number of free molecules. Because temperature (*T*, K) is always positive, the term $-T\Delta S$ is always larger than zero when adsorption occurs. To make Δ*G* negative, the enthalpy of the adsorption reaction (Δ*H*, J/mol) should be negative enough so that its absolute value exceeds *T*Δ*S*. Therefore, adsorption occurs only when it is an exothermic reaction and the enthalpy term exceeds the entropy term (Atkins and de Paula 2009), that is,

$$\Delta G = \Delta H - T\Delta S \tag{4.33}$$

According to Equation 4.33, large molecules have advantages in being adsorbed because the large contact area is subject to the strong van der Waals force between the molecule and the carbon surface. The strong attraction causes a large enthalpy loss when they are adsorbed. Meanwhile, entropy loss from the adsorption is more dependent on the number of free molecules lost than the size of the molecules. In the example shown in Figure 4.23, the macromolecule diffused into the

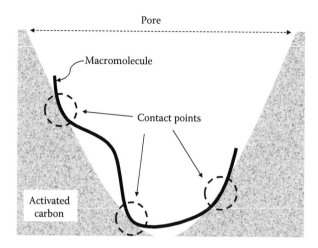

FIGURE 4.23 Adsorption of adsorbate on the pore wall of activated carbon.

pore contacts with the carbon surface at three different locations, which enhances the adsorption by increasing the enthalpy loss. For small molecules, enthalpy loss in the adsorption process is not as great as that of large molecules because of the smaller contact area, but entropy loss is about the same as that of large molecules because entropy loss is mainly dependent on the number of molecules adsorbed. Hence, small molecules do not adsorb on activated carbon as much as large molecules in general.

Charged molecules have disadvantages because the charges are accumulated in the small two-dimensional spaces by being adsorbed. Because they repulse one another in the small spaces, the energy status of the adsorbed molecules increases. In general, nonpolar compounds with lower solubility tend to adsorb activated carbon better than polar compounds. Smaller pore sizes are more effective to adsorb small organic molecules because of the larger contact area with the molecules provided by the internal curvatures of the pore shown in Figure 4.23. The molecules with double bonds or aromatic rings are better adsorbed than halogenated methanes and the molecules with saturated bonds. The electron-rich π-orbitals of adsorbents and adsorbates interact with one another and can form stable physical bonding. The surface property of the carbon can be modified by adding oxidizing agents during the manufacturing process. The surface charges and the functional groups formed on carbon surface affect the adsorption properties and the preferred adsorbates.

Adsorption is a thermodynamic equilibrium phenomenon between dissolved solutes and adsorbed solutes. The higher the solute concentration in water phase, the higher the amount of adsorbed solutes. Therefore, the lower the target organic concentration in the effluent, the lower the specific adsorption capacity. Because more organics must be removed while less amount of organics are adsorbed on unit amount of carbon, the amount of carbon required increases sharply as the target water quality becomes more stringent.

4.7.3.2 Application for the Secondary Effluent Treatment

In MBR, effluent is free of suspended solids and BOD rarely exceeds discharge limit due to the food stringent condition in the aeration tank, but COD may exceed the discharge limit, if influent contains high levels of nonbiodegradable COD. GAC columns can be used to reduce organic contents in the MBR effluent. GAC removes most of trace organics at high efficiencies when it is used to treat MBR effluent (Nguyen et al. 2012). The removal efficiencies of most trace organics were found at 97% to 100%, but the removal efficiency of acetaminophen (Tylenol®) was only at approximately 85% (Figure 6.33). PAC is not commonly used for the posttreatment of the secondary effluent because additional separation processes are required to remove the PAC before discharging the effluent. Optionally, PAC can be added to the aeration tank directly in MBR. The suspended

PAC in the aeration tank can improve the trace organic removal not only by adsorbing them but also by increasing the contact time with the microorganisms in the aeration tank. The spent PAC is discharged with excess sludge and undergoes dewatering process. PAC may scratch the membrane if aeration is excessive, especially in small-scale experiments, but it has been commercialized as MACarrier (GE), as discussed in Section 5.4.1.

If GAC is used to treat secondary effluent, the spent GAC should be regenerated and reused for economic reasons. There are several methods to regenerate GAC: (1) oxidation of the adsorbed organics using oxidation agents, (2) steaming to drive the adsorbates off from the carbon, (3) solvent extraction of the adsorbate, and (4) biological degradation of the adsorbates. GAC can also be reactivated using the same process used to produce GAC, that is, heating the carbon at 800°C to 900°C under the presence of steam, CO_2, etc., to burn off the residual organics occupying the active sites. Typically, 4% to 10% of the adsorption capacity is lost in each regeneration cycle.

4.8 RHEOLOGY OF MIXED LIQUOR

The rheological property of fluid is a consequence of complex molecular interactions in the fluid. The viscosity of fluid is defined as a proportionality coefficient, as shown in Equation 4.34. In Figure 4.24, where the force F (N or kg·m/s²), which is required to move the top plate with an area of A at a velocity v (m/s) against the parallel static plate placed with a distance y (m), is proportional to the fluid viscosity, μ (kg/m/s). Shear stress is defined as F/A (kg/m/s²). For pure water, μ is constant for a wide range of shear rate, v/y. Depending on the composition of the fluid, the molecular properties of the components, the interactions among the molecules, etc., various rheological properties are realized.

$$F = \mu A \frac{v}{y} \quad \text{or} \quad \frac{F}{A} = \mu \frac{v}{y} \tag{4.34}$$

In pure water, the fluid viscosity μ is fairly constant at a wide range of shear rate. However, if salts, organic molecules, particles, etc., are present, the extent of the momentum transfer pattern among water molecules can drastically change; hence, the fluid viscosity becomes variable depending on the shear rate. For instance, if salts exist in water, the polar water molecules are aligned around the salt ions to counter the electrical field. Because of the formation of the large ion–water clusters, momentum can transfer farther and increase water viscosity as a result. Because the size of the ion–water cluster decreases as shear rate increases, viscosity becomes a function of shear rate.

Viscosity can change depending on the magnitude of shear rate and/or the duration of the shear rate. In some cases, molecule alignment changes depending on the magnitude of shear stress so that momentum transfer is affected. If the momentum of one molecule transfers to more molecules, viscosity increases or vice versa. In other cases, molecules are aligned more or less orderly over time

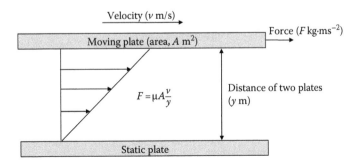

FIGURE 4.24 Definition of fluid viscosity.

when shear stress is applied. Because the new alignment takes place over time, viscosity changes gradually in this case. The aqueous solutions that consist of various molecules show a combination of varying degrees of time and shear stress dependences. Depending on the behaviors of viscosity, many different types of fluids are defined. Table 4.18 and Figure 4.25 summarize some of the basic rheological types of fluids.

The viscosity of the biological sludge is largely caused by the molecular and particular alignments that also play a role in the floc formation. In the mixed liquor with aligned constituents, the movement of one constituent causes a serial movement of other constituents connected in the same network. At an increasing shear rate, the network is disrupted and hence the effect of one constituent's movement to another becomes less significant. As a result of the smaller area of influence at a high shear rate, the constituents of the mixed liquor tend to move independently, and the force required to move the fluid decreases. As a consequence, the viscosity of biological sludge decreases as shear rate increases, and its rheological property is classified as pseudoplastic (or shear thinning) fluid, as shown in Figure 4.25a (Rosenberger et al. 2002; Mori et al. 2006; Ratkovich et al. 2013). Another aspect of the networking effect is that if shear rate changes, it takes some time for the particulates/molecular networks to reach a new steady state. The viscosity of biological sludge tends to decrease over time until it reaches a steady state when shear rate increases, shown as thixotropic fluid in Figure 4.25b (Honey and Pretorius 2000).

Figure 4.26 shows the relationship between shear rate and apparent viscosity. It is noticeable that viscosity of the mixed liquor drops from near 10,000 cP to below 10 cP when shear rate increases from 0.1 to 100 s^{-1}. In the figure, the increase of viscosity at the very high shear rate is a result of a turbulent viscosity phenomenon caused by the Taylor vortex occuring in viscometer (Mori et al. 2006). The turbulent viscosity effect may be relevant to the side stream membranes that run at very high shear rates. The thixotropic behavior of the mixed liquor is discussed in detail in other reference (Ratkovich et al. 2013), but the initial viscosity of approximately 35 cP declines to approximately 20 cP within 300 s and continues to approach an asymptotic line when a mixed liquor with 10 g/L MLSS was sheared at a constant shear rate of 1.83 s^{-1}.

TABLE 4.18
Classification of Fluid in Terms of Rheological Property

Classification		Description	Example
Newtonian fluids		Constant viscosity regardless of the shear	Pure water
Shear dependent	Dilatant (shear thickening)	Viscosity increases as shear increases	Suspensions of cornstarch (Oobleck), sand in water
	Pseudoplastic (shear thinning)	Viscosity decreases as shear increases	Paper pulp in water, latex paint, ice, blood, syrup, molasses, shampoo, biological sludge
	Bingham plastic	Minimum shear is required to make the fluid flow	Toothpaste, margarine
Time dependent	Rheopectic	Apparent viscosity increases with duration of shear	Some lubricants, whipped cream
	Thixotropic	Apparent viscosity decreases with duration of shear	Some clays, some drilling mud, paints, cerebrospinal fluid, organic sludges

Source: Modified from Wikipedia. *Non-Newtonian Fluid.* Available at http://en.wikipedia.org/wiki/Non -Newtonian_fluid (accessed December 3, 2012.)

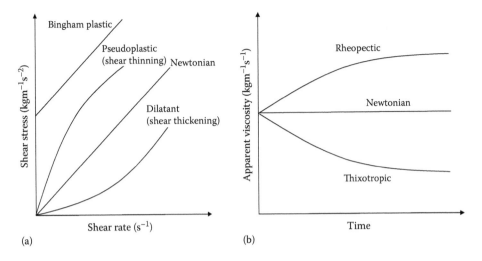

FIGURE 4.25 Various rheological behavior of liquid: (a) as a function of shear rate and (b) as a function of time.

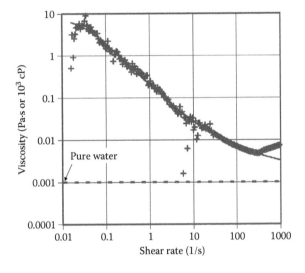

FIGURE 4.26 Shear thinning nature of the biological mixed liquor at an MLSS of 8 g/L. (From Brannock, M. et al., *Water Res* 44(10):3181–3191, 2010.)

The shear thinning (or pseudoplastic) behavior of the activated sludge can hamper the sludge mixing in MBR. The effective viscosity is the lowest near the membrane modules, where the shear rate is the highest. By contrast, the levels of effective viscosity in the corners and the sides are the highest due to the slow movement of the mixed liquor. If there is any space in the reactor without having a proper mixing due to plugged/fouled diffusers or improper diffuser placement, the shear thinning rheological effect drives the viscosity to a few thousands cP according to Figure 4.26. This dead zone effect plays a role in any of the activated sludge process, but it can be more significant in MBR because of the high MLSS. Insufficient DO in the dead zone can promote the growth of filamentous, sulfate-reducing bacteria and the production of biopolymers, which can cause operational issues.

5 Membrane Fouling in Membrane Bioreactor

5.1 OVERVIEW

The performance loss caused by membrane fouling is the fundamental drawback of membrane filtration. In the membrane bioreactor (MBR), all wastewater coming into the process must pass the absolute barrier of the membrane because inability to do so can directly lead to the necessity of bypassing a portion of untreated wastewater to river, lake, or ocean during the peak flow. It has been observed that small particles and macromolecules classified as either soluble microbial products (SMP) or extracellular polymeric substances (EPS) play a major role in membrane fouling rather than the microbial floc particles that consist of the majority of mixed liquor suspended solids (MLSS). This observation is well supported theoretically because the SMPs and the EPSs have higher tendencies to deposit on the membrane surface because of the low back-transport velocity, as discussed in Section 1.2.5. SMP and EPS are also subject to strong interaction with the membrane surface through charge and van der Waals interactions.

With the findings of SMP and EPS as the majority constituent of the cake layer formed on the membrane surface in MBR in the early 1990s, it had been believed that the membrane fouling rate could be correlated with the concentrations of such biopolymers. However, many researchers have found independently that the membrane fouling rates are not simply correlated with the SMP and EPS concentrations (Drews 2010). This suggests that the property of SMP and EPS is as important as their concentration, although further studies are required to understand what kinds of molecular properties are essential to interact with the membrane in the molecular level. In the meantime, MLSS is known to be a poor indicator of the membrane fouling potential as long as it is in a moderate range, for example, 6 to 15 g/L, in general. Meanwhile, many other factors have been identified, such as organic loading rate (OLR); solids retention time (SRT); dissolved oxygen (DO); fat, oil, and grease (FOG) contents; unsteady operating condition; salinity of feed water; nutrient balance in feed water; and inorganic scaling. It has also been known that membrane fouling is a consequence of the complex interaction among such parameters rather than it is dominated by one or two major factors.

In this chapter, the effect of each parameter affecting membrane fouling is discussed along with the relevant fouling mitigation methods, if available. Various mixed liquor characterization methods developed to estimate the membrane fouling potential are also discussed. The complex relations among the influencing factors are also discussed using the data collected from the laboratory to the field-scale studies.

5.2 FACTORS AFFECTING MEMBRANE PERFORMANCE

5.2.1 MIXED LIQUOR SUSPENDED SOLIDS

In the early days of MBR technology, MLSS was considered a primary factor determining membrane fouling. On the basis of this assumption, it was suspected that the performance of the submerged membrane could be improved by lowering MLSS in the membrane tank. In some studies, settling tanks were placed in between the aeration tank and the membrane tank to partially or fully reduce the MLSS before the mixed liquor enters the membrane tank. However, this hypothesis turned out to be incorrect based on the numerous observations made in various experimental conditions.

Chang and Kim (2005) performed side-by-side comparisons with and without a clarifier in between the aeration tank and the membrane tank. The MLSS of the settled and unsettled mixed liquors that go into the membrane tank were 100 mg/L and 6000 mg/L, respectively. Submerged hollow fiber membranes made of polysulfone were used for both trains, and scouring air was supplied from the bottom of the membrane bundles. On the contrary to the initial expectation, higher permeability loss was observed in the low MLSS train during the 7-day experiment. The fouling and cake resistance ($R_c + R_f$ in Equation 1.4) was estimated at 94.5×10^{12}/m and 46.2×10^{12}/m, respectively, with and without clarifier. In another study (Hong et al. 2002), flux was not affected by MLSS in the range of 3.6 to 8.4 g/L.

In the study of Lee et al. (2001a), the effect of MLSS was investigated by performing side-by-side tests with two MBRs with suspended culture and attached culture, respectively. The MLSS of suspended culture MBR was approximately 3000 mg/L, but that of attached culture MBR was only approximately 100 mg/L. The average suspended particle size was smaller in attached growth. It was observed that the transmembrane pressure (TMP) increase turned out to be seven times faster in the attached culture MBR. Although the TMP increasing rate of the attached culture MBR was much greater, scanning electron microscopy (SEM) images showed the cake layer was much thinner than that in suspended culture MBR. This observation suggested that the cake layer was much denser at the low MLSS condition. The exact cause of this unexpected result is not known, but there are at least two explanations. First, the attached growth MBR might be operated at a much higher food-to-microorganism (F/M) ratio because the reactor likely retained much less solids, even including the attached microorganisms. As will be discussed in detail in Section 5.2.3, membrane fouling tends to expedite at a high F/M ratio because of the elevated SMP/EPS concentration and/or the specific characteristics SMP/EPS has under the condition. Second, the unusual relationship between MLSS and flux can be explained based on dynamic membrane theory. When wastewater with organic particles was filtered using tubular membranes, the feed water with more suspended solids resulted in greater fluxes in some cases (Imasaka et al. 1993). It was explained that the initially formed cake layer with large particles formed porous cake layers that hindered further particle deposition.

As a result of the numerous observations made during the last few decades in various conditions, MLSS itself is now considered only weakly correlated with membrane fouling, if it is in a moderate range such as 6 to 15 g/L and membrane scouring is performed reasonably well. However, the excessively high MLSS beyond the moderate range still can cause accelerated membrane fouling. For example, excessively high MLSS can dramatically deteriorate the membrane scouring efficiency by hampering the bubble rising patterns due to the increasing mixed liquor viscosity. In other situations, high MLSS can lead to low oxygen transfer efficiency (OTE) that in turn reduces DO concentration. Eventually, the low DO exacerbates the membrane fouling problem, as will be discussed in Section 5.2.4. Excessive MLSS can also promote the formation of dead zones in the places without sufficient mixing. On the contrary, excessively low MLSS outside the design guideline reduces SRT and increases F/M ratio, which can trigger an accelerated membrane fouling, as discussed in Section 5.2.3.

The weak correlation between MLSS and membrane fouling can be explained by the critical flux theory discussed in Section 1.2.6. According to this theory, primarily submicron-sized particles and macromolecules deposit on the membrane surface whereas large flocs are transported away because of the particle back-transport phenomena under the cross-flow condition. Because fine particles/macromolecules take only a very small portion of MLSS, MLSS itself is not strongly correlated with membrane fouling. However, excessive MLSS can trigger unfavorable conditions for membrane filtration such as slow mixing, low DO, dead zones, etc., if it is combined with poor system design and operating condition.

Because of the complex interactions among the operational parameters, the effect of MLSS on membrane fouling can be erroneously interpreted. It is especially true when other operating parameters than MLSS are overlooked or those parameters are not controlled properly. Some of the common reasons causing improper conclusions on MLSS effect are as follows:

- In a hypothetical side-by-side experiment, identical operating conditions are applied to two reactors, including feed flow rate, aeration rate, and flux, except MLSS. In this condition, the reactor with low MLSS has a lower SRT and a higher F/M ratio than its counterpart. As a result, membrane fouling is likely more significant in the reactor with a low MLSS. Although the low SRT and the high F/M ratio are more likely causes of the greater membrane fouling rate, one may conclude that membrane fouling is more significant at a lower MLSS.

- In another hypothetical example, an experiment is performed using one MBR train while increasing MLSS gradually to see the effect of MLSS on membrane fouling. Excess sludge removal may be paused or reduced to enable MLSS increase. In this dynamic condition, the apparent SRT calculated by dividing the total sludge mass in the reactor by the sludge removed each day is longer than the effective SRT. Meanwhile, F/M ratio is dynamically decreasing because of the increasing MLSS over time. Because the two important factors affecting membrane fouling are dynamically changing, any observation made in this condition is irrelevant to the effect of MLSS. If the DO and the mixing pattern change as MLSS increases, no observation made in the experiment will be valid for anything.

- Conversely, if MLSS is decreased gradually by increasing biosolid removal, the effective SRT at low MLSS condition is longer than the steady-state SRT at the MLSS. If membrane fouling rates are measured under this environment, the likely conclusion is that membrane fouling increases as MLSS decreases.

- As MLSS increases, OTE tends to decrease because of the physical hindrance of the solids against the oxygen transfer to liquid in the gas–liquid interface. As a consequence, DO can drop, which in turn causes nonideal conditions for microorganisms and expedite membrane fouling. If the effect of MLSS on DO is overlooked, high MLSS seems a cause of the high membrane fouling rates.

5.2.2 SMP AND EPS

5.2.2.1 Definition and Quantification

EPS are the biopolymers that bind microorganisms together to form floc particles in the activated sludge process. They also enable microorganisms to attach on the surface and to provide a capability of binding with the trace metals required for metabolism. SMPs have many different origins, that is, the free biomolecules secreted by microorganisms, the debris of ruptured/hydrolyzed cells, etc. SMP mainly consists of enzymes, nucleic acids, cell walls, lipids, etc. SMP and EPS are also called biopolymers to emphasize their origin and are quantified as proteins and polysaccharides equivalents. The quantities of SMP and EPS are positively correlated with each other in a typical condition because a large portion of SMP is detached EPS.

SMP and EPS are the generic terms referring all the organic molecules originated from the microorganism. The accurate quantification of each of them is not readily possible because of the difficulties in separating them from the rest of the constituents of mixed liquor. The various chemical properties of the biomolecules with different configurations also make the accurate quantification hard. Thus, SMP and EPS are operationally defined, where any molecules, colloids, or particles are determined as EPS or SMP as long as they are detected by the quantification method used. In addition to the rather fuzzy boundary of SMP and EPS, there are no standardized quantification methods; thereby, the measured SMP and EPS can vary widely for the same mixed liquor sample depending on the experimental method used (Drews 2010). For example, the measured EPS contents are strongly dependent on the intensity of the shearing method used to extract the EPS from biomass (Liu and Fang 2002). Therefore, direct comparisons of the data from two different studies are not valid in general.

To quantify SMP and EPS, the mixed liquor sample is centrifuged and the supernatant is obtained, as illustrated in Figure 5.1. Various revolutions per minute, g-value, and duration have been used, for example, 5 min centrifuge at $5000g$ followed by filtration using a 0.2 μm filter (Germain et al. 2007), 20 min centrifuge at 20,000 rpm followed by filtration using a 0.2 μm filter (Hernandez Rojas 2005), 30 min centrifuge at 3200 rpm without filtration (Chang and Lee 1998), 10,000 rpm for 10 min at 4°C followed by filtration using a 0.45 μm filter (Ng et al. 2006), 3500 rpm (or 6000g) for 10 min (Zhang et al. 1999), 3000 rpm for 15 min (Sheng and Yu 2006), etc. In some cases, the mixed liquor sample was filtered, and the permeate was used to quantify SMP instead of centrifuging it (Drews et al. 2007).

The quantities measured from the supernatant are called SMP by definition, which is soluble EPS. The protein and polysaccharide contents are quantified most popularly by the Lawry and the Dubois methods, respectively (Lowry et al. 1951; Dubois et al. 1956). However, proteins and polysaccharides can also be analyzed by more sophisticated chromatographic methods that enable the identification of more specific molecules. Protein and polysaccharide contents are calibrated against typically bovine serum albumin (BSA) and glucose, respectively. Therefore, the quantified SMP and EPS do not represent the actual masses, but BSA and glucose equivalent masses.

The settled solids in the centrifugation step in Figure 5.1 are resuspended using either DI water or isotonic NaCl solution (0.9%) before extracting bound EPS. Heat treatment is most commonly used to detach proteins and carbohydrates from cell surface. The resuspended solids are heated in water bath at various temperatures at various durations, for example, 10 min at 80°C (Judd 2011) and 60 min at 100°C (Chang and Lee 1998). Alternatively, cation exchange resins (DOWEX Marathon C, 20–50 mesh) are mixed with the resuspended solids at a ratio of 70 g resin/g VSS and stirred for 1.5 h at 600 rpm (Frølund et al. 1996; Ng et al. 2006; Choi et al. 2013). The same resins are also used at different conditions, for example, a dose of 60 g resin/g TSS followed by mixing at 200 rpm for 12 h at 4°C (Sheng and Yu 2006). It is believed that the cation exchange resin scavenges the cations crucial for the electrostatic bondings of EPS on microbial cells. Hence, the loosely bound EPS can be easily detached from the microorganisms in the shear field generated by the mixing. In some occasions, EDTA has also been added to sequester cations from the mixed liquor sample instead

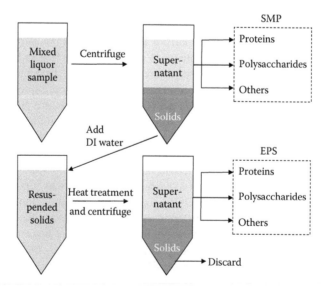

FIGURE 5.1 Flow diagram of SMP and EPS quantification.

of heat treatment or cation exchange resins (Liu and Fang 2002). The NaOH solution containing formaldehyde can also be used (Jia et al. 1996; Zhang et al. 1999). Depending on the strength and the effectiveness of the EPS detachment method, the quantified EPS can vary widely even for the same mixed liquor sample.

The treated resuspended solids are centrifuged to obtain the supernatant from the solids. The revolutions per minute of the centrifuge used for this procedure are typically higher than that for the SMP separation because this sample contains finer debris than the original mixed liquor sample. A filtration step may be followed to refine the supernatant using syringe filters. The protein and polysaccharide contents in the supernatant are analyzed using the same methods used for SMP. The measured proteins and polysaccharides are considered EPS.

5.2.2.2 Effect on Membrane Fouling

In the early days of MBR, MLSS was highlighted as a major factor affecting membrane fouling (Yamamoto et al. 1989). However, the focus moved to SMP later after finding that the biopolymers took the majority of the cake layer mass formed on the membrane surface in typical operating conditions. Soon after, the focus was expanded to the bound portion of the biopolymers, that is, EPS, whose occurrence was positively correlated with the SMP contents of the sludge in general (Chang and Lee 1998; Chang et al. 1999). It was also observed that the SMP/EPS contents of sludge appeared closely related with the microbial physiology rather than the MLSS. Since then, numerous studies confirmed the positive relationship between the SMP/EPS content and the membrane fouling rate (Cha et al. 2003; Tarnacki et al. 2005; Yoon et al. 2005; Fan et al. 2006; Grelier et al. 2006; Judd 2006). In majority of the cases, polysaccharide-based SMP appeared as a major contributor for membrane fouling rather than protein-based SMP (Judd 2004; Lesjean et al. 2004; Rosenberger et al. 2006; Yigit et al. 2008).

The positive correlations between the membrane fouling rate and the SMP/EPS are in fact quite convincing. According to Choi et al. (2013), the increasing rate of the TMP was directly related with SMP and EPS concentrations. Colloidal or total organic carbon (TOC)/chemical oxygen demand (COD) of the supernatant was also identified as indicators of the membrane fouling potential because they tend to be proportional to the SMP contents. In other studies (Cha et al. 2003; Fan et al. 2006), the mixed liquor was filtered by 0.45 μm membrane filter and the filtrate TOC was regarded as total SMP. The use of colloidal TOC as an indicator of the membrane fouling potential is discussed in Section 5.3.3.2. An exceptionally linear relationship between the SMP and the membrane fouling rate was observed in pilot MBR tests, as shown in Figure 5.2. In this study, two MBRs equipped with vertically mounted hollow fiber membranes (Memcor® of Evoqua) with a total surface area of 7 m^2 submerged in 2 m^3 reactor were used to treat municipal wastewater.

The biopolymers in mixed liquor can be also quantified using the criterion of transparent exopolymer particles (TEP) instead of SMP. TEP consist of mainly polysaccharides, but they also include proteins, lipids, nucleic acids, various sugars, etc. TEP are largely overlapped with SMP in nature and are transparent sticky gel particles. It is known to be formed by long-chain polysaccharides backbone trapping other minor constituents listed earlier. The size ranges from nanometer to millimeter. Multivalent ions play a role in stabilizing the structure by bridging molecular chains inside the gel particles. In some cases, TEP concentration in the supernatant can be used as an indicator of the membrane fouling potential (de la Torre et al. 2008).

SMP are major constituents of membrane fouling layer in MBR. EPS can also foul the membrane, but they need to be detached to become SMP. Hence, the concentration of SMP was considered a direct indicator of the membrane fouling potential. However, the dependence of the membrane fouling rate on SMP concentration has been challenged by many researchers since the mid-2000s. In fact, many experimental observations suggest that the membrane fouling rates cannot be simply correlated with the quantity of SMP (Drews 2010). According to the study performed by Drews et al. (2008), no apparent correlations are found between the SMP content and the membrane fouling rate, as shown in Figure 5.3, although identical analytical methods were used in all the experimental

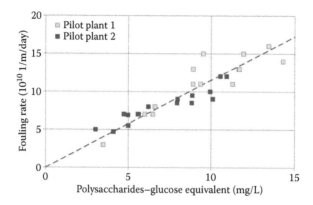

FIGURE 5.2 Fouling rate as a function of polysaccharides in the activated sludge liquid phase at an SRT of 8 days. (From Rosenberger, S. et al., *J. Membrane Sci.* 263(1–2):113–126, 2005.)

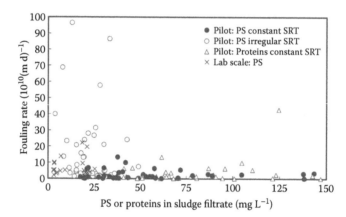

FIGURE 5.3 The relationship between the membrane fouling rate and the polysaccharides (PS) and protein portions of SMP at constant or variable SRT. (Modified from Drews, A. et al., *Desalination* 231:141–149, 2008.)

sets. As can be seen in Figure 5.3, membrane fouling can surge even at a low polysaccharides concentration. At the same time, high polysaccharide and protein concentrations do not necessarily cause fast membrane fouling.

The fuzzy relationship between the SMP content and the membrane fouling rate is not surprising because the notion started simply because SMP were the major constituent of the fouling layer formed on membrane. It should be noticed that this hypothesis is based on one dubious assumption that the SMP produced in different conditions have an equal tendency to foul membrane. However, the fuzzy relationship between SMP concentration and membrane fouling has disproven the hypothesis. Another potential reason for the poor correlation is that the analytical methods used to quantify SMP do not necessarily detect all the polysaccharides and the proteins accurately. At the same time, the actual membrane foulants are not necessarily detected by the analytical methods used to quantify SMP. In fact, the SMP measured by the analytical methods currently available also detect terrestrial humic substances, nonbiological polymeric substances, etc., in addition to polysaccharides and proteins (Judd 2008). The relationship among detectable SMP, nondetectable SMP, and SMPs actually causing membrane fouling can be illustrated as the conceptual diagram, as shown in Figure 5.4.

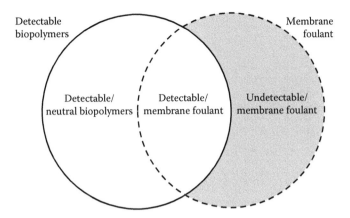

FIGURE 5.4 Conceptual diagram of the relationship between detectable biopolymers and actual membrane foulant.

5.2.3 ORGANIC LOADING

5.2.3.1 Effect of F/M Ratio on Membrane Fouling

The specific OLR can be quantified by the F/M ratio that is in turn directly related with SRT and hydraulic retention time (HRT) in a given tank. If the specific OLR (or F/M ratio) declines, excess sludge removal should be reduced to maintain a constant MLSS, which causes longer SRT. Likewise, if the influent flow rate increases in a given system (or HRT decreases), the F/M ratio increases accordingly. To maintain a constant MLSS, more excess sludge should be removed from the system, and in turn SRT decreases. Overall, if MLSS and feed strength stay constant in a given system, high SRT means low F/M ratio and low HRT, whereas low SRT means high F/M ratio and low HRT. Because all three parameters, that is, SRT, F/M ratio, and HRT, are closely interconnected, SRT will be used in this section to discuss the effect of the organic loading on membrane fouling.

It has been reported that membrane fouling tends to decrease as SRT increases and F/M ratio decreases. To elucidate the SRT effect on membrane fouling, pilot tests were performed at three different SRT at 10, 30, and 50 days (van den Broeck et al. 2012). Feed flow rates were identical to the three trains, and the SRT was controlled by differentiating the excess sludge removal rates. Although MLSS was nearly five times higher at the 50-day SRT than at the 10-day SRT (14 vs. 3 g/L), TMP rose 58 times slower (0.187 mbar/day vs. 10.8 mbar/day). The high membrane fouling rate at the 10-day SRT was attributed to the deflocculation of microbial floc. Similar results were obtained when the TMP increasing rates were compared at three different SRTs at 8, 15, and 40 days (Grelier et al. 2006). It was apparent that the fouling rate of hollow fiber membrane was the highest at the lowest SRT (8 days) and the lowest at the highest SRT (40 days). In this study, MLSS were 3.2, 4.9, and 7.7 g/L for the SRT of 8, 15, and 40 days, respectively. The fast membrane fouling at low SRT was attributed to the high polysaccharide-based SMP concentration in the mixed liquor. In another study, the same trend was found with submerged flat sheet membranes at SRT of 3 to 20 days, where low SRT is synonymous with a high F/M ratio (Ng et al. 2006). When F/M ratio was maintained identically at 0.13 g COD/g MLSS/day, it was observed that the membrane permeability was rarely affected at an MLSS range of 8 to 18 g/L (Zsirai et al. 2014).

The most rigorous experiments were performed by Trussell et al. (2006). MLSS was maintained identical at 8 g/L by adjusting excess sludge removal. The influent flow rate was controlled to obtain the target F/M ratio. Filtration was performed using three submerged hollow fiber modules (ZW500c), and the flux was maintained at 30 LMH throughout the experiment. The excess permeate was returned back to the aeration tank to maintain the water level. As a result of the sophisticated experimental setup, the parameters related with OLR (F/M ratio, SRT, and HRT) could be isolated from the MLSS and flux effects. This experiment clearly showed that the membrane fouling rate

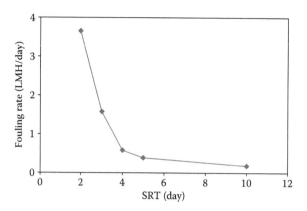

FIGURE 5.5 Effect of SRT on the membrane fouling rate measured by the time derivative of resistance increase. (From Trussell, R.S. et al., *Water Res.* 40:2675–2683, 2006.)

decreased as SRT increased. In this study, it was found that there was a critical SRT below which membrane fouling dramatically rose. Under the condition used, the TMP rate increased abruptly when SRT was less than 4 to 5 days (Figure 5.5).

Although the longer SRT has been known to enable less membrane fouling, it is not straightforward to determine the optimum SRT that provides the minimum overall costs because there are many conflicting financial and nonfinancial factors. The long SRT incurs a requirement of large biological tanks and a large footprint at a given MLSS. In addition, the long SRT increases the oxygen demand for endogenous respiration. However, the gains in OTE at a long SRT at least partially compensate the incremental increase in oxygen demand, as discussed in Section 6.5.2.6. The large aeration tanks also provide a buffer capacity to hold more water during the peak flow or to store more sludge when the sludge handling process fails. Currently, most municipal MBRs are designed targeting the SRT of approximately 20 days during the average organic loading, but the target SRT can be reduced to 12 to 15 days during the peak organic loading (Diamond 2010).

5.2.3.2 Effect of F/M Ratio on SMP/EPS

It has been consistently observed that membrane fouls slower at low F/M ratio (or at high SRT) in various settings regardless of the scale of the experiment (Cicek et al. 2001; Ng et al. 2006; Al-Halbouni et al. 2007; Cao et al. 2008; Patsios and Karabelas 2011; van den Broeck et al. 2012). However, the attempt to correlate the membrane fouling rate with the quantity of SMP and EPS in the mixed liquor has not been always successful, as discussed in Section 5.2.2.2.

In a side-by-side test, two laboratory-scale MBRs were maintained at F/M ratios of 0.17 and 0.50 g COD/g MLSS/day, respectively (Wu et al. 2013). The corresponding SRTs were 45 and 7 days, respectively. While monitoring TMP as an indicator of membrane fouling, the portion of SMP and EPS with polysaccharides and proteins was tracked to characterize the mixed liquor. The SMP and EPS contents of the two mixed liquors are compared in Figure 5.6, where SMP is marked as soluble EPS. Although the TMP increase was 5 to 20 times faster in the MBR with a short SRT (or high F/M ratio), SMP/EPS contents were not necessarily higher. This observation suggests that the quality of the SMP/EPS played a crucial role in membrane fouling rather than the quantity. Therefore, the SMP/EPS quantity alone is not considered as an indicator of the membrane fouling potential in MBR.

5.2.3.3 Effect of F/M Ratio on Membrane Fouling in Anaerobic MBR

As in aerobic MBR, the membrane fouling rate in anaerobic MBR (AnMBR) is positively correlated with F/M ratio. In a laboratory study, two AnMBRs equipped with submerged hollow fiber membranes were fed with a synthetic feed with 500 mg/L COD to obtain F/M ratios of 3.8 and 0.1 g COD/g MLSS/day, respectively (Liu et al. 2012b). It was very apparent that membranes were fouled

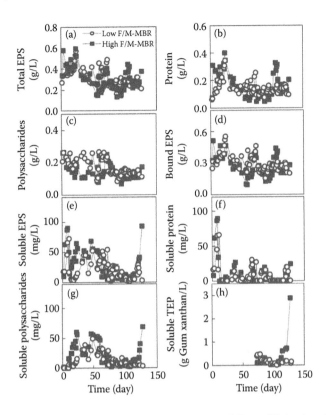

FIGURE 5.6 EPS characteristics and soluble TEP in the MBRs at different F/M ratios: (a) total EPS, (b) total protein, (c) total polysaccharides, (d) bound EPS, (e) soluble EPS, (f) soluble protein, (g) soluble polysaccharides, and (h) soluble TEP. (From Wu, B. et al., *Separ. Sci. Technol.* 48(6):840–848, 2013.)

much quickly at a higher F/M ratio (or lower SRT), as shown in Figure 5.7, where HAnMBR and LAnMBR indicate AnMBR with high and low organic loadings. It was also noticed that the cake layer formation was responsible for 99% of the permeability loss in both HAnMBR and LAnMBR when the filtration resistance was analyzed based on resistance in series model (Equation 1.4). Same has been confirmed in pilot-scale AnMBR for various industrial wastewaters, but no sufficient data exist to determine the optimum F/M ratio or SRT partly because AnMBRs are rare to date and partly because AnMBR is mainly for industrial high-strength wastewaters having variety of different sources and hence it is difficult to draw universal design parameters.

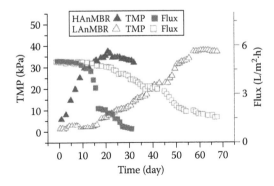

FIGURE 5.7 TMP increase and flux decline in high loading AnMBR (HAnMBR) and low loading AnMBR (LAnMBR). (From Liu, L. et al., *J. Membrane Sci.* 394–395:202–208, 2012b.)

5.2.3.4 Effect on Subsequent Reverse Osmosis Membrane Fouling in Water Recycle

The microbial physiology and the associated SMP/EPS characteristics not only affect the membrane performance in MBR but also affect the subsequent reverse osmosis (RO) membrane performance in water recycle. In a laboratory-scale study, two laboratory-scale MBRs were operated at an F/M ratio of 0.17 and 0.50 g COD/g MLSS/day, respectively, using synthetic feed. The MBR effluent was treated by subsequent RO membrane (UTC-70, Toray) with 0.0186 m². It was observed that the RO membrane treating the effluent of high F/M MBR was fouled four times quicker in terms of the increasing rate of the TMP at a constant flux (Kitade et al. 2013; Wu et al. 2013). It was also found that the concentrations of dissolved organic carbon (DOC), polysaccharide, protein, and TEP were higher in the effluent of high F/M MBR (Figure 5.8). When a cartridge filter with 0.45 μm nominal pore size was inserted in between MBR and RO, the fouling rate of RO decreased significantly especially for the permeate from the high F/M MBR. This suggested that the MBR permeate contains a sizable amount of filterable particles and/or macromolecules that could be captured by random collision and adhesion on cartridge medium.

5.2.4 Dissolved Oxygen

It has been known that DO can be one of the major factors affecting membrane fouling in MBR, if it is below a critical level, for example, 1 to 2 mg/L. At low DO, the chance of TMP surge increases by the stressed microorganisms. Even if DO is overall sufficient, a similar low DO effect can be observed if mixing is not sufficient and dead zones exist in the system. On the contrary, no firm evidence suggests that excessively high DO itself affects membrane fouling either negatively or positively. However, if high DO is obtained with excessive aeration, the floc particles exposed to the high shear can be broken down to smaller pieces, and thus, membrane fouling may be affected negatively.

Kang et al. (2003) performed experiments using a laboratory-scale MBR that had a capability of being bubbled either by air or nitrogen. Vertically mounted polyethylene membranes with 0.1 μm pore sizes were used at 20 LMH. In this study, air was used to scour membrane surface and supply oxygen to the system simultaneously. The DO effect on membrane fouling was investigated by switching air to nitrogen gas without affecting the strength of membrane scouring. It was observed

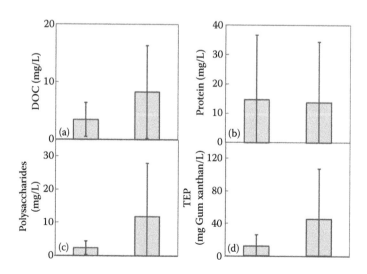

FIGURE 5.8 Effect of MBR permeate quality on RO fouling rate. (a) DOC in the MBR permeate. (b) Protein in the MBR permeate. (c) Polysaccharides in the MBR permeate. (d) TEP in the MBR permeate. (From Wu, B. et al., *Desalination* 311:37–45.)

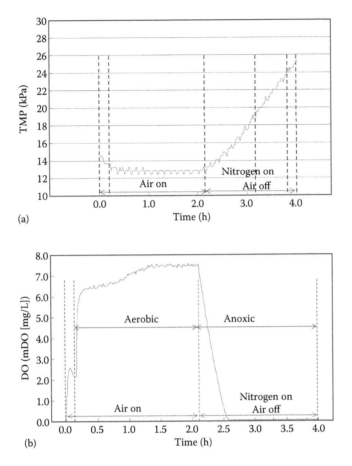

FIGURE 5.9   (a) TMP and (b) DO profiles in laboratory-scale MBR. (From Kang, I.-J. et al., *Water Res.* 37:1192–1197, 2003.)

that TMP almost immediately increased when DO started to drop, as shown in Figure 5.9. The immediate increase in TMP suggests the existence of an indirect effect of insufficient DO on membrane fouling. Meanwhile, the supernatant COD, which was closely linked with SMP, increased. Overall, the supernatant COD was in a inverse correlation with DO as it increased significantly as DO drops from above 2.0 to 0.2 mg/L. But, supernatant COD did not decrease significantly when DO increased above 1.0 to 2.0 mg/L in the lab scale experiment. A similar observation was also made in another study (Jin et al. 2006), where air and nitrogen mixtures were supplied to two reactors at two different ratios to obtain low DO (<0.1 ppm) and high DO (>3.0 mg/L). At an identical flux, the increasing rate of the TMP was much smaller at the high DO reactor.

5.2.5   FAT, OIL, AND GREASE

The FOG content of wastewater is measured typically by the n-hexane extraction method following the Method 1664 of the U.S. Environmental Protection Agency. FOG is known to attract to the membrane by hydrophobic–hydrophobic interaction because the membrane surface is more hydrophobic than water. FOG forms a dense gel layer that acts as a filtration resistance. Although there are no universally applicable FOG limits in feed water, many factors affect the practical limit. If the oil is emulsified in feed water, the tolerance should be much higher than that of free oil because oil particles do not directly interact with membrane. The FOG tolerance of membrane also increases as MLSS increases because a large surface area of floc particles can adsorb FOG. Biodegradability is another factor affecting the limit because it determines the steady-state FOG concentration in the membrane tank along with other hydraulic parameters.

There are no broadly recognized FOG limits in feed water because the characteristics of FOG are site specific and its effect is system specific. However, the FOG in feed water measured by the n-hexane extraction method are typically controlled at levels less than 50 to 100 mg/L for bio-degradable oils and less than 5 to 10 mg/L for nonbiodegradable oils. If the FOG level is higher, additional pretreatment processes may be required to remove oil from the feed water, for example, settling tank, dissolved air floatation (DAF), and aerated grit chamber.

5.2.6 ANTIFOAM

Antifoams have been used to control foam depth in MBR. The defoaming mechanisms are discussed in Section 4.3.1, but if antifoams are misused, they can suddenly increase the hydrophobic material concentration in the mixed liquor by returning the foam constituents back to the mixed liquor. The surging hydrophobic materials can not only increase the membrane fouling rate but also compromise OTE according to the mechanism discussed in detail in Section 6.5.2.7. The low OTE in turn hampers DO concentration and biological health. Eventually, the low DO leads to the expedited membrane fouling, as discussed in the previous section. It is noteworthy that the negative effect of antifoam on membrane performance is not a direct consequence of antifoam but an indirect consequence of using antifoam improperly. To prevent the antifoam-induced membrane fouling, foam depth should be maintained at a constant level by not allowing it to grow too high. Abrupt defoaming should also be avoided to prevent OTE drop.

It has been widely perceived that silicone oil and hydrocarbon-based antifoams aggravate membrane fouling whereas alcohol-based antifoams do not. This notion seems to be developed in the biotechnology industry because of the concerns over the hydrophobic interaction between the oil-based antifoams and the membrane surface (Kloosterman et al. 1988; Russotti and Goklen 2001). However, no definitive supporting empirical evidence exists as far as MBR is concerned. In fact, it is questionable that a small quantity of antifoam dosed typically less than 10 mg/L as product significantly affects membrane performance under the presence of 6 to 15 g/L biosolids that provides hundreds or thousands times larger surface areas than the membrane. In fact, there is anecdotal evidence that some silicone oil-based antifoams can be used in MBR without a noticeable membrane performance loss.

Antifoams consist of one or more of hydrophobic molecules or particles (e.g., long-chain alcohols, glycols, vegetable oil, petroleum oil, and silicone oil) and/or hydrophobic colloids, such as fumed silica/silicone, polydimethylsiloxanes dispersed in water or oil, and N,N'-ethylene *bis*-stearamide wax dispersed in oil (Wikipedia 2013a). On the basis of short-term pilot tests, "approved" and "banned" antifoams were determined by one membrane supplier, as summarized in Table 5.1 (Zenon 2003). Basically, all the antifoams containing emulsified silicone oil, petroleum hydrocarbons, or long-chain polymers are banned, whereas most of the water-soluble antifoams containing glycerins, short-chain polyether polyols, polyglycols, etc., are approved. Many membrane manufacturers adapt the similar approval criteria as a precautionary measure. However, the list is based more on the notion of hydrophobic interaction between the chemicals and the membrane surface rather than long-term experimental studies in the full-scale MBR treating real wastewater. The exact effects of the banned/approved antifoams on membrane performance in MBR are yet to be discovered.

5.2.7 EFFECT OF UNSTEADY OPERATION

5.2.7.1 Normal Fluctuation of F/M Ratio

The operating condition of MBR inevitably fluctuates as a consequence of the variations in hydraulic loading and wastewater strength depending on the timing of the day, seasons, weather, human activities occurring in the area, leak or inflow in the intake pipeline, etc. For example, two hydraulic peaks

TABLE 5.1
Approved and Banned Antifoams

Approved	Banned
Nalco IL08[a]	Betz Foamtrol AF1660
Nalco 7465	Betz Foamtrol AF3550
Nalco 7471[b]	Betz Foamtrol AF3551
Nalco 76028	Surpass Chemical Co. NOFOME AK
Dow Polyglycol 45-200	Ultra Additives Inc. FOAMTROL WT-2
Dow Polyglycol FR-530	Ultra Additives Inc. FOAMTROL WT-73
Dow Polyglycol P-1200	Ultra Additives Inc. FOAMBAN MS-5
Dow Polyglycol 112-2	

Source: Zenon Environmental Inc. 2003. A Zenon design and pilot report.
Available at http://www.gov.mb.ca/conservation/eal/registries
/brandonwastewater/eia/append-b.pdf (accessed November 30, 2010).
[a] Discontinued. Replaced with Nalco 60096.
[b] Alternative to IL08.

are apparent in the morning and in the evening in most municipal MBR. In addition, the short-term peaks by rainfall and the long-term seasonal peaks also occur at various extents and durations depending on the weather. Figure 5.10 shows yearly profiles of daily flow rate and influent BOD of a municipal wastewater treatment plant in the northeast of the United States. The peak daily flow and the peak daily OLR are 250% and 185% of the annual average in the example, respectively. Meanwhile, industrial MBR typically face more severe loading fluctuations depending on the manufacturing schedule. The hydraulic and organic loading fluctuations are buffered by flow equalization tanks to spread out the peak loadings and to minimize the capital investment for MBR, as discussed in Section 6.7.

Any OLR fluctuation in MBR has been deemed to negatively affect membrane performance by stressing microorganisms and promoting biopolymer production. However, the magnitude of acceptable fluctuation and the effect of moderate fluctuations are not clearly known with respect to membrane fouling. It has also been discovered that, if the OLR fluctuation causes low DO conditions at its peak, membranes are more prone to be fouled (see Section 5.2.7.4). Therefore, the tolerance of membrane fouling to OLR fluctuation is partially dependent on the oxygen supplying capacity.

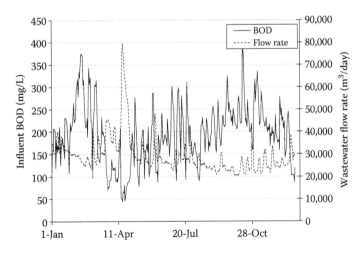

FIGURE 5.10 Daily flow rate and BOD of a municipal wastewater treatment plant in the Northeast United States.

In fact, the high OLR causes low DO not only because of the high oxygen demand but also because of the plunging OTE, as discussed in Section 6.5.2.6. For instance, α-factor, which is a linear indicator of the relative OTE to the standard OTE in clean water, can vary widely even with a moderate daily organic loading fluctuation in the aeration tank (Leu et al. 2008). In a field survey, α-factor decreased from 0.52 to 0.36 in the same spot in an aeration tank when daily peak loading occurred, which corresponds to the 31% less oxygen dissolution at a constant airflow. In fact, maintaining a constant and sufficient DO in the aeration tank is very crucial in minimizing the negative effect of organic loading fluctuation in MBR, as discussed in Section 5.2.4.

The infrequent removal of excess sludge can also elevate the membrane fouling rate. In this case, MLSS can swing in a wide range because of the infrequent sludge removal. As a result, the specific OLR, that is, the F/M ratio, fluctuates in a wide range. When around 50% of sludge in the biological system was removed periodically and freshwater was added to fill the spaces, TMP increased abruptly right after the sludge removal despite the 50% low MLSS (Drews et al. 2006). It was attributed to the abrupt increase of F/M ratio that stressed microorganisms. Meanwhile, polysaccharides based SMP concentration increased while MLSS built up, but it did not decrease proportionally to the MLSS after 50% of the sludge was removed. This suggested that the high F/M ratio stimulated microorganisms to secrete extra polysaccharides.

A laboratory-scale experiment was performed simulating the diurnal organic loading pattern commonly observed in municipal MBR (Zhang et al. 2010b). Feed TOC increased 100% twice a day for 2 h each in the treatment reactor (feed variable), whereas it was maintained nearly constant in the control reactor (feed constant). The average organic loading for the two reactors was maintained identically by reducing feed TOC later in the day for the treatment reactor. The level of DO was maintained at more than 2 ppm throughout the experiment regardless of the organic loading. It was observed that soluble TOC and polysaccharides of the supernatant, which typically indicate the membrane fouling tendency, were slightly lower with constant organic loading in the first 80 days in phase I. However, the trends were reversed in the next 80 days in phase II. Simultaneously, initially smaller particle sizes in the treatment reactor in phase I became larger in phase II, as shown in Figure 5.11. As a result, the membrane fouling rate measured by the increasing rate of the TMP (dTMP/dt) was higher in the treatment reactor in phase I, but it became lower in phase II, as summarized in Figure 5.12. It is noteworthy that this experiment was performed under sufficient DO conditions regardless of the level of OLR at higher than 2 mg/L. But, the constant DO regardless of the OLR is not easily achieved in full scale MBR. If the experiments were performed under low DO conditions during the peak loading just like in full-scale MBR, the result might have been different.

5.2.7.2 Shock Hydraulic Loading

Beyond the normal daily hydraulic fluctuations, the hydraulic shocks occur when rainwater is mixed into raw wastewater. This is not completely avoidable in municipal MBR despite the efforts on

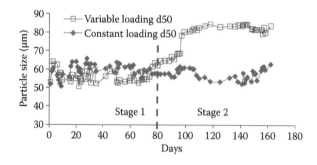

FIGURE 5.11 The median particle diameter as a volume% at constant and variable OLRs. (From Zhang, J. et al., *Water Res.* 44:5407–5413, 2010b.)

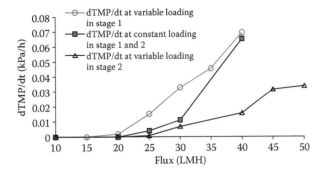

FIGURE 5.12 TMP profile over time at 20 LMH for constant loading and variable loading. (From Zhang, J. et al., *Water Res.* 44:5407–5413, 2010b.)

securing the tight collection systems. When MBR are shocked by a surging hydraulic loading by the rainfall, the OLR and the F/M ratio also surge temporarily because the settled sludges in the collection pipelines can be swept by the surging flow. Meanwhile, the ionic strengths of feed water drops because of the dilution of rainwater. Once the settled sludges in the collection pipelines are depleted, the OLR returns to near presurging condition. It has been known that hydraulic loading shock has a negative effect on membrane fouling, but it is not as significant as organic loading shocks, which will be discussed in the next sections.

Syed et al. (2009) simulated 100% hydraulic loading shock using a Z10™ module (GE) with 2.7 m² surface area submerged in a 120 L membrane tank. Feed water was diluted with an equal amount of clean water, whereas flux was raised 100% during the shock hydraulic loading condition. The mixed liquor was circulated between the membrane tank and the 380 L aeration tank. Before the test, the experimental setup was run for around 3 months to achieve a steady-state condition. Experimental conditions are summarized in Table 5.2. TMP should be increased significantly to obtain 100% higher flux during the hydraulic loading shocks that lasted 24 h each. However, the TMP was mostly recovered back to the presurge condition immediately after the hydraulic surge condition ended, as shown in Figure 5.13. Meanwhile, DO, MLSS, and pH did not change much. It was found that SMP decreased during the hydraulic shock (first 24 h) perhaps because of the expedited permeation of the small SMP at high flux condition. By contrast, bound EPS, which is attached on microorganisms, stayed at almost same level. It was also observed that floc size decreased slightly perhaps because of the low ionic strength of the diluted feed water.

TABLE 5.2
Experimental Condition of Hydraulic Shock Loading Test

Parameters	Unit	Steady State (Before Peaking)	Hydraulic Peaking Condition
F/M ratio	g COD/g VSS/day	0.26	0.26
DO	mg/L	2.5–3.5	2.5–3.0
SRT	days	15	15
HRT	hours	8	4
MLSS	g/L	8	~8
Test duration	hours	~2100	24

Source: Data from Syed, W. et al., Effects of hydraulic and organic loading shocks on sludge characteristics and its effects on membrane bioreactor performance. *Proceedings of WEFTEC.* Oct. 17–21, 2009, Orlando, Florida.

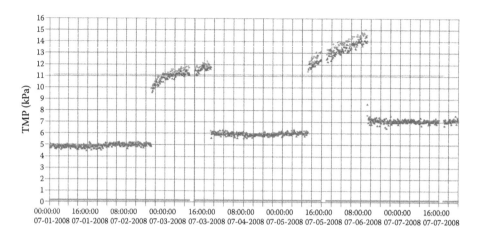

FIGURE 5.13   Effect of hydraulic loading shock on membrane fouling measured by TMP. (From Syed, W. et al., Effects of hydraulic and organic loading shocks on sludge characteristics and its effects on membrane bioreactor performance. *Proceedings of WEFTEC*. October 17–21, 2009, Orlando, Florida.)

5.2.7.3   Shock Organic Loading at Low DO

Abruptly surging organic loading causes a depletion of DO in mixed liquor because of the lagging aeration adjustment and/or insufficient aeration capacities. The low DO stresses microorganisms and stimulates biopolymer production, which in turn mitigates oxygen transfer rate in the air–liquid interface. The low oxygen transfer in turn causes lower DO and put more pressure on microorganisms to produce more biopolymers that hampers oxygen transfer. It is difficult to escape from this self-accelerating vicious cycle once the system enters into it without cutting OLR substantially.

Syed et al. (2009) investigated the effect of shock organic loading on membrane fouling under the similar condition discussed in Section 5.2.7.2. During the shock organic loading period of 8 h, F/M ratio was quadrupled from 0.26 to 1.04 g COD/g VSS/day by adding molasses solution to the screened municipal wastewater. Aeration system could not keep up with the high oxygen demand; thereby, DO immediately plunged from 3 to 0.2 mg/L. Meanwhile, TMP immediately jumped from 13 to 21 kPa at a constant flux. TMP did not decrease much even after the surge event. This suggested that a substantial irreversible membrane fouling occurred during the shock organic loading. It was found that SMP concentration increased nearly threefolds for both polysaccharides and proteins during the 8 h event, as shown in Figure 5.14. After 16 h from the end of the shock loading,

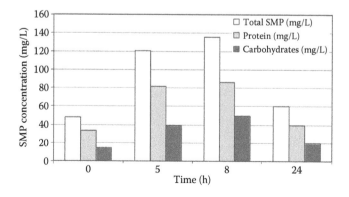

FIGURE 5.14   Effect of organic peaking on SMPs. (From Syed, W. et al., Effects of hydraulic and organic loading shocks on sludge characteristics and its effects on membrane bioreactor performance. *Proceedings of WEFTEC*. October 17–21, 2009. Orlando, Florida.)

TABLE 5.3

Experimental Condition of Organic Shock Loading Tests

Parameters	Unit	Steady State	Organic Peaking Condition
F/M ratio	g COD/g VSS/day	0.26	1.04
DO	mg/L	2	0.20–3.0
SRT	days	15	
HRT	hours	8	8
MLSS	g/L	10 ± 0.5	
Test duration	hours	~1400	8

Source: Syed, W. et al., Effects of hydraulic and organic loading shocks on sludge characteristics and its effects on membrane bioreactor performance. *Proceedings of WEFTEC.* Oct. 17–21, 2009, Orlando, Florida.

SMP levels still did not fully return to the presurge condition. The plunged DO to approximately 0.2 mg/L gradually recovered, but it took nearly 50 h to fully return to the original level at 3 mg/L (Table 5.3).

5.2.7.4 Shock Organic Loading at High DO

It is apparent that shock organic loading accelerates irreversible membrane fouling in the field condition, where dropping DO is unavoidable at least to some extent. However, it is not completely clear whether the similar irreversible membrane fouling occurs, if DO can be maintained at a constant and sufficient level. An experiment was performed with two bench-scale MBRs equipped with flat sheet Kubota membranes under a well-controlled laboratory condition (Zhang et al. 2010a). The control MBR was run at a constant organic loading, whereas the treatment MBR was run at variable OLR. Feed TOC for the treatment train was doubled twice a day for 2 h each time. By reducing feed TOC in later part of the day, average daily TOC loadings were maintained identical in the two MBRs. DO was maintained at levels higher than 2 mg/L throughout the experiment, and flux was set at 20 LMH for both MBRs. In the first 80 days, the increase in TMP was somewhat faster in the treatment MBR, but the trend was reversed in the next 80 days. Overall, the organic loading shock twice a day at 200% of the normal loading rate did not significantly affect membrane fouling. This observation suggests that temporary shock organic loadings at 2× peaking factor may not cause irreversible membrane fouling as long as DO is maintained high enough. Therefore, it is important to secure sufficient aeration capacity in full-scale MBR to avoid irreversible membrane fouling during the peak organic loading condition.

5.2.8 WATER TEMPERATURE

The membrane performance measured by permeability naturally decreases as water temperature decreases because the permeate passage through membrane pores becomes harder with the rising water viscosity, as discussed in Section 1.2.9. The long-term permeability recorded for several years shows a cyclical curve that well coincides with the water temperature, as shown in Figure 4.22. It is remarkable that the amplitude of the permeability curve is much larger than that predicted by the water viscosity. For instance, water viscosity increases from 0.89 to 1.31 cP when temperature drops from 25°C to 10°C, which is a 47% increase of viscosity that can reduce permeability by 32% according to Equations 1.13 and 1.14. However, the actual permeability declines 40% to 60% from 250 LMH/bar to 100 to 150 LMH/bar in the figure.

The more severe membrane permeability loss than the predicted by viscosity is well presented in Figure 5.15 (van den Brink et al. 2011). The membrane fouling rates at different water temperatures

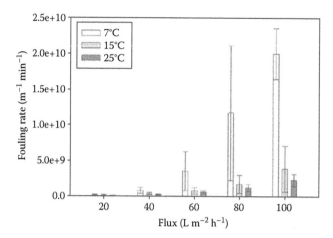

FIGURE 5.15 Average fouling rates for different flux steps at three different temperatures. (From van den Brink, P. et al., *Water Res.* 45(15):4491–4500, 2011.)

were measured using a small-scale MBR with 85 L total volume. Municipal wastewater was used as feed, and flat sheet Kubota membranes were submerged in the aeration tank. Figure 5.15 clearly shows that the membrane fouling measured by $R_m + R_f$ occurs much faster at low temperature regardless of the flux. The temperature effect on membrane fouling rate is more clear especially when the flux is above the normally accepted sustainable flux, for example 20–30 LMH. Due to the faster membrane fouling at low temperature, permeability tends to be lower than predicted by viscosity based correction equations. Therefore, flux correction against a reference temperature is valid only in a narrow range around the reference temperature.

The inaccuracy of viscosity based correction equations also can be explained by the partial deflocculation of microbial floc at low temperature. In a field survey, the supernatant TOC contents, which represents free fine particles not flocculated, increased during the winter to spring time (Wu et al. 2011). A similar result was also found in another pilot MBR study (van den Brink et al. 2011). In this study, the TOC increased in the supernatant mainly because of the increasing polysaccharide concentrations. The excess permeability loss at low temperature is also explained by particle dynamics in the membrane tank. First, the shear stress acting on the membrane surface decreases because of the decreasing upflow velocity if temperature decreases. Second, the particle back transport decreases because of the high viscosity, as discussed in Section 1.2.5. Third, the low particle diffusivity at low temperature causes a slow particle back transport and a thicker concentration polarization layer.

5.2.9 SALINITY

There is no direct effect of salinity on membrane performance because the membranes used in MBR have big enough pores that can pass inorganic ions freely and do not cause osmotic pressure. However, salinity indirectly affects membrane performance by affecting the floc size, the fine particle concentration, and the quality and quantity of biopolymers.

The effect of short-term salinity hike was investigated using a pilot-scale setup (Reid et al. 2006). When chloride (Cl⁻) concentration was increased up to 4.5 g/L using NaCl, membrane permeability dropped almost immediately, as shown in Figure 5.16 (seawater contains 19.4 g/L chloride). The recovery of flux did not occur quickly after the salinity went back to the original level. Although particle size appeared not changed, both protein and carbohydrate levels increased substantially, especially in SMP rather than EPS. Similar observations were made in a short-term side-by-side test using flat sheet submerged membranes (Singh et al. 2008). In this study, 30 g/L NaCl caused elevated polysaccharide EPS, smaller flocs, higher effluent COD, and faster increase of TMP in

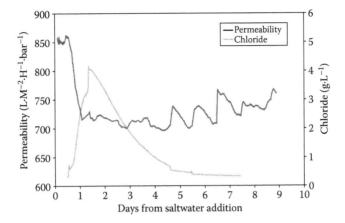

FIGURE 5.16 Permeability drop due to a salinity shock. HRT = 72 h, SRT = 54 days, flux = 8 LMH. (From Reid, E. et al., *J. Membrane Sci.* 283:164–171, 2006.)

midterm. In another side-by-side test, the surge of salinity immediately causes a shift of particle size distribution (PSD) toward a smaller direction by breaking down the over-micro-sized particles and accelerated membrane fouling (de Temmerman et al. 2014). In this study, the surge of SMP was observed with a lag time of a few hours, but powdered activated carbon (PAC) added to the mixed liquor was effective in reducing the SMP concentration. The deflocculation of microbial floc at high salinity has been explained by the ion exchange reaction by monovalent ions (Park 2002). In fact, monovalent ions replace the divalent ions that bridge organic molecules when the concentration is high enough. Because monovalent ions are less effective in bridging macromolecules, defloccula-tion becomes more prevalent when monovalent ions outnumber divalent ions by a factor of 2 to 3 or more.

The microbial deflocculation and the elevated SMP expedite membrane fouling when salinity swings in short-term, but the long-term salinity effect can be milder because of microbial adapta-tion. Microbial adaptation occurs either by changing the metabolism of the same microorganisms or by the proliferation of the most fit species under the given environment. It can be postulated that the salinity effect on membrane fouling can vary depending on the duration and the frequency of the salinity swing. In addition, the threshold salinity that triggers expedited membrane fouling can also be higher for long-term operations.

There is no established guideline for the maximum allowable salinity without causing a signifi-cant membrane fouling, but it can be somewhere around 3.5 to 7.0 g/L as NaCl or 5 to 10 mS/cm as conductivity in long-term operations, assuming the salinity does not swing abruptly. MBR is still operable above the range, but membrane surface areas larger than those normally required are necessary to treat the same flow at a lower flux. In another study (Lay et al. 2010), nitrogen removal was not affected at up to 30 g/L NaCl, but it started to be inhibited at 40 g/L.

The design flux of an MBR treating saline wastewater from a seaweed processing plant was set at around a half of that for municipal MBR, that is, 11 LMH (Lala et al. 2014). The salinity of waste-water was comparable with the seawater, and MLSS was maintained at 30 to 70 g/L. No significant troubles were experienced in the 7-year operation at the actual daily average flux of 6 to 12 LMH.

Meanwhile, BOD/COD removal efficiency is not affected much by salinity in long-term operation as long as the operating conditions are controlled stably. In one occasion, COD removal appeared not affected even at a salinity of approximately five times higher than seawater, for example, 160 g/L as NaCl, in a controlled environment (Lay et al. 2010). In a full-scale MBR treating waste-water from seaweed processing, 99% BOD removal was achieved consistently at a BOD loading rate (F/V ratio) of 1.4 kg BOD/m³/day (Lala et al. 2014). It has also been observed that the high salinity does not affect nitrogen removal significantly as long as the process condition remains stable. In one study (Guan et al. 2014), no significant effect on nitrification and denitrification was observed when NaCl concentration was approximately 11 g/L with additional multivalent ions added to the system in a long-term MBR operation. In another study (Bassin et al. 2011), nitrogen removal was not severely affected at up to 33 g/L NaCl in the long-term test. However, nitrite tended to accumulate in mixed liquor as a consequence of the disappearance of *Nitrospira* sp.

The phosphorus removal efficiency is known to be not affected significantly at low salinity as long as it maintains stably. In one study, salinity did not affect phosphorus removal significantly in an A2/O system at levels less than 7 g/L NaCl (Zhao and Guan 2011). However, the excessively high salinity can be detrimental to biological phosphorous removal. In a long-term laboratory-scale study performed at 33 g/L NaCl for 449 days, the population of phosphorus-accumulating organisms (PAO) was wiped out, and no phosphorus removal occurred beyond the level obtainable with the ordinary microbial assimilation (Bassin et al. 2011). In a batch test, it was found that phosphorus removal nearly stopped in 2 h when microorganisms were suddenly exposed to 30 g/L NaCl. In another study (Lay et al. 2010), it was found that the phosphorus removal efficiency decreased from 84% to 22% as salt concentration increased from 9 to 60 g/L. The low phosphorus removal of 22% could be naturally achieved by microbial assimilation, and it suggested PAO population was nearly wiped out.

High salinity negatively affects membrane performance by hampering floc formation. The charges of particle/molecules are more effectively shielded by the counterions. Thus, the charge-induced flocculation can be largely inhibited. In addition, the divalent cations that bridge molecules are replaced by the abundant monovalent cations (Na^+) that cannot do the same. As shown in Figure 5.17, the mixed liquor from a full-scale MBR treating wastewater from a chemical process is completely deflocculated because of the extremely high salinity at 62 mS/cm at 20°C (seawater salinity is 48 mS/cm at 20°C). The sustainable flux of membrane in this application is no more than 5 LMH because of the severe membrane fouling caused by the deflocculated sludge containing a large amount of fine particles and soluble biopolymers.

As salinity increases, more water molecules are organized around the ions to counter the charge because of the ion–dipole interaction forming larger water–ion clusters. Because the momentum transfer among the larger clusters is easier than among small water molecules, moving the water becomes harder as salinity increases. As a result, water viscosity tends to increase as salinity increases; thereby, OTE deceases. Although bubble coalescence is somewhat discouraged because of the slower bubble movement at high liquid viscosity, the net effect of rising salinity is typically dropping OTE (Lay et al. 2010). The salinity effect on viscosity is factored in using β-factor when OTE is calculated, as shown in Equation 6.20. When NaCl concentration increases from 10 to 15 to 50 g/L, β decreases from 0.94 to 0.92 to 0.74. In addition, the saturated oxygen concentration decreases as salinity increases, as shown in Figure 5.18. Combining the β-factor and the saturated DO, OTE decreases as salinity increases.

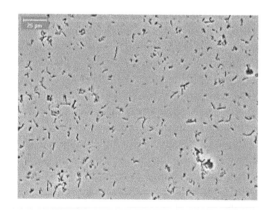

FIGURE 5.17   Microscopic image of MBR mixed liquor with a conductivity of 62 mS/cm (×400).

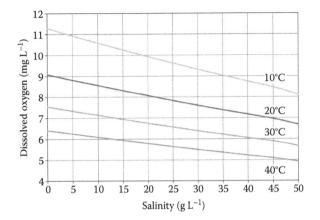

FIGURE 5.18 DO as functions of water temperature and salinity. (From Lay, W.C.L. et al., *Water Res.* 44:21–40, 2010.)

5.2.10 SHEAR STRESS

5.2.10.1 MBR with Submerged Membrane

Vigorous aeration is performed to control membrane fouling in the MBR with submerged membranes. The strong turbulence generated by the aeration not only mitigates membrane fouling but also breaks up microbial floc particles, creating fine particles. As a result, more microorganisms are exposed to the altered environment facing liquid phase with more abundant oxygen and nutrients. This may affect the physiology of microorganisms, which in turn affects membrane fouling. In addition, the high fines concentration may also contribute to membrane fouling. No extensive studies have been performed except few.

Side-by-side tests were performed to elucidate the effect of high scouring airflow using two MBRs equipped with one submerged flat sheet membrane panel each (0.48 m² surface area and 0.2 μm pore size) (Menniti and Morgenroth 2009, 2010). The aeration intensity for low shear MBR was in the high end of the typical aeration rate for flat sheet membranes at 15 L/min, but it was three times higher for high shear MBR at 45 L/min. It was found that the chord length measured by FBRM® technique (Mettler Toledo International Inc., USA) was shorter in high shear MBR as a consequence of floc breakages. Higher life-forms were observed only in the low shear MBR. There were no definitive distinctions, but bound EPS contents tended to be greater in the low shear MBR whereas SMP concentration was greater in high shear MBR. Despite the high SMP level, TMP increased much slower in the high shear MBR. This suggests that the net effect of high scouring airflow is positive on membrane performance despite the potential negative effect from the altered microbiology and the finer floc particles.

5.2.10.2 MBR with Sidestream Membranes

In the MBR with sidestream membrane, membrane units are placed outside the aeration tank, and mixed liquor is circulated between the aeration tank and the membrane tank. The high shear stress acting on the mixed liquor in the circulation pump can cause dramatic changes in the biology depending on how frequently mixed liquor is circulated. In fact, in the sidestream membrane process, mixed liquor must be circulated through the membrane system 20 to 100 times to be filtered because only a small fraction of the feed is recovered as permeate in one cycle, for example, less than 5%, as discussed in Section 2.4.1. In the laboratory-scale experiment, recovery is less than 1% because of the short channel length available in the laboratory-scale filtration equipment, and hence mixed liquor is circulated more than 100 times to be filtered.

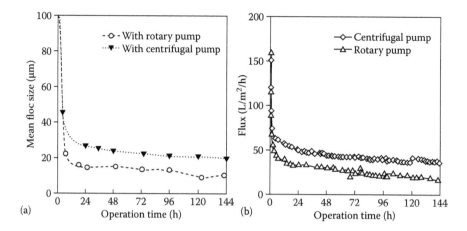

FIGURE 5.19 Effect of pump shear on floc particle size and membrane fouling in the laboratory-scale experiment: (a) mean floc size and (b) flux. (From Kim, J. et al., *Water Res.* 35(9):2137–2144, 2001.)

Kim et al. (2001) performed side-by-side studies using two laboratory-scale MBR trains equipped with sidestream tubular membranes. To investigate the pump shear effect on flux and sludge yield, mixed liquor was circulated using a rotary vane pump and a centrifugal pump, respectively. In a rotary vane pump, vanes physically touch with the casing that can grind a portion of the microorganisms passing the pump, but the hydrodynamic shear stress on floc particles is relatively low. By contrast, there is no physical grinding effect in centrifugal pump because impellers do not contact with the casing, but all the microorganisms passing the pump are exposed to the high hydrodynamic shear stress. As shown in Figure 5.19a, the mean particle size was smaller with rotary vane pump than with centrifugal pump. Because of the smaller particle sizes, much lower flux was obtained with rotary vane pump in sidestream MBR (Figure 5.19b). Meanwhile, COD in mixed liquor supernatant and permeate were much higher with rotary vane pump as a result of more severe grinding effect. The unintentional sludge disintegration by the circulation pump caused a reduced excess biosolid production. It is because the weight of new biomass generated from the disintegrated sludge is always less than the weight of the original biomass. For example, if 1 g of biomass is disintegrated, only 0.5 to 0.8 g of new biomass is generated depending on SRT. In the study, the biosolid yields of the MBR with rotary vane pump and centrifugal pump were estimated at 0.2 and 0.3 g MLVSS/g COD, respectively, and these values were substantially lower than the conventionally observed biosolid yield, that is, 0.4 to 0.5 g MLVSS/g COD.

Because of the unintentional sludge disintegration by circulation pump, projecting the sludge yield in full-scale MBR based on pilot study is tricky. Pilot tests typically use smaller-scale membrane systems with a shorter channel length when it is compared with the full-scale MBR. Because of the lower recovery from the shorter channel, mixed liquor should be circulated more frequently to be filtered. Therefore, the sludge yield of pilot-scale MBR tends to be lower than that of full-scale MBR. Extra redundancies should be reserved in the sludge treatment capacity for full-scale MBR to handle the larger than expected amount of sludge based on pilot-scale tests.

5.2.11 Nutrient Balance

The major micronutrients in the activated sludge process are nitrogen and phosphorus. Nitrogen is an essential constituent of protein/enzyme, nucleic acids, etc., and phosphorus is also an essential constituent of ATP required for energy metabolism, nucleic acids, etc. Total Kjeldahl nitrogen (TKN) is used as a measure of the nitrogen content instead of total nitrogen (TN) because most microorganisms need nitrogen in reduced form, such as free ammonia or organic nitrogen bound in amino acids or proteins. In general, the ratio of BOD:TKN:TP is recommended at 100:5:1 in raw wastewater in the activated sludge process, but the ratio varies somewhat depending on SRT and other process parameters. Deficiency in one of the two major micronutrients can lead to undesirable microbial conditions such as viscous and/or filamentous bulking (Jenkins et al. 2004). SMP and EPS levels increase when bulking occurs in many cases in activated sludge process. Viscous bulking occurs along with the surge of polysaccharides level in mixed liquor. Figure 5.20 shows the mixed liquor suffering from the viscous bulking, where hydrophilic india ink stains all hydrophilic microbial surfaces, leaving hydrophobic polysaccharide lumps as white spots.

It has been known that membrane fouling is aggravated when nutrients are deficient. In one study, side-by-side tests were performed to investigate the nutrient effect using two laboratory-scale MBRs equipped with submerged flat sheet membranes (Singh et al. 2008). One MBR was fed with a phosphorous-deficient synthetic feed, whereas the control MBR was fed with a nutrient-rich feed. It was observed that floc particles were smaller and polysaccharide concentrations were greater in the nutrient-deficient MBR. Consequently, permeate COD was high and sludge filterability was low when it was compared with those of the control reactor fed with nutrient-rich feed. Most of all, the membrane fouling was much more severe with phosphorus-deficient feed.

Metal ions are also necessary to grow healthy sludge. Monovalent cations are used to balance osmotic pressure in and out of the cell and to counter the charges of biomolecules. Divalent ions play a role in floc formation by bridging biomolecules. In fact, floc size tends to increase and sludge settling property improves when the ratio of divalent to monovalent ions is high (Park 2002). Figure 5.21 shows an example of divalent ion effect on sludge settling and supernatant clarity. In this case, the mixed liquor was deficient of divalent ions because wastewater from a chemical plant contained little divalent ions at less than 10 mg/L. The small debris or unflocculated cells did not settle easily, but 100 mg/L of calcium ions greatly improved sludge flocculation, as shown

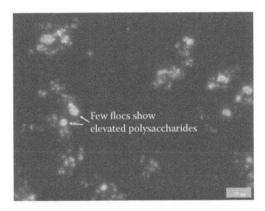

FIGURE 5.20 Polysaccharide lumps observed through a microscope at 100× after dying with india ink.

FIGURE 5.21 Effect of divalent ion on sludge settling and supernatant clarity in an activated sludge process treating wastewater from a chemical plant (100 mg/L Ca^{2+} with 30 min settling time).

in Figure 5.21. If the insufficient divalent ions (or hardness) are a cause of deflocculation, calcium and/or magnesium supplements may help reduce membrane fouling in industrial MBR. Minor transitional metal ions such as Fe, Mo, Co, Ni, and Se also play important role in the activated sludge process. Some of them are used as a cofactor in enzyme, where transitional metals provide catalytic active site for chemical reactions. In fact, it has been claimed that additional transitional metals added to feed water in a few ppm, for example, Mo, Co, and Ni, can improve microbial activities in certain environment (Burgess et al. 2000). However, the benefit of adding additional "transitional metal" ions is not fully proven in the field condition because wastewaters tend to have trace amounts of all transitional metals although the known requirements are tiny. There are anecdotal observations suggesting that adding certain transitional metals improves microbial activities, but the effect of each transitional metal seems site specific. There are no known transitional metals that improve microbial activity in most or all conditions. Although minor transitional metals are found in a healthy municipal activated sludge, for example, Zn (200–360 mg/kg), Cu (170–210 mg/kg), Pb (30–50 mg/kg), Ni (20–30 mg/kg), Cr (5–7 mg/kg), Cd (2–5 mg/kg), Hg (0.1–0.4 mg/kg), As (0.1–0.2 mg/kg), and B (20–35 mg/kg) (Tao et al. 2012), it is not certain whether all of these metals are required in the range specified in parentheses to ensure the healthy metabolism or these metals are absorbed by activated sludge just because they are available in wastewater. An independent field survey was performed to find out the potential relationship between the inorganic composition of the activated sludge and the BOD treatment, but no apparent correlation was found among six different plants running without a significant trouble. It was found that micronutrient contents varied site by site in a wide range, and no apparent commonality was observed (data not shown).

5.2.12 Scaling

Membrane fouling by inorganic precipitates is not commonly observed in municipal MBR as long as calcium concentration is not excessive, for example, less than 200 mg/L (Cornel and Krause 2008). However, inorganic fouling by $CaCO_3$ and other precipitates has been experienced in some industrial MBR treating wastewater with high calcium, sulfate, and phosphate contents. Ferric hydroxide, silica, and manganese oxide can also be a cause of membrane fouling in theory, but magnesium hydroxides cannot precipitate at near-neutral pH because of the high solubility. If aluminum or ferric salts are used to enhance phosphorus removal, metal hydroxides or phosphate can cause membrane fouling. However, hard evidences supporting the notion are scarce. Instead, many studies have shown that aluminum- and iron-based coagulants are beneficial for membrane performance as long as the pH of mixed liquor is controlled properly, as discussed in Section 5.4.2 (Lee et al. 2001b; Iversen 2009). It is perhaps because inorganic scaling (or precipitation) occurs in the bulk sludge rather than on the membrane surface. The scale particles hardly deposit on the membrane surface because of air scouring or cross flow. Meanwhile, inorganic membrane fouling is much more prevalent in AnMBR because of the formation of struvite ($MgNH_4PO_4$) under the reducing environment, as discussed in Section 7.4.4. In contrast to the porous membrane, in RO, scaling predominantly occurs on the membrane surface inside the concentration polarization layer because of the high ionic concentration. The attached scale particles on RO membrane grow and cover the membrane surface, resulting in flux decline.

Although scaling is not a common phenomenon in municipal MBR, it can become a significant factor affecting membrane performance in certain conditions. If membrane scouring is not sufficient, the scale particles formed in the bulk sludge can deposit on the membrane and grow. One autopsy study showed that $CaCO_3$ precipitation on the membrane surface significantly compromised membrane permeability (Kim and Yoon 2010). Scaling predominantly occurred on the membrane surface rather than in internal pore spaces. Therefore, it can be postulated that scale particles are formed in the bulk before they deposit and grow using the oversaturated ions in the mixed liquor. As a consequence, 90% of the calcium ions in raw wastewater were removed by the porous membrane in the study. Scales can be cleaned more efficiently after the fouled membranes are cleaned by NaOCl. By removing organics first, scale particles are better exposed to the acid used to dissolve the scales. More details on acid cleaning are discussed in Section 3.2.5.3.

Calcium carbonate scale can be minimized by maintaining a low pH (e.g., 6–7) because the solubility increases sharply as the solution pH enters into the acidic condition. Although the nitrification rate may decline as the pH decreases, it can be recovered mostly, if the low pH is maintained at a constant level for a sufficient period, as discussed in Section 4.4.1.2. Therefore, accurate pH control at a constant level is essential for a successful calcium carbonate scale control without affecting nitrification. The solubility of calcium carbonate also increases as water temperature decreases, but modifying water temperature is not feasible in the field in general. Antiscalants and crystal modifiers are successfully used in cooling tower management and pulp processing to delay scale formation and/or to discourage scale adhesion on RO membrane surface, but no sufficient information exists as far as MBR is concerned.

5.3 MIXED LIQUOR CHARACTERIZATION

5.3.1 OVERVIEW

Various forms of mixed liquor filterability have been used to track the mixed liquor quality from the perspective of the membrane fouling potential. Ideally, the measured filterability alerts the operators with upcoming accelerated membrane fouling. Free drainage, vacuum drainage, cross-flow filtration, etc., are classified as direct test methods, whereas capillary suction time (CST), PSD, etc., are classified as indirect test methods. Although one or more of such methods may be useful for the purpose in a given condition, none of them are universally reliable in all conditions. It is because of the vastly different filtration mechanisms occurring in the test methods and the actual membrane filtration. Some of the most commonly used filterability test methods are discussed in the following sections.

5.3.2 DIRECT METHOD

5.3.2.1 Free Drainage Test

The free drainage test is the simplest method commonly practiced. A filter paper, for example, Whatman 42, is folded and placed in a funnel. After pouring 50 mL of the mixed liquor sample to the funnel, the filtrate volume is measured after a predetermined time, for example, 5 min. This method does not require special equipment.

Although this method is useful to obtain a rough idea on the membrane fouling potential, the following limitations are apparent:

- Unlike actual filtration in the membrane tank, there is only a very slight driving force to filter mixed liquor. Cake compaction is much milder under this condition than that in the actual filtration. Therefore, the results obtained are nearly free of cake layer compaction effect.
- No cross flow exists in the funnel unlike actual membrane filtration. Therefore, all particles and macromolecules in the sample contribute to the cake layer resistance. By contrast, large particles are transported back to the bulk and do not contribute to cake layer in actual membrane filtration in MBR.
- Solid particles are the major constituent of the cake layer in the free drainage test, but macromolecules such as SMP and EPS are the major constituent in actual membrane filtration.
- Filterability is affected by MLSS because of the lack of normalization against MLSS.
- The results are affected by the atmospheric temperature.

5.3.2.2 Time to Filter

Time to filter (TTF) is determined by using an apparatus shown in Figure 5.22. A 90 mm Buchner funnel is used with Whatman 1, 2, or equivalent filter papers (Standard Method 2710H, APHA 1998). After pouring 200 mL of mixed liquor, the time required to obtain 100 mL of filtrate is measured at the vacuum pressure of 51 kPa (or 7.4 psi). The cake layer is formed immediately after the filtration starts, and it undergoes a compaction process because of the dynamic pressure loss through it. TTF can be normalized against MLSS, but it may be valid only at a narrow MLSS range. It has been suggested that a TTF less than 100 s indicates an easily filterable mixed liquor, 100 to 200 s indicates an intermediate mixed liquor, and more than 300 s indicates a marginal mixed liquor (Côté 2007).

5.3.2.3 Modified Free Drainage Test

To mitigate the limitation of the free drainage-based filterability test, the mixed liquor poured in a Buchner funnel can be mixed by a flat blade, as shown in Figure 5.23a. A large amount of mixed liquor (500 mL) is heated or cooled to 20°C in water bath before it is poured into the funnel to reduce the temperature effect. Filter papers with 150 mm diameter and 0.6 μm pore size are used. The mixed liquor poured in the funnel is mixed by a clamp blade agitator at a height of 1 mm above the filter paper at 40 rpm. The elapsed time (Δt) to obtain 100 to 150 mL of filtrate is measured. Finally, the sludge filterability index (SFI) is calculated by normalizing the time against MLSS, as shown in Equation 5.1 (Thiemig 2011):

$$SFI = \Delta t \ (s)/MLSS \ (\%). \tag{5.1}$$

5.3.2.4 Modified Fouling Index

Assuming that the cake layer resistance is the sole cause of flux loss and it is proportional to the cake layer weight, the modified fouling index (MFI) can be defined. The weight of the cake layer proportionally increases as the filtrate volume increases under the dead-end filtration mode used by the MFI test protocol. Equation 5.2 is the modification of Equation 1.4, where fouling resistance R_f is assumed zero and the flux J is written in a more intrinsic form. According to the initial assumption, the cake resistance R_c (m^{-1}) should be proportional to the amount of the particles deposit. The amount of the particles deposit is calculated by the product of particle concentration C (kg/m^3) and filtrate volume V (m^3), as shown in Equation 5.3, where α (m/kg) is a proportional constant. By inserting Equation 5.3 to Equation 5.2, Equation 5.4 is obtained. Applying the boundary conditions

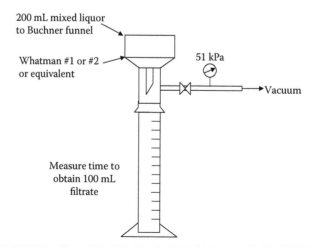

200 mL mixed liquor
to Buchner funnel

Whatman #1 or #2
or equivalent

51 kPa

Vacuum

Measure time to
obtain 100 mL
filtrate

FIGURE 5.22   Apparatus to measure TTF.

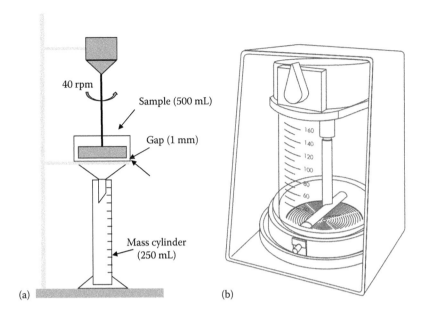

FIGURE 5.23 Experimental setups: (a) modified free drainage and (b) stirred cell. (b: Courtesy of EMD Millipore.)

of (0, 0) and (t, V), Equation 5.4 can be solved to Equation 5.5. Finally, Equation 5.5 is simplified to Equation 5.6, where MFI is defined as $\alpha\mu C/2\Delta PA^2$. The equations are presented as follows:

$$\frac{1}{A}\frac{dV}{dt} = \frac{\Delta P}{\mu(R_m + R_c)}. \tag{5.2}$$

$$R_c = \frac{\alpha CV}{A}. \tag{5.3}$$

$$\int_0^V (R_m A + \alpha CV)\,dV = \frac{\Delta PA^2}{\mu}\int_0^t dt. \tag{5.4}$$

$$\frac{t}{V} = \frac{\mu R_m A}{\Delta PA^2} + \frac{\alpha\mu C}{2\Delta PA^2}V. \tag{5.5}$$

$$\frac{t}{V} = a + MFI \cdot V. \tag{5.6}$$

According to the early method, MFI is measured using a stirred cell shown in Figure 5.23b without stirring under a dead-end filtration mode at 207 kPa (Schippers and Verdouw 1980). MFI can be denoted as either $MFI_{0.45}$ or MFI_{UF} to remark the membranes used, that is, 0.45 μm filter and ultrafilter. Typically, permeate volume is recorded more frequent than every 30 s with time stamps. Many variations exist for MFI in terms of stirring speed, membrane pore size, pressure, etc., depending on the purpose of the test and the sample characteristics (Jang et al. 2006; Khirani et al. 2006; Choi et al. 2009).

Other filterability test methods also exist. Tarnacki et al. (2005) estimated the relative membrane fouling potential of mixed liquor by measuring the time to obtain certain volume of permeate under a constant pressure. In the experiment performed with a custom-made stirred cell, membranes with 38 cm² were used at 1 bar, 20°C, and 400 rpm. Filterability and fouling index are obtained as follows:

1. Filterability, L_{15} (LMH/bar), is estimated after a 15 mL filtrate is obtained by dividing the flux by pressure, as shown in Equation 5.7. L_{15} cannot be normalized against wide range of pressure because flux is not proportional to the pressure when the cake layer compaction dominates the filterability decline under the condition as described in Section 1.2.7. Thus, it is the best practice to perform this filtration test at a fixed TMP every time it is performed:

$$L_{15} = \frac{J_{15}}{\Delta P}. \tag{5.7}$$

2. Fouling index, FI_{30}, is calculated by dividing the observed flux after 30 min of filtration, J_{30}, by the initial water flux, $J_{w,0}$ (Equation 5.8). FI_{30} is affected by TMP for the same reason as L_{15}. Therefore, FI_{30} should be measured at a fixed TMP every time the test is performed:

$$FI_{30} = \frac{J_{30}}{J_{w,0}}. \tag{5.8}$$

Despite the effort to develop fouling indices that reasonably represent the fouling potential, only limited success has been accomplished. The assumptions used in developing the previously mentioned equations do not match with the reality because of one or more of the following reasons:

- First, the weight of the cake layer is not proportional to the filtrate volume in actual membrane filtration because of the existence of the particle back transport. Depending on the course of the flux profile during the filtration, not only the amount of deposit, but also the structure of cake layer varies, as discussed in Section 1.2.5.
- Second, cake layer compression plays an important role in cake resistance, but MFI equations do not consider it. As discussed in Section 1.2.3, depending on pressure and flux during the stirred cell test, the extent of the cake layer compaction varies.
- Third, whichever laboratory test methods are used, the hydrodynamics on the membrane surface cannot be same as that in real filtration. Hydrodynamic conditions determine particle back-transport velocity and eventually the membrane fouling rates.
- Fourth, MFI is measured under an expedited membrane fouling condition typically at higher TMP than in actual filtration. Cake layer formation and aging patterns in short-term filtration cannot be same as those in the long-term filtration.

5.3.2.5 Delft Filtration Characterization Method

The Delft filtration characterization method (DFCm) relies on a single tubular UF membrane with 8 mm internal diameter and 0.03 μm nominal pore size (X-flow F5385) under inside-out filtration mode. Cross-flow velocity is maintained at 1 m/s by using a peristaltic pump and flux is maintained at 80 LMH. The pressure at the feed, concentrate, and permeate, mixed liquor temperature, pH, flux, DO concentration are recorded during the test (Evenblij 2006; Geilvoet 2010; van den Broeck et al. 2011).

From the recorded data, the profile of R_{total} (= $R_m + R_c + R_f$) is plotted using Equation 1.4, as shown in Figure 5.24. The R_{total} (or ΔR_{20}) obtained when the cumulative specific permeate volume is 20 L/m² is used to compare the membrane fouling propensities of different mixed liquors.

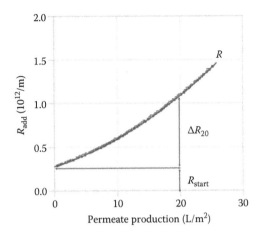

FIGURE 5.24 Typical curves obtained in DFCm. (From van den Broeck, R. et al., *J. Membrane Sci.* 378: 330–338, 2011.)

Filterability is considered poor when ΔR_{20} is higher than 1×10^{12} m^{-1}, moderate when 0.1×10^{12} m$^{-1} <$ $\Delta R_{20} < 1 \times 10^{12}$ m^{-1}, and good when ΔR_{20} values are less than 0.1×10^{12} m^{-1}.

This method provides a closer imitation of real membrane filtration than other filterability test methods, but it is still not exactly same. The flux used, that is, 80 LMH, however, is much higher than the flux commonly used in submerged membrane processes, that is, 15 to 25 LMH. The high flux is inevitable to expedite membrane fouling to complete the test within a reasonable time, but it causes an overcontribution of large particles in filtration resistance. In fact, at high flux condition, more large particles deposit on the membrane surface than at the lower flux. In addition, cake layer compaction is much more significant at the high flux; thereby, this test differs from the actual filtration occurring in the real MBR. The high shear stress in the circulation pump also breaks up floc particles and make them smaller in DFCm.

5.3.3 INDIRECT METHOD

5.3.3.1 Capillary Suction Time

The CST method has been used to measure the dewatering properties of the activated sludge. When sludge is contacted with filter paper, the water contained in the sludge starts to wet the paper because of the capillary suction phenomena caused by the hydrophilic fibers that consist of the filter paper. As the water in the sludge–paper interface is lost to the paper, the sludge that contacts with the paper becomes more compact. The dense sludge layer in the interface acts as a barrier for further water seepage to the filter paper. Therefore, the filterability of the compacted sludge in the sludge–paper interface determines the CST. If macromolecules and fine particles, which are also called SMP/EPS, are abundant in the sludge, more compact sludge layers are formed in the interface, thereby wetting speed decreases and CST increases.

According to the Standard Method 2710G (APHA 1998), 10 mL of mixed liquor is poured in the test cell reservoir (18 mm ID and 25 mm height) located in the center of the measuring apparatus, as shown in Figure 5.25. As soon as mixed liquor contacts with a filter paper, it starts to wet the filter paper underneath the reservoir and proceeds radially, as shown in the diagram (bottom). The time required for water to proceed from r1 (15.9 mm) to r2 (22.2 mm) is measured by conductivity sensors and is called CST. CST can be normalized by dividing it by MLSS (g/L), but this normalization is effective only in a narrow MLSS range.

In a laboratory study, CST was reasonably correlated with the increasing rate of the TMP (Ng et al. 2006). It was observed that the normalized CST against MLSS decreased as SRT increased as

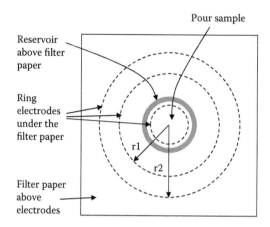

FIGURE 5.25 Apparatus to measure CST.

can be deduced intuitively from the mechanisms involving CST measurement and the membrane fouling by biopolymers. The lowering CST suggested that fine particles and macromolecule concentrations decreased as SRT increased.

5.3.3.2 Colloidal TOC

The concentration of colloidal particles in mixed liquor can be used as an indicator of the membrane fouling potential. The colloidal TOC is defined as the difference between the filtrate TOC of 1.5 μm filter paper (934-AH, Whatman, USA) and the permeate TOC of MBR (Fan et al. 2006). Colloidal TOC has been known as a good indicator of the membrane fouling potential in MBR in many different locations. In a pilot study performed in a municipal wastewater treatment plant (WWTP), critical flux was measured using a flux stepping method described in Section 1.2.6.2. The results are plotted against the synchronously measured colloidal TOC, as shown in Figure 5.26. It was observed that the critical flux was inversely correlated with the colloidal TOC. In the same study, TTF, MLSS, and permeate TOC appeared less correlated with the flux than the colloidal TOC. It has been suggested that the colloidal TOC of less than 10 mg/L indicates a good mixed liquor condition, 10 to 20 mg/L indicates an intermediate condition, and more than 20 mg/L indicates a marginal condition (Côté 2007).

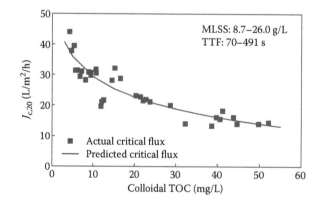

FIGURE 5.26 Relationship between the colloidal particle concentration and the critical flux corrected at 20°C. (From Fan, F. et al., *Water Res.* 40(2):205–212, 2006.)

5.3.3.3 Particle Size Analysis

The membrane fouling rate increases as the population of submicron particles increases in MBR. It is partly because of the slow particle back-transport velocity of such particles, partly because of the stronger interaction with surfaces through charge and van der Waals interactions. Although the instruments based on laser light scattering are not accurate in measuring such small particles, the shifting PSD curve to the smaller direction can be deemed an indication of the increase of submicron particles. In fact, in a side-by-side test using laboratory-scale MBR equipped with sidestream tubular membranes, the membrane fouling rate increased as the average particle size measured by a particle size analyzer based on laser light scattering decreased (Kim et al. 2001).

It has been observed that deflocculation and accelerated membrane fouling are caused by various undesirable biological conditions, for example, low DO concentration (Kang et al. 2003), high F/M ratio (Syed et al. 2009), and low SRT (Ahmed et al. 2007). It must be noted that the dynamically decreasing average particle size by external causes accelerates membrane fouling, but the small average particle size staying at a stable level is not necessarily a problem. An example is shown in Figure 5.27a, where PSDs measured at four different SRTs are plotted. According to Figure 5.27a, the average particle size is the largest when SRT is 40 days and the smallest when SRT is 20 days. Although particle size declines when SRT is extended beyond 40 days, the membrane fouling rate declines as SRT increases. Perhaps the lowering SMP content at high SRT was a more dominant factor than particle size under the condition.

5.3.3.4 Hydrophobicity of Floc

Relative hydrophobicity (RH) has been used in biotechnology as an indicator of the physiological state of microorganisms. RH is measured by a test called *microbial adhesion to hydrocarbons* (MATH) test. The MATH test is performed with solvents such as n-hexane, n-octane, n-octanol, n-dodecane, etc. After pouring a sample to a separation funnel, a selected solvent is added before shaking the separation funnel. After the water and solvent phases are separated, the ratio of microbes in the aqueous phase and in the solvent phase are measured based on light absorbance.

In the original version of the MATH test, 4 mL sample is mixed with 1 mL of n-dodecane and vortexed for 2 min followed by resting for 15 min to enable the phase separation. After taking 0.75 mL of aqueous sample, absorbance (S_e) is measured at 400 to 600 nm. The absorbance of the

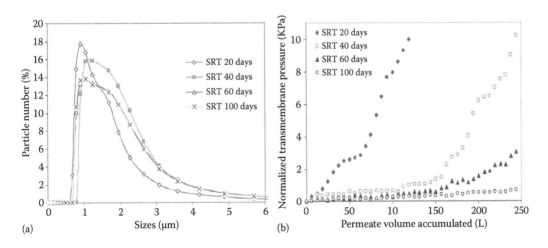

FIGURE 5.27 Relationship among SRT, PSD, and the membrane fouling rate, where flat sheet Kubota membrane was used in laboratory-scale systems fed with synthetic feed. (a) Effect of SRT on PSD. (b) Effect of SRT on membrane fouling. (From Ahmed, Z. et al., *J. Membrane Sci.* 287:211–218, 2007.)

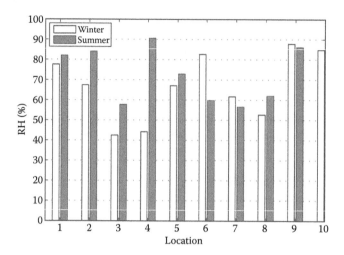

FIGURE 5.28 RH of activated sludge samples. (From van den Broeck, R. et al., *J. Membrane Sci.* 378: 330–338, 2011.)

original aqueous sample (S_i) is also measured at the same wavelength. Finally, RH is calculated using the following equation (Saini 2010):

$$\text{RH}\,(\%) = \left(1 - \frac{S_e}{S_i}\right) \times 100. \tag{5.9}$$

Because membranes are prone to be fouled by the hydrophobic components of sludge through hydrophobic–hydrophobic interaction, the RH of sludge (or biosolids) can be used as an indicator of the membrane fouling potential. There are many variations in the MATH test depending on the solvent used, the voltexing method and its duration, the wave length used, etc. In some cases, the amount of dissolved materials in solvent phase is used as an indicator of RH after evaporating the solvent.

After all, the RH value is one indicative parameter for membrane fouling, but it alone cannot be used to judge the membrane fouling propensity. Depending on other sludge characteristics, the relationship between RH and membrane fouling propensity can turn out to be opposite in many cases. According to a survey performed in 10 different locations, there was no universal trend between the RH and the filtration resistance (van den Broeck et al. 2011). In one instance, The RH of the mixed liquor in MBR was lower in summer than that in winter in site 6 of Figure 5.28, that is, 60% in summer versus 83% in winter. Thereby, the filtration resistance was lower in summer at 0.07×10^{12} m^{-1} and was 0.71×10^{12} m^{-1} in winter. On the contrary, the relationship turned out to be opposite in some other locations. For example, the mixed liquor of MBR was more hydrophobic in summer than that in winter, for example, 62% versus 53% in site 8, but the filtration resistance was lower in summer (6.28×10^{12} m^{-1}) than that in winter (9.90×10^{12} m^{-1}).

5.4 METHODS TO REDUCE FOULING

5.4.1 USE OF BIOCARRIER (BIOFILM-MBR)

It has been reported that the biocarriers added to MBR can reduce membrane fouling in some cases. The direct physical membrane scouring by biocarriers and the reduction of SMP in aqueous phase are considered the two plausible mechanisms. Lee (2002) and Lee et al. (2006) performed experiments to elucidate the direct contact effect of biocarriers on membrane fouling. In the laboratory-scale experiment, two membrane modules with vertically mounted hollow fiber membranes (Mitsubishi Rayon, Japan) were submerged in one aeration tank. One module was allowed to contact with the floating biocarriers, and the other module was guarded by an iron net to prevent the membrane bundle from directly contacting with the floating biocarriers. The biocarriers were the sponge cubes (13 mm × 13 mm × 13 mm) coated with activated carbon. It was observed that the TMP of the guarded module reached 30 kPa five times faster than that of the unguarded module. It was found that only the guarded modules were covered by slimes and sludge cakes around the fiber bundle. Similar biocarrier effects were observed in other experiments performed with porous flexible biocarriers (Yang et al. 2006). In this experiment, the PSD curve slightly moved to smaller direction with biocarriers, but the membrane fouling rate decreased. It appeared that the direct membrane scouring by biocarriers was effective in scouring the membrane surface. In another study, it was also confirmed that plastic biocarriers (AnoxKaldnes™, K1 carrier, Veolia) improved the performance of submerged flat sheet ceramic membranes when they were allowed to contact with the membrane (Jin et al. 2013).

If biocarriers are dosed excessively, they may affect membrane performance negatively (Wei et al. 2006). When polyethylene-based hollow fiber membranes (Mitsubishi Rayon, Japan) were horizontally mounted and the biocarriers with cylindrical geometry (3 mm length × 3 mm diameter) were added to the membrane tank, a positive effect on membrane fouling was observed only when biocarrier dosage was low at 1 v/v%. At a higher biocarrier dosage (5 v/v%), TMP increased faster than control. The faster membrane fouling was attributed to the smaller particle sizes than those in the control reactor without biocarrier. It was postulated that large flocs were broken down to smaller pieces because of the collisions with rigid biocarriers.

In another study (Kimura et al. 2013), granular media (or biocarrier) were added to the MBR equipped with flat sheet submerged membrane (Toray, Japan) and allowed to contact with membrane. The result showed granular media reduced membrane fouling dramatically at a same scouring airflow. It was estimated that around 50% air savings were possible if the similar membrane fouling rate is targeted. In a bench-scale MBR equipped with sidestream ceramic tubular membranes, flux increased 16% to 17% when PAC dosage was at 2 to 5 mg/L in the mixed liquor (Torretta et al. 2013).

Rigid biocarriers can damage membranes if they are allowed to contact membranes. When PAC was added to the MBR treating oily refinery wastewater, the membrane fouling rate decreased remarkably (Conner 2011). The membrane in the control reactor developed a dark color because of oil adsorption, but it was recovered by chemical cleaning. After the cleaning, the membrane surface remained as clean as the initial based on the visual inspection in both control and treatment reactors. However, the SEM picture elucidated the abrasions occurred on the membrane used with PAC within the first 30 days of operation (pictures are not shown). Membrane life was expected to be compromised substantially by up to 40% because of the abrasion. As a consequence, the granular activated carbon (GAC) column was attached to the aeration tank instead of suspending the PAC into the aeration tank.

If biocarriers are not allowed to contact membrane, membrane fouling can be more severe in BF-MBR than that in MBR depending on the conditions (Lee et al. 2002; Ivanovic and Leiknes 2011). According to a side-by-side test of BF-MBR and MBR (Yang et al. 2014), the concentrations of SMP, colloidal TOC, and TEP were similar in the mixed liquor of both processes, but membrane fouling was much more significant in BF-MBR. This suggested membrane fouling was affected not only by the quantity of the biopolymer but also by the chemical properties of the biopolymer.

It has been claimed that BF-MBR retains more biomass in the aeration tank in the forms of attached and suspended biomass. It has also been claimed that the high biomass retention enables smaller aeration tank size, whereas the low MLSS enables high OTE (Leiknes and Ødegaard 2006; Sombatsompop et al. 2006). However, the existence of biocarriers in the aeration tank has been known to hamper OTE substantially and increase the operational costs (Rosso et al. 2011). More details on the effect of biocarrier on oxygen transfer are discussed in Section 6.5.2.8. The high volumetric oxygen demand and the low OTE require an intensive aeration, but it is well known that the intensive aeration hampers OTE. Therefore, BF-MBR may enable smaller aeration tanks by retaining more biomass, but it incurs extra operating costs for the biological air.

The BF-MBR with floating GAC was commercialized by Siemens, aiming wastewater treatment in gas and oil industry. In the process called EcoRight™ MBR, GAC is added to the aeration tank, but it is filtered by strainers before the mixed liquor is transferred to the membrane tank. It is claimed that the GAC adsorbs recalcitrant organics and increases the contact time with microorganisms; thereby, the organics removal efficiency increases. Later, GE also introduced a MACarrier technology that uses PAC. In this technology, PAC can travel anywhere in the system and is allowed to contact with the membrane. The property of PAC selected as MACarrier is supposedly optimized to avoid scratching the membrane surface. The low specific aeration rate in full-scale MBR may help avoid the membrane scratch issue observed in the laboratory study discussed earlier. The periodic makeup of PAC is necessary to compensate the lost PAC through the excess sludge removal. Better membrane fouling control and better removal of recalcitrant organics are claimed in refinery wastewater treatment.

Unlike EcoRight™ and MACarrier technologies, BIO-CEL® MCP technology is relying on physical scouring effect rather than adsorption or biofilm effect. The plastic beads (diameter, 4–5 mm; specific gravity, 1.05) added to the membrane tank bombard the flat sheet membranes interrupting the cake layer formation. Because the beads are designed specifically for the purpose, they are not supposed to damage the membrane surface. The gap screens with large surface area installed in the top of the membrane tank prevent beads from escaping the tank. The rejected beads move down to the floor by current and the specific gravity and eventually drawn again to the membrane cassette by current. This method can be particularly effective in reducing membrane fouling during the peak flow in MBR, where the cake layer formation is accelerated more than proportionally predicted based on the operating flux, as explained using Figure 1.15. Because the membrane system is sized targeting the peak flow, the better peak flow performance can result in capital cost savings.

5.4.2 Use of Inorganic Coagulants and Adsorbents

It has been known that inorganic coagulants and adsorbents reduce the level of SMP and thus reduces membrane fouling rates (Lee et al. 2001b; Fan et al. 2007; Iversen et al. 2008; Iversen et al. 2009). It is believed that the metal hydroxides with net positive charge neutralize the negative charges of SMP and make them coagulated. In addition, the networks of polymerized metal hydroxide enclose SMP to form large flocs that hardly deposit on membrane. In one study, alum and ferric chlorides reduced soluble protein concentrations in mixed liquor by more than 90% from the initial amount of approximately 100 mg/L at dosages of approximately 200 mg/L (Mishima and Nakajima 2009).

Meanwhile, adsorbents such as zeolite and PAC can reduce membrane fouling by adsorbing SMP and physically scouring the membrane surface. However, a long-term use of adsorbents may not be practical because a large amount of PAC is required to treat the SMP produced continuously and to compensate the lost PAC through the excess sludge. The spent PAC also increases excess sludge production that may overload to the sludge handling process. The necessity of a large amount of PAC and the increased sludge production increase with respect to the effluent quality, adsorbents are not easily justifiable in general. Apart from the economic constraint, PAC can damage the membrane surface depending on condition, as discussed in the previous section.

Lee et al. (2001b) performed a laboratory study using hollow fiber membranes made of polyethylene (Mitsubishi Rayon Co., Japan) at 15 LMH. Alum dosage was controlled to maintain a molar ratio of Al/P at 1.5, whereas zeolite concentration was controlled at 1000 mg/L. As shown in Figure 5.29, both alum and zeolite delayed increase in TMP. In this short-term test, alum decreased nitrification rate from 92% to 76%, whereas zeolite did not affect nitrification efficiency. In another study, when polyaluminum chlorides (PACl) were added at 12.5 mg/g MLSS

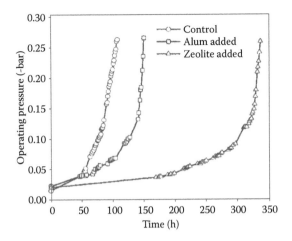

FIGURE 5.29 Effect of inorganic additives on TMP increase. (From Lee, J.C. et al., *Water Sci. Technol.* 43(11):59–66, 2001b.)

and maintained at the level by compensating the loss through excess sludge removal, the increasing rate of the TMP was significantly reduced (Teli et al. 2012). No adverse effect on COD and nitrogen removal was observed in this case, whereas phosphorus and color removal efficiencies increased significantly in the MLE-based MBR. It is noteworthy that PACl does not reduce pH as much as alum does when it is applied to mixed liquor because it is a partially neutralized aluminum chloride in nature. Thus, the pH effect on nitrification should be less for this case than the case in the previous paragraph.

The adverse effect of alum on nitrification has been observed primarily in short-term laboratory tests. The same negative effect has been rarely reported from the full-scale municipal MBR although the same alum or aluminum chlorides are used, as discussed in Section 4.5.2.6. One likely cause of the performance loss in laboratory tests is the abrupt drop in the pH with the injection of strongly acidic inorganic coagulants. Because of the acute toxicity of the low pH, the pH-sensitive nitrifiers can be partially inhibited in a small-scale experiment. On the contrary, it is easier to add inorganic coagulant gradually for longer period in full-scale MBR while stably monitoring and controlling the pH. In addition, microorganisms can adapt with the condition for longer period and become more resilient to the chemicals used.

The added inorganic coagulants, however, are known to be toxic to some filamentous bacteria in the mixed liquor at varying degrees and can alter the microbial community, as discussed in Section 4.5.2.6. Aluminum salts have been known to be toxic to some filamentous microorganisms and have been used to control foaming and bulking caused by *Microthrix parvicella* in the activated sludge process (Roels et al. 2002; Nielsen et al. 2005). However, it is suspected that the inorganic salts lose toxicity in the long run and the filamentous organisms grow back eventually.

5.4.3 USE OF WATER-SOLUBLE POLYMERS

This patented technology of using water-soluble polymers has been used to mitigate membrane fouling in MBR. The commercial products called MPE30 and MPE50 have been available from Ecolab Inc. (formerly Nalco Company) since 2004. The polyelectrolytes with net positive charge coagulate not only small debris but also SMP. It is believed that those chemicals reduce membrane fouling by reducing fine particle and SMP concentrations in mixed liquor, which leads to higher cake layer porosity (Yoon et al. 2004a, 2005; Yoon and Collins 2006; Hwang et al. 2007; Lee et al. 2007; Dizge et al. 2011). Figure 5.30 shows the efficacy of MPE30 in controlling membrane fouling in an MBR equipped with submerged flat sheet membranes (Yoon et al. 2005). The dosage of

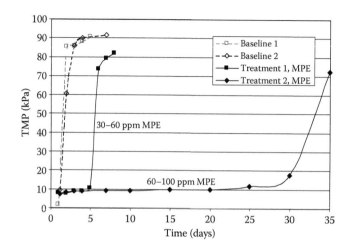

FIGURE 5.30 Effect of MPE on membrane fouling prevention. Pilot MBRs equipped with flat sheet membranes were used, and no relaxation time was applied. Flux = 22 LMH, TS = 12,000–15,000 mg/L, influent = 40 L/day, sludge temperature = 18°C–20°C. (From Yoon, S.-H. et al., *Water Sci. Technol.* 51(6):151–157, 2005.)

MPE30 was maintained at 30 to 100 mg/L in the mixed liquor by compensating the loss through the excess sludge removal. When 30 kPa was set as a threshold for membrane cleaning, the filtration duration without MPE30 was no more than a half day. However, it was extended to 5 to 30 days with MPE30. Other organic polyelectrolytes such as chitosan and starch-based polymers can also improve membrane performance (le Roux et al. 2005; Koseoglu et al. 2008), but they are hardly practical because of the high dosages required and/or the biodegradability.

In one experiment (Hwang et al. 2007), 16 mg/L of MPE50 based on reactor volume was added daily directly to the aeration tank. As a result, the average particle size increased from 101 to 179 μm. In batch tests, the soluble COD and SMP levels in mixed liquor were found the lowest when MPE50 dosage was 25 mg/g MLSS under the experimental condition. It was postulated that the zeta potential of the floc particles was reversed at the dosage higher than 25 mg/L and deflocculation started to occur. In the field condition, the optimum dosage of MPE50 varies depending on the mixed liquor condition and the source of wastewater, but maintaining 100 mg/L of MPE50 per every 3000 mg/L of MLSS seems the optimum in most cases (Guo et al. 2008). For example, if MLSS in MBR is 9000 mg/L on average, the initial dosage would be 300 mg/L. The lost MPE50 through the excess sludge should be compensated daily.

When MPE50 is applied, permeate COD tends to decrease because of the lowered SMP level in mixed liquor. In one occasion, permeate COD decreased by 50% when 200 mg/L of MPE50 was added to the aeration tank of a pilot MBR treating municipal wastewater (Yoon and Collins 2006). In another study, the activity of microorganisms measured by the specific oxygen uptake rate (SOUR) was not noticeably affected by MPE50 (Hwang et al. 2007). When 16 mg/L of MPE50 was added daily, SOUR of mixed liquor was measured at 53.6 mg O_2/MLVSS/h, whereas it was at 56.4 mg O_2/MLVSS/h in a control MBR. In the same experiment, permeate COD decreased with MPE50 from 18.2 to 14.8 mg/L. Meanwhile, permeate TN and TP remained virtually unchanged considering the measurement error, as summarized in Table 5.4.

MPE products are also effective in reducing some types of foaming from the biological tanks. The mechanism of foam reduction is not completely clear, but it might be connected with the lowered SMP and fine particles that stabilize the foam particles. In fact, fine particles and polymers with hydrophobic moieties tend to gather in the air–liquid interfaces of foam and interrupt water drainage. This leads to the extension of the foam life. As shown in Figure 5.31, the thick brown form layer on the surface of anoxic tank disappeared completely in a few hours after adding 400 mg/L of MPE50 to the membrane tank with 12,000 mg/L of MLSS (Yoon et al. 2004a). A

TABLE 5.4

COD, TN, and TP Removal Efficiency with and without MPE50

	COD			TN			TP		
	Influent (mg/L)	Permeate (mg/L)	Removal (%)	Influent (mg/L)	Permeate (mg/L)	Removal (%)	Influent (mg/L)	Permeate (mg/L)	Removal (%)
MPE50 reactor	280.2 ± 12.8	18.2 ± 5.4	93.5 ± 3.3	29.5 ± 8.3	16.0 ± 3.8	45.7 ± 4.2	4.8 ± 2.2	3.8 ± 1.3	20.8 ± 4.5
Control reactor		14.8 ± 3.2	94.7 ± 2.5		15.2 ± 2.4	48.5 ± 4.4		4.0 ± 1.8	16.6 ± 5.5

Source: Hwang, B.-K. et al., *J. Membrane Sci.* 288:149–156, 2007.

(a)   (b)

FIGURE 5.31 Effect of MPE50 on biological foam. Thick brown foam layer disappeared from anoxic tank in a municipal MBR. (a) Before MPE50 addition. (b) After MPE50 addition (400 mg/L). (From Yoon, S.-H. et al., Application of membrane performance enhancer MPE for full scale membrane bioreactors. *IWA's Water Environment Membrane Technology (WEMT) Conference*, Seoul, Korea, 2004a.)

similar foam reduction was observed when cationic flocculants were injected to the activated sludge process (Shao et al. 1997).

Recently, the efficacy of MPE30, MPE50, and MPE51 was studied in AnMBR (Díaz et al. 2014). A laboratory-scale AnMBR (4.5 L) equipped with a 70 cm long external tubular membrane was used for the study. The reactor was operated continuously by feeding synthetic wastewater at an OLR of 7 kg COD/m³/day. Headspace gas was injected in the bottom of the vertically mounted tubular membrane to circulate mixed liquor by generating the gas lift pump effect. The superficial gas flow velocity in the lumen was estimated at 0.3 m/s. The three MPE products were dosed at 300 mg/L, and critical flux was measured using the flux stepping method. MPE50 was the most effective, increasing the critical flux by approximately 100% to 23 to 27 LMH, and the high critical flux was maintained for approximately 10 days.

5.4.4 Experimental Methods

5.4.4.1 Quorum Quenching

As briefed in Section 1.3.2.3, quorum sensing is a system of stimulus and response used by microorganisms to express group behavior, for example, bioluminescence, virulence, biofilm formation, and population control. The signal molecules called autoinducer are used to communicate, for example,

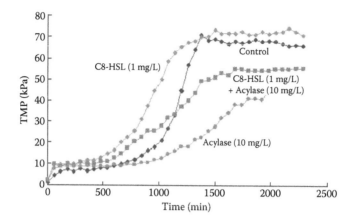

FIGURE 5.32   Effect of adding autoinducer (*N*-octanoyl homoserine lactone, C8-HSL) and acylase on membrane fouling at 15 LMH. (From Yeon, K.-M. et al., *Env. Sci. Technol.* 43(2):380–385, 2009.)

N-acyl homoserine lactones (AHL) by Gram-negative bacterial, oligopeptides by Gram-positive bacteria, and a group of molecules known as autoinducer 2 used by both Gram-negative and Gram-positive bacteria, but a group of molecules that fall into the AHL category have been primarily identified in MBR (Yeon et al. 2009a).

There are three different ways to control quorum sensing: (i) blocking AHL production, (ii) blocking the signal receptor, and (iii) inactivation of AHL signal molecules, but the third method has been successfully demonstrated (Yeon et al. 2009a,b). When acylase was added at 10 mg/L to the mixed liquor, TMP rose much slower compared with the control reactor, as shown in Figure 5.32. However, when an autoinducer (*N*-octanoyl homoserine lactone, C8-HSL) was added to a laboratory-scale MBR at 1 mg/L, biofilm formation was enhanced and TMP increased faster than the control reactor. When 10 mg/L of acylase was added simultaneously with 1 mg/L of C8-HSL, TMP increased slower than the control reactor perhaps because the enzyme quenched most of the added and naturally occurring autoinducers.

In another study, to prolong the efficacy of quorum quenching, quorum quenching bacteria were encapsulated and submerged in mixed liquor instead of using enzymes (Jahangir et al. 2012). Indigenous quorum quenching bacteria (*Rhodococcus* sp. BH4) were isolated from a full-scale MBR treating building wastewater using a medium containing only AHL as a carbon source, and they were encapsulated in polyethylene-based porous hollow fiber membranes. The necessary nutrients and the AHL produced by the bacteria in mixed liquor freely diffuse through the porous membrane capsule. By submerging the porous vessel containing quorum quenching bacteria in mixed liquor, the membrane fouling rate was significantly reduced. The similar method has also been applied for nanofiltration (NF). When acylase was immobilized on NF membrane, the formation of mushroom-shaped mature biofilm was inhibited because of the reduced EPS secretion by bacteria. As a result, flux was maintained at 90% of the initial level after 38 h of operation. Meanwhile, the flux in the control train declined to 60% of the initial (Kim et al. 2011a). Despite its potential, quorum quenching has not been fully commercialized, and there is no known full-scale MBR practicing it to date.

5.4.4.2   Vibration of Membrane or Mixed Liquor

The vibration of membrane or feed water has been studied for many decades to control membrane fouling. The most prominent commercial process is the vibratory shear enhanced process (VSEP®) that has been sold for nearly two decades primarily for removing ions from the difficult waters not suitable for the conventional spiral wound modules (Culkin 1989; Beck 2010). In VSEP, membrane discs are stacked in a pressure vessel, and the whole vessel is vibrated at around 53 Hz by an electrical motor. Recent studies have shown that the same principle can be applied for submerged membranes in MBR to mitigate the membrane fouling without air scouring. Laboratory tests have demonstrated the possibility of reducing the energy consumption to the level the existing coarse air-based systems are capable of, if it is optimized in a large scale.

Conceptually, vibration can be generated either by shaking the liquid in the membrane tank or vibrating or rotating membrane modules (Beier and Jonsson 2008; Kola et al. 2012). In a laboratory study, the whole membrane tank can be shaken on a mechanical shaker, but the same is hardly applicable for a full-scale system. Module vibration can be performed by attaching either vibration motor (Kola et al. 2014) or magnetically induced linear vibrator (Bilad et al. 2012) on the bars/frames extended from the membrane module above the water level. In vibration motor, an eccentrically positioned weight on motor shaft generates the vibration with the same frequency as the rotational speed (or rpm) of the motor. In magnetically induced linear vibrator, the electrical pulse generated by a dedicated frequency generator alternate the magnetic poles to generate linear vibration.

The linear velocity of the membrane is determined by the frequency and the amplitude of the vibration, and hence they are the primary parameters that need to be controlled in vibratory membrane filtration. Bilad et al. (2012) compared total membrane resistances in a laboratory-scale MBR by using either coarse air or magnetically driven linear vibration at 50 Hz with 2 mm amplitude. The vibration was performed intermittently for 50% of the time and the cycle time was 4 min.

As shown in Figure 5.33, substantial reductions in total filtration resistance were observed with vibration compared with the membrane with coarse air scouring. The energy demand in this small-scale test was estimated at 2.3 kWh/m^3 when six 0.016 m^2 membranes were used. Although the energy demand is substantially higher than those for the state-of-the-art submerged membranes based on coarse air scouring, vibration showed its potential as an alternative membrane scouring method.

In a pilot test, a belt-driven reciprocating frame attached to a membrane cage was submerged in the membrane tank. Self-lubricating plane bearings were used for the reciprocating frame. The membrane cage was vibrated using a low-speed variable speed gear motor running at 23 to 111 rpm without air scouring. TMP was stably maintained below 5 psi at 40 LMH at 10 to 12 g/L MLSS. The specific energy required for filtration was estimated at approximately 0.15 kWh/m^3, when flux was 20 LMH and reciprocating frequency was 0.43 Hz (Ho et al. 2013).

There are several hurdles for the vibration method to be a commercially viable option. Module design should be optimized at a larger scale, and the maximum shear yields must be secured by minimizing the physical contacts of the module with the stagnant structure in the membrane tank. Vibration frequency, amplitude, and flux must be optimized simultaneously to maximize the specific energy

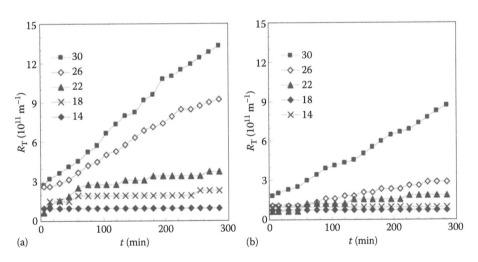

FIGURE 5.33 Effect of operational flux on filtration resistance in laboratory-scale MBR: (a) air scouring and (b) vibration. (From Bilad, M.R. et al., *Water Res.* 46(1):63–72, 2012.)

efficiency. Securing the mechanical reliability at the operating condition would be another critical goal. Some extent of aeration may be unavoidable in the membrane tank to maintain the DO at a sufficient level (1–2 mg/L) to prevent the accelerated membrane fouling. Consequently, vibration strength and aeration intensities should be optimized together to obtain the overall high-energy efficiency.

5.4.4.3 Electrical Field

To mitigate the membrane fouling caused by charged macromolecules, electrical fields have been tested in various conditions for many decades (Yukawa et al. 1983). Because the macromolecules and fine particles are charged negatively, cathodes with negative electrical potential are inserted in the opposite side of the feed water across the membrane (or permeate side) to repulse them from the membrane surface. Anodes are located in the middle of the feed side in the form of metallic meshes. In the electrical field across the membrane, negatively charged macromolecules migrate away from the membrane and move to the anode. Monovalent ions with high diffusivities such as Na^+ and Cl^- are much less affected by the electrical field than macromolecules, thereby playing a role in establishing the local charge balance.

Although positive results have been obtained in laboratory-scale experiments, this principle has not been applied in large-scale membrane filtration primarily because of the complexities in the module configuration and the long-term fouling of electrodes. Akamatsu et al. (2010) tested flat sheet membranes in a plate and frame module equipped with platinum meshes in both sides of the membrane as a cathode and an anode. When the activated sludge was filtered, flux improved nearly 100% at 3 to 10 kPa with an electric field of 6 V/cm. When the electrical field was activated at an intermittent mode, for example, 90 s on and 90 s off, the lost flux during the period of no electrical field was immediately recovered. Liu et al. (2012a) tested a conductive flat sheet submerged membrane doped with pyrroles as a cathode. Because the membrane itself was an electrode, very low electrical field was sufficient to mitigate the fouling, for example, 0.2 V/cm. In this study, two stainless steel meshes were placed 5 cm away from the membrane in both sides of the flat sheet membrane as anodes whereas the pyrrole-doped membrane acted as cathode. The electrical field reduced membrane fouling, and the flux increased around 100% from the baseline at a vacuum pressure of 12 kPa.

5.4.4.4 Ultrasonication

Ultrasonication has been tested not only to enhance membrane performance but also to enhance membrane cleaning in MF/UF processes. It has been known that ultrasonication helps improve membrane performance by hindering particle deposition due to the vibrational motion of both particles and membrane. If ultrasound is used for membrane cleaning, it can improve the cleaning efficiency by enhancing particle detachment from the membrane. Several factors are known to affect the performance of sonication, for example, frequency, strength, and duration. When polystyrene latex particles were filtered in dead-end filtration mode, it was observed that the initial flux was maintained better at the lower end of the frequency range tested, that is, 20 to 100 kHz (Lamminen et al. 2004). When ultrasonication was performed to clean fouled membranes, membrane cleaning efficiency increased as the power density (W/cm²) increased. However, membranes are damaged with excessive exposure to the ultrasounds, depending on the membrane materials and the fabrication methods. When PVDF-based hollow fiber membranes were used, the maximum allowable ultrasonic dose (E_U, Wh/cm²) calculated by multiplying ultrasonic power density (P_U, W/cm²) and its duration (h) declined as ultrasonic strength increased. The relationship shown in Equation 5.10 was found when power density ranged between 0.2 and 0.9 W/cm² (data from Jin et al. 2008). The maximum allowable exposure time (t_{max}, h) varied depending on power density, as shown in Equation 5.11, in the same power density range:

$$E_U = 2.71P_U^2 - 4.97P_U + 2.36, \tag{5.10}$$

$$t_{max} = \exp(3.26 - 6.40\,P_U). \tag{5.11}$$

Ultrasonic membrane cleaning is also affected by the distance between the tip and the membrane (Chen et al. 2006). When ultrasonic tip was placed at varying distances from the membrane at 3.5, 2.6, or 1.7 cm, respectively, it was observed that the membrane cleaning efficiency was the highest with the shortest distance. However, when membranes were placed too close to the tips within the ultrasonic cavitation region (1.3 cm from the tip), membranes were damaged easily. Ultrasonication is effective to remove cake layer but is not considered effective to remove gel layer (Wang et al. 2014c).

Despite the promising results, ultrasonic fouling mitigation and membrane cleaning have not been implemented in the large-scale membrane process. It is due to the cost associated with the complex membrane system design. It is also difficult to apply ultrasound-based fouling mitigation system in submerged membrane-based MBR because of the concerns over the floc breakages and the elevated fine particle concentrations in mixed liquor.

5.4.4.5 Intermittent Ozone Sparging

In order to control membrane fouling, membranes can be contacted with ozone intermittently. In one experiment (Takizawa et al. 1996), a synthetic water with kaolinite suspended in tap water was filtered by vertically mounted hollow fiber membranes submerged in a membrane tank. The flux was maintained at a constant level (60 LMH), while either air or ozone-containing air (0.0435 mg O_3/L) was sparged to scour the PE-based membrane. When air and ozone-containing air were sparged alternatively every 10 s, TMP rise delayed substantially.

The similar ozonation has been applied for MBR in order to reduce sludge production in MBR as will be discussed in detail in Section 5.6.5. In this process, ozone is not directly sparged to the membrane, but it is reacted with mixed liquor in order to disintegrate microorganisms before returning the treated mixed liquor back to the MBR system. It has been observed that the ozonation of mixed liquor reduces not only excess sludge production but also membrane fouling by reducing SMP and fine particle concentration in mixed liquor.

5.5 EFFECT OF FLOW BALANCING ON MEMBRANE PERFORMANCE

5.5.1 Scouring Air

5.5.1.1 Cause of Unbalanced Scouring Air

As the number of membrane cassette increases, it becomes harder to balance the airflow rate among air nozzles underneath the cassettes. The maldistribution of scouring air is a common problem that causes many secondary problems in MBR. There are many causes of imbalanced airflow, but the following causes have been commonly observed:

- The total head loss from the blower to the air nozzle can vary significantly depending on the number of tees/crosses and the length of pipes in between the two. As the system size increases, air distribution system becomes more complicated. Thus, the air must pass different numbers of tees, crosses, and valves and travel different distances, depending on the location of the nozzle. As a result, the chance of having unbalanced airflow in each air nozzle grows as the system size increases. This problem can be dramatically aggravated when the dynamic pressure loss is excessive in the main air pipe because of the small diameter.
- In the pipelines shown in Figure 5.34a, the small main air pipe causes excessive pressure loss and in turn causes gradually declining airflow in the branch pipes toward the downstream.
- On the contrary, Figure 5.34b shows well-designed pipelines, where the main pipe diameter is three times larger than the branches (or nine times larger cross-sectional areas). Because of the negligible head loss in the main pipe, airflow and pressure in the branch pipes are nearly identical.

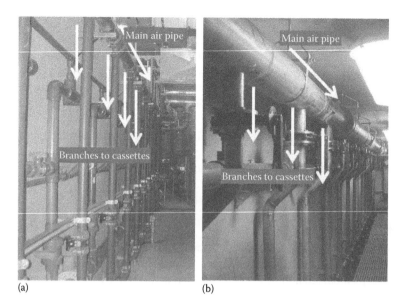

(a)   (b)

FIGURE 5.34 Scouring air distribution systems in two different MBR sites. (a) Air distribution system with issues. (b) Well-balanced air distribution system.

- Differences in the length of branch pipes can cause varying degree of pressure head loss in each pipe. In fact, the airflow rate through a pipe is inversely proportional to the square root of the pipe length, if the total pressure losses in branch pipes are identical. Therefore, if one branch is two times longer than the other branch, the airflow through the pipe will be only 71% of the other shorter pipe (= 1/sqrt(2)).
- The random fouling of air nozzles can cause high-pressure head losses in some air nozzles. Air nozzle fouling should be minimized to obtain reasonable airflow distribution, but there are no known methods that guarantee the clean air nozzles. To monitor actual airflow to the cassettes, airflow meters are required in all the membrane cassettes, but it is often constrained by the capital budgets.
- Even if airflows in nozzle are identical, air can be maldistributed among membrane modules and panels, if fibrous materials interrupt the airflow. This happens when the prescreen of wastewater fails and a large amount of fibrous materials captured in the bottom of membrane cassettes interrupt the airflow, as shown in Figure 5.35.

FIGURE 5.35 Fibrous materials can block the air–liquid channel, if screen fails regardless of membrane types. (Courtesy of Mott MacDonald.)

- Nonuniform air nozzle submergence can cause the maldistribution of scouring air. It is important to place all the air nozzles with an identical submergence. However, as shown in Figure 6.13, diffuser pipe can be installed slightly slanted to compensate the internal pressure loss and secure constant effective air pressure along the pipe.

5.5.1.2　Consequence

The maldistribution of air causes uneven airflow pattern, as shown in Figure 5.36a, where aeration is overly vigorous in the center cassette whereas it is not vigorous enough in the right cassette and a portion of the left cassette. The membranes located in the low airflow zone can develop sludge cake in between two membrane panels because of the insufficient mixed liquor supply. Eventually, the sludge cakes can merge in the middle of the upflow channel and plug it up. As more water is vacuumed out through the cake layer, cake layer becomes denser and cannot be removed without disassembling the cassette. Depending on the amount of the sludge cake built in each channel, some membrane panels can bulge and deform. The bulging membrane panels need to be replaced after disassembling the cassette (Manson et al. 2010). Figure 5.36b shows a membrane panel heavily damaged by sludge cake.

　　The membrane fouling started from the underaerated zone tends to spread because the flux loss in the affected cassette must be compensated by other cassettes in a constant flux mode. It is crucial to maintain evenly distributed scouring air for stable and reliable operation. All the potential causes of imbalanced aeration listed in the previous section must be carefully reviewed during the design stage. From the operation perspective, the most common causes of uneven air distribution are the insufficient removal or accidental leak of fibrous materials from the pretreatment process as well as the air nozzle/diffuser fouling.

5.5.2　Permeate Drawing

In submerged membrane filtration, all membranes are supposed to run below the sustainable flux to avoid fast membrane fouling. If the flux in a portion of membranes is above the sustainable flux, membrane fouling occurs quickly in that area. Then the flux in other areas must increase under the constant flux mode to compensate the flux loss in the fouled area. To avoid the uneven flux among membrane modules and cassettes, vacuum pressure in all membrane modules should be controlled evenly. However, as system size grows, it becomes harder to maintain flux evenly in all membrane surfaces.

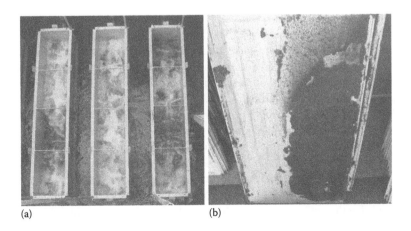

(a)　　　　　　　　　　　　　　　　　　　(b)

FIGURE 5.36 Consequence of uneven scouring air distribution. (a) Scouring air maldistribution. (b) Cake formation on membrane. (Courtesy of Mott MacDonald.)

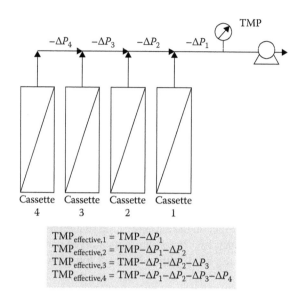

$$TMP_{effective,1} = TMP-\Delta P_1$$
$$TMP_{effective,2} = TMP-\Delta P_1-\Delta P_2$$
$$TMP_{effective,3} = TMP-\Delta P_1-\Delta P_2-\Delta P_3$$
$$TMP_{effective,4} = TMP-\Delta P_1-\Delta P_2-\Delta P_3-\Delta P_4$$

FIGURE 5.37   Effect of pressure loss in permeate pipe on effective TMP.

Figure 5.37 shows an example with four membrane cassettes. Pressure losses largely occur in tees, joints, valves, elbows, etc., but it can be significant even in straight pipes in poorly designed system. Because of the pressure loss in the permeate pipe, effective TMP decreases as the distance of cassette from the permeate pump increases. As a consequence, the flux of cassette 1 is the greatest whereas that of cassette 4 is the smallest in Figure 5.37. One way of avoiding maldistribution of permeate flow is metering permeate flow in each module or cassettes, but it is not always possible in a full-scale system because of the budget constraints. The other common cause of imbalanced permeate flow is the imbalanced scouring airflows among cassettes, as discussed in the previous section.

5.5.3 MIXED LIQUOR CIRCULATION

In the MBR system with multiple aeration and membrane tanks, liquid flow distribution among the tanks must be rigorously controlled to maintain reasonable MLSS in each tank. The evenly distributed solid levels are essential for ensuring not only the optimum biological performance but also the optimum membrane performance. If mixed liquor distribution is not performed properly, the MLSS level can increase excessively in some membrane tanks. It may cause fast membrane fouling in those tanks. If a membrane tank fails to produce a design flow because of the excessive MLSS, the lost flow must be compensated by other membrane tanks. It eventually causes accelerated membrane fouling across all membrane tanks.

Figure 5.38 shows an example, where mixed liquors from the three different aeration tanks are distributed among four membrane tanks. After extracting equal amounts of permeate from the four membrane tanks, the concentrated mixed liquor merges again with the concentrate channel before recycling back to the front end of the aeration tanks. In this system, the inlet and the outlet of each membrane tank are open to the sludge channels to enable free mixed liquor flow driven by the water level difference between aeration and membrane tanks. In this system, the mixed liquor flow through tank 1 is the highest because of the short hydraulic distances and the favorably located mixed liquor channels between aeration and membrane tanks. By contrast, the mixed liquor flow through tank 4 is the lowest. As a result, MLSS in tank 4 was found approximately 100% greater than that in tank 1. The imbalanced flow distribution problem occurs more often than not, especially

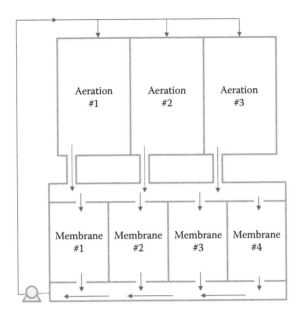

FIGURE 5.38   A hypothetical example of imbalanced mixed liquor circulation.

in retrofit plants where existing tanks are reused to save capital costs. Appropriate attention must be taken to avoid unintended imbalance in mixed liquor circulation.

5.6   EXCESS SLUDGE REDUCTION PROCESS

5.6.1   PRINCIPLE

In biological processes, microbial population grows by consuming food in the influent, but simultaneously endogenous respiration occurs when old cells die and are consumed by live cells. In natural condition, sludge (or biosolids) yield can be controlled by limiting food supply and extending sludge age (or SRT) to allow more time for endogenous respiration. However, this method requires a large aeration tank to store extra amount of biomass and more aeration to suspend the large amount of biomass stored. In addition, the endogenous respiration rate slows down over time as nonbiodegradable portion of dead cells accumulate in the sludge. The accumulating nonbiodegradable solids cause inefficient use of the aeration tank and the aeration energy. In fact, it has been known that the cell walls are refractory in the aeration tank because they are designed to protect cells from the hydrolysis and the enzymatic attack in the same condition. According to the activated sludge model number 1 (ASM#1) of the International Water Association (IWA), nonbiodegradable materials take approximately 8% of the total live biomass (Henz et al. 1987). Therefore, approximately 8% of the biomass accumulates in the mixed liquor as nonbiodegradable solids every time live microorganisms die. As SRT increases, more nonbiodegradable cell debris accumulates in the mixed liquor while the live microbial population declines under a constant MLSS condition. As nonbiodegradable solids accumulate, reducing excess sludge production becomes increasingly hard.

The sludge production process can be written as Equation 5.12. In this equation, X is the MLSS (mg/L), S_e is the COD in mixed liquor (mg/L), K_S is the half-reaction constant (mg/L), μ_m is the maximum growth rate constant (mg/L/day), and k_d is the cell decay rate (/day). If there is no sludge removal, X (MLSS) increases over time because the growth rate is greater than the death rate (Yoon 2003; Yoon et al. 2004b). However, if the death rate, k_d, is raised artificially, the MLSS growth rate, dX/dt, can be controlled at zero or even negative. In fact, all methods used

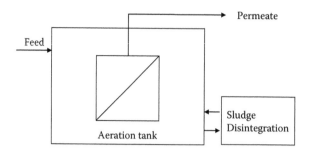

FIGURE 5.39 Principle of sludge (or biosolids) reduction by partial sludge disintegration and the recycle to the aeration tank.

for sludge reduction are about how to artificially raise k_d economically and reliably. The equation is as follows:

Apparent growth rate = growth rate – death rate.

$$\frac{dX}{dt} = \frac{\mu_m S_e}{K_S + S_e} - k_d X. \tag{5.12}$$

As shown in Figure 5.39, the mixed liquor from the aeration tank is disintegrated in a sidestream device/tank and recycled back to the aeration tank. The debris of disintegrated cells are consumed by the live cells in the aeration tank and reproduced as new cells. Some carbons are lost as CO_2 during the reproduction process. By controlling the rate of sludge disintegration (or k_d), the desired sludge reduction rate can be obtained. If the sludge disintegration method used can convert the non-biodegradable cell debris to biodegradable, the net sludge production can reach zero.

5.6.2 SLUDGE REDUCTION BY EXTENDING SRT: ECONOMIC ASPECT

The amount of sludge produced can be reduced by simply extending SRT, as discussed in Section 6.3.4. The long SRT condition can be obtained by combining high MLSS and large aeration tanks. As SRT increases, the relative portion of the endogenous respiration in the total respiration increases and the observed sludge yield decreases. Meanwhile, dead cells and cell debris concentration increase in the sludge while live microorganism population decreases under the constant MLSS. Because of the accumulation of nonbiodegradable sludge, the effect of SRT on sludge reduction gradually decreases as SRT increases, as shown by the curve based on ASM#1 in Figure 6.5. Because of the lowering sludge reduction rate at high SRT, the specific capital and operating costs to obtain additional sludge reduction increases as target sludge reduction efficiency increases. In general, reducing sludge production by extending SRT is generally not considered economical.

Raising MLSS beyond the ordinary level causes several design and operational challenges. Whereas oxygen dissolution rate should be increased to meet the high oxygen demand to oxidize extra biosolids, a partial loss of OTE is unavoidable because of the high aeration rate at the high MLSS, as discussed in Section 6.5.2. Therefore, disproportionally more scouring air is required than the linearly projected aeration rate based on the oxygen demand. In addition, the high MLSS tends to make sludge more viscous and can increase membrane fouling by hampering membrane scouring efficiency. With respect to MBR operation, the high MLSS raises a risk of operational failure when excess sludge treatment stops due to mechanical issues because there is no sufficient buffer to store more sludge in the biological tanks. In ordinary MBR, MLSS can be allowed to increase to some extent while the excess sludge treatment systems are repaired. If sludge reduction is desired without raising MLSS, larger aeration tanks must be used, but this incurs the increase in operating cost for suspending more biomass in the larger tanks in addition to the increase in capital cost.

Reducing excess sludge by extending SRT is not attractive economically in typical conditions. If the specific energy requirement for oxygen dissolution is approximately 1 kg O_2/kWh, the operating cost of reducing 1 ton of dry sludge would be $142, assuming $0.1/kWh and 1.42 kg COD/kg VSS. If the oxygen costs required for nitrification are included, operating costs will be close to approximately $200/ton. Additional costs may be required to supply sufficient air to suspend solids in the large aeration tanks. If the long SRT is achieved by high MLSS instead of larger aeration tanks, the specific energy required to dissolve oxygen will be much more than $200/ton due to low oxygen transfer efficiency at high MLSS. Because the cost of operation exceeds the cost of sludge disposal in most geographical areas, sludge reduction by extending SRT is not commonly practiced.

Meanwhile, the MBR with extra-high MLSS are commonly used for excess sludge thickening, as discussed in Section 6.13.3. The secondary biological sludge is thickened by an MBR-like system equipped with submerged membranes. Design fluxes are typically less than half of the ordinary MBR, for example, 4 to 6 LMH. In one study, flux was gradually reduced from 7.75 to 2.5 LMH during the thickening process to mitigate membrane fouling when the waste activated sludge was thickened from 3 to 8 g/L to 45 g/L using submerged hollow fiber membranes (ZW500d) (Natvik et al. 2009). In another study, target MLSS was set at 40 g/L with the same membranes (Cantor et al. 2000).

5.6.3 Sludge Disintegration Methods

5.6.3.1 Biological Method

In the 1940s, it was found that the net excess sludge production from a municipal WWTP could be reduced by adding anaerobically digested excess sludge to the aeration tank (Torpey et al. 1984). This led to the idea that the alternate exposure of sludge to different environments could substantially reduce the quantity of the sludge to be disposed. Later in the early 1980s, sludge disintegration was performed by iterating excess sludge in between the aeration tank and the thermophilic digester in a municipal WWTP. However, these biological methods failed to get wide acceptance because of the dwindling sludge reduction performance over time (Prakasam et al. 1990). It was considered that the alternating exposure of the sludge to two different conditions encourages the growth of microorganisms that can survive in both conditions. Once those versatile microorganisms replace the microorganisms present in the original sludge, alternating environmental conditions does not significantly affect the survivability of microorganisms (or k_d) as it should to make the process economical. There are a few commercial or experimental processes that fall into this category. The use of anoxic (Cannibal®, Evoqua), anaerobic (Chon et al. 2010), and thermophilic (S-TE, Kobelco Eco-solutions) reactors have been attempted to treat the excess sludge grown in aerobic condition in surrounding temperatures.

Some of the excess sludge treatment methods used for sludge stabilization reduce the sludge weight, as discussed in more detail in Section 6.13, for example, aerobic digestion, autothermal thermophilic aerobic digestion, and anaerobic digestions. The primary goal of the sludge stabilization is to qualify the sludge for land application or landfill in a typical situation. The savings from the sludge reduction alone are usually not sufficient to justify the costs of running the process.

5.6.3.2 Chemical Method

The most successful and proven sludge reduction method is based on ozone oxidation and was commercialized as Bioleader™ (Kurita) in the 1990s (Yasui 2000; Yasui and Shibata 1994; Yasui et al. 1996, 2005). In this method, a portion or the entire amount of RAS is passed through the ozone contactor. Each time the sludge passes the ozone contactor, the sludge is dosed with ozone at approximately 0.05 g O_3/g MLSS. Overall, around three times more sludge should be treated than the sludge reduced (Sakai et al. 1997), and hence the specific ozone demand is calculated at around 0.15 g O_3 per 1 g of biosolids reduced. In this process, excess sludge is passed through closed bubble columns in which pressurized oxygen with 10 to 12 w/w% ozone is sparged from the bottom. The exhaust gas containing residual ozone and pure oxygen is directed to the aeration tank to be sparged. Sparging the exhaust gas in the aeration tank is not only for safely treating the residual ozone gas but also for supplying extra oxygen to meet the additional oxygen demand caused by the solubilized sludge. With the ozone treatment, sludge settling properties and effluent quality improve perhaps because the ozone tends to react preferentially with small suspended particles with a large specific surface area rather than large floc. The other benefit of this method is that ozone can convert nonbiodegradable cell debris to biodegradable, preventing the gradual decline of sludge reduction efficiencies at high SRT observed in the biological sludge disintegration. As a result, 100% biosolid reduction is possible, except the small amount of inert materials contained in influent, for example, silts and cellulosic and plastic debris (Sakai et al. 1997). In addition, the sludge reduction efficiency does not decline over time in long-term applications because microorganisms cannot adapt with such oxidants that chemically attack the cell constituents.

The original Bioleader™ has been modified by using various ozone dosages, various ozone contactor configurations, ozone injection points, etc., but as the ozone dosages in the ozone contactor decreases, the amount of sludge that should be treated to obtain a unit sludge reduction increases (Ahn et al. 2002; Park et al. 2003). Other commercial processes using ozone include Lyso™ (Praxair), Halia™ (Air Product), AspalSLUDGE™ System (Air Liquide), etc. Unlike other methods, Lyso™ uses a plug-flow reactor to maximize the efficiency of ozone use.

The use of chlorine was also attempted instead of ozone (Chen et al. 2001; Saby et al. 2002). By treating a portion of sludge in the aeration tank at a chlorine dosage of 0.133 g/g MLSS every day, excess sludge production was reduced by 65%. However, the sludge settling properties (or SVI) were severely compromised, and the method did not appear practical in activated sludge process. In addition, there were concerns over the production of chlorinated organics such as trihalomethanes, which may make the effluent disqualified for discharging to the nature.

Alkaline treatment is an effective method to hydrolyze microbial cells and can be used alone for sludge disintegration. Alternatively, alkaline treatment can be combined with ozone treatment. In a pilot test performed in a municipal WWTP, a portion of mixed liquor in MBR equivalent to 1.5% of the influent flow rate was treated by NaOH at 22.3 mEq/L (or 0.89 g/L) for 3 h, followed by ozone treatment at a dosage of 0.02 g O_3/g MLSS. It was observed that MLSS stayed stably at 10 to 11 g/L for 200 days, whereas the MLSS of the control MBR increased to 25 g/L (Oh et al. 2007). When the same alkaline–ozone treatment was applied before excess sludge is sent to an aerobic digester, a fast sludge reduction occurred in the digester (Yeom et al. 2005). In this study, around 70% of sludge reduction was observed at a retention time of only 3.1 days in the digester.

Metabolic decoupling agents can also be used. These toxic agents to microorganisms work by decoupling the metabolic pathway of microorganisms, causing inefficiencies in cell growth/division. It has been known that some chemicals interrupt biomass production by uncoupling the oxidative phosphorylation and disrupting the direct energetic link between catabolism and anabolism, for example, 2,4-dinitrophenol (2,4-DNP), *o*-chlorophenol, 2,4-dichlorophenol, rotenone, dicoumarol, trichlorophenol, 3,3′,4′,5-tetrachlorosalicylanilide (TCS), and *p*-nitrophenol (Mayhew and Stephenson 1998; Ye et al. 2003). When 100 mg/L of *para*-nitrophenol was added to the aeration tank, the oxygen uptake rate increased simultaneously with a 30% lower sludge yield (Low et al. 2000). Despite the promising observations, some microorganisms in sludge are suspected to adapt with the decoupling agents eventually and form a resistant sludge. In addition, the use of decoupling agent compromise the effluent quality by forming diffused microbial flocs that do not settle in clarifier.

5.6.3.3 Physicochemical Method

Physicochemical methods such as ultrasound, ball mill, plasma, electrolysis, thermal hydrolysis, high shear disintegration, etc., can be used to disintegrate sludge. In a study performed using a laboratory-scale MBR with submerged membranes in the combined aeration and membrane tanks, a portion of the sludge (1 L out of 8.5 L) was taken out every day and returned back to the aeration tank after treating it with ultrasound. The energy input in the sludge disintegration process was 216 kJ/g MLSS (or 0.06 kWh/g MLSS), and the OLR of the MBR was 0.91 kg BOD/m³/day. As shown in Figure 5.40, the MLSS of the MBR with sludge disintegration stayed at 6 to 7 g/L for 28 days, whereas the MLSS of the control MBR without sludge disintegration increased to 14 g/L. As soon as the sludge disintegration stopped, MLSS started to increase at a similar rate with the control MBR (Yoon et al. 2004b).

High-pressure homogenizers have been used to mix liquids and powders in the food and beverage industry. The similar equipment can be used to disintegrate cells. In the patented process called MicroSludge™, sludge is chemically pretreated using alkaline to make the cells easier to burst before it is fed to homogenizer (Stephenson and Dhaliwal 2000; Rabinowitz and Stephenson 2006). The pretreated sludge is pressurized to more than 800 bar and released to the atmosphere through a nozzle. When the cells are depressurized abruptly, dissolved gases in the cell mass expand suddenly and the cells burst releasing easily degradable cytoplasmic materials. This method has been used to pretreat excess sludge being fed to an anaerobic digester.

Hybrid processes have also been attempted and commercialized, such as thermophilic digestion and chemical sludge disintegration using H_2O_2 (AFC™ process by PMC BioTec), thermal hydrolysis and anaerobic digestion (Biothelys™ by Veolia), sludge homogenizer and anaerobic digestion (Crown™ process by Evoqua), and thermochemical process (Banu et al. 2011).

Mechanical sludge disintegration occurs inadvertently in MBR equipped with sidestream tubular membranes, where microorganisms are ruptured by the high shear stress exerted by the circulation pump impeller (Kim et al. 2001). In small-scale MBR systems, sludge yields were observed at only 0.3 and 0.2 g MLVSS/g COD with a centrifugal pump and a vane pump, respectively. The observed sludge yields were substantially lower than those observed without a circulation pump, that is, 0.4 to 0.5 g MLVSS/g COD. The sludge reduction by circulation pump is less significant in large-scale MBR because feed water circulates less frequently through the membrane system than that in small systems. The long membrane channel length enables a higher water recovery when the

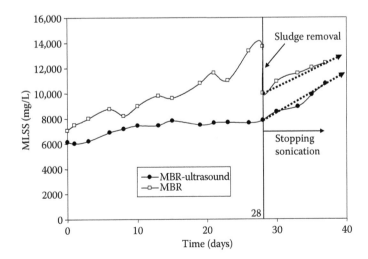

FIGURE 5.40 Comparison of MLSS development with and without a partial sludge disintegration by ultrasound. (From Yoon, S.-H. et al., *Process Biochem.* 39(12):1923–1929, 2004b.)

TABLE 5.5
Sludge Disintegration Methods

Methods		Remark
Biological	High SRT	Efficiency decreases as SRT increases
	Anoxic treatment	Store excess sludge in anoxic tank for 7–10 days (e.g., Cannibal™)
	Anaerobic treatment	Recycle anaerobically digested sludge to the aeration tank
	Thermophilic treatment	Expose sludge to thermophilic condition (e.g., S-TE™)
Chemical	Ozone	Oxidize a portion of RAS using ozone (e.g., Bioleader™)
	Chlorine or H_2O_2	Substitute of ozone
	Alkaline treatment	Hydrolyze sludge using NaOH and return it to the aeration tank
	Decoupling agents	Add metabolic decoupling agents to the aeration tank
Physicochemical	Ultrasonication	Rupture cells using ultrasound
	Ball mill	Grind the sludge and return it to the aeration tank
	Homogenizer	Compress sludge with air and abruptly release the pressure to rupture the cells
	Heating	Heat sludge using microwave, steam, etc., to deactivate microorganisms

pressurized mixed liquor passes the membrane system. Thereby, less amount of feed water needs to be circulated by pump to obtain same amount of permeate (Table 5.5).

5.6.4 Effect of Sludge Reduction Process on Nutrient Removal

One intrinsic drawback of sludge (or biosolids) reduction processes is the compromised biological phosphorus removal efficiency. It is simply because biological phosphorous removal is dependent on the amount of phosphorous carried out by the excess sludge. On the contrary, nitrogen removal efficiency can be retained after implementing the sludge reduction process. Because the disintegrated sludge recycled back to the biological system carries a sufficient amount of COD compared with TKN, for example, COD/TKN ≈ 10, denitrification can be accomplished at a high efficiency. In some cases, the excess COD carried by the disintegrated sludge can even improve the nitrogen removal efficiency, especially when the COD/TKN ratio of raw water is low (<3), and it is a bottleneck in the denitrification. In the MLE process shown in Figure 6.28, the disintegrated sludge is fed to the anoxic tank to use the COD in reducing the nitrate recirculated from the aeration tank to nitrogen gas. The remaining TKN goes to the aeration tank and is oxidized before recirculated back to the anoxic tank.

It has been reported that the sludge disintegration by ozone does not affect nitrogen removal efficiencies under the presence of anoxic tanks in full-scale tests in a municipal WWTP (Yasui et al. 1996; Sakai et al. 1997). This observation was also confirmed mathematically based on IWA's ASM#1 (Yoon and Lee 2005). According to the study, total nitrogen (TN), including NH_4-N and NO_3-N, in the effluent only marginally increases in an A/O process combined with sludge disintegration (A/O-SD), as shown in Figure 5.41. In this study, the volumes of anoxic and aeration tanks were assumed 5 and 10 L, respectively, and the influent flow rate was 30 L/day. The COD of the influent was assumed 300 mg/L and the TKN 30 mg/L. The parameter β in Figure 5.41 indicates the conversion efficiency of nonbiodegradable matters to biodegradable during the disintegration process. Perhaps β is higher with chemical sludge disintegration methods than with mechanical methods. Until the sludge disintegration rate reaches 4 L/day in the system with 15 L total volume, TN in the effluent only moderately increases regardless of β. However, once the sludge disintegration rate exceeds the threshold at 4 L/day, TN in the effluent increases rather sharply, especially with low β. The sharp increase of effluent TN is attributed to the declining nitrifier population as a consequence of increasing sludge disintegration rate. Meanwhile, TN in the effluent is much higher in the system

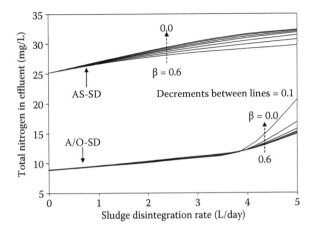

FIGURE 5.41 Effect of the sludge disintegration rate on total nitrogen (TN) in effluent, where A/O-SD indicates the activated sludge process with anoxic and aeration tanks combined with sludge disintegration. (From Yoon, S.-H., Lee, S. *Water Res.* 39:3738–3754, 2005.)

without anoxic tank. In the activated sludge with an add-on sludge disintegration system (AS-SD), TN in the effluent increases gradually as the sludge disintegration rate increases.

The intrinsic drawback of sludge reduction processes in the phosphorous removal efficiency can be partially covered by using the enhanced biological phosphorus removal process. On the basis of the ASM#2 model, in an anaerobic–anoxic–oxic (A$_2$O) process with a sidestream sludge disintegration process, effluent phosphorus concentration can be controlled 20% to 30% higher than the baseline without sludge disintegration while the sludge reduction efficiency is maintained at 50% to 70%. The enriched PAOs play a role in buffering the artificially raised specific phosphorus loading by the disintegrated sludge. Inorganic coagulants can also be used to enhance the phosphorus removal, but it hampers the sludge reduction efficiencies.

5.6.5 EFFECT OF SLUDGE REDUCTION PROCESS ON MEMBRANE FOULING

It has been well known that the ozonation of mixed liquor tends to increase the average floc size. It is perhaps because the highly reactive ozone oxidizes most organic substances on its way; thereby, small particles with large specific surface areas are more vulnerable to be solubilized than large particles. The other potential cause of the particle size increase is that the zeta-potential of microbial floc tends to change to more neutral with ozonation. The neutrally charged particles tend to coagulate better due to the lack of charge repulsion. As a consequence of the growing particle sizes, it has been observed that sludge settling improves in the secondary clarifier in CAS when sidestream ozonation was performed for CAS to reduce excess sludge production (Yasui et al. 1996).

Hwang et al. (2010a,b) performed side-by-side tests with and without ozone injection. In this study, ozone was injected through a Venturi injector called the *turbulent jet flow ozone contactor* (TJC). When the Venturi was operated without ozone, the cavitation generated by the Venturi injector was not sufficient to make any difference in sludge yield. However, the sludge yield decreased significantly with an ozone injection. Meanwhile, the average zeta potential of particles increased from −20 mV to −8 mV with ozonation, and the average particle size grew from 167 to 213 μm on average. As a result, the membrane fouling rate measured by the increasing rate of the TMP decreased (TJC-MBR), as shown in Figure 5.42. In another study (Song et al. 2003), sludge was taken from the membrane tank and contacted with ozone in a stirred tank. The specific ozone dosage was set at 0.1 g O$_3$/g SS. The treated sludge was returned to the anoxic tank to be used as a carbon source. Although no noticeable membrane fouling was observed under the condition, nitrogen removal improved slightly. MLSS stayed at the initial level with ozone for during the

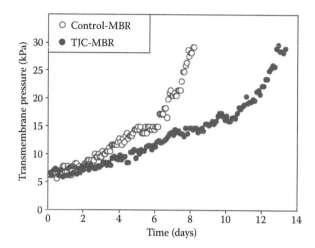

FIGURE 5.42 Effect of ozonation on membrane fouling. (From Hwang, B.-K. et al., *J. Industrial Eng. Chem.* 16:602–608, 2010b.)

35-day operation, whereas it increased gradually in the control reactor. In this study, the total EPS remained nearly unchanged, but loosely bound EPS (LB-EPS) defined in another literature (Li and Yang 2007) was significantly reduced with ozone.

Except ozonation, most other sludge disintegration methods have a potential to aggravate membrane fouling. Most chemical methods such as alkaline/acid treatment, metabolic decoupling agents, etc., increase the concentration of cell debris in mixed liquor. Such small cell debris and macromolecules that originated from the dead cells very likely accelerate membrane fouling in MBR. In fact, when alkali-thermal treatment was performed to disintegrate sludge, EPS concentrations increased and as a result membrane fouling accelerated (Banu et al. 2011). Meanwhile, mechanical methods also create cell debris and macromolecules that raise the membrane fouling potential, for example, ball mill, electrolysis (Kim et al. 2011b), ultrasound (Yoon et al. 2004b), cavitation by Venturi injector (Hwang et al. 2010a,b), homogenizer (Rabinowitz and Stephenson 2006), plasma treatment (Chang et al. 2011), and thermal biosolid treatment (Bougrier et al. 2006).

5.6.6 Economics of Biosolid Reduction Process

When sludge reduction is performed by iterating disintegrated sludge according to the methods discussed in Section 5.6.5, the economics of the process is affected mainly by the following two factors:

- *Cost of sludge disintegration*—The capital and operating costs of sludge disintegration process vary widely, depending on the sludge disintegration method. Sludge disintegration using biological methods such as anoxic, anaerobic, or thermophilic reactors are less costly than chemical or physical methods, but the efficacy is limited because of microbial adaptation in the main reactor and the sidestream reactor. Ozonation provides reliable sludge reduction efficiencies up to nearly 100%, but the operating costs required for sludge disintegration are more than $200 per one ton of dry biosolid reduced. Substantial capital costs are also required for ozone contactor, ozone generator, pure oxygen generator or liquid oxygen storage tank, and additional aeration capacity in the aeration tank.
- *Target biosolid reduction efficiency*—The cost of biosolid reduction increases faster than proportionally predicted as the target biosolid reduction efficiency increases. For example, the specific cost of reducing 80% of sludge production is much greater than the specific cost of reducing 50% of sludge production. It is mainly because refractory solids accumulate

in the mixed liquor and consume extra chemicals or energy as the target sludge reduction efficiency increases. Meanwhile, the cost efficiency of sludge disintegration also decreases, if the target disintegration (or solubilization) efficiency increases. For example, the specific ozone dosage increases as the target solubilization efficiency increases because increasingly more ozone molecules react with already ruptured cells and their debris.

The concept of sludge disintegration number (SDN) has been developed as an indicator of biosolid disintegration efficiency. It was defined as the ratio of the sludge disintegrated to the sludge reduced (Yoon 2003; Yoon and Lee 2005). For example, if 10 kg of dry solids is treated while 2 kg is actually reduced, SDN becomes 5. SDN varies depending on how much thoroughly the biosolids are disintegrated. When ozone dosage is 0.05 g O_3/g MLSS or higher, SDN was observed at approximately 3, and the total ozone consumption was approximately 0.15 g O_3/g MLSS removed (Yasui et al. 1996; Sakai et al. 1997). A similar SDN was observed when ultrasonic cell disintegration was performed at an energy input of 2.16 kJ/g MLSS (Yoon et al. 2004b). However, SDN can be much larger than 10, if chemical/energy dosages are low, for example, less than 0.01 to 0.02 g O_3/g MLSS, because only partial cell damages occur each time sludge is treated (Fabiyi and Novak 2007).

The specific cost of reducing excess biosolids using ozone is calculated at \$357/ton MLVSS reduced, as shown in Table 5.6, assuming the target biosolid reduction is 80% to 90% or higher at a specific ozone consumption of 0.15 g O_3/g MLVSS reduced. It should be noted that the cost is before considering the equipment and labor costs. In this calculation, it is assumed that ozone is produced from pure oxygen because it is less costly than using air due to the much lower specific energy consumption for ozone production. The specific cost can decrease substantially if the target sludge reduction is low, for example, 20% to 30%, instead of more than 70% to 80%.

The energy cost for ultrasonic cell disintegration can also be calculated using the similar method used for ozonation. If the energy input is 2.16 GJ/ton as observed by Yoon et al. (2003) and the SDN is 3, the total energy required to run the sonicator is 6.48 GJ/ton of dry biosolid removed. If the unit electricity cost is \$0.072/kWh as assumed for ozonation, the specific energy costs become \$130/ton of dry biosolids removed. By adding the costs to dissolve additional oxygen, that is, \$102/ton, as shown in Table 5.6, the total operating costs become \$232/ton of MLVSS (or dry biosolids) removed before considering the costs to remove the additional nitrogen created from the disintegrated sludge. In addition, the capital costs for the sonication devices, diffusers to supply extra air to the aeration tank, labor costs, etc., should be considered. Overall, the total cost of sludge reduction can reach near \$400/dry ton.

TABLE 5.6
Approximate Operating Costs to Reduce 1 Ton of Dry Biosolids Using Ozone

Item	Consumption/Cost	Assumptions
Liquid oxygen, required	1.25 ton O_2 $125	O_3 requirement = 150 kg/ton dry biosolids
		O_3 in ozone stream = 0.12 kg O_3/kg mixture
		O_2 price = $100/ton
Power cost for ozone	1800 kWh $130	Unit power cost = $0.072/kWh (US average)
		Specific power consumption = 12 kWh/kg O_3
Power cost for additional aeration	1420 kWh $102	Added O_2 demand by disintegrated sludge = 1420 kg COD
		Specific energy demand to dissolve oxygen = 1 kWh/kg O_2 dissolved
		Oxygen credit from sparging O_2-rich off-gas in the aeration tank and oxygen requirement for nitrification is not considered
Total operating costs	$357	Equipment costs for ozone generator not considered

5.7   SCALABILITY OF MBR FROM LABORATORY SCALE TO FULL SCALE

The scalability of laboratory-scale system to larger-scale systems is poor in MBR. The vastly different specific aeration rates per footprint cause differences not only in microbiology but also in hydrodynamic patterns on the membrane surface. As a result, observations from laboratory-scale system cannot be directly used to project the design parameters of larger-scale systems. Some of the differences between small- and large-scale systems are as follows:

- *Different hydrodynamic conditions on the membrane surface*—Full-scale membrane cassettes with 2 to 3 m of height spare large spaces underneath the membrane modules to develop a uniform upflow before the air hits the membrane. On the contrary, small-scale membrane modules with 0.2 to 1.0 m height can hardly have such spaces. More vigorous aeration (or greater specific aeration rate) is often used in small-scale membranes to compensate the drawback. In the full-scale system, the bubbles have more chances to coalesce together while rising causing potentially different membrane scouring mechanisms in the top and the bottom of the membrane. In addition, the longer upflow channels cause a larger hydraulic resistance for the flow.
- *Different shear effects on floc*—In the aeration tank, the OTE is generally proportional to the tank depth. Thus, less aeration is necessary to transfer the same amount of oxygen in deeper tank. Meanwhile, larger-scale modules use the scouring air more efficiently because the rising bubbles cover larger surface areas. In addition, larger-scale modules tend to have greater packing densities that require less specific airflow. Overall, the specific airflow per footprint can be even an order of magnitude lower in full-scale systems. This directly affects the shear stress acting on the microbial floc and in turn affects the floc size. If PACs are added to the membrane tank, it can scratch the membrane surface in the laboratory-scale experiment because of the high specific aeration rate, but the membrane scratch effect can be milder in large-scale MBR.
- *Different internal pressure loss*—Full-scale hollow fiber membranes with 1 to 2 m effective channel length suffer more from the internal pressure loss than pilot or laboratory-scale membranes with less than 0.5 m effective length. Thus, the negative effect of the internal pressure loss on flux is more significant in full-scale modules in general. Meanwhile, the longer fibers have larger amplitudes, which can at least partially compensate the negative effect of the internal pressure loss. The net effect of fiber length on flux is module and operating condition specific and is difficult to be projected based on the data obtained with different modules.
- *Different biological conditions*—As explained earlier, the specific aeration rate tends to be higher in small-scale experiments than that in larger-scale experiments. Under the high specific airflows, more carbon dioxides are stripped out from the mixed liquor; thereby, the carbon dioxide concentration tends to be low. The low carbon dioxide concentration increases pH, which in turn can affect the physiology of the microorganisms. In addition, the carbon dioxide concentration in mixed liquor might be a factor affecting the growth rate of autotrophic microorganisms that require carbon dioxide as a carbon source (Denecke and Liebig 2003). This raises a possibility of having different nitrification efficiencies depending on experimental scale.
- *Different mixing conditions*—Although a thorough mixing is readily possible in a small-scale experiment by carefully designing the system and supplying sufficient mixing energy, the same degree of mixing is rarely possible in full-scale systems. Some degree of DO gradient is unavoidable in a large-scale system and the microbial physiology may be affected. Sampling of mixed liquor can be a significant disruption in the small scale, but it is not in the large scale.

- *Different feed water qualities*—Even if the feed water taken from a full-scale WWTP is used as a feed for small-scale tests, it is not possible to exactly mimic the fluctuating water quality and quantity occurring in full-scale system. In addition, the solids contained in the raw water sample tend to settle in the feed tank in a small-scale experiment. Thereby, the water fed to the reactor is not exactly same as the raw water. If the raw water is mixed continuously to keep it homogeneous, oxygen is dissolved into the water and biodegradations occur before the raw water is fed to MBR.
- *Different experimental durations*—Laboratory-scale experiments are often performed under expedited fouling condition at a high flux under a well-controlled environment to complete the experiment in a reasonable time frame. However, full-scale systems undergo small and large fluctuations in operating parameters and the filtration cycles last a few months to even a year. The experimental duration may also affect the degree of biological adaptation to the operating condition. For example, the low pH (<6.5–7.0) may diminish the nitrification rate in short-term experiment, but the microorganisms may adapt to the low pH in full-scale system if the low pH is sustained for extended period. The different biological conditions likely affect the membrane fouling rate too.
- *Different buffer capacities for environmental fluctuations*—If any environmental condition changes, the effect is more immediate and profound in small-scale systems. For example, if acid is released accidently to the reactor, the pH drops more severely and quickly in small systems compared with large systems. If the surrounding temperature fluctuates, water temperature also fluctuates more abruptly and widely in a small-scale system. MLSS can also change abruptly in small system when excess sludge is drawn or sludge sample is taken. MLSS changes very slowly in full-scale MBR because of the limited sludge handling capacity.

The differences of laboratory-, pilot-, and full-scale MBR with respect to their design and operating parameters are summarized in Table 5.7.

TABLE 5.7
Typical Ranges of Operating Parameters and Conditions for Different Plant Sizes

Parameter	Laboratory Scale	Pilot Plant	Full Scale
Volume	1–10 L	100–1000 L	>10 m^3
Operating time	Hours–Months	Weeks–Months	Years
Wastewater	Synthetic/grab/stored samples	Synthetic/real wastewater	Real wastewater
Temperature/hydraulic load/ feed composition	Constant	Constant/fluctuating	Fluctuating
Specific aeration demand (SAD$_p$)	150–500 m^3/m^3	10–100 m^3/m^3	10–30 m^3/m^3
Dissolved CO$_2$	Low	Medium	High
Energy input	2–20 kWh/m^3	1–10 kWh/m^3	0.5–1 kWh/m^3
MLSS fluctuation	High	Middle	Low
Polysaccharide concentration	20–100 mg/L	20–100 mg/L	<20 mg/L

Source: Kraume, M. et al., *Desalination* 236(1–3):94–103, 2009.

6 MBR Design

6.1 LESSONS LEARNED FROM HISTORY

The current membrane bioreactor (MBR) technology based on submerged membranes technology is the product of numerous trials and errors throughout the last two or three decades. Many misunderstandings, underexpectations and overexpectations, and misguidances have been corrected. The following are some of the remarkable findings during this time:

- *The necessity for fine screens to remove debris was often underestimated in the early days.* This was partly because membranes were erroneously considered compatible with debris. The clogging/ragging of membranes by fibrous materials is detrimental for the membrane system. The debris caught by hollow fiber bundles can only be removed manually after disassembling the cassette. The sludge cake plugging the channels in flat sheet membranes can be removed by mechanical surface scraping after taking all membrane panels out from the cassette. As discussed in Section 6.2, membrane system operation can be hampered by debris when mechanical prescreens do not function properly regardless of the membrane type. Although prescreens with 1 to 3 mm mesh size have been routinely used, this problem is still one of the most common issue along with excessive membrane fouling due to the selection of improper screen types, improper installation that allows leaking debris, inherent limitation of screens in removing fibrous materials, etc. (Stefanski et al. 2011).
- *MBR were often designed for excessively high mixed liquor suspended solids (MLSS), such as 20 to 30 g/L.* This was partly because a small footprint was the biggest selling point along with excellent effluent quality in the early days. However, excessively high MLSS interferes with membrane scouring by slowing the upflow. In addition, it can cause low dissolved oxygen (DO) by hampering oxygen transfer in mixed liquor, which in turn triggers excess biopolymer secretion and accelerated membrane fouling. If the high MLSS is combined with poor membrane scouring system design, the chance of module clogging also increases, especially in hollow fiber membranes. As a consequence, the target MLSS has decreased significantly since the late 1990s. Now, the optimum MLSS in aeration basins is considered 8 to 12 g/L, whereas some MBR are designed at a lower MLSS such as 6 g/L.
- *MBR was thought to be compatible with high food-to-microorganisms (F/M) ratio.* Solids settling in the clarifier were no longer a concern in MBR because membrane rejected 100% of the solids. As a result, the F/M ratio, which was kept low in conventional activated sludge (CAS) for good sludge settling, was not considered a significant factor in MBR. This misperception was combined with the desire of saving footprints and resulted in overly compact aeration basins. Later, it turned out that the high F/M ratio was not only the cause of low oxygen transfer efficiency (OTE) but was also the cause of expedited membrane fouling. Nowadays, the optimum F/M ratio of MBR is considered to be 1/3 to 1/2 of the optimum F/M ratio for CAS, as summarized in Table 6.2
- *MBR was considered tolerable to organic loading shock.* In CAS, organic loading rates must be maintained stably as much as possible to obtain good sludge settling properties by not disturbing the process. However, it was thought to be an obsolete concept due to the lack of clarifier in the new MBR process. As a result, reducing or eliminating the holding (or equalization) tank seemed plausible. It is somewhat arguable, but widely varying F/M

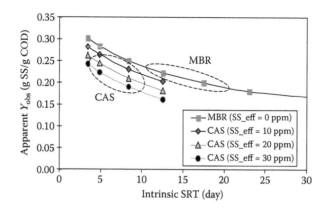

FIGURE 6.1 Comparison of observed sludge yield in CAS and MBR.

expedites membrane fouling in many situations by stressing microorganisms to produce more soluble microbial products (SMP) or by causing insufficient DO at high organic loading conditions.

- *Biosolids yield was considered much lower in MBR than in CAS.* This claim was based on some laboratory-scale or pilot-scale experiments performed at the dawn of MBR technology in the early 1990s or earlier, where unrealistically high solids retention times (SRT) such as 50 to 100 days or longer were employed. However, field engineers soon realized that there were no meaningful differences in the apparent sludge yields between MBR and CAS. Although MBR produces somewhat less sludge due to its long SRT (12–30 days for MBR vs. 5–10 days for CAS), it does not lose sludge through the effluent. As a result, the apparent biosolids yields are not noticeably different in typical conditions. Figure 6.1 shows the simulation data based on the International Water Association's (IWA) activated sludge model #1 (ASM#1) assuming varying degrees of suspended solids in CAS effluent. If the CAS effluent carries out 10 to 30 mg/L of suspended solids, apparent sludge yields (Y_{obs}) are about the same in both CAS and MBR at the common SRT ranges, as indicated in the figure.

6.2 PRETREATMENT

Proper pretreatment is crucial for successful MBR operation, especially for MBR with submerged membranes. A great deal of operational failures are directly or indirectly related to improper pretreatment. The clogging/ragging of membranes by fibrous materials and large debris can be detrimental for hollow fiber membranes (Figure 6.2a), but it also hampers flat sheet membrane operation by blocking the entrance of upflow channels in membrane cassettes (Figure 6.2b). The fibrous materials and debris caught by hollow fiber bundles and air pipes can only be cleaned manually after taking the modules out from the tank or after draining the tank. The upflow channel of flat sheet membranes can be completely plugged up by the sludge cake formed in between two membrane panels. The affected membrane panels must be taken out from the cassettes and the sludge cake needs to be scraped off before washing with water jet. Fibrous materials also can wrap the air pipes and diffusers, causing zones with no or little air in the space above. The cleaning process is not only very costly by itself but it also causes a substantial downtime that can be detrimental from the perspective of handling peak flow and meet the regulatory requirements. Moreover, the debris can physically damage the membrane by nibbling on the membranes in the turbulent flow.

Traditional settling tanks are an efficient means to remove debris and fibrous materials from wastewater (Moustafa 2011). In addition, settling tanks can remove more biological oxygen demand (BOD) than mechanical screens by removing some organic particles with less than 1 to 2 mm

(a) (b)

FIGURE 6.2 Ragging in MBR when fine screens fail to remove debris in municipal MBR plants: (a) in hollow fiber bundles and (b) on air pipes.

diameter. The low BOD in the MBR feed can mitigate membrane fouling in MBR by lowering the F/M ratio, as discussed in Section 5.2.3. However, clarifiers do not guarantee the same BOD removal performance during peak flow due to the high superficial velocity in the clarifier. In addition, their bulkiness often contradicts the purpose of having compact MBR systems. As a result, a vast majority of newly built MBR are equipped with mechanical screens with self-cleaning capabilities such as rotating belt sieve, rotary drum sieve, conveyor belt sieve, etc. Membrane manufacturers typically recommend mesh screens with less than 2 mm mesh size for hollow fibers and less than 3 mm for flat sheet membranes. However, it is not uncommon to find screens with smaller mesh sizes in MBR. For gap screens with linear slits, slit size should be smaller by 1 mm than the recommended mesh sizes. For perforated screens with holes, slightly larger holes are allowed.

The problems caused by fibrous materials and debris are still commonly experienced in the field due to mechanical defects, improper screen types, or improper screen installations (Stefanski et al. 2011). Once fibrous materials are caught on membrane modules and frames, turbulence near the affected area decreases and as a result more fibrous materials such as cloth/paper fibers, hairs, etc., can be caught on the initially caught materials. Short fibers and particles can reconstruct into large rags in mixed liquor and eventually clog membranes. Therefore, a high concentration of large particles, perhaps at 1 to 4 mm, can be used as an indicator of potential membrane clogging although such particles are, by themselves, too small to clog membranes.

The quality and quantity of debris varies widely depending on the geographic location and the season. Therefore, choosing mechanical screens with proper functions and capacities is site specific to some extent. When six different municipal sites in North America were surveyed, captured solids ranged from 4 to 95 L/1000 m³, although all of the surveyed sites were using 2 mm screens (Côté et al. 2006). If grits and grease are a concern, additional removal processes can be added in between coarse and fine screens, as shown in Figure 6.3. Aerated grit chambers and American Petroleum Institute (API)/corrugated plate interceptor (CPI) separators are commonly used.

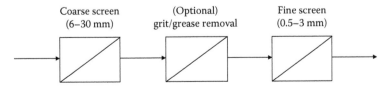

FIGURE 6.3 Typical pretreatment process.

TABLE 6.1

Screening Contents of Typical Sewage with Different Screen Configurations

Feed	Screen	Captured Solids (mg/L)	Comment
Raw WW	Bar screen (7.2 mm slots)	<1.0	Very little removal
Raw WW	Vibrating screen (0.75 mm holes)	14	Removal of essentially all trash (hair, seeds, etc.)
Raw WW	Brush screen (0.75 mm holes)	23	Removal of essentially all trash (hair, seeds, etc.)
Raw WW	Rotary drum screen (0.5 mm holes)	94	Significant removal of paper fibers that could be degraded in the MBR
Settled WW	Rotary drum screen (0.5 mm holes)	2.8	Primary clarification removes most trash. Screen is for membrane protection

Source: Côté, P. et al., Pretreatment requirements for membrane bioreactors. *Proceedings of WEFTEC.* October 21–25, 2006. Dallas, Texas.

If tree leaves fall into MBR basins, they undergo an abrasion process in the turbulence and eventually the structural components such as leafstalks with midribs and large veins remain whereas the rest of the leaves break apart. Just like other hard debris (plastics, woods, and metals), the moving leafstalks with midribs can scratch off the membrane surface and possibly compromise membrane integrity. Basin covers may be necessary if aeration and membrane tanks are exposed to falling leaves or other debris.

As the hole size of the screen decreases, larger screen surface area is required to process the same flow rate (Côté et al. 2006; Frechen et al. 2008), thereby the capital costs are inversely related with the screen pore size. Meanwhile, the amount of captured solids by the screen increases as the screen becomes finer. In one study, the quantity of captured solids increases dramatically from 23 to 94 mg/L when the screen hole size decreases from 0.75 to 0.5 mm as shown in Table 6.1. The more solids removal leads to less BOD loading and lower F/M ratio of MBR. Consequently, aeration costs for both biological and membrane scouring can decrease due to the lower BOD and the lower membrane fouling propensities. The optimum screen pore size must be decided considering the costs of the screen and the savings in the operation.

In other surveys, chemical oxygen demand (COD) and total suspended solids (TSS) of the raw water were reduced by 20% and 33%, respectively, when 1 mm mesh screens were used in a municipal MBR in Kaarst, Germany (Schier et al. 2009; Frechen et al. 2010). In another location in Swanage, United Kingdom, COD and TSS removal efficiencies were 13% and 23%, respectively, when 2 mm hole screens were used. When gap screens with slot openings were used, removal efficiencies were lower than those of mesh or hole screens with the same size openings. When 1 mm gap screens were used, COD and TSS removal were 8% and 34%, respectively, in Monheim, Germany.

6.3 BIOLOGICAL SYSTEM DESIGN

6.3.1 ORGANIC LOADING RATE OR F/M RATIO

Organic loading rate is one of the principal design and controlling parameters in biological wastewater treatment process. It is measured by the amount of food provided to a unit amount of biomass or reactor volume for a unit of time. The F/M ratio is based on the amount of biomass whereas the food-to-volume (F/V) ratio is based on reactor volume.

The F/M and the F/V are calculated using the following equations:

$$F/M = \frac{QS_0}{XV} \tag{6.1}$$

$$\text{F/V} = \frac{QS_0}{V} \tag{6.2}$$

where
F/M food-to-microorganism ratio (g BOD/g MVLSS/day)
F/V food-to-volume ratio (g BOD/g MLVSS/day)
Q influent flow rate (m³/day)
S_0 influent BOD or COD (mg/L)
X MLSS in aeration tank (mg/L)
V tank volume (m³)

As summarized in Table 6.2, four different F/M ratios exist depending on the unit used for organic contents in the influent and for the biomass content in the mixed liquor. Those four are interconvertible using the ratio of COD/BOD and mixed liquor volatile suspended solids (MLVSS)/MLSS. COD/BOD and MLVSS/MLSS vary depending on the wastewater source and the SRT, but are typically 2.0 and 0.8, respectively, for municipal wastewater.

MBR are designed at lower F/M ratios than CAS to avoid the high membrane fouling rate and low oxygen transfer efficiencies. The preferred F/M ratio of MBR is approximately a third to a half of that in CAS, that is, 0.05 to 0.15 g BOD/g MVLSS/day or 0.1 to 0.3 g COD/g MLVSS/day in municipal MBR.

It is noticeable that nitrogen contents are not taken into consideration in the F/M and F/V ratio calculation even though the oxygen demand for nitrogen oxidation (or nitrification) can be quite significant. It is partly because the nitrogen contents of wastewater are largely proportional to the organic contents (BOD/COD) in municipal wastewater, and partly because nitrification is performed independently by autotrophic microorganisms rather than heterotrophic microorganisms. Therefore, if tank sizes are proper for organic treatment, it is very likely suitable for nitrogen treatment unless operating conditions are significantly unfavorable for nitrification in municipal

TABLE 6.2
Comparison of Preferred Ranges of Operating Parameters in MBR and CAS

Design Parameter	Unit	MBR	CAS
F/M	g BOD/g MLSS/day	0.04–0.12	0.16–0.24
	g COD/g MLSS/day	0.08–0.24	0.32–0.48
	g BOD/g MLVSS/day	0.05–0.15[a,b]	0.2–0.3
	g COD/g MLVSS/day	0.1–0.30[c]	0.4–0.6
F/V	g BOD/L/day	0.5–1.5	0.6–0.9
	g COD/L/day	1.0–3.0	1.2–1.8
MLSS	g/L	8–12[a]	2–4
MLVSS	g/L	6–10	1.7–3.4
SRT	days	10–30	5–10
SOUR	mg O₂/g MVLSS/h	2–5[d]	6–12
OUR	mg O₂/L/h	15–60	20–40
DO	mg/L	1–2	1–2

[a] From Judd, S., *MBR Book*. The Netherlands: Elsevier, 2006, pp. 165–205.
[b] From Brepols, C., *Operating Large Scale Membrane Bioreactors for Municipal Wastewater Treatment*. London: IWA Publishing, 2011.
[c] Converted from the BOD based F/M assuming COD/BOD = 2.0 for municipal wastewater.
[d] From Ng, H.Y. et al., *Environ. Sci. Technol.* 40:2706–2713, 2006.

wastewater treatment. In the meantime, industrial wastewater often contains excessive nitrogen compared with the organic content and it becomes a determining factor in reactor sizing.

6.3.2   HRT/SRT

The hydraulic retention time (HRT) of a system is calculated using Equation 6.3. The sum of all reactor volumes $\left(\sum_i V_i\right)$ is divided by influent flow rate (Q)

$$\text{HRT} = \frac{\sum_i V_i}{Q} \tag{6.3}$$

For municipal MBR, HRT typically ranges from 2 to 4 h in aeration tanks and from 1 to 2 h in membrane tanks. Total HRT in aeration and membrane tanks ranges from 3 to 6 h, if there are no anoxic or anaerobic tanks attached. For modified Ludzack–Ettinger (MLE)-based MBR, anoxic tank proceeds aeration tank and the HRT of the anoxic tank ranges from 1 to 2 h in general. Overall, HRT ranges from 5 to 8 h in A/O based municipal MBR (Section 4.6.2). If multiple anaerobic and anoxic tanks are added to enhance nutrient removal, as shown in Section 4.6.5, overall HRT can exceed 12 h.

The SRT is calculated as Equation 6.4. The term SRT is often used interchangeably with sludge age and mean cell residence time (MCRT) although the definitions of the terms are somewhat different. Oftentimes, MCRT includes the SRT in the sludge blanket of clarifier, but SRT does not include it in CAS. These two terms become identical in MBR due to the lack of a clarifier.

$$\text{SRT} = \frac{\sum_i X_i V_i}{Q_X X_X} \text{ or } \frac{XV}{Q_X X_X} \tag{6.4}$$

where
   Q_X excess biosolids removal rate (m³/day)
   V total reactor volume (m³)
   V_i individual reactor volume (m³)
   X average MLSS (mg/L)
   X_i MLSS in each reactor (mg/L)
   X_X MLSS in the excess biosolids flow (mg/L)

The typical ranges of SRT for municipal wastewater is discussed in Section 5.2.3.1, but it has been known that the lower the SRT is, the higher the membrane fouling is. In general, 12 to 15 days has been deemed the minimum SRT that does not incur excessive membrane fouling. No strong evidences exist, which indicate accelerated membrane fouling at an extended SRT beyond the typical range of 12 to 30 days. However, such a long SRT is not preferred because it is accompanied by high capital costs to build large bioreactors.

6.3.3   Oxygen Uptake Rate

Oxygen uptake rate (OUR) is a measure of oxygen consumption rate per mixed liquor volume by definition. In ordinary conditions with a reasonable microbial health, OUR varies depending on the F/V ratio because the amount of the substrate to be oxidized is closely proportional to the amount of oxygen required at steady state. Meanwhile, the specific activity of microorganisms is measured by specific OUR (SOUR) and can be obtained by dividing OUR by MLVSS or MLSS. As OUR is related with F/V, SOUR is directly related with F/M. The low SOUR in MBR is a natural

$$F/V = \frac{QS_0}{V} \qquad (6.2)$$

consequence of the low F/M ratio employed in MBR. Typical design and operational parameters are summarized in Table 6.2.

Two different measurement methods exist, one is called a non–steady state method, which relies on the rate of DO declining, and the other is called a steady state method, which relies on the oxygen demand at a constant DO.

- In the non–steady state method, clean water is saturated with oxygen by bubbling it with air and then filled into a test bottle near to the top according to Method 1683 (USEPA 2001b). After adding a mixed liquor sample with known volume and MLSS/MLVSS, the test bottle is topped off with additional water. A DO probe equipped with a propeller mixer at the end is inserted into the bottle. The stopper in the middle of the probe seals the bottle. No air bubble should remain in the bottle. The declining DO in the mixed liquor sample is a direct consequence of the oxygen uptake by the microorganisms in it. If the DO is plotted against time, a graph very similar to Figure 6.22 is obtained, where the slope of the DO curve is equivalent to the OUR by definition. By multiplying the dilution factor to the measured OUR, the OUR of the mixed liquor is obtained. The non–steady state desorption method discussed in Section 6.5.3.2 relies on the same principle to estimate the OTE in aeration tank in situ.
- In the steady-state method, OUR is measured while DO remains constant. Under this condition, oxygen dissolution rates exactly match with OUR by definition. OUR is calculated based on the oxygen dissolution rate. For example, in respirometry test, a known amount of mixed liquor sample is poured into a test bottle immersed in a water bath and mixed with stirring bars or paddles to dissolve oxygen gas in the headspace to the mixed liquor. While absorbing the CO_2 produced from the biological reaction using solid NaOH or KOH in the headspace, pure oxygen is supplied to the test bottle to maintain constant headspace pressure. Under this condition, OUR is equivalent to the volumetric oxygen consumption rate of the mixed liquor by definition.

In MBR conditions, although microorganisms are capable of very high SOUR (e.g., >50 mg O_2/g MLVSS/h), oxygen dissolution cannot keep up with such high demand due to the lowering OTE at high aeration rates. As aeration intensity increases, more air bubbles coalesce and surface area does not increase proportionally to the airflow rate. At a certain point, additional airflow does not create additional surface areas. A maximum OUR that can be supported by aeration is perhaps approximately 100 mg/L/h when MBR is aerated by diffusers. In CAS, it is approximately 150 mg/L/h due to the low MLSS.

It must be noted that OUR and SOUR do not necessarily indicate the health of the microorganisms because they are more likely a consequence of the organic loading rate. For instance, SOUR decreases as organic loading rate decreases just because microbial activity decreases with a lower amount of food. In a laboratory-scale MBR study (Maeng et al. 2013), the specific ATP concentration per biomass weight was tracked as an indicator of microbial activity at a few different SRT. Here, a long SRT is synonymous with a low specific organic loading rate. It was observed that the specific ATP concentration sharply declined as SRT increased, as shown in Figure 6.4. Therefore, OUR/SOUR alone cannot be deemed as a performance indicator of microorganisms in biological systems.

6.3.4 SLUDGE YIELD

Microorganisms stay in the bioreactor for a longer period than the life cycle of most microorganisms on average before they are discharged as excess sludge. In the meantime, dead cells are consumed by live cells while losing a portion of the carbon to the atmosphere as carbon dioxides. This is the so-called endogenous respiration that reduces the amount of excess biosolids depending on the amount of time the cells spend in the biological tanks before being discharged. The observed sludge yield, Y_{obs}, is the ratio between the total sludge (or biosolids) produced and the total substrate

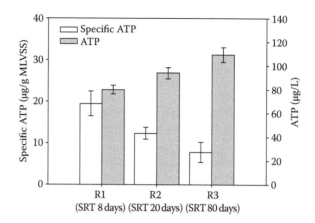

FIGURE 6.4 Total and specific ATP content in mixed liquor in laboratory-scale MBR treating synthetic feed. (From Maeng, S.K. et al., *Water Res* 47(9):3151–3162, 2013.)

consumed, and is written as Equation 6.5. Due to endogenous respiration, Y_{obs} is always smaller than that of the intrinsic sludge yield, Y.

$$Y_{obs} = \frac{Q_X X_X}{Q(S_0 - S_e)}$$

(6.5)

where

Q influent flow rate (L/day)
Q_X mixed liquor removal rate (L/day)
S_0 influent COD or BOD (mg/L)
S_e effluent COD or BOD (mg/L)
X_X MLSS or MLVSS of removed mixed liquor (mg/L)
Y_{obs} observed biosolids yield (g MLSS/g COD or B)

The Y_{obs} can also be estimated from the kinetic parameters. According to the mass conservation law, the rate of biosolid accumulation in the system is equal to the difference between the rate of biosolids produced from the substrate and the rate of excess biosolids removed from the system. The rate of biosolids produced is again equal to the difference between the intrinsic biosolid production rate and its decay rate. These relations can be written as Equation 6.6.

Rate of biosolids accumulation = Rate of biosolids production from substrate − Rate of biosolids decay − Rate of biosolids removal

$$\frac{dX}{dt} = Y\frac{Q(S_0 - S_e)}{V} - k_d X - \frac{Q_X X_X}{V}$$

(6.6)

where

Y intrinsic sludge yield (g MLSS/g COD or BOD)
k_d first-order biosolids decay constant (/day)

At steady state, dX/dt equals zero because X in the system is constant. The right side of the equation can be rearranged as Equation 6.7.

$$Y = k_d \frac{VX}{Q(S_0 - S_e)} - \frac{Q_X X_X}{Q(S_0 - S_e)}$$

(6.7)

By inserting Equation 6.4 and Equation 6.5 into Equation 6.7, the final equation is obtained as follow.

$$Y_{obs} = \frac{Y}{1 + k_d \cdot SRT} \tag{6.8}$$

Typical ranges of Y and k_d are summarized in Table 6.3. Four different Y values exist depending on the unit, but they can be converted each other using the ratio of MLVSS/MLSS and COD/BOD. The COD/BOD ratio is higher than 1.0 in municipal wastewater samples because a portion of substrate converts to microorganisms and does not completely oxidize to CO_2 in the BOD measurement. Nonbiodegradable COD increases the COD/BOD ratio in some industrial wastewaters. For example, some tannery wastewaters are reported to have high COD/BOD ratio at approximately 4 (Jenkins et al. 2004).

It should be noted that Equation 6.8 is applicable only for typical SRT in the CAS process from which Y and k_d were obtained. This is because Equation 6.6, from which Equation 6.8 is derived, does not count on the accumulation of nonbiodegradable solids. Thus, Equation 6.8 relies on a fixed k_d regardless of the SRT. However, the effective biosolids decay rate, k_d, decreases as SRT increases due to the gradual accumulation of nonbiodegradable solids that do not decay biologically under the activated sludge conditions. Therefore, if Y_{obs} is calculated for high SRT (e.g., >10 days), using Equation 6.8, an underestimated Y_{obs} will be obtained due to the overestimated k_d. It is apparent that Equation 6.8, which is commonly used for CAS, is not valid for MBR in which SRT is at least 12 days or longer, more likely around 20 days.

The Y_{obs} of MBR can be more accurately predicted by either empirical equations or IWA's ASM, which takes nonbiodegradable solids into consideration. Figure 6.5 compares two empirical curves, one theoretical curve based on ASM#1, and one curve based on Equation 6.8 for the same hypothetical condition. The Y_{obs} predicted by Equation 6.8 agrees reasonably with other Y_{obs} when SRT is less than 10 days, but it deviates substantially from others at higher SRT. ASM#1 predicts higher Y_{obs} than Equation 6.8 because of the nonbiodegradable materials that accumulate in the mixed liquor. The most direct and accurate correlation might be the empirical curve obtained from well-controlled laboratory experiments (Macomber et al. 2005). The correlation based on field data (Gildemeister 2003) results in the highest Y_{obs} in the typical SRT range of 10 to 30 days in MBR. Table 6.4 summarizes the relation among SRT, Y_{obs}, and F/M in well-controlled laboratory experiments.

If the phosphorous discharge limit is tight, inorganic coagulants are often used with or without the help of biological nutrient removal (BNR; Lee et al. 2001b). The added aluminum (Al^{3+}) or ferric (Fe^{3+}) ions form various insoluble inorganic salts in mixed liquor such as $AlPO_4$, $Al(OH)_3$, $AlO(OH)$, $FePO_4 \cdot 2H_2O$, $Fe(OH)_3$, $Ca_5PO_4(OH)_2$, etc. Therefore, apparent biosolid production increases and all the correlations in Figure 6.5 become invalid.

TABLE 6.3

Typical Kinetic Parameters for CAS for Municipal Wastewater

Coefficient	Unit	Range	Typical	Remark
Y	g MLSS/g COD	0.4–0.6	0.5	Convertible each other assuming COD/BOD = 2 and
	g MLVSS/g COD	0.3–0.5	0.4	MLVSS/MLSS = 0.8 for municipal wastewater
	g MLSS/g BOD	0.8–1.2	1.0	
	g MLVSS/g BOD	0.6–1.0	0.8	
k_d	/day	0.025–0.075	0.06	Used with MLVSS-based Y
		0.02–0.06	0.05	Used with MLSS-based Y
MLVSS/MLSS	–	0.7–0.9	0.8	No inorganic coagulant addition was assumed
COD/BOD	–	1.25–2.5	2.0	

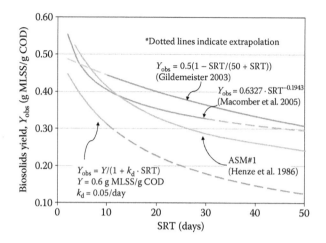

FIGURE 6.5 Sludge (or biosolids) yields based on different models and experiments for the process without inorganic coagulant addition.

TABLE 6.4
Observed Sludge Yield and F/M Ratio as a Function of SRT

	Y_{obs}			F/M
SRT (days)	g VSS/g COD	g MLSS/g COD	MLVSS/MLSS	COD/g VSS/day
2	0.477	0.558	0.855	1.048
5	0.384	0.461	0.833	0.521
10	0.329	0.396	0.831	0.304
20	0.298	0.358	0.832	0.168
30	0.268	0.328	0.817	0.124

Source: Data from Cicek, N. et al., *Water Sci Technol* 43(11):43–50, 2001; Macomber, J. et al., *J Environ Eng* 131(4)579–586, 2005.

6.3.5 Oxygen Demand

6.3.5.1 Overall Reaction

In biological wastewater treatment, oxygen is required mainly to meet carbonaceous and nitrogenous oxygen demands, although a small amount might be required to oxidize inorganic ions such as ferrous ions (Fe^{2+}), manganese ions (Mn^{2+}), sulfides, etc. Figure 6.6 illustrates the oxygen, carbon, and nitrogen flows in a biological reaction. Oxygen is consumed when carbons convert to heterotrophs and CO_2, which is the most dominant mechanism in carbon-rich wastewater treatment. The nitrogen measured as total Kjeldahl nitrogen (TKN) is oxidized to NO_3-N by the autotrophs that use the energy from this nitrogen oxidation reaction to fix CO_2 and produce a new cell mass. A portion of TKN contained in raw wastewater is also used to build the new cell mass of heterotrophs and autotrophs instead of being oxidized.

If there is an anoxic stage in the process, a portion of the NO_3-N can be used as an oxygen source. As a consequence, the NO_3-N is reduced to nitrogen gas by heterotrophs and this process is called denitrification. Excess sludge includes both heterotrophs and autotrophs, but typically the autotrophs take less than 10% of the total biomass due to the low TKN content in wastewater as well as the low sludge yields. As a result, the production of autotrophs from TKN is often neglected when biosolids yields, Y_{obs}, are estimated.

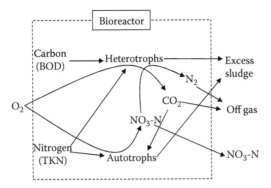

FIGURE 6.6 Mass flow chart illustrating biological reaction in bioreactors.

6.3.5.2 Oxygen Demand Calculation

There are three important components contributing to oxygen demand in activated sludge process, that is, (1) carbonaceous O_2 demand, (2) nitrogenous O_2 demand, and (3) O_2 credit from denitrification.

1. *Carbonaceous O_2 demand*

   The carbonaceous O_2 demand is estimated based on the difference between the amount of COD removed from the raw wastewater and the COD stored in excess biosolids. The chromium-based COD is suitable for this calculation because hexavalent chromium reacts with a vast majority of the carbonaceous components, but they do not react with the nitrogenous components. The total effective carbonaceous O_2 consumption rate is calculated by subtracting the rate of COD storage in excess biosolids from the apparent COD removal rate as follows (Equation 6.9):

   Carbonaceous O_2 demand = COD removed − COD assimilated to excess biosolids

$$O_{2,\text{Carbon}} = Q(S_0 - S_e) - \lambda Y_{\text{obs}} Q(S_0 - S_e) \tag{6.9}$$

   where

$O_{2,\text{Carbon}}$	oxygen consumption rate (kg/s)
Q	influent flow rate (m³/s or L/s)
S_0	COD of influent (kg/m³ or g/L)
S_e	total COD of effluent (kg/m³ or g/L)
λ	COD/MLSS (g COD/g MLSS)
Y_{obs}	observed sludge yield (g MLSS/g COD)

   In CAS that runs without inorganic coagulant addition, λ ranges between 1.1 and 1.2, but it tends to be in the low end of the range if SRT is high due to the accumulation of inorganic materials in sludge. In MBR, λ can be assumed 1.1 or slightly lower as MBR runs at substantially longer SRT than CAS. The observed sludge yield can be calculated from the empirical equation that will be discussed in Section 6.3.4.

2. *Nitrogenous O_2 demand*

   The nitrogen in reduced form needs oxygen to be oxidized to nitrate $\left(NO_3^-\right)$. The reduced nitrogen exists mainly as ammonium $\left(NH_4^+\right)$, amino acids, and proteins, but it also exists as nitrite $\left(NO_2^-\right)$ and the constituent of nucleic acids. Typically, nitrite concentrations are neglected because they are negligible (<1 mg/L) in wastewater compared with other components. The nitrogen in reduced form is quantified as TKN.

The nitrogen oxidation reaction can be written as the following equation. One nitrogen atom in ammonia needs four oxygen atoms to be completely oxidized. By dividing the weight of four oxygen atoms (64 g/mol) by the weight of one nitrogen atom (14 g/mol), a multiplying factor of 4.57 g O_2/g TKN is obtained.

$$NH_3 + 2O_2 \rightarrow H^+ + NO_{3-} + H_2O$$

It must be noticed that 1 mol of acid is produced when 1 mol of nitrogen is oxidized in the above equation. Because 1 mol of acid is equivalent to 50 g of alkalinity as $CaCO_3$, the alkalinity consumption during nitrification becomes 3.57 g alkalinity/g TKN oxidized (= 50/14).

The nitrogenous O_2 consumption rate is calculated by multiplying the rate of TKN oxidization by the multiplication factor, 4.57, as shown in Equation 6.10. Ammonia stripping rarely occurs at near-neutral pH in activated sludge and is neglected.

Nitrogenous O_2 consumption = 4.57 × (Apparent TKN removed − TKN in excess biosolids)

$$O_{2,Nitrogen} = 4.57[Q(S_{TKN,0} - S_{TKN,e}) - i_{XB}Y_{obs}Q(S_0 - S_e)] \qquad (6.10)$$

where

$O_{2,Nitrogen}$	oxygen consumption rate for nitrification (kg/s)
$S_{TKN,0}$	TKN in influent (kg/m^3 or g/L)
$S_{TKN,e}$	TKN in effluent (kg/m^3 or g/L)
i_{XB}	TKN content in biosolids (g TKN/g MLSS)

According to the ASM#1 (Henze et al. 1986), i_{XB} was proposed as 0.086 g N/g COD, where MLSS is expressed as COD. Considering the COD to MLSS ratio (i.e., 1.1), the original i_{XB} can be converted to 0.095 g N/g MLSS.

3. *Oxygen credit from denitrification*

If DO is not sufficient and ORP is 50 mV or below, heterotrophs start to consume the combined oxygen taken from NO_3-N because molecular oxygens are scarce. By definition, the amount of combined oxygen consumed in this process is equivalent to the amount of COD oxidized, which again equals to the amount of molecular oxygen saved. In the following chemical equation, when four NO_3-N are reduced to nitrogen gas, ten oxygen atoms are produced. The ratio of oxygen produced to the nitrogen reduced is 2.86 g O_2/g NO_3-N (= 10 × 16/4/14). It is also noticeable that 1 mol of acid disappears when 1 mol of NO_3-N is reduced, which is opposite from nitrification reaction.

$$4HNO_3 \rightarrow 2N_2 + 2H_2O + 5O_2$$

The total amount of nitrogen nitrified can be calculated by subtracting the amount of TKN assimilated to microorganisms from the TKN removed by the system. A portion of the nitrate produced from the nitrification reaction is denitrified while providing oxygen to the biological system. The amount of nitrogen denitrified can be obtained by subtracting the amount of NO_3-N discharged from the total amount of nitrogen nitrified based on the mass balance. From the amount of denitrified nitrogen, the oxygen credit can be calculated by multiplying the conversion factor, that is, 2.86 g O_2/g NO_3-N. Finally, the oxygen credit is described as Equation 6.11, where the first two terms in parentheses are adopted from Equation 6.10.

O_2 credit = -2.86(TKN removal rate − TKN assimilated − NO_3-N discharged through effluent)

$$O_{2,Denit} = -2.86[Q(S_{TKN,0} - S_{TKN,e}) - i_{XB}Y_{obs}Q(S_0 - S_e) - Q(S_{NO_3,e} - S_{NO_3,0})] \qquad (6.11)$$

The total O_2 consumption rate, $O_{2,Total}$ (kg/m³ or g/L), is calculated by adding all three components discussed above.

$$O_{2,Total} = O_{2,Carbon} + O_{2,Nitrogen} + O_{2,Denit} \qquad (6.12)$$

$$O_{2,Total} = Q[(1 - \lambda Y_{obs} - 1.71i_{XB})(S_0 - S_e) + 171(S_{TKN,0} - S_{TKN,e}) - 2.86(S_{NO_3,0} - S_{NO_3,e})] \qquad (6.13)$$

Conceptually, the specific O_2 consumption rate per reactor volume ($O_{2,Total}/V$, kg O_2/m³/s) is equivalent to OUR in the steady state. Specific oxygen demand (SOD) per wastewater volume (kg O_2/m³) is calculated by dividing $O_{2,Total}$ by wastewater flow rate as follows:

$$SOD = \frac{O_{2,Total}}{Q} \qquad (6.14)$$

6.3.6   COMPARISON OF CSTR WITH PFR

6.3.6.1   Mixing Patterns in CSTR and PFR

In continuously stirred tank reactor (CSTR) shown in Figure 6.7, the substrates entering into a tank are quickly mixed with the liquid in the tank. Due to the diluting substrates in the tank, a reaction occurs at low substrate concentrations. The mass balance equation can be written as Equation 6.15, assuming instantaneous complete mixing. In this equation, the substrate accumulation rate in the tank is calculated by multiplying the reactor volume (V, m³) by the substrate concentration changing rate (dS/dt) and it is in turn equated to the substrate accumulating rate calculated by multiplying flow rate (Q) and the difference between incoming (S_0, mg/L) and outgoing (S_t, mg/L) substrate concentrations at time, t. This equation can be solved for the boundary conditions of (0, S_1) and (t_1, S_t) to obtain Equation 6.18. If the initial substrate concentration was zero, it can be simplified as Equation 6.19.

$$V\frac{dS}{dt} = Q(S_0 - S_t) \qquad (6.15)$$

$$\int_{S_1}^{S_i} \frac{dS}{S_0 - S_t} = \frac{Q}{V}\int_0^t dt \qquad (6.16)$$

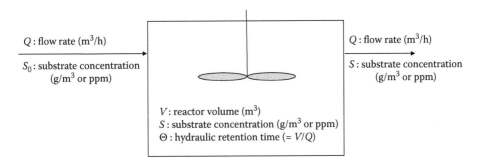

FIGURE 6.7   Continuously stirred tank reactor.

$$\ln\frac{S_0 - S_t}{S_0 - S_1} = -\frac{t}{\theta} \tag{6.17}$$

$$S = S_0 - (S_0 - S_1)e^{-\frac{t}{\theta}} \tag{6.18}$$

$$S = S_0\left(1 - e^{-\frac{t}{\theta}}\right) \tag{6.19}$$

If feed water with a substrate concentration of S_0 starts to be fed at a flow rate of Q to a CSTR with a volume of V, the normalized substrate concentration against feed concentration (S/S_0) increases as Figure 6.8. When the normalized time against HRT (t/θ) reaches 1, S/S_0 reaches 63% of the maximum level. At $t/\theta = 5$, S/S_0 reaches 99%. Conversely, once substrate feeding stops at $t/\theta = 10$, S/S_0 decreases to 37% of the maximum at $t/\theta = 11$ and it decreases again to 1% of the maximum at $t/\theta = 15$.

In ideal plug flow reactors (PFR), anything that enters into the system comes out from the system in the order of its entrance. In the real world, however, some extent of forward and backward mixing is inevitable due to molecular diffusion and the lagging liquid flow near the reactor wall. Thus, true plug flow does not exist even in narrow capillary channels. As illustrated in Figure 6.9, when a

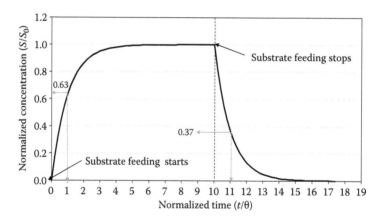

FIGURE 6.8 Normalized concentration in CSTR as a function of normalized time where feed water with a concentration, S_0, comes in at time zero and the feed concentration drops to zero at $t/\theta = 10$.

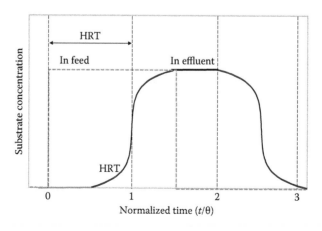

FIGURE 6.9 Relation between feed concentration and effluent concentration in PFR.

step change occurs in feed concentration under a constant flow rate, effluent concentration starts to increase before t/θ reaches 1 due to the forward mixing or lagging movement of the existing fluid in the channel. It needs more time than HRT to reach the feed concentration due to the substrate lost through forward mixing. Once the substrate concentration in the feed drops to zero, effluent concentration starts to decrease some time later, but earlier than the HRT due to backward mixing. Because of the tailing effect, the substrate concentration in the effluent does not drop to zero even after the HRT.

6.3.6.2   Advantages and Disadvantages of PFR in the CAS Process

The primary design and operational goal of CAS is obtaining sludge with good settling properties in a secondary clarifier. Low BOD and low TSS in the effluent are nearly automatically obtained with good sludge settling because BOD removal is not a bottleneck of the process in typical conditions, and the effluent TSS is a direct consequence of good sludge settling. The F/M ratio is limited to 0.2 to 0.4 g BOD/g MLSS/day in a typical CAS for these purposes, although microorganisms can accommodate with a much greater F/M ratio in essence.

Despite the efforts to achieve good sludge settling properties, CAS often suffers from poor sludge settling in the secondary clarifier due to abnormal biological conditions when excessive filamentous bacteria grow or excessive biopolymers are produced. The exact causes of the filamentous bloom are not clear, but one popular theory suggests that the filamentous bloom is linked to the excessively low F/M ratio, where filamentous bacteria may have an edge over floc-forming bacteria because of its large specific surface areas that provide a better capability to absorb food. Sludge settling issues are more commonly experienced with CSTR-type aeration tanks in municipal wastewater treatment. In CSTR, the COD provided by wastewater is quickly diluted in the large volume of mixed liquor and thereby all the microorganisms grow under starving conditions. The filamentous bloom in CSTR is often attributed to this dilution effect that creates a favorable environment for filamentous bacteria.

Meanwhile, PFR-based aeration tanks are often considered an option to discourage filamentous blooms. In a PFR-based system, aeration tanks are designed as long narrow channels through which the mixture of fresh wastewater and return-activated sludge flows. Because the fast-growing, floc-forming microorganisms quickly absorb most of the food in the upstream, little food is left for filamentous microorganisms in the downstream. Alternatively, small aeration tanks can be serially connected to maximize the plug flow pattern by minimizing backward and forward mixing in the channel. The drawbacks of plug flow-type aeration tanks include: (1) the high oxygen demand in the upstream makes it hard to dissolve enough oxygen to support biological COD degradation; (2) OTE is low in the upstream due to the high airflow and the high F/M ratio; and (3) the high F/M ratio causes greater diffuser fouling in the upstream, which again makes it harder to dissolve enough oxygen.

6.3.6.3   Advantages and Disadvantages of PFR in the MBR Process

PFR has been adopted for CAS to obtain good sludge settling properties in the secondary clarifier. However, the PFR concept is nearly obsolete in MBR not only because sludge settling is not a part of the MBR operation but also because PFR increases membrane fouling potential due to the high F/M ratio near the entrance. In addition, MBR sludge typically contains high levels of filamentous organisms likely due to the low F/M ratio, as discussed in Section 4.3.5, but filamentous bacteria themselves are arguably not a primary concern for membrane fouling as long as SMP/extracellular polymeric substance (EPS) levels stay low at a constant operating condition.

One potential advantage of PFR in MBR is that effluent TKN can be better controlled by preventing them from short-circuiting. TKN more readily leaks out in CSTR without being nitrified especially when aeration and membrane tanks are merged, but the same hardly occurs in PFR because the influent moves through aeration and membrane tanks sequentially. Because COD is quickly absorbed by heterotrophs, as can be seen in contact stabilization processes, COD short-circuiting is not noticeable regardless of the reactor configuration.

A strict plug flow design can cause expedited membrane fouling in MBR due to the excessively high local F/M ratio in the upstream. Figure 6.10 shows an extreme case of PFR, in which small aeration tanks are serially connected. This type of reactor arrangement is commonly used in chemical processes to maximize the reaction rates in the upstream. At the same time, reaction efficiency is also boosted by allowing the reaction of residual substrates in the downstream reactors. However, the serially connected reactor design can hamper the MBR operation especially when F/M is excessively high in the first reactor. The excessively high F/M causes not only high OUR but also low OTE. Therefore, it becomes very hard to maintain a sufficient DO level in the first reactor and the stressed microorganisms produce excessive biopolymers in the first reactor, as discussed in Sections 5.2.3 and 5.2.4. The biopolymers produced in the first reactor undergo a degradation process in the following reactors, but the biopolymer degradation is not fast enough to be completed before the mixed liquor reaches the membrane.

In one full-scale MBR, the average F/M ratio was surveyed at a proper level at 0.1 g BOD/g MVLSS/day, but the local F/M in the first tank was 0.7 g BOD/g MLSS/day in the serially connected aeration tanks similar to that in Figure 6.10. The high F/M ratio not only increased OUR but also reduced OTE. As a result, the DO level in the first tank was below the detection limit (e.g., <0.1 mg/L), which again stimulated microorganisms in the tank to produce more SMP, which negatively affected OTE. With a combined effect of high SMP and vigorous aeration, severe foaming was observed in the first aeration tank, as shown in Figure 6.11. Excessive membrane fouling was observed in this municipal MBR due to the SMP produced in the first aeration tank.

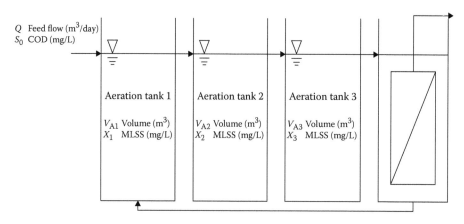

FIGURE 6.10   Plug flow MBR design using serial multiple CSTR.

FIGURE 6.11   Excessive foaming in the first aeration tank in serially connected CSTR-based MBR, where the volume of the first aeration tank takes 1/7 of the total system volume.

6.4   MEMBRANE SYSTEM

6.4.1   Flux and Membrane Area

Design fluxes for membranes are given by membrane manufacturers for typical municipal waste-waters with a set of conditions that need to be met including minimum and maximum MLSS, minimum temperature, scouring airflow rate, F/M ratio, etc. There are multiple design fluxes depending on the duration, for example, monthly or weekly average flux, and daily or hourly peak fluxes. The number of membrane modules required to treat influent is determined to avoid violating any of the constraints given by the manufacturer in terms of the flux and the durations at the given water temperature. However, determining the required membrane area is not straightforward in the real world due to the following uncertainties:

- Calculating the flow rate to be treated in each moment is not straightforward due to the complexities in the flow pattern combined with the equalization tank capacity. In fact, the flow rate varies depending on the time of the day, temporary weather conditions, season, types of residences around the plant, existence of industrial sources, etc. In addition, the size and the operation strategy of the equalization tank also affect the flow pattern to MBR. It is not only laborious to obtain rigorous information on the flow rate pattern but it is also time-consuming. As a consequence, the MBR plant often has to be designed without having all required hydraulic information.
- Membrane fouling occurs in any flux, but it occurs increasingly faster as the flux increases beyond the sustainable flux, as discussed in Section 1.2.6. The sustainable flux itself is dependent on the biological conditions, which vary depending on the timing. Therefore, running a membrane system under the manufacturer's guidelines does not guarantee successful operation. As a consequence of the varying flow rate and the varying sustainable flux, it is not completely certain how well membranes can perform during peak flow. It is always true that the lower the flux is, the lower the membrane fouling is. Therefore, determining the membrane area is a matter of how much risk we would like to or must take under the given budget constraint.
- Figure 6.12 shows an example in which daily average flux varies between 15 and 73 LMH whereas the yearly average is 25 LMH. Assuming hourly flow variations in a day are handled by equalization tanks, the membrane system must be operated at more than 40 LMH

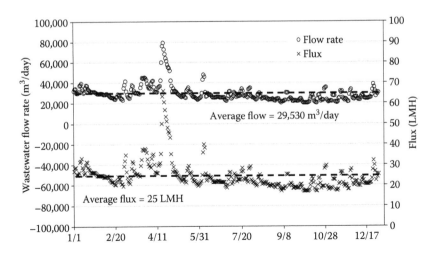

FIGURE 6.12   Daily average flux and flow profile in a municipal wastewater plant in North America, where total membrane area was 50,000 m².

for more than 10 days in a row in April. If 50% more membrane modules are installed to handle the peak flow, yearly average flux and peak daily flux will decrease to 17 and 50 LMH, respectively, but significant capital and operating costs will be required.

• As discussed previously, overlapping the membrane manufacturer's flux guidelines with the hydraulic profile and finding out the required membrane area to comply with the flux guideline is the standard procedure. However, given the many uncontrollable natural conditions, budget constraints, and the uncertainties in hydraulic data, rigorous follow-up of such a logical process is not always readily possible. Thus, design engineers also rely on the experiences obtained from prior sites with similar conditions to estimate the number of membrane modules required.

6.4.2 Specific Air Demand

The efficiencies of submerged membranes are often benchmarked based on specific air demand (SAD). There are two ways of estimating SAD depending on how it is normalized, for example, SAD per permeate volume (SAD_p) or SAD per membrane surface area (SAD_m). Because SAD_m is directly affected by the module compactness, which varies widely depending on membrane configuration, it is not relevant for estimating the efficiency of the aeration. On the other hand, SAD_p is a direct indicator of the aeration efficiency, which indicates the specific air volume required per permeate volume. Although the actual energy consumption is also dependent on the water depth in the membrane tank, SAD_p is used as a crude measure of the energy efficiency assuming that the depths of the membrane tanks are approximately equal. SAD_p tends to be lower for hollow fiber membranes than for flat sheet membranes due to the greater packing density as discussed in more detail in Section 3.2.3.

It has been reported that SAD_p can be as low as approximately 7 with intermittent aeration, but it should be noted that the low SAD_p claimed by membrane manufacturers are valid only when membranes run at the design flux and the other operating conditions are in favorable ranges. If membranes run below the design flux, SAD_p increases because the air scouring rate does not proportionally decrease with the flux. In addition, every major operating parameter must be monitored to maintain good biological conditions and to keep the filterability high. However, full-scale MBR inevitably have fluctuating hydraulic and organic loading (or F/M), fluctuating DO level, MLSS swings, potential nutritional imbalance, etc. After all, the field SAD_p has been surveyed at 10 to 20 or higher depending on the membrane type and site.

6.4.3 Air Scouring System

Designing an air scouring system is one of the key technical challenges in MBR. Airflow rate must be uniform in all nozzles to maximize the membrane performance by controlling the membrane fouling rate evenly. Otherwise, localized membrane fouling can occur in places without sufficient scouring airflow. The fouled membrane area can expand because the flux in the unaffected area must increase to compensate for the lost permeate flow. Because the membrane fouling rates exponentially increase with the flux, as shown in Figure 1.17, membrane fouling can quickly spread across entire membrane cassettes. On the contrary, membrane modules with excessive scouring air can be damaged by broken fibers or bent membrane panels. Therefore, maintaining uniform aeration underneath the membrane modules is crucial for successful MBR operation.

However, maintaining uniform airflow rates is very challenging in field conditions as the system scale increases. First, pressure losses in air pipes make it hard to equalize the total pressure drop to all nozzles. Second, air nozzles tend to be fouled at different rates, causing unequal pressure drops among nozzles. Third, the submergence of air nozzles may not be completely identical. The first and the third issues can be solved by rigorous engineering practices, but no firm solution exists for the second issue despite the countless different attempts that have been made by numerous membrane and engineering companies.

The major cause of coarse bubble diffuser fouling is known to be the dried sludge inside and outside of the diffuser pores (Kim and Boyle 1993; Hung and Boyle 2001). Because the air passing the diffuser pores is heated by adiabatic compression occurring in the blower, it has a very low relative humidity. Therefore, if there is any sludge in the passage of the dry hot air, it quickly dries. In coarse bubble nozzle used for membrane scouring, wet sludge intrudes the nozzle by the static pressure when aeration is paused for intermittent aeration and is dried when aeration resumes. As the aeration cycles repeat over an extended period, the dry sludge cake builds up in the nozzle. The plugged pores cause higher airflow through the unplugged pores and make the bubbles larger. Simultaneously, the increasing back pressure causes declining airflow rate.

One way of preventing the dry sludge from plugging air nozzles is by periodically flushing the nozzles with mixed liquor. Some of Kubota Corp.'s MBRs are built with self-cleaning nozzle systems, as illustrated in Figure 6.13. In this system, the main air pipe underneath the membrane panels is slightly inclined toward the outlet to compensate for the air pressure loss in the main pipe. As a result, the effective air pressure in the entrances of the smaller branch pipes are kept approximately identical. In the meantime, large air bubbles are generated through the macropores facing upward with 8 to 10 mm diameters in the branch pipes. The left over air is discharged through the end of the branch pipes as illustrated in the figure.

The cleaning valve in the end of the main air pipe is closed during normal operation, but it is opened either manually or automatically during the nozzle-cleaning period. If the valve is open, the internal air pressure drops in the main air pipe and the mixed liquor pours into it through the branch pipes by the static pressure. Then, airflow sweeps the mixed liquor through the main pipe and pushes the mixed liquor up through the vertical sludge discharge pipe by the airlift pump effect. Sufficient sweeping airflow rate is necessary to drive the mixed liquor in the main air pipe out to the drain above the water level. The dry sludge in the pipe is wetted by the mixed liquor and can be discharged during the cleaning cycle. This system can be robust, if it is combined with a robust pretreatment system that does not allow large debris or fibrous materials to enter the MBR. However, if the pretreatment is not sufficient enough, diffuser plugging problems can be aggravated by the fibrous materials passing the pores during diffuser cleaning.

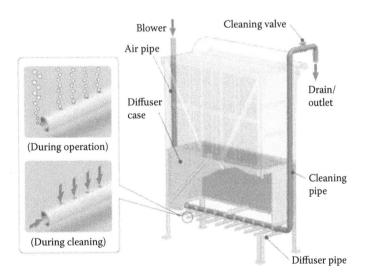

FIGURE 6.13 Coarse air diffuser design (left) and diffuser cleaning method. (Courtesy of Kubota Membrane USA Corporation.)

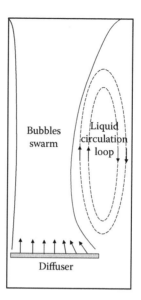

FIGURE 6.14 Bubble rising pattern in membrane tank equipped with a submerged flat sheet membrane when diffusers are off-centered. (Reproduced from Nguyen Cong Duc, E. et al., *J Membr Sci* 321:264–271, 2008.)

The upflow pattern of two-phase flow can change dramatically by subtle changes in airflow distribution, flow rate, dimension and configuration of the tank, location of the air nozzles, MLSS, etc. (Gresch et al. 2010, 2011). Even biological conditions can change the upflow patterns by affecting sludge rheology and the consequential liquid movement. For example, if air nozzles are placed slightly off-centered in the left side of the tank, as shown in Figure 6.14, single-phase liquid circulation loop develops in the right side while air bubbles rise quickly through the bypass channel formed in the left. If membrane modules are placed in this tank, those on the left side are exposed to excess airflow whereas those on the right side are not scoured enough. Because membrane scouring by liquid flow is not as efficient as by air–liquid mixture, membranes on the right side will be preferentially fouled. Even if the diffuser pipe is placed in the center, if a few nozzles on the right side are plugged up, similar problems can occur. Due to the complexities of the flow dynamics and mixing patterns, empirical experiences are an important part of the decision making on top of the theory in determining the configurations of aeration and membrane tanks, including the placement of membrane modules and diffusers.

6.5 OXYGEN TRANSFER EFFICIENCY

6.5.1 THEORY

6.5.1.1 Standardized Oxygen Transfer Efficiency

Standardized OTE (SOTE) is defined as the OTE in clean water at the standard condition in which no DO and no salinity is present at 20°C. SOTE is not a fixed property of a given reactor and diffuser configurations but is variable depending on airflow rates. As airflow rate increases, bubbles rise faster through the bubble plume strengthened by the increasing number of bubbles. Meanwhile, the specific surface area shrinks due to the expedited coalescence at the denser bubble population. As a consequence of the shorter contact time and the smaller specific surface area, SOTE decreases. Therefore, the diffuser's SOTE is usually reported as a function of the airflow rate.

Because SOTE is valid only for the water depth that the diffuser was tested for, SOTE is often reported as specific SOTE or $SOTE_S$ by dividing SOTE by diffuse submergence. $SOTE_S$ decreases

TABLE 6.5
Comparison of Clean Water OTE of Selected Diffusers

Diffuser Type Place in Grid	SOTE at 4.5 m Submergence (%)
Ceramic discs	26–33
Ceramic domes	27–29
Porous plastic tubes/discs	28–32
Jet aeration	15–24
Flexible sheath tubes	22–29
Nonrigid porous plastic tubes	26–36
Single spiral roll	19–37
Perforated membrane tubes	22–29
Nonporous diffusers (coarse bubble)	
Dual-spiral roll	12–13
Mid-width	10–13
Single-spiral roll	9–12

Source: Data from Shammas, N. Fine pore aeration of water and wastewater. In *Advanced Physic-Chemical Treatment*, edited by L.K. Wang, Y-T. Hung, and N.K. Shammas. Humana Press, Totowa, NJ, 2007.

as bubbles rise because oxygen partial pressure in air bubbles decreases as a consequence of the loss of oxygen to water and the decreasing static pressure. However, the effect of depleting oxygen in bubbles is typically not significant in biological wastewater treatment due to the low OTE at approximately 10% in 4 to 5 m diffuser submergence. In addition, the negative effect of the reduced static pressure is largely compensated for by the increasing bubble surface area. Thus, SOTE is calculated by multiplying diffuser submergence and $SOTE_S$ assuming $SOTE_S$ is constant in the process with a 4 to 5 m diffuser submergence. The typical SOTE and $SOTE_S$ of various diffusers are summarized in Table 6.5. More accurate values for a specific diffuser can be obtained from the manufacturer.

6.5.1.2 OTE Estimation in Process Water/Mixed Liquor

In process water (or mixed liquor), oxygen transfer is hampered by dissolved materials and suspended solids. The extent of the effect from impurities is expressed using α-factor, which is a relative OTE measured in process water to the SOTE under a standard condition (20°C, no DO, no salinity). The α-factor ranges between 0 and 1 depending on the interference of impurities in process water. The field OTE is calculated by multiplying the parameters affecting oxygen transfer to the SOTE as shown in Equation 6.20. The typical range of each parameter is summarized in Table 6.6.

TABLE 6.6
Typical Ranges of Factors Affecting OTE

	Typical Ranges	Median
α	0.2–0.6	0.4
β	0.95–0.99	0.97
θ	1.022–1.026	1.024
f	0.7–0.9	0.8
h	3–6 m (standard)	4.5 m
	6–12 m (deep)	
$SOTE_S$	0.05–0.07/m (fine pore)	0.06/m
	0.02–0.03/m (coarse pore)	0.025/m

Inorganic ions in the water also interfere with OTE. If inorganic ions are present in water, the polar water molecules are aligned around the ion to form water-ion clusters. Under these conditions, the movement of one cluster affects a large number of water molecules associated with other clusters. Therefore, to move a portion of water, a stronger force is necessary and this directly means higher water viscosity by definition. At high viscosity, diffusion of oxygen molecules from the air–water interface to the bulk liquid slows down and hence the OTE decreases. The effect of soluble ions on OTE is factored in using β-factor in Equation 6.20. By definition, the β-factor is 1 in pure water and ranges from 0 to 1 in process water.

As diffusers are fouled by inorganic and organic deposits in mixed liquor, the number of open pores on the diffuser surface gradually declines. Under constant aeration conditions, the airflow to the open pores should increase along with air pressure increase. The high effective airflow rate through the diffuser pore causes low OTE by producing large bubbles. The loss of efficiency by diffuser fouling is considered in the equation by fouling factor, f, which is 1.0 for new diffusers. As f declines, energy efficiency loss is inevitable due to the combined effect of low OTE and high air pressure required to overcome high back pressure.

Temperature is another factor affecting OTE by affecting the viscosity of water and the diffusivity of oxygen. As temperature increases, oxygen dissolves in water quicker due to the increasing diffusivity. In addition, oxygen molecules dissolved in the air–water interface dissipate to the bulk quicker due to the lower water viscosity. The quick dissipation of oxygen from the interface expedites oxygen dissolution by reducing the oxygen concentration (C) and increasing the driving force. However, the saturated DO (C_0^*) declines as temperature increases and this acts as a drawback in the driving force. The temperature effect on the oxygen dissolution kinetics is considered using the θ-factor as shown in Equation 6.20. Combining all these effects together, the OTE increases slightly as temperature rises as found in the thermophilic biological processes discussed in Section 7.7.

$$\text{OTE} = \alpha\theta^{T-20}hf\,\text{SOTE}_\text{S}\left[\frac{\beta C_0^* - C}{C_{0,20}^*}\right] \tag{6.20}$$

where

α	alpha factor (–)
β	beta factor (–)
θ	theta factor (–)
f	fouling factor of diffuser (–)
h	diffuser submergence (m H$_2$O)
C	average DO in aeration tank at a given condition (mg/L)
C_0^*	average saturated DO at a given temperature and zero salinity (mg/L)
$C_{0,20}^*$	average saturated DO at 20°C, 1 atm, and zero salinity (mg/L)
SOTE$_\text{S}$	specific SOTE at 20°C, 1 atm, and zero salinity (/m H$_2$O)
T	water temperature (°C)

OTE in the process water (or mixed liquor) at the standard condition is often written as αSOTE, which is a normalized OTE against 20°C, zero salinity, and zero DO (Stenstrom et al. 2006). The αSOTE can also be written as αfSOTE including the diffuser fouling effect. Unless the water temperature significantly deviates from 20°C or salinity is exceptionally high, β- and θ-factors are often neglected because α-factor dominates the OTE whereas f-factor may have a sizable impact.

The salinity factor (β), and average saturated DO (C_0^*) are calculated using Equations 6.21 and 6.22, Equation 6.21 is based on the data taken from Weiss (1970).

$$\beta = 1 + (2.297 \times 10^{-8}T - 3.976 \times 10^{-6})S \tag{6.21}$$

$$C_0^* = [-5.823 \times 10^{-5} T^3 + 7.049 \times 10^{-3} T^2 - 3.929 \times 10^{-1} T + 14.56]$$

$$\times \left[1 + \frac{h}{20.67} - 1.099 \times 10^{-4} A_L \right] \tag{6.22}$$

where

T	water temperature (°C)
h	diffuser submergence (m H_2O)
S	conductivity of wastewater at 25°C (µS/cm)
A_L	altitude of the plant (m)

The average saturated DO at sea level at 20°C $\left(C_{0,20}^* \right)$ is calculated as Equation 6.23.

$$C_{0,20}^* = 9.1 \left[1 + \frac{h}{20.67} \right] \tag{6.23}$$

6.5.2 FACTORS AFFECTING OTE

6.5.2.1 DIFFUSER PORE SIZE

Diffusers with small pore sizes generate a large number of small bubbles. The surface area of each bubble is proportional to the second power of the bubble size whereas the number of bubbles is inversely proportional to the third power of the bubble size at a given airflow rate. Overall, total available surface area is inversely proportional to the bubble size. For example, if the bubble size decreases to half, the total surface area increases twofold. Consequently, the total surface area of the bubble is inversely proportional to the bubble size at a given airflow rate. In general, fine bubble diffusers have pores smaller than 2 mm and produce bubbles that are smaller than 5 mm. The fine bubble diffusers generally incur higher pressure loss than coarse bubble diffusers, but the pressure loss can be kept low by decreasing the specific airflow rate per diffuser. Perforated membrane diffusers in disc/dome configuration are commonly used, but perforated tubes are also used. As materials, thermoplastic or elastomeric materials such as plasticized PVC, EPDM rubber, and neoprene rubber with patterned pores or gaps (1–2 mm) are commonly used. These diffusers are also called flexible diffusers because pores can expand when airflow is high, producing larger bubbles. Sintered HDPE and ceramic disc diffusers are also commonly used.

Coarse bubble (or nonporous) diffusers typically have holes or slots that are bigger than 6 mm. Coarse bubbles not only coalesce when they rise but also break down to smaller bubbles due to strong turbulence. Thus, the relative sizes of the coarse bubbles do not increase as much as fine bubbles while they are rising. The steady state bubble size is 6 to 10 mm under typical conditions and does not change drastically when airflow rate increases. Therefore, the OTE of coarse bubble diffusers does not fluctuate much as airflow rate changes. In some cases, OTE even slightly increases as airflow rate rises due to turbulence (Shammas 2007). Compared with fine bubble diffusers, coarse bubble diffusers are less prone to diffuser fouling in nature due to their large pore openings. Therefore, the efficiency gap between the fine bubble diffusers and the coarse bubble diffusers is not as big as what bubble size and SOTE suggest in field conditions. The advantages and disadvantages of fine bubble diffusers are summarized in Table 6.7.

TABLE 6.7

Comparison of Fine Bubble Diffusers and Coarse Bubble Diffusers

	Fine Bubble Diffuser	Coarse Bubble Diffuser
Pore size	<2 mm	>6 mm
Bubble size	2–5 mm	6–10 mm
Pressure loss	0.1–0.2 m H_2O (new) 0.2–0.5 m H_2O (used)	0.1–0.5 m H_2O (new and used)
Relative advantage	Higher OTE and lower power costs Less volatile organic compound emissions $SOTE_S$: 4%–6%/m	Less susceptible to fouling Less susceptible to chemical attack Less efficiency loss at high airflow Less susceptible to uneven air distribution Less maintenance costs $SOTE_S$: 2%–3%/m

6.5.2.2 Airflow Rate

OTE from the same diffuser can vary widely depending on airflow rate. In fact, as airflow rate increases in a given diffuser system, bubble size increases and thereby the total bubble surface area does not increase proportionally. In addition, at high airflow rates, the rising bubbles have greater chances to collide with each other and coalesce due to the high population density. As a result, small diffuser pores do not necessarily create small bubbles when they are overloaded.

The effect of bubble coalescence is less significant in coarse bubble diffusers because coarse bubbles generate stronger turbulence that can keep the bubbles from coalescence in the event of collision. As a result, α-factor of coarse bubble diffusers does not decrease as much as that of fine bubble diffusers at high airflow rates. Figure 6.15 shows the strong airflow effect on α-factor in five different full-scale plants. As the performance of fine bubble diffusers is significantly affected by airflow rate, diffuser manufacturers often provide SOTE curves as a function of airflow rate.

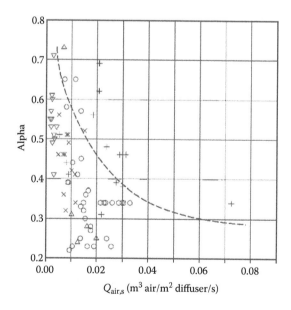

FIGURE 6.15 Effect of airflow rate on α-factor in field, in which five groups of symbols indicate five different full-scale plants. (From Rosso, D. et al., *Water Env Res* 77(3):266–273, 2005.)

6.5.2.3 Spatial Distribution of Diffuser

OTE is affected by diffuser placement on the floor of the aeration tank. Fine bubble diffusers are more sensitive to diffuser placement than coarse bubble diffusers because the specific airflow rates are lower. If specific airflow per footprint is high in a section of the aeration tank, air plumes can develop in the area through which bubbles bypass quickly. Inadequate diffuser placement and improper airflow distribution cause the short-circuiting of air, which promote bubble coalescence in the high airflow zones and decreases the contact time by increasing the bubble rising velocity. To prevent air short-circuiting (or channeling), the entire floor should have evenly spaced diffuser coverage. If airflows are distributed evenly across the reactor floor, bubble rising velocity is minimized due to the lack of large columns serving as downcomers.

6.5.2.4 Diffuser Fouling

In aeration basin, diffusers are fouled over time by inorganic precipitates, dry biomass, biofilm, etc. Diffuser fouling can be classified into three categories (Kim and Boyle 1993). Inorganic fouling is classified as type A fouling and the organic fouling as type B. The mixture of inorganic and organic fouling is classified as type C. Type A fouling increases head loss, but it can increase OTE in some cases by decreasing the effective pore size of the diffuser. Type B fouling occurs mainly by dried or wet biomass. It increases head loss, but decreases OTE because the biofilm forms bigger secondary pores on the diffuser surface. In type C fouling, the mixture may increase OTE slightly by decreasing effective pore size in some cases, but it certainly decreases energy efficiency by increasing the head loss.

In biological wastewater treatment, type B fouling by organic deposits is most commonly observed. Thus, diffuser fouling almost always causes large bubbles with low OTE in MBR. Based on a wastewater plants survey (Rosso et al. 2007), it was found that the average α-factor of new diffusers was 0.5, but it decreased to 0.4 on average after 2 to 24 months and to 0.35 afterward as shown in Figure 6.16. In this figure, the fouling factor (f) of 2- to 24-month-old diffusers is calculated at 0.8 (= 0.4/0.5) and that of more than 24-month-old diffusers is 0.7 (= 0.35/0.5).

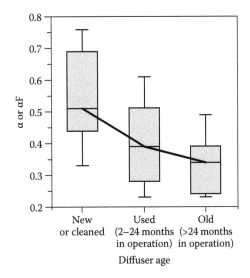

FIGURE 6.16 Effect of diffuser fouling on α- and f-factor in CAS. (From Rosso, D. et al., Energy-conservation in fine pore diffuser installations in activated sludge processes: Final report to California government. Southern California Edison Project 500-03-001, 2007.)

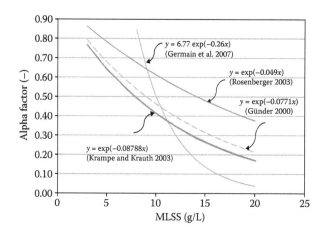

FIGURE 6.17 Alpha-factor of fine bubble diffusers as a function of MLSS.

6.5.2.5 MLSS and Viscosity

The suspended solids measured by MLSS are one of the factors affecting α-factor (Germain et al. 2007). In general, α-factor declines as MLSS increases, if all other conditions remain identical, but it may appear opposite in certain conditions due to the complex connection between the two through intermediate parameters as will be discussed later. Various correlations between MLSS and α-factor are available in the literature as illustrated in Figure 6.17. The dramatic differences among the curves suggest that there are other factors strongly influencing the α-factor. In fact, organic loading rate (or F/M ratio), SRT, specific aeration rate per footprint, etc., are factors affecting OTE, but some of these factors are often overlooked during the OTE measurement. For instance, it was found that the α-factor fluctuates significantly throughout the day even in the same location in the same aeration tank depending on the organic loading and the specific aeration rate (Cornel et al. 2003). Therefore, the absolute values that the curves suggest in the figure are valid only for the condition from which the values were obtained.

Although α-factor is highly variable depending on the process condition, the two curves (Günder 2000; Krampe and Krauth 2003) providing moderate α-values in the most common MLSS range of 8 to 12 g/L can be considered as guidelines for municipal MBR with approximately 20-day SRT. Based on these curves, α-factor is approximately 0.4 ± 0.1 in typical MBR. For industrial MBR, α-factor can be very different from those in municipal MBR especially when the wastewater contains components that hamper biological health and floc formation. The uncertainties in α-factor are a significant hurdle in process design because the aeration and diffuser capacities are proportionally affected by it. Therefore, prior experiences with wastewater with similar compositions are a very important part of the process design.

The α-factor can also be correlated with mixed liquor viscosity. In fact, oxygen must first dissolve into the air–liquid interface and then migrate into the bulk liquid by Brownian diffusion and convective flow. According to the Einstein–Stokes equation, the diffusivity of oxygen is inversely proportional to the viscosity of the medium. In addition, convective oxygen transportation from the air–liquid interface to the bulk also slows down as medium viscosity increases. As a result, if oxygen transportation to the bulk liquid slows down, oxygen dissolution in the air–liquid interface also slows down, thereby OTE and α-factor decrease. As shown in Figure 6.18, α-factor is inversely proportional to mixed liquor viscosity at a shear rate of 10 s^{-1}.

6.5.2.6 SRT and F/M Ratio

It has been known that α-factor is strongly influenced by SRT (Rosso et al. 2005, 2007; Rosso and Stenstrom 2007; Leu et al. 2010). The exact causes are not clear, but the varying quantity and quality

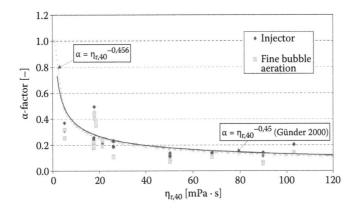

FIGURE 6.18 Connection between representative viscosity at a shear rate of 10/s and the determined α-factor. (From Krampe, J., and Krauth, K., *Water Sci Technol* 47(12):297–303, 2003.)

of SMP depending on SRT seems to be the cause. As shown in Figure 6.19, when SRT increases from 2 days to 20 days, the α-factor increases threefold from 0.2 to 0.6 in activated sludge processes based on the survey of four different CAS plants. It is also known that α-factor tends to be greater in BNR processes than in CAS due to the longer SRT caused by the extra tanks used for denitrification (Rosso et al. 2005; Leu et al. 2010). As a result of the high α-factor and the oxygen credit from the denitrification process, the overall aeration costs of BNR are not necessarily greater than those of CAS despite the larger tank size.

SRT is reciprocally linked with the specific organic loading rate, that is, F/M ratio. At a low F/M ratio, fewer excess biosolids are produced and, as a consequence, SRT is long. Therefore, it can be generalized that α-factor increases as F/M ratio decreases. The α-factor is not only affected by the long-term F/M ratio but is also affected by the short-term fluctuation of F/M ratio (Rosso and Stenstrom 2007). When α-factor was tracked at a fixed location in an aeration tank, it fluctuated widely in a day between 0.36 and 0.53, depending on the instantaneous

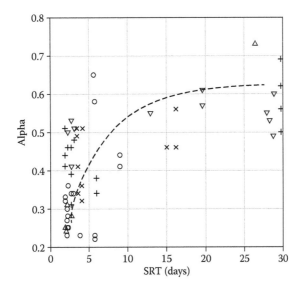

FIGURE 6.19 Effect of SRT on α-factor in full-scale CAS with fine bubble diffusers. Each dot indicates data from one full-scale plant. (From Rosso, D. et al., *Water Env Res* 77(3):266–273, 2005.)

organic loading and the aeration rate at the time of the α-factor measurement. A similar trend was observed in another study performed in a full-scale municipal MBR using off-gas analysis method, where α-factor varied between 0.4 and 0.6 depending on organic loading in a day (Cornel et al. 2003).

The quick response of α-factor to organic loading can be partially explained by the surfactants and surfactant-like substrates carried by the feed wastewater. When feed flow rate increases, more surfactants are fed to the aeration basin and are attracted to the air–liquid interface of rising bubbles due to their hydrophobic moieties. The surfactant molecules covering the bubble surface physically interrupt oxygen transfer and reduce the α-factor (Stenstrom and Masutani 1990). The quick response of the α-factor to the F/M ratio can also be linked to the changes in the quantity or quality (or both) of SMP. In fact, it has been observed in MBR that transmembrane pressure (TMP) increases as F/M ratio increases with only a few hours' gap at a constant flux (Syed et al. 2009). The cause of the increasing TMP at high F/M ratio is not completely clear, but the increasing SMP concentration, potentially with changing molecular properties, is believed to be a culprit. If any change occurs in the quantity or the quality of SMP, mass transfer rate in the air–liquid interface can change due to modifications in the air–liquid interface.

If airflow rate is adjusted to meet the oxygen demand at varying organic loading rates, the α-factor fluctuates according to the varying airflow rate. As airflow rate increases, more bubbles coalesce with each other and the contact time decreases due to the bubbles rising faster. OTE and α-factor decrease as a consequence.

It is noteworthy that the α-factors in CAS and MBR are not in very different ranges despite their vastly different MLSS in the aeration tank. As can be compared in Figures 6.17 and 6.19, α-factors of MBR are only slightly lower than those of CAS. Except some high α-factors in CAS, α-factors are mostly in the same range (0.3–0.5) at the common operational conditions. The relatively high α-factor of MBR for the high MLSS can be explained by two factors. First, the much lower F/M ratio of MBR at 1/3 to 1/2 of those of CAS positively affects α-factor. Second, diffuser fouling is potentially lower in MBR due to the low F/M ratio. As discussed in Section 6.3.6.2, diffusers are fouled faster at high F/M ratio perhaps due to the high SMP concentration and the specific properties of the SMP secreted at the condition. It has been well known that the diffusers in the upstream foul quicker than those in the downstream in PFR-based systems.

6.5.2.7 Foaming

One common cause of foaming in activated sludge is the surfactant-like molecules. Artificially manufactured surfactants are the most efficient foam-causing molecules, but the naturally occurring polysaccharides and proteins with hydrophobic and hydrophilic moieties can also function as inefficient surfactants. The hydrophobic moiety prefers to stick out to the air whereas hydrophilic moiety roots into the water in the air–liquid interface of the bubble. Surfactant molecules can cause foams at a sufficiently high concentration by aligning side by side on the foam's surface. Microorganisms and cell debris with hydrophobic surfaces also accumulate on the thin film of foam particles and stabilize the foam by interfering with water drainage.

The surfactant molecules and hydrophobic particles also tend to accumulate on rising bubble surfaces in the aeration tank. The accumulated molecules and particles on the bubble surface interrupt oxygen transfer by physically screening the air–liquid interface. In activated sludge process, light foams can be formed by synthetic surfactants carried by raw wastewater, but brown (or heavy) foams are formed by the biologically produced surfactant-like molecules and hydrophobic cell debris. Some microorganisms with hydrophobic surfaces, for example, *Nocardia*, *Microthrix parvicella*, and type 1863 also contribute to foam stabilization by accumulating on the foam's surface and interfering with water drainage from the foam. Meanwhile, the biopolymers that cause or stabilize foams can also expedite membrane fouling. Therefore, foaming, low OTE, and high membrane fouling are closely interconnected. Depending on the condition, these three could occur simultaneously or sequentially.

Improper use of antifoam chemicals may damage the OTE dramatically by returning the hydrophobic materials accumulated in the foam layer back to the water. The foaming and defoaming can be split into four stages as explained in the following paragraphs. Figure 6.20 illustrates the four-stage cycle.

Stage 1—In normal conditions, foaming and defoaming balance each other and only a thin foam layer exists. The foam-causing materials such as proteins, polysaccharides, filamentous microorganisms with hydrophobic surfaces, and hydrophobic cell debris move up to the top and cause foaming, but those hydrophobic materials return to the water when the foams collapse.

Stage 2—If artificial or biological surfactants and hydrophobic particles become more prevalent in mixed liquors, the balance between foaming and defoaming is broken. Then, more hydrophobic materials move to the foam layer than those returning to the bulk. The OTE may not be affected because the hydrophobic materials that interrupt oxygen transfer are scavenged to the foam layer. If the surfactant scavenging by the rising bubbles is fast enough, OTE can even increase at this stage.

Stage 3—If antifoam is dosed and the foam layer collapses, a large amount of accumulated hydrophobic materials in the foam layer return to the mixed liquor within a short period. This can cause a dramatic decrease in OTE by increasing the concentration of hydrophobic materials in the mixed liquor. DO decreases as a result of low OTE, which in turn increases the SMP secretion or changes the SMP property by stressing microorganisms. High SMP increases the chances of a second surge of foaming.

Stage 4—As the dissolved hydrophobic materials are discharged from the bubble column with effluent and excess biosolids, OTE gradually recovers.

The dramatic effect of foam layer collapse on gas transfer efficiency (GTE) has been observed in syngas fermentation. Using an in-line mass spectrometer, the real-time GTE was tracked. As shown in Figure 6.21, the initially stable GTE in stage 1 started to increase as foam height increased rapidly in stage 2. In this stage, microorganisms with hydrophobic surfaces accumulated to the foam layer

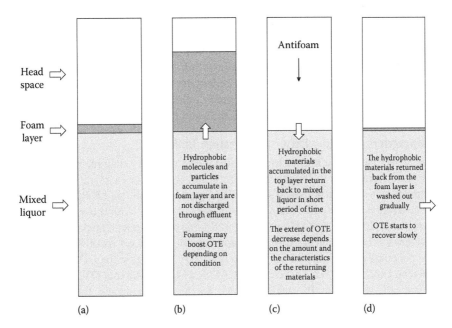

FIGURE 6.20 Mechanism of OTE collapse: (a) normal condition, (b) foam accumulation, (c) foam collapse, and (d) recovery of OTE.

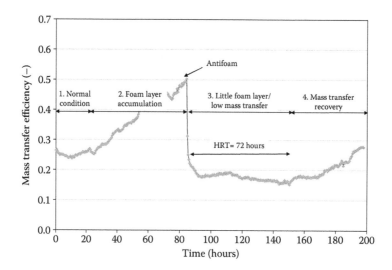

FIGURE 6.21 Effect of foam layer collapse on mass transfer efficiency in syngas fermentation using bubble column at 72 h HRT.

and as a consequence suspended solids in the fermentation broth decreased. However, as soon as the antifoam chemicals were added and the foam layer collapsed, the GTE sharply decreased from 50% to 18%. Because the amount of antifoam used was much less than 1 mg/L based on the reactor volume, it is unlikely that the antifoam itself was the cause of such a huge GTE decrease. In addition, it was difficult to consider that such instantaneous GTE drop was caused by antifoam toxicity. Therefore, the sudden GTE decrease can be attributed to the large amount of hydrophobic foam constituents returned back to the liquid. After the fermentation broth was purged for one cycle of SRT (i.e., 72 h), GTE started to increase again.

The above observation suggests that the foam layer should not be killed abruptly to not disrupt OTE in biological processes. If foam depth is excessive, it is more desirable to reduce it gradually while allowing hydrophobic materials to exit from the system with excess sludge and effluent. Most desirably, foam depth should be controlled at a low manageable level (e.g., <20–30 cm), without allowing it to grow excessively. In MBR, if foam layer is killed abruptly, the hydrophobic materials returned to the mixed liquor not only hamper OTE but also expedite membrane fouling. Moreover, the low OTE can indirectly enhance the membrane fouling by causing low DO, as discussed in Section 5.2.4. The preferred antifoams for MBR are listed in Table 5.1.

6.5.2.8 Biocarriers and Flow Pattern in Aeration Tank

Floating biocarriers have been known to reduce OTE although they may increase the contact time of bubbles by interfering with the bubble rising in certain conditions. In a field survey, the OTE in a full-scale integrated fixed film activated sludge (IFAS) process with AnoxKaldnes™ biocarrier and coarse bubble aeration were studied and the results were compared with those of CAS without biocarriers (Rosso et al. 2010, 2011). OTE was measured using the American Society of Civil Engineers (ASCE) standard method, which is a form of steady state method based on off-gas analysis. Because oxygen must penetrate through the biofilm, it was necessary to increase the DO to more than 3 mg/L in IFAS whereas 1 to 2 mg/L was sufficient for CAS. It was observed that OTE of IFAS were much lower than that of CAS despite the much lower MLSS. As a consequence, the aeration demand of IFAS was up to 50% greater than that of CAS during the course of the study (Rosso et al. 2010). The low OTE and α-factor of IFAS were attributed to the large bubble sizes caused by the enhanced bubble coalescence by the floating biocarriers at high airflow rates. In addition, the high DO of IFAS also contributed to the low OTE by compromising the driving force for oxygen dissolution.

OTE is not uniform even in the same aeration tank depending on the location in the tank. The short-circuiting of airflow caused by nonideal air distribution is one reason, but F/M ratio gradient along the aeration tank is also responsible for the OTE variation, especially when the aeration tank is designed as a PFR. The gradual decline of substrate concentration along the streamline can cause OTE gradient. In a field survey, α-factors were observed at 0.3 and 0.55 in the inlet and the outlet of a 300-m-long plug flow tank in a municipal CAS process (Rosso and Stenstrom 2007). This observation suggests that a great deal of tapering is required in aeration rate toward the downstream in PFR because not only does oxygen demand decrease but OTE also increases toward the downstream.

6.5.3 OTE Measurement

6.5.3.1 Non–Steady State Absorption Method

The method employing changing DO concentrations during the OTE measurement is called the "non–steady state" method. If DO increases during the measurement, it is further classified as the "absorption" method. Otherwise, it is classified as the "desorption" method.

Non–steady state absorption methods are widely used to measure OTE and SOTE mainly due to its simplicity. According to Method 02-84 of ASCE (1991), SOTE is measured in the aeration tank filled with clean water. Sodium bisulfite and cobalt catalysts are added to remove the DO before bubbling starts. Typically, sodium bisulfite ($NaSO_3$) is added in the amount equivalent to 125% to 175% of the stoichiometric requirement and approximately 0.05 mg/L of cobalt chloride is also added as cobaltous ions (Stenstrom et al. 2006). Subsequently, air bubbling starts while real-time DO is traced using DO probes. SOTE is mathematically calculated from the DO curve and normalized against zero DO and zero salinity at 20°C. Apparent mass transfer coefficient (k_La_0) is also calculated from the DO curve as shown in the later part of this section (Equation 6.25).

OTE is measured in the aeration tank filled with mixed liquor. When DO stays at a constant level, the oxygen dissolution rate is equivalent to the oxygen consumption rate. The oxygen consumption rate by microorganisms is estimated by the OUR in a test bottle as briefed in Section 6.3.3. The total oxygen dissolution rate in the tank is obtained from the product of the OUR and the reactor volume. OTE is calculated by dividing the total oxygen dissolution rate by the oxygen supply rate calculated from the airflow rate. Apparent mass transfer coefficient (k_La) is calculated from the OUR and the DO, as shown in the later part of this section (Equation 6.26). The α-factor is calculated from the ratio between k_La and k_La_0 (Equation 6.29).

Despite the convenience, the accuracy of this method is not high because the accuracy of measuring OUR in a test bottle is not high and hence the estimated k_La from OUR is not very accurate. In one study, for example, the OUR measured in a test bottle was always higher (by up to 40%) than the OUR measured *in situ* in the aeration tank using a steady state method (Krause et al. 2003). The high OUR can be attributed to the high shear rate caused by the micromixer in the OUR test bottle. Under high shear conditions, floc particles are broken down and more microorganisms are exposed to the DO. Due to the overestimated OUR, the estimated OTE also tends to be overestimated in non–steady state adsorption methods. The following are the procedures for calculating OTE, SOTE, and α-factor based on the non–steady state absorption method.

Step 1—Measurement of apparent mass transfer coefficient in clean water, k_La_0

Oxygen transfer (or dissolution) rate in clean water can be described by Equation 6.24, where DO increasing rate is proportional to the apparent mass transfer coefficient in clean water, k_La_0, and the difference between saturated DO (C_0^*) and the DO at time t, $C_{0,t}$.

$$\frac{dC_{0,t}}{dt} = k_La_0(C_0^* - C_{0,t})$$

(6.24)

where

$C_{0,t}$ DO at time t in clean water (mg/L)

C_0^* average saturated DO in clean water (mg/L)

t elapsed time (h)

$k_L a_0$ apparent mass transfer coefficient in clean water (/h)

In this equation, C_0^* represents an average saturated DO in the tank and can be calculated using Equation 6.22. Equation 6.24 can be integrated using two boundary conditions, that is, $C_{0,t} = C_i$ at $t = 0$ and $C_{0,t} = C_{0,t}$ at $t = t$.

$$\ln \frac{C_0^* - C_{0,t}}{C_0^* - C_i} = -k_L a_0 t \tag{6.25}$$

The log-term on the left side of the equation can be plotted against t to obtain $k_L a_0$ from the slope.

Step 2—Measurement of apparent mass transfer coefficient in process water, $k_L a$

In the activated sludge process, volumetric oxygen consumption rate is measured by OUR. At a hypothetical steady state, the oxygen dissolution rate should be equal to the oxygen consumption rate (OUR) as described by Equation 6.26.

$$k_L a(C^* - C_{SS}) = OUR \tag{6.26}$$

where

C_{SS} DO in mixed liquor at steady state (mg/L)

C^* Concentrated DO in mixed liquor (mg/L)

$k_L a$ apparent mass transfer coefficient in mixed liquor (/h)

By definition, C^* is equivalent to βC_0^* and can be calculated using Equations 6.21 and 6.22 for the experimental conditions. OUR is experimentally measured according to Method 2710 (APHA 1998). Finally, $k_L a$ can be estimated using Equation 6.26.

Step 3—Calculation of oxygen transfer rate, OTR

Standard OTR in clean water at the standard condition (20°C, 1 atm, zero DO, and zero salinity), OTR_0, can be calculated using Equation 6.27. OTR at a given DO (C_{SS}) in process water (OTR, kg O_2/h) can also be calculated using a similar equation, where V indicates reactor volume (m³).

$$OTR_0 = \frac{k_L a_0 C_0^* V}{1000} \text{ or } OTR = \frac{k_L a(C^* - C_{SS})V}{1000} \tag{6.27}$$

Step 4—Calculation of SOTE and OTE

Oxygen transfer efficiency in clean water (SOTE) and in mixed liquor (OTE) are estimated by the following equation:

$$SOTE = \frac{OTR_0}{0.279 Q_{air}} \text{ or } OTE = \frac{OTR}{0.279 Q_{air}} \tag{6.28}$$

where Q_{air} is airflow rate in Nm³/h at 20°C and 1 atm. The factor in the denominator, 0.279, is the oxygen density at 20°C, 1 atm, and 0% relative humidity, but it becomes 0.275 at 20°C, 1 atm, and 36% relative humidity where most blowers are rated for.

By definition, α-factor can be calculated as follows from the ratio between $k_L a$ and $k_L a_0$. It represents the relative ratio of oxygen dissolution in process water against that in clean water.

$$\alpha = \frac{k_L a}{k_L a_0}.$$

(6.29)

6.5.3.2 Non–Steady State Desorption Method

In this method, OUR is measured *in situ* from the DO decay curve obtained while pausing the oxygenation of aeration tank (Wagner et al. 2002). The underlying assumption is that OUR is constant at the DO range employed in the test, for example, 2 to 10 mg/L. This assumption is valid because the activities of heterotrophic microorganisms are hardly affected by DO, if DO is higher than 1 to 2 mg/L. In fact, the half reaction constant (K_S) of DO concentration for BOD oxidation is no more than 0.5 mg/L in activated sludge processes. According to the Monod equation (Equation 6.30), the reaction constant is not linearly proportional to DO (or S in the equation). If S is equivalent to K_S, the reaction constant, k, becomes half of the maximum reaction constant, k_{max}. As S increases beyond three to four times that of K_S, the reaction constant, k, approaches to the maximum and becomes nearly constant.

$$k = \frac{S}{K_S + S} k_{max}$$

(6.30)

In one study, a mechanical mixer was used to dissolve pure oxygen into mixed liquor. Oxygen was supplied intermittently to the continuously running mechanical mixer and as a result DO swung periodically between 2.6 and 4.0 mg/L. Organic loading rate was kept constant during the experiment. Figure 6.22 shows a time curve of DO in an aeration tank. This graph clearly shows DO drops linearly during the two oxygenation cycles shown in the figure. The linearly declining curve indicates that oxygen consumption rate is not a function of DO in this range. Finally, the OUR can be calculated at 88 mg/L/h from the slope.

Oxygen consumption rate in the aeration basin can be estimated by multiplying OUR by the basin volume, which is equivalent to the oxygen dissolution rate. OTE can be calculated by dividing the oxygen dissolution rate by the amount of oxygen supplied to the water (Equation 6.31), where Q_{air} is airflow rate at 20°C and 1 atm (Nm^3/h).

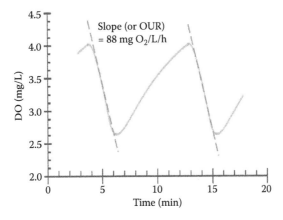

FIGURE 6.22 Time curve of DO when intermittent oxygenation is performed in aeration basin. (From Irizar, I. et al., *Water Sci Technol* 60(2):459–466, 2009.)

$$\text{OTE} = \frac{\text{OUR} \cdot V}{0.279 Q_{\text{air}}} \tag{6.31}$$

The apparent mass transfer coefficient, $k_L a$, in the aeration tank can also be calculated using Equation 6.32, where C_{SS} is the steady state DO at the process condition for which $k_L a$ is calculated.

$$k_L a = \frac{\text{OUR}}{C^* - C_{SS}} \tag{6.32}$$

The non–steady state desorption method is simple and accurate, but it is not commonly used in air-based biological treatment processes because it is difficult to increase DO to more than 4 mg/L using the given aeration capacity. In addition, sludge settling during the no-aeration period can cause errors. Another potential drawback is that biological flocs can be broken down when airflow is increased to increase DO. The sheared sludge can artificially increase OUR by exposing more microorganisms to the substrates and oxygen. However, if mechanical mixers are used with pure oxygen to oxygenate the aeration basin, DO can be readily increased close to or above 10 mg/L just by increasing pure oxygen flow without changing the mechanical mixing speed (Irizar et al. 2009). Because biological flocs are exposed to a constant shear field in this condition regardless of the oxygen supply, floc breakages can hardly affect OUR and a high measuring accuracy can be obtained.

6.5.3.3 Steady State Method

Off-gas analysis is an *in situ* method that provides the most accurate real-time estimation and can be applied for most aeration tanks regardless of the aeration method being used. It estimates the OTE by analyzing the off-gas composition using a floating off-gas collector on the aeration basin surface as shown in Figure 6.23. By comparing the O_2 contents in the off-gas and in the fresh air, OTE is calculated. The details of the method are well described in the literature (Leu et al. 2009).

The off-gas flow rate is slightly lower than the inlet airflow rate in typical activated sludge processes due to O_2 dissolution. Although a portion of biologically produced CO_2 is stripped out from mixed liquor and added to the off-gas, it is less than the amount of the oxygen lost to the mixed liquor. In the steady state method, off-gas flow rate is assumed to be the same as that of fresh airflow, but the error from this assumption is generally not significant due to the low OTE. If OTE is

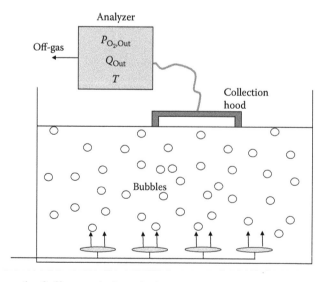

FIGURE 6.23 Schematic of off-gas analysis method.

10%, the maximum error from the airflow rate can be only up to 2.1% (10% of 21% O_2 in dry air), if no CO_2 is stripped out from the mixed liquor. Moisture contents in the air and in the off-gas also play a role, but the error is only up to 2.3% and 4.2% at a water temperature of 20°C and 30°C, respectively. As a result, OTE can be calculated using the simplified equation below assuming Q_{Out} is equivalent to Q_{In} (Redmon et al. 1983). If all the partial oxygen pressures are normalized against 0% humidity, the accuracy of the calculation can improve.

$$\text{OTE} = 1 - \frac{Q_{Out} P_{O_2,Out}}{Q_{In} P_{O_2,In}} \approx 1 - \frac{P_{O_2,Out}}{P_{O_2,In}} \tag{6.33}$$

where
Q_{In} airflow to aeration tank (m³/min at 20°C, 1 atm)
Q_{Out} off-gas flow rate from aeration tank (m³/h at 20°C, 1 atm)
$P_{O_2,In}$ partial O_2 pressure in fresh dry air (bar)
$P_{O_2,Out}$ partial O_2 pressure in dry off-gas (bar)

OTR (kg O_2/h) can be estimated assuming local off-gas flows are uniform across the aeration basin surface such as in Equation 6.34 (Leu et al. 2009). The OTE measurement can be performed at multiple points on the water surface to obtain more accurate average OTE.

$$\text{OTR} = 0.279 \, Q_{Out} \, \text{OTE} \, \frac{A_{Tank}}{A_{Hood}} \tag{6.34}$$

where
A_{Tank} surface area of aeration tank (m²)
A_{Hood} the area covered by hood for off-gas collection (m³)
Q_{Out} off-gas flow rate from the floating hood (m³/h at 20°C, 1 atm)

6.6   OXYGEN BALANCE IN MBR

In a typical MBR, more air is supplied to the membrane tank than to the aeration tank, but the majority of biological degradation (or oxygen consumption) occurs in the aeration tank. In a field survey, it was observed that 78% of compressed air was used for membrane scouring whereas only 22% was used for biological degradation, as shown in Figure 6.24. Nonetheless, the majority of oxygen dissolution took place in the aeration tank, that is, 67%. It was also observed that the OTE was estimated 7.2 times greater in the aeration tank than in the membrane tank. Given the fact that the OTE in the aeration tank with 8 to 12 g/L MLSS is less than 10%, OTE in the membrane tank can be approximately 1%. The low OTE in the membrane tank is attributed to the low driving force at high DO (4–8 ppm) and the coarse bubbles used for membrane scouring.

A substantial amount of excess DO in the membrane tank (4–8 mg/L) is carried over to the aeration tank by internal recycling whereas only a small amount of oxygen is transferred to the membrane tank from the aeration tank (1–2 mg/L). A net oxygen transfer occurs to the aeration tank due to the difference in DO in the two streams. The amount of oxygen transferred from the membrane tank to the aeration tank corresponds to approximately 10% of the total oxygen consumption in the system (Giesen et al. 2008). Overall, 76% of the total DO was consumed in an aeration tank whereas 23% was consumed in a membrane tank. The amount lost through permeate was estimated at 1%. Aeration demand for membrane scouring was 3.5 times greater than that for biological aeration on average, but the ratio significantly dropped during peak flow conditions due to the increase of biological aeration.

It is noteworthy that a significant amount of oxygen consumption occurs in the membrane tank although wastewater is not fed to it. This is partly because of endogenous respiration of

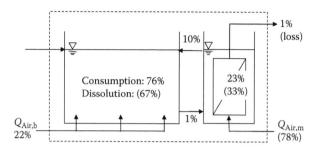

FIGURE 6.24 Oxygen balance of MBR equipped with vertically mounted hollow fiber membranes (GE Water) based on a 1-year study in Varsseveld, The Netherlands. (From Giesen, A. et al., Lessons learnt in facility design, tendering and operation of MBR's for industrial and municipal wastewater treatment. *Proceedings of the Water Institute of South Africa (WISA) Biennial Conference and Exhibition.* May 18–22, 2008. Sun City, South Africa; Courtesy of http://www.royalhaskoningdhv.com.)

microorganisms that requires oxygen in the membrane tank. If SOUR of biosolids in the membrane tank is 3 mg O_2/g MLSS/h and MLSS is 12 g/L, OUR in the membrane tank becomes 36 mg/L/h, which is comparable to or not much less than that in aeration tank. In the meantime, the high DO in the membrane tank can be used for nitrification. A sizable amount of TKN can pass the aeration tank without being oxidized due to the slow nitrification rate depending on DO, HRT, TKN loading, existence of toxic compounds, sludge age, etc. Wastewater channeling in the aeration tank can also contribute to the TKN leak to the membrane tank, but the leaked TKN is rapidly oxidized at the high DO (4–8 mg/L) environment in the membrane tank. As a result, it has been observed in many locations that the TKN concentration in the plant effluent can be consistently low even when varying amount of TKN is carried over to membrane tank (Figure 3.6 in Brepols 2011).

6.7 PEAK FLOW HANDLING

MBR produces high-quality effluent at a wide range of hydraulic and organic loading rates. However, the absolute particle filtration performed by the membrane makes the system prone to fail at peak flow conditions due to the inability to produce partially compromised permeate at high flow. Therefore, rigorous planning is required to prepare for the peak flow in the system design stage. The most desirable solution is having sufficient membrane surface areas to treat the peak flow, but it is economically constrained. Therefore, equalization tanks are used to buffer the short-term diurnal variation of flow. This allows capital cost savings by increasing the system utilization efficiency. The optimum size of the equalization tank can be determined using various methods, but the economics, space availability, the maximum holding time without causing biological problem by the partially degraded wastewater in the subsequent biological processes, etc., are some of the constraints in sizing the equalization tank.

An example of a graphical method to determine the required equalization tank size is shown in Figure 6.25. Cumulative wastewater volume is plotted for a day or two in the season with the highest flow variations. A straight line is drawn from the first to the last data point. The slope of the line indicates the average flow rate in the period for which the equalization tank targets to provide the buffer capacity, for example, 340 m³/h in the figure. Then, two tangent lines parallel to the average flow line are drawn, where one line must be farthest to the left and the other must be farthest to the right from the center line. The vertical distance of the two tangent lines indicates the equalization tank volume required for the MBR treating the average flow rate. A moderate safety factor can be multiplied to the estimated equalization tank volume to be prepared for unprecedented flow fluctuations.

To calculate the minimum membrane area required, daily average fluxes are plotted assuming a few different membrane surface areas as shown in Figure 6.26. In this example, the daily flow rates were assumed to be the same as those in Figure 5.10. Then, the flux curves were drawn for the three different membrane areas, that is, 50,000, 75,000, and 100,000 m². The membrane area required is

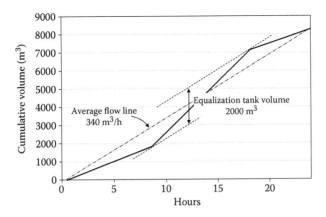

FIGURE 6.25 Graphical method to determine equalization tank volume to hand diurnal flow.

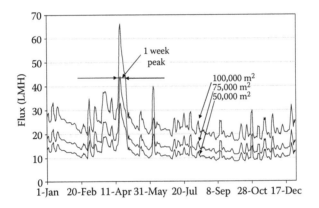

FIGURE 6.26 Determination of membrane surface area required for a municipal WWTP in the northeast United States.

dominated by peak fluxes, which occurred in April in this case. If the daily and weekly peak fluxes allowed are 40 and 30 LMH, respectively, the peak flux for 50,000 m² in April is above the limit. On the contrary, the daily and weekly peak fluxes for 100,000 m² are much lower than the limits. When the membrane surface area is 75,000 m², daily peak flux exceeds 40 LMH slightly for a few days. Therefore, the membrane area can be decided at somewhat larger than 75,000 m².

Additional safety factors can be considered to prepare for the uncertainties in weather, population growth, emergency system maintenance, and so on. However, installing sufficient membrane surface areas for short-term events is often not cost-effective, especially when peaking factor (peak flow divided by average flow) is high. Although there is no absolute solution that applies for every condition, a number of different methods have been devised to save excessive capital costs.

As one option, the CAS process can be built parallel to the MBR. During normal times, incoming wastewater is split between the two processes depending on the need for meeting the effluent quality for discharge or reuse. During peak events, excess wastewater that cannot be treated by MBR can be directed to CAS temporarily. If excessive TSS leaks from the clarifier during the peak flow event, inorganic or organic coagulants can be added temporarily. Having parallel CAS processes with MBR is particularly feasible when only a portion of the wastewater needs to be recycled as a high quality recycled water. In this case, MBR produces feed water for RO whereas CAS produces the effluent to be discharged.

As discussed in Section 5.2.7.2, surging organic loading causes much more significant membrane fouling than surging hydraulic loading during the peak flow event. Therefore, peak flow can be handled better by reducing BOD of the influent during the event. It has been known that at least 20%–30% or up to 50% of BOD can be removed from raw wastewater by removing suspended solids. This method can be readily applied when primary clarifier is used as pretreatment. Coagulants can be added to the injection well of the primary clarification to enhance the removal of BOD particulates. If inorganic coagulants are used, pH should be carefully controlled to not disrupt nitrification. If F/M ratio is controlled at or below the normal level, DO will stay at the normal level and membrane fouling rate can be kept low during the peak event. The airflow for membrane scouring can also be increased to the maximum to delay membrane fouling, as discussed in Section 3.3.2.

Alternatively, membrane performance-enhancing agents such as MPE50® from Ecolab (formerly Nalco) can be added to the membrane tank during the peak flow. As discussed in Section 5.4.3, the transparent yellowish water-soluble polymer with a net positive charge coagulates SMP and reduces membrane fouling rate. The dosage of MPE50® varies depending on mixed liquor property, but the standard dosage is known to be 100 mg/L based on reactor volume for every 3 g/L of MLSS. The chemicals lost through excess sludge removal need to be compensated for by adding additional amounts to maintain efficacy. Inorganic coagulants such as aluminum and ferric salts also can be used, but their efficacies are not as high as MPE50® and careful pH monitoring is required to not disrupt the biological nitrification process.

6.8 AERATION OPTIMIZATION

In activated sludge process, maintaining DO at a proper level, for example, 1–2 mg/L or higher, is crucial for successful operation. To minimize the energy costs while achieving the target DO, airflow rates can be controlled depending on the oxygen demand. A few options exist in controlling aeration rate, for example, variable frequency drive (VFD), on/off control of multiple small blowers as needed, or throttling the inlet side of the blowers. Fine bubble diffusers are often preferred despite the higher maintenance requirements. Saving aeration energy is more critical in MBR than in CAS because MBR requires two to five times stronger aeration than CAS in general mainly due to the aeration needed for membrane scouring. In fact, membrane scouring takes the majority of the total airflow in most municipal MBR, as discussed in Section 6.6. Therefore, the scouring air minimization is a critical part of the overall energy optimization.

Airflow is adjusted stepwise according to the influent flow rate. In practice, scouring airflow rates are adjusted in two to four steps using multiple blowers depending on the hydraulic flow rate MBR should treat. For example, if four blowers are available, airflow can be changed from 100% to 150%, to 200% by using two, three, or four blowers, respectively, depending on the feed flow rate. If there is only a very low or no hydraulic loading, air scouring can be performed either intermittently to suspend the mixed liquor or performed by using one blower. Controlling the scouring airflow rate linearly proportional to the wastewater flow rate is not a preferred strategy except for small plants with one or two blowers in total. This is because such linear airflow control requires costly blowers equipped with variable speed drives (VFD) and the efficiencies of the blower and the VFD are not high under varying flow rates. In addition, there are no strong evidences that such linearly varying airflow rate is superior to the stepwise airflow rate in terms of membrane scouring effect.

In one example presented by Ovivo Water, the daily flow curve can be segmented to no-flow, low-flow (<0.5 Q), medium-flow (0.5–1.5 Q), and high-flow (2 Q) periods and different aeration rates are applied to minimize the energy consumption (Codianne 2012). As shown in Figure 6.27, the diurnal flow curve is split into six different periods considering the capacity of the equalization tank. At low-, medium-, and high-flow periods, 3.5, 5.25, and 7 L/m²/min of scouring air is supplied per membrane panel, respectively. At no-flow period, only intermittent aeration is performed in aeration and membrane tanks to prevent sludge from settling and from being septic.

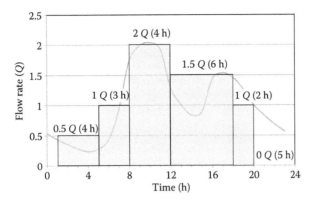

FIGURE 6.27 Daily flow rate pattern, where Q indicates maximum monthly average flow.

In a plant with three membrane tanks, the overall plant operation strategy is decided using PC-based software, for example, EnergyPro™. At no-flow condition, aeration is performed only occasionally in the membrane tanks to prevent sludge settling. At low-flow conditions (<0.5 Q), two of the three membrane tanks are used while the standby tank is aerated intermittently. If the net flow rate to each train is lower than 0.5 Q, airflow rate to each membrane panel becomes 3.5 LPM. Otherwise, it can be 5.25 LPM per membrane panel at medium flow (0.5–1.5 Q). It is automatically decided whether using two trains at high airflow would be better or if using three trains at medium airflow would be better in terms of energy consumption. At peak flow at 2.0 Q, all trains are used at high airflow rate (7.0 LPM per cartridge).

6.9 PLANNING HYDRAULIC GRADE LINE

Optimizing the hydraulic grade line is one important step to minimize the capital and operating costs of MBR. The total hydraulic head required must be calculated first to decide the elevation of the first biological tank. Flow distribution among aeration and membrane tanks, flow rates at each point during peak-, average-, and low-flow conditions, and the elevation required for gravity suction in some cases (Section 3.2.1.2) need to be considered (Qasim 1999). The natural grade line and the effect of underlying plant geology on structural costs also need to be considered.

Design decision is a tradeoff involving structural costs, pumping costs, piping costs, consideration for future expansion, footprint restrictions, etc. (WEF 2012). There might be structural cost-saving opportunities by compromising pumping costs depending on underlying geology or piping and pumping costs can be saved by compromising structural costs. Figure 6.28 shows one example of grade line profile for an MBR based on A/O process. The elevation of the equalization tank can

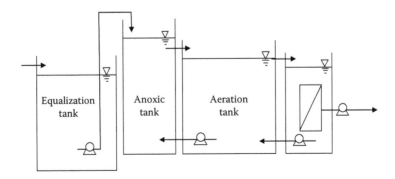

FIGURE 6.28 Hydraulic gradeline of an MBR plant based on MLE process.

be decided considering the costs of ground work depending on geology, pumping costs to anoxic tank at an average water level in the tank, storm water level, etc. The levels of anoxic, aerobic, and membrane tanks are decided considering head loss at the flow rates among tanks, the head loss caused by flow distribution, etc. By switching the elevations of aeration and membrane tanks with each other, energy costs can be saved further due to the smaller pumping costs by 1 Q as shown in Figure 2.11c.

6.10   MEMBRANE LIFE SPAN

6.10.1   FACTORS AFFECTING MEMBRANE LIFE

Exposure of membranes to free chlorine during periodic membrane cleaning makes the membranes prone to breakage by being oxidized. Membrane chemistry and manufacturing methods are the two major factors affecting chlorine tolerance. Polyvinyl difluorides (PVDF) and blends of polyethersulfone (PES) and polyvinylpyrrolidone (PVP) are known to be highly chlorine tolerant, for example, 500,000 ppm·h for PVDF (Fenu et al. 2012) and 250,000 ppm·h for PES/PVP blends (Pilutti and Nemeth 2003). There is a perception that polyethylene (PE) membranes are vulnerable to free chlorine, but this misperception perhaps started when the first PE-based hollow fiber membranes commercialized as submerged membranes in the 1990s had a low chlorine tolerance. In fact, PE itself is highly tolerant to chlorine due to the lack of functional groups that can solicit a chemical attack on the polymeric backbone. For example, 5% to 10% household bleaches are sold in PE bottles. One membrane manufacturer claims its hollow fiber membranes have a chlorine tolerance of 1,000,000 ppm·h, which is higher than that of PVDF membranes (Econity Inc.).

As shown above, chlorine tolerance of membranes is reported as a time concentration by multiplying chlorine concentration by exposure time (e.g., ppm·h), which is also called CT value. The underlying assumption is that exposure time and chlorine concentration are equally detrimental to membranes or that the two parameters are linearly influential. However, it has been found that exposure time has a much more significant effect on membrane degradation than chlorine concentration (Abdullah and Bérubé 2013). In one study, two membrane coupons were exposed to free chlorine solutions with different strengths, that is, 3600 and 44,300 mg/L. By controlling the exposure time at 556 and 45 h, respectively, the two membrane coupons were exposed to chlorine at identical time concentrations of 2,000,000 ppm·h. Interestingly, the membrane coupon exposed to 44,300 mg/L for 45 h (or 2,000,000 ppm·h) was less damaged than the other membrane coupon exposed to 3600 ppm NaOCl for 278 h (or 1,000,000 ppm·h). It was apparent that the exposure time had a more significant effect on membrane degradation than the chlorine concentration. Therefore, the popular chlorine exposure scale, that is, ppm·h, can be better expressed as ppm·h^a, where $a > 1$.

All the factors affecting membrane fouling including HRT, SRT, toxic or harmful components in raw wastewater, etc., indirectly affect membrane lifespan by affecting the membrane cleaning frequency that again determines the chlorine exposure of membrane. Prescreening of wastewater is another factor affecting membrane lifespan. The solids and fibrous materials originating from fabrics, hairs, papers, etc., can clog membranes and diffusers, and thereby induce membrane fouling. Hard debris such as leaves, plastics, small branches, and packing materials can cause a persistent abrasion on the membrane surface and can lead to a failure of membrane integrity in the long run.

Scouring aeration must cover the entire footprint evenly to prevent air channeling to small areas. If airflow is excessive to a portion of membrane panels, they can be bent and the gaps among membrane panels become irregular. On the contrary, in areas without sufficient aeration, upflow channels (or gaps) can be filled up with sludge cake that pushes the membrane panels away from each other, thereby membrane panels can be bent. Meanwhile, flat sheet membrane panels must be tightly held in a cassette to prevent them from rubbing against the cassette frame during turbulence. As shown in Figure 6.29, the top and bottom corners of the membrane panels have been damaged as a result of abrasion with the rubber holders in the cassette during its 10 years of service.

FIGURE 6.29 Abrasion seen at the top and bottom corners of membrane panels (left: top corner, right: bottom corner, scale in mm). (From Nishimori, K. et al., *Water Sci Technol* 62(3):518–524, 2010.)

In theory, the organic solvents contained in industrial wastewater can damage membrane modules especially when the solvent dissolves any of the membrane and module components. However, the solvent damage is unlikely because the effective solvent concentration in steady state is not high enough to damage membrane modules in MBR operated under reasonable conditions. This is because solvent concentrations in mixed liquor are negligible at steady state for the most commonly used solvents in industry, for example, benzene, acetone, hexane, DMSO, short-chain alcohols, and ethers due to their high biodegradabilities. Even the solvents with more complex molecular structures are also readily biodegradable, for example, methylethyl ketone (MEK), dimethyl amine (DMA), dimethyl formamide (DMF), *p*-nitrophenol (PNP), *o*-chlorophenol (OCP), trichlorophenol (TCP), dichlorodiethyl ether (DCDEE), and fluorescent whitening agents (FWA) based on stilbene-cyanuric type (Dojlido 1979). MBR has an extra advantage over CAS regarding the removal of slowly degrading organics due to the high SRT (12–30 days), which allows for the enrichment of microorganisms specific for difficult solvent molecules. In a field survey of an electronic parts manufacturing plant, DMSO and MEK contents in the influent was approximately 150 and 65 mg/L, respectively, on average, but both solvents were found to be below the detection limit at less than 0.1 mg/L each in the MBR effluent. Because the membrane used in MBR does not reject solvents, those found in the effluent are deemed as steady state concentrations in the mixed liquor.

6.10.2 Manufacturer's Estimation

Major membrane companies provide conditional guarantees on membrane life, typically 4 to 7 years. However, many evidences found recently suggest that the lifespan of commercial membranes from reputable manufacturers are close to 10 years or longer, if the membranes are properly installed and used under reasonable operating conditions. For flat sheet membranes, only 7.8% of the original panels had been replaced with new ones in the first 10 years of operation in one of the first MBR plants built in 1999 by Kubota in Japan. In another plant built in 1998 in Porlock, United Kingdom, only 6.4% of the original panels had been replaced as of 2008. The average replacement ratio in the all plants surveyed was 6.4% within the first 5 years in Europe (Nishimori et al. 2010). In an industrial MBR treating saline food wastewater with high BOD (1100 mg/L), membrane failure rate was reported at 1.25% after 7 years of operation, where membranes were used at low flux at 4 to 12 LMH (Lala et al. 2014).

Hollow fiber membranes from GE also have a similar lifespan (Côté et al. 2012). In Figure 6.30, the three cumulative surface areas are plotted for installed membrane, replaced membrane due to the end of service life, and replaced membrane due to premature failure covered by warranty. In addition, hypothetical membrane replacements were also plotted assuming 5-year and 10-year life spans. The 5-year and 10-year curves move behind the total installed area curve with 5- and 10-year lags, respectively. In the graph, it is not clear how many years of service life were assumed normal for the membranes replaced due to the end of service life. It is also not clear how premature failure was defined because there is no clear-cut classification in judging premature failure as will

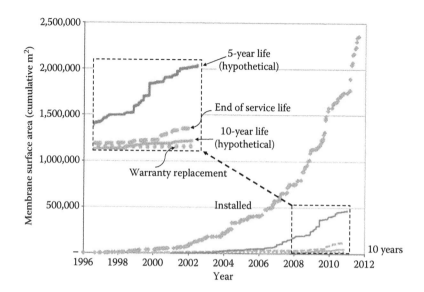

FIGURE 6.30 Cumulative membrane installations, failure replacement, warranty replacement, and hypothetical lives in North America for all generations of ZeeWeed® membranes. (From Côté, P. et al., *Desalination* 288(1):145–151, 2012.)

be discussed in the next sections. Nonetheless, it is reasonable to consider the average membrane life is close to 10 years because the sum of the cumulative areas for "warranty replacement" and "end of service life" are still close to the 10-year curve. Considering the improving durability of membrane materials and installation/operating technologies, membrane service life is indeed considered to be approximately 10 years for new membranes manufactured by top-tier suppliers in municipal MBR.

6.10.3 THIRD PARTIES' ESTIMATION

For membrane manufacturers, it is easy to draw a line between live and dead membranes because it can be judged by the membrane replacement. However, there is no clear border between live and dead membranes for membrane users. From a practical point of view, membrane life ends when it cannot treat peak flow due to decreasing permeability over time. However, membrane life can be extended by performing more frequent recovery cleanings with chlorine or organic acids. More attention on equalization tank operation and effluent production schedule also help with membrane life extension. Oversized membrane capacities in the design stage often enable partially compromised membranes to treat the flow. Membrane performance can be improved by removing extra COD/TSS in the pretreatment using coagulant. In some cases, nearby wastewater plants may take excess wastewater that cannot be handled by MBR. Because the decision on the end of membrane life is tied up with the extra resources available in most cases, determining membrane life is rather arbitrary unless the membrane integrity is severely compromised.

Figure 6.31 shows the breakdown of the causes of membrane replacement. It is noticeable that the major cause is damage occurring during the installation process by sparks from the welding equipment. The next common cause is gradual membrane seal failure due to turbulence in the membrane tank.

In a study performed in Schilde, Belgium, where GE ZW500c® membranes were employed, MBR failed to produce the design flow of 220 m³/day in winter (Fenu et al. 2012). However, the MBR treated almost the design flow in the next winter with more frequent recovery cleaning. In this case, if membrane life was judged based on the capability of producing the design flow rate,

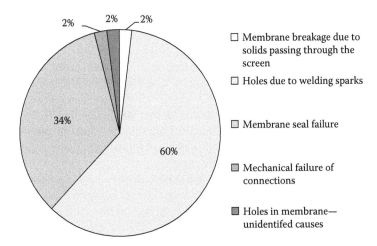

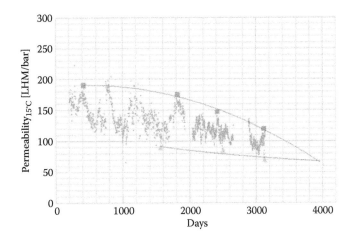

FIGURE 6.31 The causes of membrane replacement. (Data from Ayala, D.F. et al., *J Membr Sci* 378:95–100, 2011.)

the membranes could be called failed at around 4.5 years (or 1650 days) and premature membrane replacements could occur. In fact, it is difficult to judge membrane failure because the low flux event could simply be caused by a temporary deterioration of mixed liquor quality. Therefore, it is hard to conclude that membranes need replacement even if membranes fail to produce design flow.

One way of predicting the membrane lifespan is by tracking the membrane permeability and extrapolating it (Fenu et al. 2012). In a study performed in Schilde, Belgium, membrane permeability was monitored and recorded ever since the MBR was commissioned. Membrane permeability fluctuated following a sine curve, where permeability increased in the summer and decreased in the winter over a period of 8 to 9 years. As shown in Figure 6.32, it is apparent that the amplitude of the permeability curve declines over time whereas absolute permeability was also trending down. This means that recovery cleaning becomes less effective due to the accumulation of irreversible membrane fouling and membrane weathering. If the trends of the highest and the lowest points in each year are extrapolated, those two lines meet at around 4000 h or 11 years. This suggests that recovery cleaning will not be effective at the 11th year and membranes should be replaced.

FIGURE 6.32 Determination of the timing of membrane replacement. (Fenu, A. et al., *J Membr Sci* 421–422:349–354, 2012.)

Membrane lifespan can also be predicted based on cumulative chlorine exposure (Fenu et al. 2012). In the Schilde MBR, maintenance cleaning was performed weekly while recovery cleaning was performed whenever TMP reaches 0.4 to 0.5 bar. If the cumulative chlorine exposure is plotted, the time required to reach the maximum chlorine tolerance of ZW500 membranes, that is, 500,000 ppm·h, can be estimated by extrapolating the curve (graph not shown). Chlorine concentration in the cleaning solution was at 600 mg/L as NaOCl. In the first 10 years, cumulative chlorine exposure increased exponentially due to the increasing needs for membrane cleaning. If the curve is extrapolated, cumulative chlorine exposure reaches the maximum limit of 500,000 ppm·h in the 15th year of the commissioning, but it seemed overly optimistic compared with the result obtained from the permeability data as mentioned in the previous paragraph.

As discussed previously, there are no decisive criteria that could trigger the end of the membrane's life. However, the capability of obtaining design flux at below the maximum TMP with the help of recovery chemical cleanings at a reasonable frequency can be used as a criterion to judge the membrane life. In Schilde MBR, membrane life estimation of the least performing train out of four trains was 8.1 to 10.0 years using this criterion. Overall, the estimated membrane life spans based on a few different criteria suggests that membrane life is close to 10 years or longer. This agrees well with the manufacturer's estimation discussed in the previous section.

6.11 TRACE ORGANICS REMOVAL

6.11.1 Removal Mechanism

In recent decades, concerns over trace organics such as pesticides, pharmaceuticals, personal care products, and steroidal hormones have been growing. Many of those trace organics are suspected of being endocrine disruptors in aquatic organisms and eventually in humans (Yoon et al. 2007). Removal mechanisms by activated sludge vary depending on the nature of the trace organics, but biodegradation, adsorption to biosolids and filtration medium, evaporation, etc., have been identified as major removal mechanisms.

Many trace organics are not biodegradable at all under the conditions provided by activated sludge. However, for slowly biodegradable trace organics, biodegradation rate is mainly dependent on the level of enrichment of the naturally occurring microorganisms that can degrade them, thereby having long SRT is beneficial for high treatment efficiencies. MBR can be superior to CAS with respect to removing slowly biodegradable trace organics because it runs at substantially longer SRT. For instance, in a side-by-side test with MBR and CAS treating municipal wastewater, the removal efficiency of phenazone, propyphenazone, and formylaminoantipyrine increased from 10%–35% to 60%–70% over a 5-month period as a consequence of microbial adaptation in a pilot MBR test (Zuehlke et al. 2007). The removal efficiency of the same trace organics by CAS remained consistently low at 15% to 40%. Meanwhile, the removal efficiencies of nonbiodegradable compounds can be similar in both MBR and CAS. For example, in the same study, the removal efficiencies of estrone and estradiol were about the same in both MBR and CAS, and this suggested that biodegradation is not the main removal mechanism for such compounds.

Adsorption on biosolids and other suspended solids can be an important removal mechanism for nonbiodegradable compounds. If multiple different trace organic species exist, they compete with each other for the limited number of adsorption sites on biosolids. If some of them are biodegradable, the adsorption equilibrium becomes more complicated. Therefore, various removal efficiencies can be observed for the same species depending on the existence of competing organics, the biodegradation rate of each component, and the contact time (or HRT). For example, nearly 100% of bisphenol A was removed in one experiment (Trinh et al. 2012), but only approximately 90% of it was removed in another experiment (Nghiem et al. 2009), although the exact causes of the difference are not traceable.

The compounds with low polarity (or high hydrophobicity) are easier to be adsorbed on biosolids than those with high polarity. Because biosolids are more hydrophobic relative to water molecules, nonpolar molecules undergo stronger interaction with biosolids compared with water. On the other hand, the molecules with a high polarity tend to be better stabilized in the water phase due to the dipole–dipole or dipole–charge interactions. For a given compound, removal efficiency tends to decline as the concentration in wastewater increases due to limited adsorption capacity. High SRT is not advantageous to obtain high removal efficiency for the nonbiodegradable trace organics because the amount of excess sludge production decreases with SRT. As a consequence, if compounds are not biodegradable at all, removal efficiency of MBR can be lower than that of CAS. Because compounds with high affinity are preferentially adsorbed, the removal efficiencies of low-affinity compounds can vary widely depending on the existence of the high-affinity compounds. Due to the complexities and various factors playing a role in trace organics removal, the removal efficiency observed in one site can hardly be applied for other sites, where the wastewater composition as well as the operating conditions are different.

Adsorption of trace organics on the membrane also plays a role in membrane process. In one study, a hybrid system of powdered activated carbon and subsequent ultrafiltration (PAC-UF) was used to remove o-dichlorobenzene (DCB) from water (Kim et al. 1996). It was observed that a significant amount of DCB was adsorbed on membranes under the competitive adsorption environment between PAC and the membrane. As a result, DCB concentration in the UF permeate was much lower than the mathematically predicted value assuming only the adsorption on PAC in the short-term experiment. Although the adsorption capacity of membrane acts as a buffer when DCB concentration fluctuates, the effect of this mechanism would not be significant in long-term operation due to the limited adsorption capacity of the membrane. In addition, the desorbed organics from the membrane during membrane cleaning must be sent back to the equalization tank and thus the net adsorption becomes zero. In theory, the trace organics with a high volatility can be removed by being evaporated from the water surface. However, most of the trace organics are heavier than 100 Da with functional groups and varying degrees of polarity essential for high boiling points. Due to the high boiling point and low vapor pressure of the surrounding temperature, it is very unlikely that evaporation contributes to removal efficiency in a meaningful way.

6.11.2 Factors Affecting Removal Efficiency

6.11.2.1 Effect of Hydrophobicity of Compound

In general, the removal efficiencies of nonbiodegradable pharmaceutical compounds are positively correlated with the hydrophobicity of the compounds. Fenoprop, ketoprofen, naproxen, diclofenac, carbamazepine, pentachlorophenol, 4-tert-butylphenol, etc., fall into this category. In a laboratory study, these and other pharmaceutical compounds were added to MBR at 5 mg/L each and the MBR was operated without sludge removal for 6 weeks. The removal efficiencies were plotted against the octanol-water distribution coefficients (K_{OW}) of the compounds. It was found that the removal efficiency was positively correlated with K_{OW} as shown in Figure 6.33. The removal efficiencies of biodegradable compounds were out of the trend line especially for the compounds with low K_{OW}. The untreated compounds were treated at 95%–100% efficiencies by passing the MBR effluent through a granular activated carbon (GAC).

The hydrophobicity of trace organics can be affected by pH if amine or carboxyl groups are present in the molecule. When four compounds with variable hydrophobicity (sulfamethoxazole, diclofenac, ibuprofen, and ketoprofen) depending on pH and two compounds with fixed hydrophobicity (bisphenol A and carbamazepine) were treated by MBR, the removal efficiencies of the first four compounds were affected by pH, but the last two compounds were not. Ibuprofen removal efficiency was particularly sensitive to pH. It was removed by almost 100% at pH 5, but nearly none of it was removed at pH 9 due to the ionization of the carboxyl group in the molecule (Tadkaew et al. 2010).

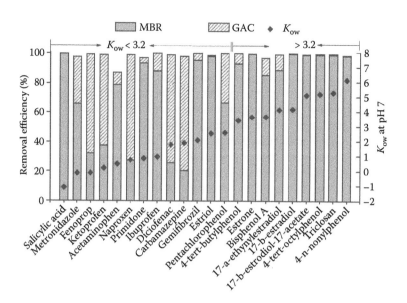

FIGURE 6.33 Removal efficiency of trace organics by MBR and GAC, where DO and pH were at 2 to 4 mg/L and 7.2 to 7.5, respectively, and empty bed contact time (EBCT) in GAC column was 7 min. (From Nguyen, L.N. et al., *Bioresource Technol* 113:169–173, 2012.)

In another study, the removal efficiency of trace organics was studied using a MBR with anoxic and aerobic tanks (Boonyaroj et al. 2012). Leachate with about 9400 ppm COD was fed for 200 days after 100 days of the acclimation period. HRT in aerobic tank was maintained at 1 day. MLSS stayed between 10 and 20 g/L without excess sludge removal. Trace organic removal efficiency ranged between 50% and 76% depending on the compounds, that is, napthalene (68%), anthracene (70%), 4-methyl-2,6-di-tert-butylphenol (74%), bisphenol A (67%), dimethylphthalate (50%), diethylphthalate (51%), di-*n*-butylphthalate (68%), benzyl butyl phthalate (72%), bis(2-ethylhexyl) phthalate (75%), and di-*n*-octyl phthalate (76%). It was observed that the removal efficiency tended to be greater for the compounds with large octanol-water partition coefficients (K_{ow}).

6.11.2.2 Effect of Temperature

The removal efficiency of some trace organics can vary depending on water temperature. In a study performed with two pilot MBR based on A/O process in a municipal WWTP, high removal efficiencies of phenazone, propyphenazone, and formylaminoantipyrine were observed in summer, but they decreased noticeably in winter. Figure 6.34 shows that the removal efficiencies of acetylaminoantipyrine (AAA) by CAS process (KW) and two pilot MBR (PP1 and PP2) are fairly proportional to water temperature. However, this observation should not be generalized for all trace organic species because the extent and the trend of temperature effect are not consistent for all trace organics.

6.11.2.3 Effect of SRT

The removal efficiency of certain pharmaceutical compounds improves as SRT increases because of microbial adaptation. According to a long-term laboratory study using synthetic feed (Maeng et al. 2013), the removal efficiency of gemfibrozil, ketoprofen, and clofibric acid increased substantially as SRT increased from 8 days to 20 days to 80 days. The rising removal efficiency with SRT suggested that those compounds were slowly biodegradable. Meanwhile, bezafibrate (Bezalip®), ibuprofen (Advil®), fenoprofen (Fenopron®), phenacetine (Vicks®), acetaminophen (Tylenol®), phentoxifyline (Trental®), and caffeine were removed at nearly 90% regardless of the SRT, which suggested that these compounds are readily biodegradable or easily adsorbed on biosolids. On the contrary, the removal efficiencies of diclofenac (Voltaren®), naproxen (Aleve®), and carbamazepine

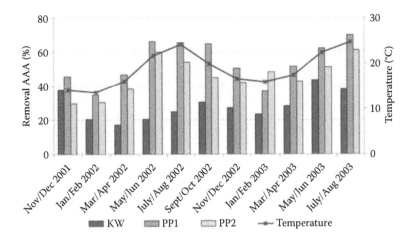

FIGURE 6.34 Removal efficiencies of AAA (acetylaminoantipyrine) by CAS process (KW) and two pilot MBR (PP1 and PP2) with anoxic tanks treating municipal wastewater. (From Zuehike, S. et al., *Water Environ Res* 78(13):2480–2486, 2007.)

(Tegretol®) were low at 5% to 30% at all SRT, perhaps due to the low biodegradability in the condition provided by MBR.

To estimate removal efficiency reliably, MBR must run for at least two to three times of the SRT until steady state is reached. For example, if SRT was 20 days, MBR must run at least 40 to 60 days from the perspective of removal mechanisms by adsorption. If the experimental duration is shorter than that, the removal efficiency can be overestimated especially for nonbiodegradable compounds. Nonetheless, the complex interactions among biological adaptation, biodegradability, adsorption, SRT, temperature, and other environmental factors make the data interpretation very hard, especially in field conditions. Due to the complex removal mechanism and the inability to collect and comprehend all the influencing parameters, the removal efficiency obtained from one plant could be very different in other plants running under similar conditions.

6.11.3 REMOVAL EFFICIENCY

As the use of chemicals proliferates for pharmaceutical, personal care products, agriculture, manufacturing, etc., any municipal wastewater stream carries many of those chemicals. The release of artificial chemicals can potentially harm the environment by affecting the metabolism of aquatic organisms and eventually humans. Generally, steroidal hormones such as androstenedione, androsterone, etiocholanolone, dihydrotestosterone, testosterone, 17β-estradiol, estriol, and estrone are biodegradable, and removal efficiencies are 90% or higher. However, most pesticides and herbicides are not biodegradable and are removed mainly by adsorption mechanisms. In addition, even for the same compound, removal efficiency can vary widely depending on the existence of competing trace organics, biological conditions employed such as MLSS, SRT, and temperature, unless the compound is readily biodegradable as discussed in the previous sections. The common ranges of removal efficiencies are summarized in Table 6.8 for the commonly found trace organics, but the actual removal efficiency in a specific environment can be outside of the range.

6.11.4 COMPARISON WITH CAS PROCESS

MBR are different from CAS with respect to the trace organic removal characteristics. This is mainly due to its longer SRT (or sludge age) and lower intrinsic excess sludge production. Although long SRT allows the enrichment of microorganisms specific for slowly degradable trace organics,

TABLE 6.8

Removal Efficiencies of the Trace Organic Contaminants by MBR

Removal Efficiency	0%	10%	20%	30%	40%	50%	60%	70%	80%	90%	100%
17β-estradiol											x
4-Tolyltriazole			x								
5-Tolyltriazole										x	
Acetaminophen										x	x
Amitriptyline											x
Androstenedione											x
Androsterone											x
Atenolol							x	x			
Atorvastatin										x	x
Atrazine	x	x	x	x	x						
Benzotriazone			x								
Bezafibrate											x
Bisphenol A									x	x	x
Caffeine										x	x
Carbamazepine	x	x	x								
Chlozapine										x	
Cimetidine									x		
Clarithromycin											x
Clofibric acid	x	x	x	x							
Codeine											x
DEET									x	x	x
Diazepam			x	x	x	x					
Dichlofenac		x	x	x	x	x	x	x	x	x	
Dihydrotestosterone											x
Dilantin	x	x									
Enalapril											x
Erythromycin									x		
Estriol											x
Estrone											x
Etiocholanolone											x
Fenoprofen										x	
Fluoxetine			x	x	x						
Gemfibrozil						x	x	x	x	x	
Hydroxyzine											x
Ibuprofen											x
2-Hydroxy-ibuprofen											x
Ketoprofen						x	x	x	x	x	x
Linuron			x	x							
Meprobamate				x							
Metformain										x	x
Naproxen					x	x	x	x	x	x	x
Nonyphenol											x
o- and p-Hydroxy atrovastatin											x
Omeprazol						x	x	x	x		
Oxycodone	x										
Paracetamol											x
Pentoxifylline										x	x

(Continued)

TABLE 6.8 (CONTINUED)
Removal Efficiencies of the Trace Organic Contaminants by MBR

Removal Efficiency	0%	10%	20%	30%	40%	50%	60%	70%	80%	90%	100%
Phenacetine										X	X
Primidone			X								
Propylparaben											X
Ranitidine								X			
Risperidone											X
Roxithromycin		X	X	X							
Sim-hydroxy acid								X	X	X	X
Simvastatin											X
Sulfamethazine										X	X
Sulfamethoxazole						X	X	X	X	X	
Testosterone											X
t-octylphenol							X	X	X	X	X
Triamterene				X							
Triclocarban										X	X
Triclosan								X	X	X	X
Trimethoprim								X	X	X	X
Verapamil										X	

Source: Data from Sahar, E. et al., *Water Sci Technol* 63(4):733–744, 2011; Trinh, T. et al., *Water Sci Technol* 66(9):1856–1863, 2012; Tadkaew, N. et al., *Water Res* 45(8):2439–2451, 2011; Choi, B.G. et al., *Desalination* 309:74–83, 2013; Kim, M. et al., *Water Sci Technol* 69(11):2221–2229, 2014.

lower sludge production can partially hamper the removal of nonbiodegradable compounds through adsorption mechanism. As a result, depending on the main removal mechanism, the removal efficiency of a specific trace organic can be greater either in MBR or in CAS.

In one study, it was observed that MBR tends to remove hydrophilic compounds better than CAS, for example, trimethoprim and sulfamethoxazole of which K_{ow} is less than 1.0 (Sahar et al. 2011). Because these compounds are not adsorbed in the biomass as much as hydrophobic compounds, they rely more on biodegradation to be removed. MBR provides a better degradation environment by providing a longer residence time for microorganisms to adapt to the compounds. On the contrary, some of the hydrophobic compounds such as clarithromycin (CLA), roxithromycin (ROX), and erythromycin (ERY) were removed better by CAS because those hydrophobic organics rely more on adsorption mechanisms to be removed with excess biosolids removal.

6.12 OILY WASTEWATER TREATMENT

6.12.1 PRETREATMENT

The success of MBR for oily wastewater treatment largely hinges on the removal of nonemulsified free oil drops before the wastewater reaches the MBR. This is just like removing debris and fibrous materials from the influent is crucial for the success of municipal MBR. The refractory oil particles must be removed thoroughly before they reach the membrane tank. Once the free oil drops leak into the membrane tank, they tend to attach to the membrane surface through hydrophobic–hydrophobic interactions because membranes are more hydrophobic than the water and the biosolids. The attached oil drops can cause not only immediate membrane permeability loss but also softened membrane materials due to the solvent effect depending on the membrane chemistry. As a consequence, membranes become more vulnerable to physical damage.

American Petroleum Institute (API) separators are commonly used to remove oil and solid particles from the water (Figure 6.35). Its design is very similar to the rectangular clarifiers used for municipal wastewater, but the design and operation of the skimmer and scraper are optimized for treating oily wastewater. API separators are designed to capture oil globules larger than 0.15 mm. In theory, if there is no turbulence and short-circuiting in the separator, the oil globules with a rising velocity greater than overflow velocity will be captured by the skimmer. The overflow velocity is calculated by dividing the influent flow rate by the surface area of the separator. API recommends having a minimum of two parallel channels so that one can be in-line while the other is out of service.

Corrugated plate interceptor (CPI) contains corrugated plates that are 19 to 51 mm apart at an angle of 45° from the vertical (Figure 6.36). Free oil globules need to rise only such distance to be captured by the slanted plate and rise along it. Heavy particles are also captured by the other side of the plates and fall down. As a result, the space efficiency of the separate can be greatly improved compared to API separator. The laminar flow condition must be maintained in the channels among plates in order not to disturb the separation process.

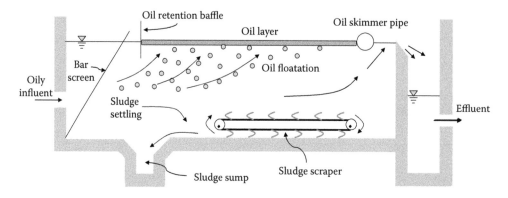

FIGURE 6.35   Side views of API separator.

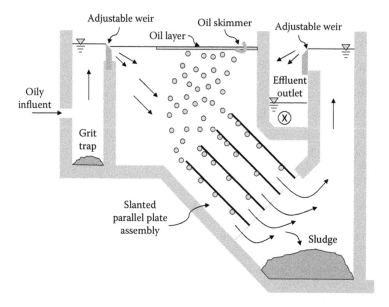

FIGURE 6.36   Corrugated plate interceptor.

Parallel plate interceptor (PPI) is similar to a plate clarifier and not extensively used in petrochemical industry because CPI are more efficient in coalescing free oils. Circular separators are also used for oily wastewater and are similar to the circular clarifier used in municipal wastewater treatment in terms of the underlying design principle.

6.12.2 Produced Water

6.12.2.1 Characteristics of Produced Water

Produced water is the oily water produced as a byproduct when oil and gas are produced. It is formed by the water that exists below the gas and oil layers in the well and the injected water with artificial chemicals. Therefore, produced water contains not only natural hydrocarbons but also various artificial chemicals added to the well to enhance the extraction process and to protect the equipment. It includes some or all combinations of gas hydrate inhibitor, corrosion inhibitor, demulsifier, antifoam, biocide, wax inhibitor, scale inhibitor, etc. Although it is highly variable depending on the source and the maturity of the gas/oil well, produced water often contains high levels of TSS, salts, COD/TOC, heavy metals, etc. Produced water typically appears as a murky brown or gray water with high turbidity. It potentially has a floating oil layer, if excess oil is contained without being emulsified. Nitrogen and phosphorous contents may not be sufficient for activated sludge process depending on the water source. Table 6.9 shows examples of produced water quality. When biological treatment is adopted, a few weeks to a few months of acclimation time may be required to obtain a steady state performance because microorganisms need extra time to adapt to the high total petroleum hydrocarbons (TPH) and high salt concentrations.

One complexity in treating produced water is that the water quality and quantity change as the well matures. In the early phase of production, not as much produced water is produced as oil, but the volume of produced water increases as oil and gas layer is depleted in the later phase of the well's lifecycle. The water-to-oil ratio is 3:1 on average over the lifetime of a typical well, but it can even be 50:1 in the end. Simultaneously, more water with chemicals is injected into the well as the pressure in the well decreases and hence produced water quality changes (Igunnu and Chen 2012).

TABLE 6.9
Quality of Produced Water from Gas Fields and Oil Fields

Parameter	Natural Gas Produced Water	Oil Field Produced Water
Oil/grease (mg/L)	40	260
pH	4.4–7.0	4.3–10
TSS (mg/L)	5500	1000
TDS (mg/L)	360,000	6554
TOC (mg/L)	37–38,000	1500
COD (mg/L)	120,000	1220
Density (kg/m³)	1020	1140
Arsenic (mg/L)	0.005–151	0.005–0.3
Lead (mg/L)	0.2–10.2	0.008–8.8
Chromium (mg/L)	0.03	0.02–1.1
Mercury (mg/L)	–	0.001–0.002
Oil droplet size (μm)	2–30	2–30

Source: Redrawn from Fakhru'l-Razi, A. et al., *J Hazard Mater* 170(2–3):530–551, 2009.

6.12.2.2　Optimum Design Parameters

No universally applicable MBR design parameters exist for oily water treatment due to the lack of sufficient field data as well as the diverse water qualities. However, biological wastewater treatment systems have been designed following the same guidelines for municipal MBR including F/M and F/V ratio, MLSS, anoxic tank design, etc. In one study, COD and oil and grease removal from produced water was 95.2% to 98.6% whereas F/V ratio ranged from 0.9 to 2.6 kg COD/m³/day, which was close to the typical F/V ratio found in municipal MBR, as shown in Table 6.2 (Sharghi and Bonakdarpour 2013). Meanwhile, flux tends to be somewhat lower than those for municipal MBR. Sustainable flux seems to be 10 to 20 LMH instead of 20 to 30 LMH for municipal MBR due to the dispersed floc particles at high salt and oil and grease contents. More research is required to establish firm guidelines for MBR design.

6.12.2.3　Treatment Efficiency

The removal efficiencies of COD, oil and grease, and hydrocarbons are high in general (Kwon et al. 2008; Zilverentant et al. 2009; Kose et al. 2012; Sharghi and Bonakdarpour 2013). As shown in Table 6.10, TPH removal is nearly complete including monoaromatic hydrocarbons, benzene, toluene, ethylbenzene, xylene, aliphatic hydrocarbons with less than 40 carbons, aliphatic acids, aromatic acids, naphthenic acids with less than 9 carbons, polycyclic aromatic hydrocarbons (PAH), and fatty acids. Most of those biodegradable hydrocarbons can be reduced to less than 1 µg/L (or ppb) by using a series of aeration tanks with an extended residence time.

Figure 6.37 shows the COD removal efficiency of laboratory-scale MBR treating produced water from TPAO basin of Turkey (Kose et al. 2012). In this laboratory-scale test, effluent COD is consistently 200 to 400 mg/L regardless of the influent COD. Fat, oil, and grease (FOG) contents in influent and effluent were 41 and 12 mg/L on average, respectively, which corresponds to 70% removal efficiency. FOG removal efficiency increased as SRT increased and this suggested that microorganisms acclimate and actually decompose a portion of the FOG. In another study, produced water with 35.5 mg/L of FOG was treated using MBR (Zhang et al. 2010a). When HRT increased from 8 to 12 h by reducing the flow rate, FOG removal efficiency increased from 65% to 83%. It is noteworthy that SRT increases automatically as HRT increases in the given MBR setup. On the contrary, in other studies, FOG removal efficiency tended to decrease as SRT increased and excess biosolids production decreased. This contradicts all previously discussed observations, and suggested that the FOG contained in the influent used in the study was removed mainly by adsorption rather than biodegradation (Sharghi and Bonakdarpour 2013; Sharghi et al. 2013).

TABLE 6.10
Removal Efficiencies of Organic Components of Produced Water in Sangachal Terminal in Azerbaijan

Component	Concentration (mg/L)		Removal (%)
	Influent	Effluent	
Monoaromatic hydrocarbons	27	0.0019	99.9
Benzene, toluene, ethylbenzene, xylenes (BTEX)	12	0.0019	99.9
C_9–C_{40} aliphatic hydrocarbons	36	0.046	99.9
Aliphatic acids	5000	0.34	99.9
Aromatic acids	580	1.8	99.7
C_5–C_9 naphthenic acids	3900	4.3	99.9
Polycyclic aromatic hydrocarbons (PAHs)	0.21	0.00021	99.9
C_2–C_7 volatile fatty acids	1707	<10	>98

Source:　Zilverentant, A. et al., *World Oil* 230(8):21–27, 2009.

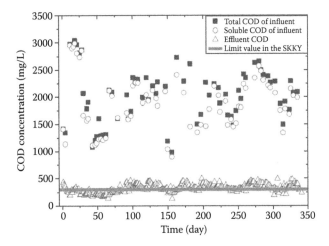

FIGURE 6.37 COD removal efficiency of MBR treating produced water when average feed COD and TDS are 2400 and 8400 mg/L, respectively. (From Kose, B. et al., *Desalination* 285:295–300, 2012.)

6.12.3 Petroleum Refinery Wastewater

Both refinery wastewater and produced water contain salts, petroleum hydrocarbons, oil and grease, suspended solids, etc., but refinery wastewater is typically much less concentrated than produced water because it is mainly from the desalting process, where crude oil comes into contact with a larger quantity of clean water to strip salt out to meet the crude oil specification for distillation tower. Wastewater is also produced from distillation, thermal cracking, catalytic conversion, etc., but it typically takes a minor portion of the total wastewater. The amount and quality of wastewater produced from the refining process varies widely site by site depending on the source/quality of the crude oil used for the refining process. Wastewater volume produced from the refining process varies at 0.4 to 1.6 times the volume of crude oil (Diya'uddeen et al. 2011). The qualities of four different refinery wastewaters are shown in Table 6.11. Supplemental nitrogen and phosphorous may be required if BOD/TKN/TP significantly deviates from 100:5:1.

Similar to produced water, removing the nonemulsified free oil particles is crucial for the successful operation of MBR. The oil can cause not only short-term membrane performance losses but also

TABLE 6.11
Petroleum Refinery Wastewater Quality

	Case 1	Case 2	Case 3	Case 4
BOD_5 (mg/L)	570	150–350	150–350	–
COD (mg/L)	850–1020	300–800	300–600	330–556
Phenol (mg/L)	98–128	20–200	–	–
Oil (mg/L)	12.7	3000	50	40–91
TSS (mg/L)	–	100	150	130–250
BTEX (mg/L)	23.9	1–100	–	–
Heavy metals (mg/L)	–	0.1–100	–	–
Chrome (mg/L)	–	0.2–10	–	–
Ammonia (mg/L)	2.1–5.1	–	10–30	4.1–33.4
pH	8.0–8.2	–	7–9	7.5–10.3
Turbidity (NTU)	22–52	–	–	10.5–159.4

Source: Ishak, S. et al., *J Sci Ind Res* 71:251–256, 2012.

long-term membrane integrity issues by compromising the physical property of the polymeric membranes. API separators are commonly used to simultaneously remove floating oil and to precipitate solids from the wastewater. Its design is very similar to the rectangular clarifiers used for nonoily wastewater, but the skimmers' and scrapers' design and operation are optimized for treating oily wastewater. Slanted parallel plates can also be inserted to save footprints just like those for municipal wastewater.

Membrane fouling seems more significant in refinery MBR than in municipal MBR due to the combination of high oil and grease contents, more dispersed floc particles at high salt concentrations, nonideal nutrient balance, presence of potentially toxic compounds that stimulate microorganisms to secrete biopolymers, etc. Daily average design flux ranges from 10 to 20 LMH (or 6–12 gfd), which is lower than that for municipal MBR, that is, 25 LMH (or 15 gfd).

Treatment efficiencies are in line with those of produced water discussed in the previous section. Petroleum hydrocarbon removal is generally very high if the operating conditions are reasonably maintained including nutrient balance, HRT, SRT, etc. For example, in a properly designed system, BTEX (benzene, toluene, ethylbenzene, and xylene) can be reduced to 1 to 10 µg/L or lower regardless of the concentration in the influent.

6.13   EXCESS SLUDGE TREATMENT

6.13.1   Overview

Post treatment of excess sludge is an important process in biological wastewater treatment, which is essential to secure the biological safety of sludge before the final disposal process or to modify the sludge property to fit it to the final treatment process such as composting, incineration, etc. The low water content of sludge obtained along with the stabilization process is also critical to minimize the leachate production in landfill sites and to minimize the heating costs for incineration. Other purpose of sludge stabilization is reducing aggressive odors, especially when sludge is exposed to open spaces in farms, forests, etc., where direct or indirect human contact is expected. The excess sludge of MBR is treated by the same methods used for CAS.

Excess sludge treatment is performed by combining the methods discussed in the following sections to achieve specific goals depending on the environment. If sludge is to be treated by incineration, a stabilization process may not be required, but the methods that provide the least moisture contents are advantageous in saving fuels. If the sludge would be sprayed on land, excess sludge should go through proper stabilization processes to reduce pathogens, but obtaining low moisture content is not a primary goal because the final product needs a proper moisture content to be sprayed. The commonly used methods for each step are summarized in Table 6.12.

6.13.2   Sludge Pretreatment

Sludge must be pretreated chemically to enhance the dewatering properties. There are two steps to chemical pretreatment, that is, coagulation and flocculation. These two terms are being used interchangeably in many cases, but coagulation is typically referred to the formation of small conglomerates (or floc) by destabilized particles through charge neutralization. Bridging of multiple particles by the polymers with opposite charges can also occur, but this is not a strong mechanism in coagulation. Inorganic coagulants, such as $AlCl_3$, $Al_2(SO_4)_3$, $FeCl_3$, $Fe_2(SO_3)_2$, or low molecular weight organic coagulants, such as polydiallyldimethylammonium chloride (pDADMAC) in dry or solution form and epichlorohydrin-dimethylamine (Epi-DMA) copolymers in solution form are popularly used. Organic coagulants are much more expensive than inorganic coagulants, but they hardly increase sludge production. On the contrary, inorganic coagulants increase sludge production by being added at high dosages and forming metal hydroxides. In addition, inorganic coagulants are strongly acidic and decrease the pH. The addition of alkalines may be necessary to prevent pH drop depending on the dosage and the alkalinity of the sludge.

TABLE 6.12

Sludge Treatment and Disposal Processes for Secondary Sludge

Category	Method	Description	Remark
Pretreatment (as needed)	Coagulation	Small flocs are formed by charge neutralization, bridging, networking, etc.	Inorganic: Al/Fe based Organic: pDADMAC in dry or solution form/Epi DMA in solution form
	Flocculation	Large cationic macromolecules bring small particles together to from large flocs	DMAEA.MCQ in dry form or in oil continuous emulsion form are the most common
Thickening	Gravity settling	Excess sludge is settled in a clarifier. Supernatant is sent back to biological tanks	2%–3% solids in 12–24 h. Simple and bulky system
	Membrane thickening	Secondary sludge is thickened in membrane tanks with submerged membranes	Up to 4%–5% solids at 4–6 LMH. Compact. Harder to operate than others
	Dissolved air floatation (DAF)	Air bubbles are attached on solids to float by oversaturating the sludge with air or bringing it into contact with water carrying oversaturated air	3%–6% solids. Compact system. Requires flocculants
	Gravity belt	Water is drained from flocculated sludge while it is on a porous filter cloth (belt) and transferred within the machine	5%–7% solids with flocculants
	Belt press	One more set of belts is added to the gravity belt and squeeze the sludge in the middle to obtain lower moisture contents in the sludge cake	10%–15% solids with flocculants
	Centrifuge	Sludge is fed to a spinning bowl typically mounted horizontally. Sludge cake near the wall of the bowl is scraped out by an internal helical scroll	15%–20% solids with flocculants. Suitable for large plants. Skilled workers required
	Filter press		
Stabilization (odor and pathogen control)	Chemical stabilization	Alkalines are mixed with sludge to obtain pH higher than 12	CaO/Ca(OH)$_2$ powders are used. Exothermic reaction
	Aerobic digestion	Excess sludges with 1%–3% solids are aerated for an extended period under endogenous respiration conditions. Alkali addition is usually required due to the pH drop by nitrification	40%–60% sludge reduction in 40–60 days. Simple system, but bulky
	ATAD	Autothermal thermophilic aerobic digestion relies on biological heat generated from the thickened sludge (4%–5%) to heat the system under aerobic conditions	Performed at 50°C–65°C. Aspirating mixers/aerators are used. 2 kg O$_2$/kWh. 30%–50% solids reduction in 5–10 days
	Anaerobic digestion	Sludge is digested to methane gas under strict anaerobic condition	45%–50% solids reduction in 20–40 days
	Composting	Dewatered sludge is piled and aerated through corrugated pipes or mechanical turnover for 21 to 28 days followed by curing for 30+ days. Core temperature in compost pile reaches 50°C to 70°C	Must meet Class A or B criteria to be qualified as soil conditioner, fertilizers for fields, forests, etc.

(Continued)

TABLE 6.12 (CONTINUED)

Sludge Treatment and Disposal Processes for Secondary Sludge

Category	Method	Description	Remark
Drying	Drying beds	Settled sludges are dried on porous media, e.g., sand, artificial media, and pavement by both drainage and evaporation with or without solar heat	Low costs and little maintenance. Large lands required. Potential odor issues
	Heat drying	Wet sludge is brought into contact with hot gas in flash, rotary, and fluidized bed driers. Proper initial moisture content, e.g., 50%–70%, is required for efficient sludge handling inside the process	Solids contents can be increased up to ~90% suitable as fuel for furnace. Small footprint, but high capital and operating costs
Disposal	Landfill	Thickened sludge is landfilled in sites with liners to hold leachate. Leachate and off-gas should be treated properly	Costs of landfill are rising in many locations due to the scarcity of land
	Incineration	Partially or fully dried sludge is fed to multiple-hearth or fluidized bed incinerators. Temperature should be higher than 760°C to ensure odor removal and complete burning	Emission controls are required in most locations. Suitable for large plants. High capital and operation costs
	Land application	Stabilized sludge qualified as Class A and B sludge can be spread to fields, forests, etc., to improve soil properties	Details are in USEPA Part 503 for biosolids rules
	Ocean dumping	Sewage sludge can be carried by barge or ship to the ocean and dumped. It is still being practiced in many parts of the world	Sewage sludge dumping has been banned since 1992 in the US

Flocculation is the reaction that brings small flocs together to form large conglomerates by bridging the small flocs. Typically, high molecular weight cationic flocculants with 30% to 80% mole charges are used for secondary sludges, where mole charges are determined by the ratio of the monomers with charges in the polymeric chain. N,N-dimethylaminoethylacrylate methyl chloride quaternary amine salts (DMAEA.MCQ) are most commonly used, but N,N-dimethylaminoethylacrylate benzyl chloride quaternary amine salts (DMAEA.BCQ), Mannich-type flocculants, etc., are also used. DMAEA.MCQ and DMAEA.BCQ have fixed positive charges regardless of pH and are typically sold as emulsion polymers, but Mannich flocculants have variable charges depending on pH due to the amino groups in the side chains and are sold as solution polymers.

The added cationic flocculants destabilize the negatively charged bioparticles by neutralizing the charges. The destabilized colloidal particles are brought together closely by long flocculant chains to form large flocs while releasing water in the spaces among particles. Emulsified flocculants encapsulated in reverse micelles in oil continuous phase are the most commonly used. These emulsion-type flocculants must be inverted to water continuous phase by being added to turbulent water at 0.3% to 1.0% before being used. Alternatively, dry flocculants in powder form can be used, but they also need to be dissolved in water at 0.1% to 0.3% using a specially designed make down system. Typical flocculant dosages are 3 to 10 kg/ton dry solids in secondary sludge; however, depending on the source and quality of the sludge, dosage can vary widely. It has been perceived that the specific flocculant demands for MBR sludge are somewhat greater than that for CAS sludge due to the smaller particle sizes with larger surface areas.

As a pretreatment for secondary sludge dewatering, coagulants are typically not necessary because cationic flocculants alone can form large enough flocs suitable for the subsequent dewatering process. However, using flocculants alone, it is difficult to form large flocs with some industrially

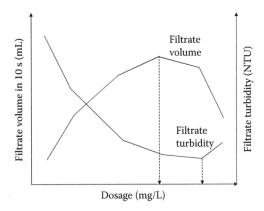

FIGURE 6.38   Determination of optimum chemical dosage from free drainage test.

sourced sludges, or they may require excessive dosages of flocculant. In these cases, the sludge can be treated using a dual chemical program, in which sludge is preconditioned by coagulants before being treated by flocculant. The sludge, partially netralized by coagulant, may allow a partial saving of more expensive flocculant and potentially improve the sludge dewatering properties.

No accurate *ex situ* exprimental methods exist to determine the optimum chemicals and their dosages, but free drainage tests similar to those described in Section 5.3.2.1 are often performed for these purposes. Instead of filter paper, a piece of press belt sample is used as a filter medium. Flocculants should be diluted to the concentrations recommened by the supplier before being used for the test. After adding flocculant to a small sludge sample (e.g., 200–500 mL), the sample is mixed with a paddle mixer for 5 to 10 s at high rpm (100–200 rpm) followed by a slow mixing for longer periods at low rpm (30–60 rpm). After pouring the mixture on the press belt, filtrate volume is measured for 5 to 30 s. Dosages, mixing conditions, and drainage times should be optimized depending on sludge properties. Filtrate turbidity is measured before it is discarded. By performing the above procedures at various flocculant dosages, the data can be plotted as the example shown in Figure 6.38. As flocculant dosage increases, filtrate volume collected in the initial 10 s increases, but it decreases after the plateau due to the charge reversal of the particles by overdosed cationic flocculants. The optimum dosage with regard to filterability is determined based on the filtrate volume. Meanwhile, the trend of filtrate turbidity generally follows filtrate volume, but the filtrate turbidity–based optimum dosage does not necessarily agree with the filtrate volume–based optimum dosages. Therefore, two different optimum dosages are determined from the perspective of dewaterability and filtrate quality. In field conditions, sludge property is not fixed, but variable with time. As a consequence, the determined optimum dosage from the free drainage is not necessarily applicable as is. Therefore, the optimum dosage determined by the free drainage test is considered an approximate projection that can be used as a starting dosage. Free drainage–based dosage estimation may be most accurately applied for belt press and filter press because little shear stresses are exerted on the sludge in the process. However, the dosages projected by free drainage test may not be indicative for screw press, centrifuge, rotary drum screens, etc., where substantial shear stress is given to the floc particles. In the field, flocculant dosages should be fine-tuned by tweaking the flocculant pump speed while checking the filtrate turbidity or the sludge cake dryness.

6.13.3   Sludge Thickening or Dewatering

Sludge thickening is performed to reduce the sludge volume by removing water. Gravity thickeners are commonly used to thicken the excess sludge before other processes are applied to thicken the settled sludge further. The excess sludge taken from the membrane tank with 0.8% to 1.5% solids is fed to a center well of a clarifier similar to the secondary clarifier in the CAS process (Figure 6.39).

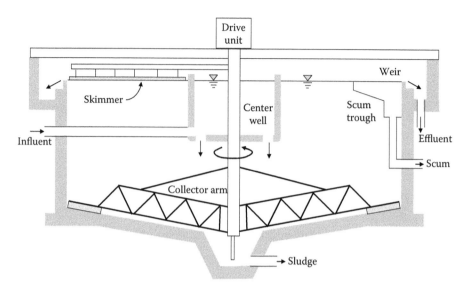

FIGURE 6.39 Schematic diagram of a gravity thickener.

The sludge is gently stirred using deep trusses or vertical pickets to open up the spaces among sludge particles. As water escapes from the sludge, the sludge becomes thicker. The thickened sludge is collected from the conical clarifier bottom. The supernatant from the tank is sent to either equalization tank or aeration tank. Gravity thickening allows solids contents of 5%–10% and 2%–3% for primary and secondary sludges, respectively. Recommended maximum hydraulic overflow rates for primary and secondary sludges are 16 to 51 m/day and 4 to 8 m/day, respectively (Tchobanoglous et al. 2003). The corresponding HRT are 2 to 6 h and 12 to 24 h for primary and secondary sludges. If the HRT is excessive in the thickener, septic conditions can develop and gases are produced, for example, CH_4, CO_2, and N_2. Sludge thickening can be hampered by the gas bubbles attached on sludge particles and float them.

Many Greenfield MBR do not have primary clarifiers. Instead, they have coarse and fine screens before the aeration tank. The collected debris and garbage are sent to landfills and the filtered wastewater is treated by MBR. The excess biosolids (or secondary sludge) from the MBR process are often thickened by submerged membranes in very similarly designed tanks with MBR. At a steady state, dilute excess sludge is fed to a membrane tank with submerged membranes while thickened sludge is removed. Flux is maintained much lower than in MBR, for example, 4 to 6 LMH, with or without coagulants (Zsirai et al. 2014). By increasing 0.8%–1.5% solids to 3%–6% solids, sludge volume decreases from 15%–30% of the original. The membrane-based thickening is inherently free of the sludge flotation problem commonly experienced in gravity-based thickeners. In addition, it requires a much smaller footprint because the HRT can be controlled much shorter at 4 to 8 h compared with gravity-based thickeners. Overall, membrane-based thickening allows substantial footprint reduction by eliminating gravity-based clarifiers and thickeners in Greenfield MBR.

Dissolved air floatation (DAF) is also used to thicken excess sludge. In this process, sludge is passed through a pressure chamber filled with pressurized air at 4 to 6 bar before it is released into a floatation tank. Theoretically, 147 g of air is dissolved in 1 ton of water at 20°C at 5 bar. If the sludge with the excess dissolved air is released into a floatation tank, the excess air forms fine bubbles on the solids' surface, where the nucleation energy is the lowest for bubble formation. Depending on the system design, all or a partial amount of incoming sludge is pressurized. In some cases, the treated effluent is passed through the pressure chamber and is mixed with the incoming wastewater to avoid mechanical troubles in the pressure chamber by solids and to preserve solids without shearing them. Alternatively, fine bubbles can be generated using diffusers or mechanical agitators and placed in contact with the sludge if the surface property of the sludge permits easy bubble attachment. As a consequence of the lowered density by fine bubbles, solids flocs float to the water's surface and are skimmed by surface scrapers. If proper coagulants/flocculants are used to pretreat the sludge, DAF can thicken secondary sludge to 3% to 6%. If solids loading rate exceeds 10 kg/m²/h, operating difficulties may arise.

In gravity belt thickening, sludge is pretreated by injecting flocculants and mixing it to form large flocs in flocculation tank. In-line flocculation also can be performed by passing the flocculent-sludge mixture through in-line mixers. Long enough pipelines are necessary to allow sufficient contact time, for example, 0.5 to 1 min. The flocculated sludge is fed to a distribution box and is evenly spread across the width of moving belt as shown in Figure 6.40. While the flocculated sludge moves toward the discharge end, the sludge is ridged and furrowed by a series of plow blades placed along the belt track, allowing the water released from the sludge to pass through the belt. After the sludge cake is removed by the blade at the end of the belt, the belt undergoes a wash cycle. This method provides 5% to 7% of solids in the sludge cake in secondary sludge thickening.

The belt press is a more advanced form of belt thickener, which allows higher solids contents in the sludge cake by physically compressing the sludge in between two belts. Sludge pretreated with flocculants is fed to the distribution box and is spread evenly at the front end of the moving belt. After passing an initial free drainage area, the sludge on the belt meets with the other belt above it. When those two belts, with partially dewatered sludge in the middle, pass a series of rollers that turn the direction of the belt, the sludge is pressed hard and additional filtrate is released through the belt. The source and the property of sludge, chemical treatment program, belt property, belt speed and tension, etc., affect the cake dryness. This method allows 10% to 15% of solids content when applied to secondary sludge. Solids loading rates are 45 to 180 kg/h per meter of belt width for the secondary sludge. Pretreatment with flocculants is necessary at a dosage of 2 to 5 kg dry flocculent/ ton or 4 to 10 kg emulsion flocculent/ton for typical secondary sludge.

Filter presses are also used to thicken the sludge. Sludge is filled up in between two filter screens attached to perforated rectangular plates and the plate perimeter is sealed. The gap between the two filter plates is closed squeezing the sludge in between. Generally, 10 to 20 min is required to fill the press and 15 to 30 min of constant pressure is required to obtain the desired cake dryness. Filter cake is discharged after the cycle. Multiple plates are stacked together and systematically moved back and forth to complete the sequential process by pneumatic pressure.

Centrifuge is a compact system relative to other sludge dewatering options and it requires no or much less flocculants, for example, 0 to 4 kg dry flocculants/ton solids. However, the relatively high capital costs and maintenance costs make it suitable mainly for large-scale plants (e.g., >20,000 m³/ day), especially with space constraints. The solid bowl centrifuge consists of a spinning long bowl with a tapered end (Figure 6.41). Sludge (or slurry) is introduced into the unit continuously from the left side of the system (through the axis in the figure). As soon as the sludge enters through discharge nozzles in the middle of the rotating scroll, solid–liquid separation starts to occur through centrifugal force. An internal helical scroll is controlled to spin at a slightly different speed from the bowl by the differential speed gear box located on the other side of the sludge inlet. Due to the slowly

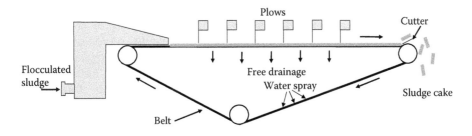

FIGURE 6.40 Gravity belt thickening.

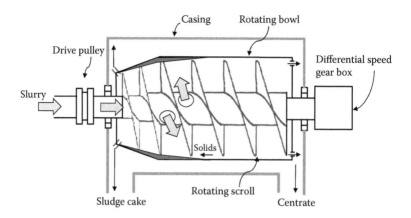

FIGURE 6.41 Schematic diagram of a centrifuge used for sludge thickening.

spinning helical scroll against the bowl, the solids accumulated in the circumference of the bowl move toward the tapered end. Additional solids concentration occurs while the sludge is transported by the scroll toward the discharge ports. Centrate is discharged through the port on the other side of the sludge inlet over the weir. The sludge cake produced contains 10% to 30% of solids depending on the source of sludge, but it is around 20% for the secondary sludge pretreated by flocculants.

Rotary drum screen is simple in structure and is relatively less expensive than other processes. Flocculated sludge is fed to the high side of a slanted perforated drum spinning slowly. While water is drained through the perforated drum surface, sludge slides down toward the other end. The sludge turning over in the spinning drum improves the dewatering process by helping expose the confined internal spaces in floc, but the high shear stress exerted on the sludge may compromise the filtrate quality. Extra flocculants may be necessary to hold the floc tight by compensating for the shear effect.

6.13.4 SLUDGE STABILIZATION

Thickened sludge is stabilized to reduce pathogens and to eliminate offensive odors. It is also required to inhibit/reduce/eliminate the potential for excessive decay during transportation to the final disposal places. The most commonly used stabilization methods include aerobic or anaerobic digestions, autothermal thermophilic aerobic digestion (ATAD), lime treatment, composting, etc.

Aerobic and anaerobic digestions are commonly used to produce qualified sludges for final disposal processes in terms of pathogenic population, but they also substantially decrease sludge quantity. Pathogenic stabilization is essential especially when the sludge is applied in forests, farmlands, deserts, etc. Aerobic digestion is similar to activated sludge, where thickened sludge is aerated in an aeration tank for extended periods. Under these condition, microorganisms in the sludge enter the endogenous respiration mode without having additional foods. As a result, a portion of the carbon in the sludge is lost to the atmosphere as CO_2. Aerobic digestion has a few benefits, for example, low BOD in supernatant, little odor issues, ability to produce odorless humus-like sludge, easy operation, and low capital costs. It has been known that up to 75% to 80% of secondary sludge (or biosolids) can be oxidized by aerobic digestion in CAS while the rest of the sludge is not easily degraded under aerobic conditions. In practice, sludge reduction is targeted at 40% to 60% because the biosolids reduction rate slows down as the target reduction rate increases. The slowing sludge reduction rate increases operating costs and diminishes the benefits of aerobic digestion. Approximately 8% to 10% of biomass consists of nitrogen and is oxidized to nitrate during aerobic digestion. Approximately 7.14 kg of alkalinity is consumed when 1 kg of nitrogen is oxidized, as discussed in Section 4.4.1.1, and as a consequence, pH decreases. Simultaneous denitrification can occur in oxygen-depleted areas such as the corner/side of the tank, inside the microbial floc, etc., especially when DO is low. If pH decreases to less than 5.5, pH adjustment may be required by feeding a base solution. With respect to pathogenic control, sludge must meet Class B requirements, at a minimum, to be land-applied in the United States. To meet Class B requirements, SRT must be at least 40 days at 20°C or 60 days at 15°C (Tchobanoglous et al. 2003).

If the solids content in the thickened sludge is 4% to 5% or greater and the heat loss from the digester is controlled properly, temperature automatically increases to 50°C to 65°C because of the biologically generated heat. This process is called ATAD. ATAD allows 30% to 50% biosolids reductions whereas HRT and SRT are maintained identically at 5 to 10 days. Approximately 20 MJ of heat is generated when 1 kg of volatile solids are destroyed to CO_2. The high reaction temperature allows rapid reduction of pathogenic viruses and parasites below the detection limit. The treated sludge by ATAD is deemed Class A sludge that can be readily land-applied. Unlike aerobic digestion, nitrification does not occur in ATAD due to the high temperature and short SRT. The nitrifiers carried over with the thickened sludge are destroyed by heat, but the SRT (5–10 days) is not sufficient to allow the nitrifiers compatible with high temperatures to grow. As a result of ammonia production from sludge, the pH of ATAD typically ranges high at 8 to 9. Diffusers are prone to fouling under these conditions due to the high scaling potential at high temperature and high MLSS, thereby mixing and oxygenation become an issue. As a result, most ATAD systems adopt aspirating aerators instead of diffusers. In aspirating diffusers, air is introduced through a hollow shaft by the aspiration effect generated by the propellers around it.

Thickened sludge can also be treated by anaerobic digestion (AD), but this process is economically feasible only for large plants due to the high capital costs. The operating costs of AD are low due to the lack of aeration and it makes AD competitive when the sludge quantity to be treated is large. The fundamental of AD is discussed in Section 7.4.2. Feed sludge must be fed either continuously or semicontinuously every 0.5 to 2 h to not disrupt the reaction. High-rate AD system is based on CSTR, where mixing is performed to homogenize the sludge in the reactor. High-rate AD is designed as either single-stage or two-stage systems. In practice, the target sludge reduction efficiency is 45% to 50%.

As a chemical treatment, hydrated lime ($Ca(OH)_2$) or quicklime (CaO) can be added to sludge to make pH high enough (i.e., >12) to inhibit microbial activities and eliminate pathogens. Sufficient lime addition is important because a portion of the lime added can be consumed by carbon dioxides and organic acids formed microbiologically. The lime reaction is basically an acid–base reaction that generates heat. Therefore, lime stabilization comes with a substantial sludge temperature increase that expedites the hydrolysis of biosolids.

Sludge composting is becoming a more popular option because finding landfill space is becoming increasingly harder. Dewatered sludge is piled and aeration is performed by injecting air through corrugated plastic drainage pipes and mechanical turning over. Bulking agents such as wood chips can be mixed with the sludge to improve aeration performance. The temperature in the core of the compost pile can reach 50°C to 70°C as a result of the biological heat. Enteric pathogens are pasteurized under these temperatures. In the initial phase of composting, fungi and acid formers play a role in mesophilic conditions. As temperature rises, thermophilic bacteria, actinomycetes, and thermophilic fungi appear and the degradation rate of biosolids reaches the maximum. As foods are depleted, the cooling stage starts and mesophilic bacteria and fungi come back. Composting typically takes 21 to 28 days with an additional 30+ days for the curing process. Throughout this process, water evaporation occurs at a high temperature and the moisture contents of the thickened sludge decrease. The final product is a humus-like stable material without an aggressive odor. Properly composted secondary sludge can be used as soil conditioners in agriculture and forestry (Tchobanoglous et al. 2003).

6.13.5 Sludge Disposal

The stabilized sludge classified as Class A or B can be land applied, if proper lands are available. In the United States, Class A sludge can be land applied without restriction but Class B sludge is allowed with some restrictions with respect to the harvest timing, animal grazing, public contact, etc. Details are found in EPA Part 503 for biosolids rules (USEPA 1994). Typically, liquid sludges are trucked out and sprayed on land. Solids sludges can be surface applied using conventional manure spreaders. It is performed for agricultural land, forests, disturbed land, and dedicated land disposal sites. Pathogens and toxic organic substances are further destroyed by sunlight, microorganisms, and desiccation. A portion of the fertilizer can be replaced by the nutrients provided by the sludge. The organic matter in the sludge helps soil to retain not only water but also potassium, calcium, and magnesium by improving the cation exchange capacity of the soil. Details on regulations associated with crop harvesting, groundwater, securing buffer zones between land application site and sensitive areas such as wells, residential areas, roads, etc., should be thoroughly reviewed before performing land application.

Landfills are the oldest and perhaps the most popular method to dispose of wastes around the globe. The landfill sites are contained with landfill liners to prevent leachate from intruding the ground water system. The methane gas produced from the depths of the site is collected by collection pipes and typically flared on a stack. Leachate should also be collected from the wells placed in the landfill site and treated properly. It may require more than 15 to 20 years of gas collection and more than 30 years of leachate treatment after the landfill sites are closed. The costs of landfills are increasing as a result of the depletion of landfill sites and the tightening environmental regulations in many countries with high population densities.

Sludge incineration is the ultimate way to reduce sludge volume that leaves minimal amount of ash to be landfilled without leaving potential hygienic issues. If the sludge cake has small enough moisture contents, net heat recovery is also possible. No stabilization process is required before incineration. However, the high capital and operating costs, including the potential necessity for emission control systems, are the hurdles in justifying this process. For complete odor removal, the temperature of the flue gas should be increased to more than 760°C. The heating value of primary sludge ranges from 23 to 29 MJ/kg, but that of anaerobically digested primary sludge ranges low from 9 to 14 MJ/kg because a portion of biodegradable carbons are lost. Secondary sludge has a heating value of 20 to 23 MJ/kg. In multiple-hearth incinerators, sludge cake is fed to the top hearth and is raked to the center hole through which the cake drops to the second hearth and so on. Temperature is the highest in the middle hearth where moisture content is low and sufficient unburnt carbons exist. Multiple-hearth reactors are mainly used in large plants due to the necessity of highly skilled workers to run the complex system. Fluidized bed incinerators are commonly used for sludge incineration, where sand particles are blown by the air injected from the bottom with dry sludge cakes at 760°C to 820°C. Due to the large surface area exposed to the hot air, sludge cake is quickly dried as soon as it enters the reactor. Ash and combustion gas escape from the top of the reactor and are scrubbed.

6.14 PRACTICAL ISSUES IN MBR DESIGN

MBR is designed based on solid theories, as discussed earlier in this chapter, but intrinsic limitations exist in the accuracy of design due to the uncertainties in input data. Although knowing BOD is the starting point of the MBR design, its accuracy is often questionable due to a number of reasons. A typical BOD test starts at a DO of approximately 9 mg/L and preferably ends at approximately 3 to 6 mg/L at 20°C for maximum accuracy. Therefore, samples are diluted to 3 to 6 mg/L in the BOD bottle. As a result, when the BOD of a high-strength sample is measured, a large dilution factor is inevitable and the error of dispensing small volume sample increases. If suspended solids exist in the high-strength sample, the BOD can vary even more widely depending on the quantity of the suspended solids taken during the sampling and the sample dispensing. The nitrification occurring in the BOD test bottle also adds errors by adding extra oxygen demand. The activity of the inoculum used for BOD testing causes additional uncertainty. The microbial diversity of the inoculum used to seed BOD bottles varies depending on the source. The low diversity may produce insufficient biodegradation and in turn underestimate BOD because some components in the wastewater may not be decomposed (Jouanneau et al. 2014). Because the accuracy of the BOD test is highly dependent on how the test is performed, comparing the data from different sources is not always straightforward. The accuracy issue in BOD tests contributes to the unreasonably wide variability of COD to BOD ratio for municipal wastewater. In addition to the intrinsic errors in BOD values, the fluctuations in BOD and flow rate make it difficult to estimate the organic loading to the system. Although sampling can be performed multiple times with time interval using an autosampler, having some error is inevitable.

Dichromate-based COD may provide better accuracy in determining organic contents in water because it requires much lower dilution factors and test conditions can be more rigorously controlled with fewer human factors. The sample COD can be measured without dilution up to 1000 to 1200 mg/L with good accuracy (i.e., ±5%). In addition, dichromate-based COD is not affected by nitrification effects because dichromates do not oxidize ammonia. However, one drawback is that dichromate-based COD affects nonbiodegradable substrates. As a result, if any industrial wastewater contains a high level of nonbiodegradable COD, the oxygen demand cannot be estimated based on COD value.

The actual oxygen consumption of wastewater can be measured by respirometry. In this method, the mixed liquor sample in a test bottle is mixed vigorously to suspend the solids and dissolve the oxygen in the headspace. The KOH solution that comes into contact with the headspace of the test bottle absorbs the CO_2 produced from the biological reaction while pure oxygen fills the bottle to maintain a constant headspace pressure. The quantity of the oxygen supplied to the test bottle is measured based on the number of oxygen bubbles passing the water column or based on the mass flow. From the time curve of the oxygen supplied to the bottle, accurate biodegradation kinetics and oxygen demand are calculated. Although this method provides a very realistic oxygen consumption pattern for the sample, it is still not the exact information applicable for full-scale processes because SRT and the associated sludge yields in the test bottle are not identical to those of real MBR. Most of all, the respirometry is only for a spot check that does not reflect the varying organic strength and flow rate in the actual site.

Wastewater characteristics are unique in every site especially in industrial wastewater. Depending on the wastewater characteristics, there might be significant differences in the biosolids yield, ratio of MLVSS/MLSS, mixed liquor characteristics that affect membrane fouling, etc. Meanwhile, the raw data required to estimate daily, weekly, and monthly flow rates are often not available in its complete form. The accuracy of the flow rate data may be questionable. Moreover, there is no guarantee that the old flow rate is an indication of the future flow rate.

In addition to the organic loading rate and the associated oxygen demand, OTE is another part of the equation in determining aeration demand. However, OTE widely varies depending on diffuser layout, specific airflow rate per diffuser, biological condition, organic loading rate, etc., as discussed in Section 6.5.2. It is well known that the α-factor that determines the OTE varies dramatically (e.g., 0.3–0.5), even in the same location in the same aeration tank because of the organic loading rate changes occurring in the same day. If biological conditions change due to sludge bulking, foaming, etc., the α-factor can vary more widely. As a consequence, sizing the aeration system becomes tricky and an oversized aeration system with a large redundancy becomes a practical solution that reduces the risk of low DO.

The inaccuracies in oxygen demand and OTE create a significant gap between theory and practice. Design engineers should fill up the gap by finding working solutions combining their own experiences obtained from existing plants. One example on oxygen demand quantification is that the total oxygen demand of municipal wastewater is often estimated by multiplying a factor to BOD as shown in the following equation. The underlying assumptions of this equation are that the wastewater characteristics, OTE, and SRT are similar regardless of the site, and no anoxic tanks exist for denitrification (Tchobanoglous et al. 2003).

$$O_{2,\text{avg}} = Q_{\text{avg}}(fS_0 + 4.57S_{\text{TKN},0}) \tag{6.35}$$

where

$O_{2,\text{avg}}$ average oxygen demand (kg/day)
f multiplication factor (–)
Q_{avg} average wastewater flow rate (m³/day)
S_0 wastewater COD (kg/m³)
$S_{\text{TKN},0}$ wastewater TKN (kg/m³)

The multiplication factor, f, ranges between 1.1 and 1.25 for municipal wastewater, but it may be out of the range for industrial wastewater. SRT is a factor affecting f by affecting the sludge production and the oxygen demand. The calculated oxygen demand does not include the oxygen credit from anoxic tanks. In practical situations, the uncertainties in oxygen demand calculation are covered by various multiplication factors, for example, a flow peaking factor of 1.5 to 2.5, a safety factor of 1.1 to 1.2, and redundancies in aeration capacity for future expansions. As a result, a degree of errors in air demand quantification can be covered in great detail.

6.15 POWER COSTS

6.15.1 BLOWER

Blowers can be categorized as either positive displacement blowers or centrifugal blowers. Positive displacement blowers provide relatively constant airflow in a wide range of discharge pressures, but they are typically more costly and require more maintenance than centrifugal types. Centrifugal blowers provide a wide range of flow rates over a narrow range of discharge pressures. The efficiency of the blower varies depending on the type of blower and the condition they are used in; their approximate ranges are listed in Table 6.13. The types of blower in each category and their operation, airflow rates, and advantages and disadvantages are summarized in Table 6.14.

The equation to calculate the power consumption of the blower can be derived by assuming adiabatic air compression in the blower chamber. Because air is compressed quickly without losing the internal energy to the surroundings, air temperature increases under the adiabatic compression condition. Two equations are available depending on the unit of gas flow as shown in Equations 6.36 and 6.37.

$$P_b = \frac{Q_{\text{air}}P_1}{17.4e_B e_M}\left[\left(\frac{P_2}{P_1}\right)^{0.286} - 1\right] \tag{6.36}$$

TABLE 6.13
Typical Efficiencies

Blower Type	Nominal Efficiency (%)	Nominal Turndown (% of Rated Flow)
Positive displacement (variable speed)	40–65	50
Multistage centrifugal (inlet throttled)	50–70	60
Multistage centrifugal (variable speed)	60–70	50
Single-stage centrifugal, integrally geared (with inlet guide vanes and variable diffuser vanes)	70–80	45
Single-stage centrifugal, gearless (high speed turbo)	79–80	50

Source: USEPA, Evaluation of energy conservation measures for wastewater treatment facilities. EPA/832-R-10-005, 2010.

Note: Values may vary beyond the range, depending on the application.

TABLE 6.14
Overview of Blower Types for Aeration of Wastewater

Category	Description and Operation	Types	Typical Airflow and Pressures	Advantages	Disadvantages
Positive displacement	Provides fixed volume of air for every shaft revolution. Operates over a wide range of discharge pressures	Most common is two counterrotating shafts (rotary) with two- or three-lobed impellers on each shaft	5–50,000 scfm, 1–14 psig	Low capital cost, economical at small scale Can achieve higher output pressure at same airflow rates Simple control scheme for constant flow applications	Difficult to operate at variable flow rates without VFD Can be noisy (enclosures are commonly used for noise control) Requires more maintenance than other types Typically least energy efficient
Centrifugal multistage	Uses a series of impellers with vanes mounted on a rotating shaft (typically 3600 rpm). Each successive impeller increases discharge pressure. Individual units operate at a narrow range of discharge pressures and at a wide range of flow rates	Number of stages dictates discharge pressure	500–30,000 scfm, 4–14 psig, can be higher with more stages	Can be more energy efficient than positive displacement Lower capital costs compared with single-stage centrifugal blowers Can be quieter than single stage units	Can be less energy efficient than single-stage centrifuge Efficiency decreases with turndown
Centrifugal single-stage integrally geared	Similar to multistage but uses a single impeller operating at high speed (typically 10,000–14,000 rpm) to provide discharge pressure. Uses gearing between motor and blower shaft	Differences are in speed and type of control (e.g., one or two sets of variable vanes)	500–70,000 scfm, 4–24 psig	Can be more energy efficient than multistage or positive displacement Can maintain good efficiency at turndown Typically come with integral control systems for surge protection	More moving parts than multistage units. Surge can be more damaging Can be noisy (enclosures are commonly used for noise control) Higher capital cost compared with multistage or positive displacement
Centrifugal single-stage gearless (high-speed turbo)	Centrifugal single-stage blower uses special low-friction bearings to support shaft (typically ~40,000 rpm). Uses a single or dual impeller	Magnetic or air bearing	400–10,000 scfm, 4–35 psig (manufacturers are currently expanding their range of offerings)	Small footprint Efficient technology for lower airflow capacity ranges Can maintain good efficiency at turndown May come with integrated control systems to modulate flow and for surge protection Can be easy to install (place, plumb, and plug in)	Typically higher capital cost compared with multistage or positive displacement blower (although likely less expensive than integrally geared) Limited experience (new technology) More units required for larger plants (will change as manufacturers expand airflow range)

Source: USEPA. Evaluation of energy conservation measures for wastewater treatment facilities. EPA/832-R-10-005. 2010.

$$P_b = \frac{wRT_1}{8.41e_B e_M} \left[\left(\frac{P_2}{P_1} \right)^{0.286} - 1 \right] \tag{6.37}$$

where

e_M motor efficiency (–)
e_B blower efficiency (–)
P_1 inlet pressure, absolute (kPa)
P_2 outlet pressure, absolute (kPa)
P_b blower power (kW)
Q_{air} airflow in ambient temperature (m³/min)
R gas constant (8.314 J/mol/K)
T_1 ambient temperature (K)
w mass flow of air (kg/s)

6.15.2 PUMP

The efficiency of pump (ε_p) is defined as the ratio of power transferred to the water to the power supplied to the pump. The brake horsepower of motor (P_M) represents the power supplied to the pump. If a gear box is used to reduce motor speed, the gear box efficiency (~95%) must be multiplied to P_M to obtain the actual power transferred to the pump.

The efficiencies of commonly used centrifugal pumps range from 75% for mixed liquor up to 85% for clean water if the pumps are used at the preferred operating ranges with respect to flow rate and differential pressure. In field conditions, however, the flow rate and the differential head pressure often do not fall into the preferred ranges. Moreover, motor speed may be reduced occasionally to accommodate the dry weather flow, which not only reduces the motor efficiency but also reduces pump efficiencies. Throttling, bypassing, recirculation, etc., are also performed to accommodate various flow demand, but these methods cause more dramatic efficiency losses. Overall, pump efficiency likely ranges from 35% to 70% in field conditions, especially when flow rate changes irregularly with frequent on and off cycles (Kaya et al. 2008).

The power consumption of pump (P_P) can be calculated by dividing the energy transferred to liquid ($\Delta P Q_L$) by pump efficiency (ε_p) and motor efficiency (ε_M) as seen in Equation 6.38.

$$P_P = \left(\frac{\Delta P}{\rho g} + \frac{v_L^2}{2g} + \Delta h \right) \frac{\rho g Q_L}{\varepsilon_P \varepsilon_M} \tag{6.38}$$

where

g gravity acceleration (9.8 m/s²)
Δh vertical distance between water intake and discharge (m)
P_P power consumption of pump (W)
ΔP differential pressure between pump inlet and outlet (Pa)
Q_L liquid flow rate (m³/s)
v_L liquid velocity in pump outlet (m/s)
ρ liquid density (kg/m³)
ε_P pump efficiency (–)
ε_M motor efficiency (–)

In practical situations, the velocity head, $v_L^2/2g$, is negligible compared with the pressure head, $\Delta P/\rho g$, unless the liquid velocity, v_L, is very high (e.g., >3–5 m/s). As a result, velocity head can be neglected in pump energy calculation in most situations in mixed liquor transportation among biological tanks.

6.15.3 Motor

Electric motors are responsible for approximately 90% of the electric energy consumption in a typical wastewater treatment plant including those for blowers, pumps, and mixers (WEF 2009; USEPA 2010). Motor efficiency (ε_M) is a measure of mechanical power output (P_m, kW) relative to electrical power input (P_e, kW).

$$\varepsilon_M = \frac{P_m}{P_e} \tag{6.39}$$

P_m is also known as motor power or brake horsepower whereas P_e is known as wire power or wire horsepower. The plant power cost is directly related to P_e.

Most electric motors are designed to run at 50% to 100% of the rated load. Maximum efficiency is obtained at usually near 75% of the rated load and preferred loads are between 60% and 90%. As a result, a 10 hp motor has an acceptable load range of 5 to 10 hp, a peak efficiency at 7.5 hp, and preferred load range at 6 to 9 hp. A motor efficiency tends to decrease dramatically when loads are approximately less than 50%, but larger motors can operate with a reasonable efficiency at loads down to the 25% range. Because motors are often used far from the 75% rated load, the practical operating efficiencies are usually less than the nominal full-load efficiency.

The range of acceptable efficiencies vary with individual motors and tends to extend over a broader range for larger motors. A motor is considered underloaded, if it is operated at an efficiency range far below the maximum. Other factors that reduce efficiency in the field include power quality (i.e., voltage, amps, and frequency) and temperature. Table 6.15 shows the mean efficiency of standard and high-efficiency motors at full load. More accurate efficiencies can be obtained from the manufacturer.

Many motors are designed with a service factor that allows occasional overloading. Service factor is a multiplier that indicates how much a motor can be overloaded under ideal ambient conditions. For example, a 10-hp motor with a service factor of 1.15 can handle an 11.5 hp load for short periods of time without incurring significant damage. VFD, also called variable speed drive (VSD), is used to closely match the speed of blowers, pumps, mechanical oxygenators, etc., to the variable demand. When variable outputs are required, the maximum pump and blower efficiencies can be achieved with VFD rather than throttling and bypassing. VFD's efficiency stays nearly constant until the relative speed decreases down to 70%, but it can decrease rapidly below that especially for the motors rated at less than 50 hp. More accurate efficiency can be obtained from the VFD manufacturer.

6.15.4 Power Factor

In a purely resistive AC motor, voltage and current waveforms exactly matches in phase-changing polarity at the same instant in each cycle. In actual AC motors, however, reactive elements such as capacitors or inductors are present. The energy storage in these reactive elements results in a time delay in the current. The stored energy is not consumed by the motor, but is returned to the power source; therefore, less power is consumed than it appears based on the apparent power consumption calculated by voltage and current. Power factor (PF) is defined as the ratio of real power used for work (P_e) to the apparent power (S) as shown in Equations 6.40 and 6.41. Because an extra transmission capacity must be prepared for the extra current that returns to the power source, fees may be applied in some locations for the devices with low PF. Typical PF for the motors with various sizes at different loads are summarized in Table 6.16.

$$PF = \frac{P_e}{S} \tag{6.40}$$

TABLE 6.15
Efficiency of Motor at Full Load

| Name Plate Power | | Mean Efficiency | |
hp	kW	Standard Motor	High Efficiency Motor
1	0.746	0.825	0.865
1.5	1.119	0.84	0.894
2	1.492	0.84	0.888
2.5	1.865	0.812	0.87
3	2.238	0.875	0.895
4	2.984	0.827	0.889
5	3.73	0.875	0.902
7.5	5.595	0.895	0.917
10	7.46	0.895	0.917
15	11.19	0.91	0.93
20	14.92	0.91	0.936
25	18.65	0.924	0.941
30	22.38	0.924	0.941
40	29.84	0.93	0.945
50	37.3	0.93	0.95
60	44.76	0.936	0.954
75	55.95	0.941	0.954
100	74.6	0.945	0.958
125	93.25	0.945	0.954
150	111.9	0.95	0.958
200	149.2	0.95	0.958
250	186.5	0.954	0.962
300	223.8	0.954	0.962

Source: USEPA, Evaluation of energy conservation measures for wastewater treatment facilities. EPA/832-R-10-005, 2010.

TABLE 6.16
Typical Power Factor of 1800 rpm Motor

| Name Plate Power (hp) | PF at Different Loads | | |
	50%	75%	100%
0–5	0.72	0.82	0.84
5–20	0.74	0.84	0.86
20–100	0.79	0.86	0.89
100–300	0.81	0.88	0.91

$$S = \sqrt{3}\, iV \tag{6.41}$$

In the above equations, i is the current (A) and V is voltage (V). The actual power consumption, P_e, is calculated by inserting the following equation to the above equation as shown in Equation 6.42. For single-phase motors, $\sqrt{3}$ is not required in the equation.

$$P_e = \sqrt{3}\, iV \cdot \text{PF} \tag{6.42}$$

6.15.5   Price of Electrical Power

Average electrical power prices for industry range from $0.07 to $0.20/kWh in most industrialized countries (with exceptions), but the actual power cost varies quite significantly depending on the following factors even in the same country:

- *Location*—Depending on geographical location, power cost varies significantly. In the United States, power costs tend to be low in the Midwest, South Central, and Pacific West at $0.04 to $0.08/kWh, whereas it is high in New England at $0.10 to $0.13/kWh (Electric Choice website 2013).
- *Timing*—The capacity of power plant required is dependent on the peak power demand, not the average demand. To reduce the fixed costs to build additional power plants, the government and power companies often imply surcharges for peak demands to discourage power usage at peak times. To save power costs, the peak water treatment time can be manipulated using redundant capacities in equalization tanks.
- *Contract power*—To make the power consumption more predictable while discouraging power demand surge, each end user may be allocated with maximum power allowance, perhaps with time variations. If power consumption is beyond the contract level, hefty surcharges may be imposed.

6.15.6   Case Example No. 1: MBR with Submerged Membranes

6.15.6.1   Condition and Assumption

An example of specific energy demand (SED) calculation is demonstrated in this section for the MBR shown in Figure 6.42 treating municipal wastewater. Excluding pumping energy in collection systems, excess biosolids handling, posttreatment, etc., there are five major processes directly related with wastewater treatment in MBR with regard to energy consumption: (1) biological aeration for organic removal, (2) aeration for membrane scouring, (3) permeate suction, (4) pumping for wastewater feeding and mixed liquor recycle, and (5) permeate drawing. All the parameters required to calculate energy consumption are listed below.

Influent
- Q Average wastewater flow rate (150 m³/h or 0.0417 m³/s)
- S_0 Influent COD (350 mg/L or 0.35 kg/m³)
- $S_{TKN,0}$ Influent TKN (60 mg/L or 0.06 kg/m³)
- $S_{NO_3-N,0}$ Influent NO₃-N (3 mg/L or 0.003 kg/m³)

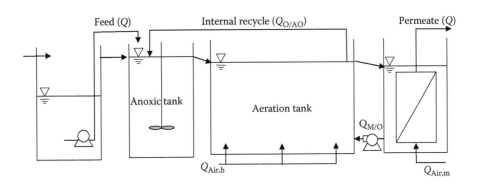

FIGURE 6.42   Flow diagram of typical MBR with anoxic tank.

Effluent

S_e Effluent COD (10 mg/L or 0.01 kg/m³)

$S_{TKN,e}$ Effluent TKN (1 mg/L or 0.001 kg/m³)

$S_{NO_3-N,e}$ Effluent NO_3-N (10 mg/L or 0.01 kg/m³)

Biosolids

SRT   Solids retention time (20 days)

Y_{obs} Observed biosolids yield (0.358 g MLSS/g COD based on Macomber's equation in Figure 6.17 elsewhere)

i_{XB} Nitrogen content in biosolids (0.086 mg N/mg MLSS)

λ Ratio of COD to MLSS (1.1 g COD/g MLSS)

X_0 MLSS in aeration tank (10 g/L)

Process condition

A_M Membrane surface area (9000 m²)

DO   Dissolved oxygen in aeration tank (2 mg/L)

H_F Average waterhead required to pump raw wastewater to anoxic tank (4 m H_2O)

H_S Submergence of scouring air nozzle (2.5 m H_2O)

$H_{L,z}$ Head loss for scouring air (1.0 m H_2O)

H_b Submergence of fine bubble diffusers (4.5 m H_2O)

$H_{L,b}$ Head loss for biological air (0.5 m H_2O)

J_{net} Net flux (25 LMH or 6.94×10^{-6} m/s)

SAD_m Net specific air demand per area (0.36 m³/m²/h)

$Q_{O/AO}$ Internal recycle, 3 Q (450 m³/h or 0.125 m³/s)

Q_{MO} Mixed liquor recycle to aeration tank, 3 Q (450 m³/h or 0.208 m³/s)

T Water temperature (25°C)

ε_B Blower efficiency (0.7)

ε_P Pump efficiency (0.3 for permeate, 0.4 for mixed liquor recycle, 0.6 for wastewater pumping)

ε_M Motor efficiency (0.7 for permeate, 0.9 for all others)

In mechanical screen operation, most of the energy is consumed by the feed pump to transport wastewater against gravity and the head loss in the pipe, but the screening operation itself does not need too much energy. Therefore, the power consumption by mechanical screen is neglected in the calculation. Pumping power (P_p) is calculated using Equation 6.38. The following are some of the assumptions made:

1. *Wastewater pumping from the equalization tank to the anoxic tank*: Due to variations in the water level in the holding tank, the differential pressure of the pump is variable. Here, the average pressure difference between the inlet and the outlet of the feed pump (ΔP_{Feed}) is assumed at 3 m H_2O. Flow rate is equivalent to the wastewater flow rate, Q. Pump efficiency (ε_p) is assumed at 0.6 whereas motor efficiency (ε_M) is assumed at 0.9.

2. *Internal recycle*: The average head pressure (ΔP) that the recycle pump must overcome is assumed at 2 m H_2O, including the head loss in pipelines. Internal recycle flow rate is assumed at 3 Q. Because the pump handles a liquid with high MLSS, pump efficiency (ε_p) is assumed somewhat low at 0.4 due to the higher viscosity of the mixed liquor than water.

3. *Recycle from membrane tank*: Due to variations in the water level in the aeration tank, the head pressure of the pump is variable. The average differential pressure of the recycle pump (ΔP) is assumed at 2 m H_2O. Recycle flow rate to membrane tank, Q_{MO}, is assumed at 2 Q. Pump efficiency (ε_p) is assumed at 0.4 as for the mixed liquor recycle pump.

4. *Anoxic tank mixing*: The energy required to mix the anoxic tank is dependent on the depth of the tank, mixed liquor viscosity, tank size and aspect ratio, etc., but it is generally assumed at 8 W/m³. Assuming the HRT of anoxic tank is 2 h, the specific denitrification rate (SDNR) required to obtain the target effluent nitrogen content is calculated as Equation 6.43, where $i_{XB}Y_{obs}(S_0-S_e)$ represents the nitrogen assimilated to biomass. MLVSS can be assumed at 80% of MLSS. V_{AO} of 300 m³ is obtained by multiplying HRT (2 h) and Q_0 (150 m³/h). The calculated SDNR, that is, 0.063 g NO₃-N/g VSS/day is within the range of the SDNR for wastewater, that is, 0.03 to 0.11 g NO₃-N/g VSS/day, as shown in Table 4.5, and thus the assumed anoxic tank volume (300 m³) is justified. Finally, the power required to mix the anoxic tank is calculated at 2400 W or 2.4 kW by multiplying the SED (8 W/m³) and the anoxic tank volume (300 m³).

$$SDNR = \frac{Q_0\left[(S_{TNK,0} + S_{NO_3-N,0} - S_{TNK,e} - S_{NO_3-N,e}) - i_{XB}Y_{obs}Q_0(S_0 - S_e)\right]}{MLVSS \cdot V_{AO}}$$

(6.43)

$$SDNR = \frac{150 \times 24 \times \left[(60 + 3 - 1 - 10) - 0.086 \times 0.358 \times (350 - 10)\right]/1000}{0.8 \times 10 \times 300}$$

$$= 0.063 \text{ g NO}_3\text{-N/g VSS/day}$$

5. *Permeate drawing*: Permeate drawing is performed intermittently in MBR. The efficiency of the permeate pump is not high due to the frequent on–off cycles and is assumed at 0.3 here. The efficiency of VFD that controls the permeate flow is assumed at 0.7. The flow rate is assumed to be similar as the influent flow rate, Q, because the loss to the excess sludge removal (Q_x) is less than 0.01 Q. Suction pressure is assumed at 2 m H₂O whereas it typically ranges from 10 to 30 kPa. It is also assumed that additional 2 m H₂O head is required to transport the permeate to a different location. Finally, the SED (kWh/m³) for each unit process is calculated using the following equation:

$$SED = \frac{24P}{Q}$$

(6.44)

6.15.6.2 Results

The aeration power for biological reaction is calculated based on the total oxygen demand and OTE. Total oxygen demand can be calculated as follows using Equation 6.13 for the condition described above.

$$O = 150[(1 - 1.1 \times 0.358 - 1.71 \times 0.086 \times 0.358)(0.35 - 0.01) + 1.71(0.06 - 0.001)$$
$$- 2.86(0.003 - 0.01)] = 46.4 \text{ kg O}_2\text{/h}$$

Specific oxygen demand per wastewater volume is calculated at 0.309 g/L by dividing the oxygen demand (O) by feed flow rate, Q (150 m³/day).

OTE can be calculated using Equation 6.20, but α-, β-, and θ-factors, average saturation DO in aeration tank, fouling factor, f, and specific SOTE are required. Alpha-factor is calculated using the Krampe and Krauth equation shown in Figure 6.17,

$$\alpha = e^{-0.08788 \times 10} = 0.415$$

Average saturated DO, C_0^*, is calculated using Equation 6.22 as follows:

$$C_0^* = \left[-5.823 \times 10^{-5} \times 20^3 + 7.049 \times 10^{-3} \times 20^2 - 3.929 \times 10^{-1} \times 20 + 14.56 \right]$$
$$\left[1 + \frac{4.5}{20.67} \right] = 11.0 \, \text{mg/L}$$

Next, OTE is calculated using Equation 6.20. The median values summarized in Table 6.6 are used for all unknown variables.

$$\text{OTE} = 0.415 \times 0.95 \times 1.024^{25-20} \times 0.8 \times 4.5 \times 0.06 \frac{11.6-2}{11.6} = 0.078 \text{ or } 7.8\%$$

Total biological airflow rate can be calculated using the oxygen demand (O) calculated above and the oxygen content of air, that is, 0.275 kg O_2/m^3 at the condition blower performances are rated for, that is, 20°C at 36% relative humidity at sea level.

$$Q_{\text{Air}} = \frac{O}{0.275 \text{OTE}} = \frac{46.4}{0.275 \times 0.078} = 2167 \, \text{m}^3 \text{ air/h or } 0.60 \, \text{m}^3/\text{s}$$

The moisture content of air is typically not a significant source of error in the calculation compared with the errors in oxygen demand and OTE calculations, except in high-temperature and high-humidity situations (e.g., >30°C and >50% relative humidity).

The head pressure required for bio-air blowing is calculated as

$$\Delta P_{\text{Air,b}} = 9.804 \, (H_b + H_{L,b}) = 9.804(4.5 + 0.5) = 49.0 \, \text{kPa}$$

The required blower power for biological aeration is calculated at 38.7 kW by inserting all values to Equation 6.36. Using Equation 6.44, SED is calculated at 0.258 kWh/m³.

The blower power (P_b) required for membrane scouring is calculated using Equation 6.36. By multiplying membrane area and SAD$_m$, scouring airflow can be calculated at 3240 m³/h or 0.9 m³/s in this case. The total head pressure required can be calculated by adding nozzle submergence and head loss in the pipe and the nozzle ($H_s + H_{L,s}$), that is, 3.5 m H_2O or 34.3 kPa. By inserting these values into Equation 6.36, the power required for membrane scouring is calculated at 42.9 kW. Subsequently, SED for membrane scouring is calculated at 0.286 kWh/m³ using Equation 6.44.

Meanwhile, the power required for mixed liquor pumping is calculated by inserting all the information shown in the previous section into Equation 6.38. The SEDs of each unit process for the examples given in this section are summarized in Figure 6.43 and Table 6.17. According to the figure, energy consumptions of non–air-related processes are minor, for example, permeate drawing (7%), internal mixed liquor recycle (7%), mixed liquor circulation between membrane and aeration tanks (4%), anoxic tank mixing (2%), and feed pumping (2%). Aeration takes around 78% of the total energy consumption, where biological aeration takes slightly less energy (37%) than membrane scouring (41%) for the given case.

6.15.7　Case Example No. 2: MBR with Crossflow Membranes

The specific energy consumption of crossflow filtration systems is widely variable depending on how well the internal pressure loss is controlled, as discussed in Section 2.4.1.1.

Figure 6.44 shows a typical tubular membrane system. In this system, feed water is distributed to multiple tubes in a module and merged again before moving to the next module. In addition, the feed

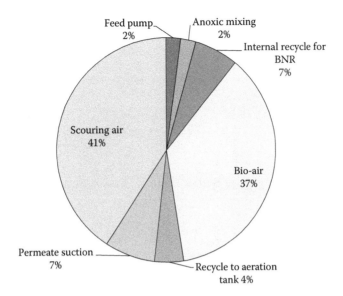

FIGURE 6.43 Breakdown of energy consumption of each unit process in MBR (Case Example No. 1).

water encounters changes in the channel cross-sectional areas that result in flow velocity varying depending on the longitudinal location in the system. As a consequence of the converging or diverging flows and the varying flow rates, a large pressure loss occurs along the channel. Single-tube tubular modules can mitigate the longitudinal pressure loss by minimizing such drawbacks, where channel diameter is maintained constant from the entrance to the exit by using connectors with an identical diameter with the membrane tube. The low pressure loss allows a long membrane channel length up to 48 m (vs. 10–15 m for typical tubular membranes) at a TMP of 6 bar. As a result, high permeate recovery is possible in one feed cycle. This case example is for the single-tube system and hence it must be noted that the estimated SED here is at the low end of the ranges found in the MBR with sidestream membranes.

The conditions are given as follows:

P_{in} inlet pressure (550 kPa)
ρ liquid density (1000 kg/m³)
v liquid velocity in membrane module (3.7 m/s)
D_i internal diameter of membrane (0.0254 m)
n number of channels (10 ea.)
ε_p pump efficiency (0.6)
ε_m motor efficiency (0.9)
J_{avg} average flux (83.3 LMH or 2.31×10^{-5} m/s)
L effective membrane length (48 m)

The liquid flow rate to membrane system, Q_{feed} (m³/s), can be calculated as follows:

$$Q_{feed} = \frac{\pi D_i^2 n v}{4} = \frac{3.14 \times 0.0254^2 \times 10 \times 3.7}{4} = 0.0187 \, \text{m}^3/\text{s or } 67.5 \, \text{m}^3/\text{h}$$

The inlet pressure already includes the pressure required to overcome longitudinal pressure loss and head loss/gain from the difference in the elevations of inlet and outlet. The power required, P (kW), can be calculated using Equation 6.38. The velocity head in the equation is not neglected here because it can be significant in tubular membranes due to the high liquid velocity.

TABLE 6.17

Breakdown of Energy Consumption in the MBR with Submerged Membrane

Process		ΔP^a m H$_2$O	ΔP^a kPa	Q_L or $Q_{Air}{}^b$ m³/h	ε_P or $\varepsilon_B{}^c$ –	Motor Efficiency, ε_M –	Power (P) kW	SED Individual kWh/m³	SED Individual %	SED Total kWh/m³	SED Total %
Biological process	Feed flow (Q)	3	29.4	150	0.6	0.9	2.3	0.015	2.2	0.334	48
	Anoxic mixing						2.4	0.016	2.3		
	Internal recycle ($Q_{O/AO}$)	2	19.6	450	0.4	0.9	6.8	0.045	6.5		
	Bio-air ($Q_{Air,b}$)	5	49.0	2127.6	0.7	0.9	38.7	0.258	36.7		
Membrane filtration	Recycle (Q_{MO})	2	19.6	300	0.4	0.9	4.5	0.030	4.3	0.368	52
	Permeate suction (Q)	4	39.2	150	0.3	0.7	7.8	0.052	7.4		
	Scouring air ($Q_{Air,m}$)	3.5	34.3	3240	0.7	0.9	42.9	0.286	40.7		
Total							105.4	0.702		0.702	100

a Total differential pressure required.
b Flow rate of liquid or air.
c Efficiency of pump or blower.

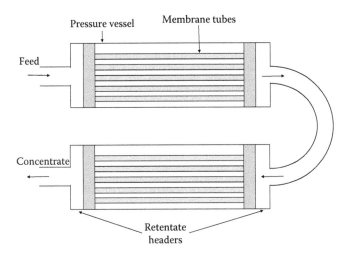

FIGURE 6.44 Membrane module with multiple tubular membranes (or multitube module).

$$P_{\mathrm{p}} = \left(550{,}000 + \frac{1}{2} \times 1000 \times 3.7^2\right) \frac{0.0187}{1000 \times 0.6 \times 0.9} = 18.4\,\mathrm{kW}$$

Total permeate flow rate, Q_{p} (m³/s) can be calculated as

$$Q_{\mathrm{p}} = (\pi D_i Ln) \cdot J = (3.14 \times 0.0254 \times 48 \times 10) \times 2.31 \times 10^{-5} = 8.86 \times 10^{-4}\,\mathrm{m}^3/\mathrm{s}\ (\text{or } 3.19\,\mathrm{m}^3/\mathrm{h})$$

Finally, SED is calculated as follows:

$$\mathrm{SED}_{\mathrm{filtration}} = \frac{P_{\mathrm{p}}}{3600\,Q_{\mathrm{p}}} = \frac{18.4}{3600 \times 8.86 \times 10^{-4}} = 5.8\,\mathrm{kWh/m}^3$$

Although exceptionally long tube channels (48 m) were assumed, the permeate recovery from a single pass of the mixed liquor is calculated at only 4.7% (= 3.19/67.5, $Q_{\mathrm{p}}/Q_{\mathrm{feed}}$). As a result, 21 times more liquid recirculation is required than the permeate obtained. The SED obtained above (5.8 kWh/m³) falls into the low end of the literature values, that is, 4 to 12 kWh/m³ (Côté and Thompson 2000). In practical conditions, the membrane channel length is only 1/4 to 1/2 of the assumption. Therefore, the SED is two to four times greater than the values calculated here.

Despite the optimistic assumption, it is very apparent that SED of sidestream filtration running under a positive pressure is more than an order of magnitude greater than that of submerged membrane system (0.09–0.5 kWh/m³). This is a significant handicap when tubular membranes are used for a process that does not create high value, for example, municipal wastewater or drinking water treatment. In general, tubular membranes have competitive advantages when submerged membranes are not suitable for the application due to the requirement for thorough pretreatment, volatile organic carbon (VOC) emission, inability to obtain high recovery, inability to handle solids that have high membrane fouling tendencies, etc.

Assuming the same biological condition shown in Section 6.15.6, the SED of MBR with a sidestream membrane system can be broken down, as shown in Figure 6.45. The energy consumption for feed circulation overwhelms all other energy consumptions. Although very optimistic assumptions were made in membrane channel length here, overall SED is very high at 6.0 kWh/m³ and

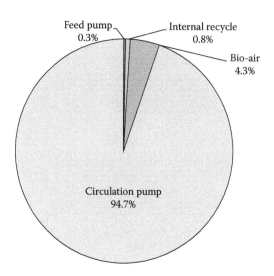

FIGURE 6.45 Breakdown of energy consumption of each unit process in MBR with sidestream membranes (Case Example No. 2).

this is almost an order of magnitude greater than that for the MBR with submerged membranes (0.69 kWh/m^3 as shown in Section 6.15.6).

6.16   CAPITAL EXPENDITURE

The prices of submerged membranes have decreased significantly in the last two decades. The price drop has slowed down since the late 2000s, but it is still in the middle of a declining trend. The membrane prices vary widely depending on the volume of order, freighting method, geographical area, market situation at the time of the order, manufacturer's marketing strategy, etc. In North America, the unit price of the membranes sold by some of the top-tier suppliers are approximately \$40 to \$50/m^2 for large-scale projects including cassette frames, but excluding shipping, handling, and tax. The membrane prices of the same reputable membranes are known to be 20% to 30% higher in Europe.

Table 6.18 summarizes approximate market size and prices of membranes for dialysis, RO, water filtration, MBR, and gas separation applications. It should be noted that the prices in the table are for large-scale projects. In comparison, the price of replacement membranes for small MBR in New Zealand was reported at US\$140 to 160/m^2 (Manson et al. 2010).

TABLE 6.18

Approximate Market Size and Price of Membranes for Various Applications

Market	Configuration	Market Size (M\$/year)	Membrane Area Sold (mm^2/year)	Approximate Price (\$/m^2)
Dialysis	Hollow fiber	1800	400	4.5
Reverse osmosis	Flat sheet	600	50	12
MF/UF filtration	Hollow fiber	350	10	35
MF/UF MBR	Hollow fiber	350	7	50
Gas separation	Hollow fiber	600	6	50

Source: Côté, P. The future of MF/UF in water treatment. *Proceedings of WETSUS Conference.* October 19, 2010. Leeuwarden, The Netherlands.

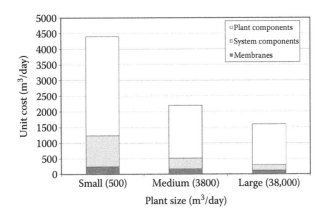

FIGURE 6.46 MBR plant cost structure. (From Côté, P. The future of MF/UF in water treatment. *Proceedings of WETSUS Conference*. October 19, 2010. Leeuwarden, The Netherlands.)

As a consequence of low membrane prices, membranes take only a very small portion of the total plant capital costs. As the system size grows, unit membrane prices decrease, but the unit costs of other components decreases faster. This is because the required membrane area increases proportionally to the flow rate whereas other capital costs increase less than proportionally. In large-scale plants, the specific membrane costs for unit flow rate is approximately $100/(m^3/day), as shown in Figure 6.46, excluding all associated piping, blowers, and civil works. Overall, membrane costs take only 5% to 10% of the capital costs regardless of the plant size.

6.17 SED OF SUBMERGED MBR

The average SED of many different municipal MBR plants were surveyed including biological air, membrane scouring air, sludge circulation, permeate suction, office lighting, etc. (Lüdicke et al. 2009). The average SED of CAS with or without tertiary filtration was also surveyed and the results were compared with those of MBR. SED was mainly affected by the type of treatment system and the treatment capacity. SED of MBR was greater than that of CAS by approximately 0.3 kWh/m^3 due to the energy required for membrane systems.

In general, SED ranges from 1.8 to 6 kWh/m^3 for plants with a capacity of less than 5000 person equivalents and 0.8 to 1.4 kWh/m^3 for the plants with a capacity of more than 5000 person equivalents, as suggested by the data summarized in Table 3.3. In a highly optimized plant with a design flow rate of 23,000 m^3/day in Ulu Pandan, Singapore, SED was observed at 0.55 kWh/m^3 in a certain period in 2009. The low energy consumption can be attributed to the intermittent aeration performed at 10 s on–10 s off mode and the existence of an anoxic tank that allowed oxygen credit from the denitrification reaction. In the same plant, SED could be reduced to 0.4 kWh/m^3 by adjusting the aeration mode to 10 s on–30 s off when organic loading rate was low (Chen et al. 2012a). It is apparent that SED tends to decrease as the flow rate or the plant capacity (or both) increases, as shown in Figure 6.47.

In a given plant, SED varies depending on influent flow and strength. For example, SED decreased from 1.66 to 0.39 kWh/m^3 when flow rate increased from 8000 to 44,000 m^3/day in Nordkanal, Germany in 2007 (Engelhardt 2008). Meanwhile, the yearly average SED was at 0.85 kWh/m^3 in the same place in 2007. Similar results were obtained in a small MBR with a 6000 person equivalent capacity in France. SED was estimated at 0.69 and 1.01 kWh/m^3 for winter and summer, respectively (Barillon et al. 2012). The low SED in winter was attributed to the higher flow rate at lower organic strength than in summer.

In a survey for two municipal MBR equipped with hollow fiber membranes (GE), total plant SED were determined at 0.90 kWh/m^3 each (De Wever et al. 2009). The SED of membrane aeration, bioreactor aeration, recirculation pumps, permeate pumps, mixers, and other miscellaneous processes

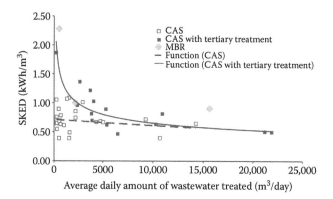

FIGURE 6.47 SED of WWTPs in the Erft region. (From Brepols, C.H. et al., *Water Sci Technol* 61(10):2461–2468, 2010.)

were estimated at 0.23, 0.30, 0.03, 0.04, 0.05, and 0.25 kWh/m³, respectively, for the MBR with 16,000 m³/day dry weather flow in Kaarst-Nordkanal, Germany. Likewise, the SED was broken down to 0.34, 0.24, 0.11, 0.12, 0.04, 0.05 kWh/m³, respectively, for the MBR with 6000 m³/day dry weather flow in Varsseveld, The Netherlands. It is noteworthy that the SED for membrane aeration in these field conditions are as high as 0.23 and 0.34 kWh/m³, although much lower SED have been claimed by many membrane manufacturers (0.07–0.15 kWh/m³). The gap between the two is explained largely by lower flow rate than design flow rate and less than ideal mixed liquor conditions with respect to filterability. In essence, the data in this table is in line with the data obtained from a case example summarized in Table 6.17. The plant SED is plotted against the membrane system capacity in Figure 6.48. It is apparent that the plant SED tends to decline as the membrane system capacity increases. In addition, the SED of submerged hollow fiber membranes appear somewhat lower than those of submerged flat sheet membranes with exceptions.

The SED of membrane systems vary depending on membrane type (Barillon et al. 2012, 2013). Hollow fiber (HF) membranes tend to require less scouring air than flat sheet (FS) membranes according to the audit performed in six different MBR plants (two FS and four HF). In Figure 6.49, SED includes the energy to run activated sludge processes and membrane systems, but it does not include the energy for indirect purposes such as building maintenance. According to this survey, flat sheet membranes consume nearly 50% to 100% more specific energy for membrane scouring than HF especially when hydraulic loadings are low at 40% to 50% of the design values.

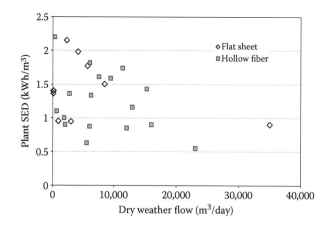

FIGURE 6.48 Plant SED of municipal MBR depending on the membrane configuration and the dry weather flow (or design flow). (Redrawn from Krzeminski, P. et al., *Water Sci Technol* 65(2):380–392, 2012.)

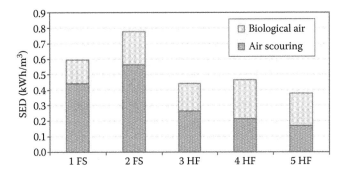

FIGURE 6.49 Energy consumption for biological air and membrane air scouring in five different MBR sites with submerged flat sheet (FS) or submerged hollow fiber (HF) membranes. (From Barillon, B. et al., *Water Sci Technol* 67(12):2685–2691, 2013.)

6.18 OPERATING EXPENDITURE

Total operating costs are dependent on the goals of the treatment, water quality, existence of BNR processes, membrane types used, age of the membrane, unit power costs, labor costs, etc. If discharge limits exist for trace organics or certain solvents, activated carbon adsorption or RO (or both) may be required. In some locations, additional alkalinity may be necessary to adjust the pH due to acidic influent or excess nitrogen in the effluent. If excess biosolids disposal costs are high due to the insufficient availability of landfill sites, extended aeration may be economical to reduce excess biosolids at the expense of aeration costs. Therefore, the total costs and the cost structure widely vary depending on the site. However, if there are no big and unusual cost items such as adsorption column, RO, etc., operating costs can range from $0.20 to $0.50/m^3 in municipal MBR including sludge disposal. Typical operating costs for structures are shown in Figure 6.50, where electrical costs, sludge dewatering and disposal costs, and labor costs are the major cost items. Depending on the location, sludge disposal costs can be much higher than those shown in the figures.

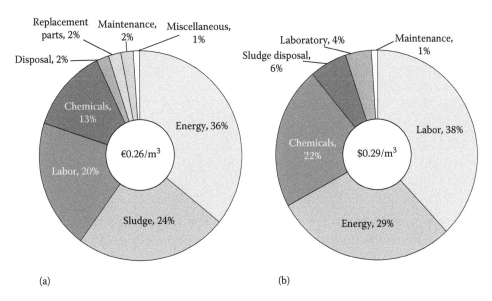

(a) (b)

FIGURE 6.50 Operating cost breakdown of municipal MBR in Nordkanal, Germany in 2006. (a) Nordkanal, Germany and (b) Cauley Creek, USA. (Based on Engelhardt, N., Long-term experience with Europe's largest membrane bioreactor plant for municipal waste water treatment. *Proceedings of IFAT*. September 23–25, 2008. Shanghai, China; Durden, K., and Dodson, D., *AMTA News* 22:10–13, 2004.)

7 Other Applications

7.1 DIRECT MEMBRANE FILTRATION OF WASTEWATER

7.1.1 Overview

Membrane bioreactor (MBR) has been widely used to treat municipal and industrial wastewater in recent decades. The main drivers of this relatively new process include superior effluent quality and smaller footprint relative to the conventional activated sludge (CAS). MBR is often combined with biological nutrient removal (BNR) processes to lower the nitrogen and phosphorus concentrations in the effluent to meet the stringent discharge limit. In the meantime, the demand of recycled wastewater for irrigation and landscaping has been increasing in many geographical regions not only due to the stringent water supply but also due to the desire to save the costs associated with water footprint. Deliberate nutrient removal from wastewater may not be required for these purposes. In fact, the ratio of nitrogen and phosphorous in typical municipal wastewater roughly falls into the ideal nitrogen:phosphorous (N:P) ratio for growing terrestrial plants, that is, 100:7.4–14.8 in a mass basis (Geider and La Roche 2002). The chemical oxygen demand (COD) removal efficiencies of direct membrane filtration (DMF) is known to be high in general at 37% to 97%, depending on the membrane types used and the initial water quality as summarized in Table 7.1. The high COD removal efficiency is attributed to the existence of particulate COD in raw water and the conversion of soluble COD to particulate COD by the biological assimilation inadvertently occurring during the filtration process.

DMF is one of the simplest forms of wastewater treatment that can produce readily recyclable water for irrigation and landscaping. Several potential advantages over CAS and regular MBR have been proposed: (1) lower net energy consumption and smaller footprint due to the lack of biological tanks, (2) less operational challenges from the biological fluctuations, (3) lower capital costs, and (4) higher nutritional value of recycled water due to the lack of BNR (Sutton et al. 2011). Contacting the filtered wastewater with zeolite (clinoptilolite) column was proposed to remove ammoniacal nitrogen, if necessary. Phosphorus removal can be performed by adding coagulant to the raw wastewater. However, most of these hypothetical advantages have been denied mainly because of the inferior membrane performance in treating raw wastewater without thorough biological pretreatment.

As discussed in the following sections, DMF is not economically feasible in treating the organic-rich wastewater exposed to the extended period of septic condition when it is applied for treating municipal wastewater. The submerged membrane–based DMF can be much more economical than sidestream membrane-based DMF, but it is still less economical than the regular MBR because of the necessity of extra membrane surface areas to compensate the low flux. As a result of the economic constraint, DMF has not been used in commercial scale in municipal wastewater treatment.

7.1.2 DMF Based on Tubular Membrane

Most of the known studies have been performed using tubular membranes (Till et al. 1998; Bendick et al. 2004, 2005). It has been found that membrane fouling is much more significant in DMF than that in the regular MBR equipped with sidestream tubular membranes. As a result, larger membrane surface areas and greater operating costs are expected. This disadvantage makes the DMF based on tubular membrane technology nearly obsolete because even the regular MBR with full aerobic treatment are not attractive in municipal wastewater treatment. The energy costs can

TABLE 7.1

Permeate Quality during the DMF of Sewage

Influent Type	Membrane Type/ Pore Size	COD (mg/L)			Suspended Solids (mg/L)
		Influent	Permeate	Efficiency (%)	
Sewage/settled	MF 0.05 μm	216	59	73	119
Gray water/screened	MF 0.1 μm	75	9	88	
Sewage/screened	MF 0.1 μm	122	9	93	91
Sewage/settled	MF 0.1 μm	70	14	80	43
Sewage/settled	MF 0.2 μm	216	76	65	69
Sewage/settled	MF 0.45 μm	194	55	72	110
Sewage/settled	MF 0.5 μm	216	35	84	71
Sewage/settled	MF 0.8 μm	216	62	71	105
Sewage/settled	MF 1.2 μm	242	84	65	99
Sewage/settled	MF 1.4 μm	216	52	76	61
Sewage/screened	NF	725	98	86	
Sewage/settled	UF	135	78	42	46
Sewage/screened	UF	218	138	37	63
Gray water/screened	UF 300 kDa	70	14	80	43
Gray water/screened	UF 300 kDa	42	6	86	
Gray water/screened	UF 150 kDa	70	12	83	43
Sewage/filtered	UF 50 kDa	416	12	97	100
Sewage/filtered	UF 50 kDa	30	9	70	24
Gray water/screened	UF 30 kDa	70	11	84	43
Gray water/screened	UF 15 kDa	41–86	7–16		

Source: Diamantis, V.I. et al., Direct membrane filtration of sewage using aerated flat sheet membranes. *Proceedings of the 11th International Conference on Environmental Science and Technology.* September 3–5, 2009. Chania, Crete, Greece.

easily reach 15 to 20 kWh/m³ or higher in DMF depending on the sustainable flux. It is because the organic and particulate contaminants in raw wastewater directly contact with the membrane without being treated biologically under the optimized condition. In fact, raw wastewater is inevitably exposed to the septic condition when it is transported through collection pipes. In the septic condition, small amounts of microorganisms are exposed to the large quantity of food (or high food-to-microorganism [F/M] ratio) under low dissolved oxygen (DO) condition. As discussed in Sections 5.2.3 and 5.2.4, the stressed microorganisms secrete the SMP extremely proficient for membrane fouling. In addition, the cake layer formed on the membrane surface in DMF is likely much denser than that in the MBR because of the lack of large floc particles.

7.1.3 DMF Based on Submerged Membrane

Submerged membranes can be more economical than tubular membranes for DMF because of not only the low capital costs for the flow but also the lack of water circulation at high pressure. The DMF based on the submerged membrane is essentially an MBR with very short hydraulic retention time (HRT) without a dedicated reactor space for biological degradation. The submerged membrane–based DMF can provide a comparable compactness with tubular membrane-based DMF in terms of footprint requirement. It has a few more extra advantages in addition to the general advantages of DMF mentioned previously: (1) the inadvertently occurring biological degradation in the membrane tank ensures low soluble COD in the effluent, (2) much less energy consumption than tubular membranes by a factor of 10 to 40, and (3) much lower capital costs than tubular membranes due to the inexpensive membrane modules. In fact, COD removal efficiencies can be well more than 80% if submerged membranes are used without a dedicated biological aeration tank (Diamantis et al. 2010). If the flux in DMF is the same as in typical MBR, the minimum HRT obtainable with submerged hollow fiber membranes will be approximately 1 to 2 hr depending on flux.

For DMF, submerged membranes are much more economical options than cross-flow tubular membranes because of the much less expensive membranes and the lower operating costs. In the submerged membrane-based DMF, the contaminants in raw wastewater are at least partially treated biologically by the inadvertently growing microorganisms in the membrane tank. However, the excessive F/M ratio at negligible DO causes expedited membrane fouling. Diamantis et al. (2009, 2010) have tested submerged flat sheet membranes to filter municipal wastewater without a separate aeration tank. It was found that membrane permeability was only 50 L/m²/h/bar, which was much lower than that of typical MBR, for example, 100 to 300 L/m²/h/bar (Brepols 2011). Mezohegyi et al. (2012) also tested submerged flat sheet membranes to concentrate municipal wastewater and observed more significant membrane fouling than that in typical MBR perhaps because of the excessive extracellular polymeric substances. As a result, only low fluxes, for example, <7 L/m²/h (LMH), were obtainable in a sustainable manner. Thereby, the projection for specific capital and operating costs was much higher for DMF than for MBR. According to a study, the total costs of treating wastewater using submerged flat sheet membranes were €2.0/m³ and €0.73/m³ for DMF and MBR, respectively, including capital and operating costs (Diamantis et al. 2010).

In submerged membrane-based DMF, HRT is no more than 1 to 2 h, and volumetric organic loading rate (OLR) is extremely high. Under this condition, excess sludge must be removed fast to maintain a low enough mixed liquor suspended solids (MLSS). Thereby, solids retention time (SRT) becomes very short, for example, 0.5 to 2 days. In such a short SRT condition, only the microorganisms with very short duplication time or those growing attached can survive in the membrane tank. Therefore, the microbial diversity becomes extremely narrow. In a laboratory study, hollow fiber membranes with 0.035 μm pore size were used to treat synthetic wastewater with 300 mg/L COD at an SRT of 1 day. As can be seen in Figure 7.1, only five morphologically distinguishable species were present in the mixed liquor. Despite very short HRT (0.5 h) and low MLSS (<500 mg/L), the effluent COD was consistently around 20 mg/L during the experiment. However, slowly growing nitrifiers could not be enriched at such a short SRT, and as a consequence, no nitrification occurred at all. In another study, when SRT was less than 2 days, very narrow genetic diversities were also observed based on 6s rRNA analyses using the polymerase chain reaction denaturing gradient gel electrophoresis (PCR-DGGE) method, but COD removal efficiency was observed very high at 93% (Teksoy Başaran et al. 2012).

In a laboratory study, it was observed that microorganisms tend to form a biofilm on reactor walls when SRT is extremely low, for example, 0.5 to 2 days. A portion of the biofilm on reactor wall is

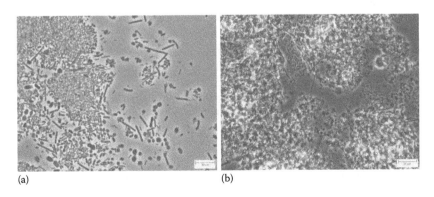

(a)   (b)

FIGURE 7.1 Comparison of microbial diversity at (a) short and (b) long SRT. (a) SRT = 1 day. (b) SRT = 20 day.

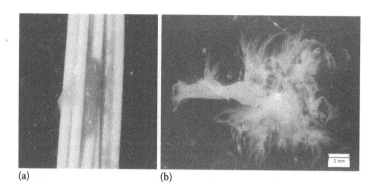

(a)   (b)

FIGURE 7.2 Jelly material attached on hollow fiber membrane and its hairy appearance when washed by DI water: (a) jelly materials on membrane and (b) washed jelly material in petri dish.

eventually detached by the turbulence. The floating large debris is captured by hollow fiber membranes causing expedited membrane fouling by interfering the fiber movement. Figure 7.2a shows biofilm debris with jelly appearance captured by hollow fiber membranes. Once this large debris is attached on the membrane, it can encourage the further deposition of other floating materials in the long run by hampering the membrane movement and by creating low flow zones nearby. The core of the jelly material consists of polyamide-based hairy fibers that act as a backbone of the biofilm supporting the structure (Figure 7.2a).

In the DMF based on the submerged membrane, capital and operating costs can be saved by increasing the membrane tank size. Under the longer HRT, contaminants in the raw wastewater are more thoroughly treated biologically, and as a consequence, membrane performance improves. This in turn reduces the membrane surface area required and the total scouring airflow. However, as membrane tank size increases to reduce membrane fouling, DMF becomes close to the regular MBR in terms of volumetric organic loading rate.

7.2 MEMBRANE CONTACTORS FOR GAS TRANSFER

7.2.1 OVERVIEW

The membrane contactor can be used to transfer gas components to and from the liquid. Membrane degassing, membrane oxygenation, humidification/dehumidification, and membrane distillation (MD) are some of the examples. Dissolved gas, solvent vapor, or any of the components in liquid and gas can cross the membrane depending on the concentration (or partial pressure) difference across the membrane. The same membrane contactors can also be used to remove biological oxygen demand (BOD), nitrate, perchlorate, etc., from contaminated water by growing the biofilm on the liquid side of the membrane. Table 7.2 summarizes various applications of membrane contactors.

Porous hydrophobic membranes and nonporous membranes can be used for the applications. Porous hydrophobic membranes provide high mass transfer rate because of the existence of the micropores on the surface, but they are inherently vulnerable to pore wetting that breaks membrane integrity. In fact, hydrophobic membranes are fouled over time by the organic and inorganic contaminants carried by the feed, and hence the contact angle. Microbial growth on the membrane surface can also compromise the contact angle. Eventually, water starts to intrude the pores with the most foulants, and the integrity is broken. The contact angle can be recovered by membrane cleaning, but the recovery of hydrophobicity is hardly complete. The surfactant used can make the membrane less hydrophobic than initial by coating the surface. Oxidizing agents (Cl_2, H_2O_2, etc.) can also compromise hydrophobicity by partially oxidizing the membrane itself. On the contrary, nonporous membranes are free of the integrity issue caused by pore wetting, but the mass transfer rate is inherently low because of the lack of pores on the surface.

TABLE 7.2

Application of Membranes for Gas Transfer

Application	Process Description	Dominant Membrane Type
Degassing	Removes dissolved gas such as O_2, CO_2, N_2, H_2S, etc., through HF membranes using vacuum. Used for ultrapure water production for chip manufacturing, water for food processing, water for analytical instruments, etc. Nonporous membranes are popularly used, but porous hydrophobic membranes also can be used.	HF Nonporous
MD	Hydrophobic porous membrane divides warm water and cold air. The water vapor diffused to the cold air condenses. Pore wetting can be expedited by the contaminant deposit or microbial growth on the membrane surface.	FS Porous
Oxygenation	Air or pure O_2 is supplied to the lumen side of the HF membrane submerged in the activated sludge tank. Simultaneous air scouring may be required to control biofilm growth. Condensate formed in the lumen hampers the efficiency of the process.	HF Nonporous
Wastewater treatment	Configuration is the same as oxygenation, but biofilm is grown on the membrane surface to treat wastewater. Condensate formed in the lumen hampers the efficiency of the oxygenation. Porous membranes suffer from pore wetting.	HF, FS Nonporous/porous
Removal of NO_3^- and ClO_4^-	Autotrophic biofilm is grown on the shell side of nonporous HF membrane and H_2 is provided from the lumen. Condensate formed in the lumen hampers the process efficiency.	HF, FS Nonporous
Humidification	Dry gas is supplied to the lumen of the hollow fiber submerged in water. Used to prehumidify the inlet air of fuel cell, humidification of respiratory gas, etc.	HF Nonporous
Dehumidification	Wet gas is supplied to the HF lumen while sweeping dry gas flows in the shell.	HF Nonporous

Note: FS, flat sheet; HF, hollow fiber.

One of the most popular applications out of all membrane contactor applications today might be vacuum degassing process as shown in Figure 7.3a. It is used to remove dissolved gases from water such as oxygen, nitrogen, carbon dioxide, and hydrogen sulfide (Leiknes and Semmens 2001). The vacuum degassing process has been adopted for the applications that do not prefer chemicals, for example, ultrapure water production for semiconductor manufacturing, water for food and beverage production, and boiler feed water treatment. Lumen plugging by condensate does not occur in vacuum degassing because the vacuum prevents the condensation of vapor in the lumen.

Hollow fiber membrane contactors can be used for the oxygenation of water, as shown in Figure 7.3b. Air or pure oxygen is supplied typically to the lumen side of the hollow fiber membrane submerged in water. Oxygen is diffused through the membrane and dissolves in water. Oxygen transfer efficiency (OTE) can be controlled by controlling the air/O_2 flow rate. OTE increases as air/O_2 flow rate decreases and contact time increases. Minimal feed gas pressure is sufficient to drive the feed gas through the membrane in this process because the gas pressure does not need to overcome the static water pressure. Inexpensive low-pressure blowers are sufficient for oxygenation. Another benefit of this process is that no bubbling, no foaming, no aerosol, and no volatile organic carbon emission are anticipated. However, this type of application suffers from lumen plugging by

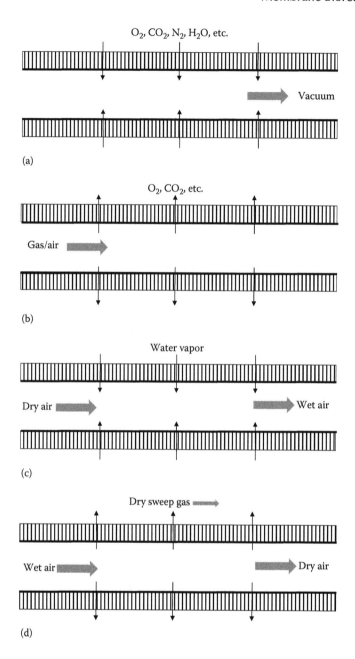

FIGURE 7.3 Application of hollow fiber for gas: (a) degassing, (b) gas dissolution, (c) humidification, and (d) dehumidification.

the condensate formed by the water vapor that intruded the lumen for both porous and nonporous membranes, as will be discussed in Section 7.2.2.1. Lumen plugging shuts down some of the fibers and reduces the overall process efficiencies in many applications.

The gas transfer rate of hollow fiber contactors can be enhanced by growing the biofilm on the waterside of the membrane surface. This process is called membrane-assisted biofilm reactor (MABR), membrane-supported biofilm reactor (MSBR), or membrane biofilm reactor (MBfR). In this process, gas consumption occurs most actively right on the interface between the biofilm and the membrane, as shown in Figure 7.4. As a consequence, gas concentration profile in the interface becomes much steeper with biofilm. According to the Fick's first law, the diffusive flux of gas components is proportional

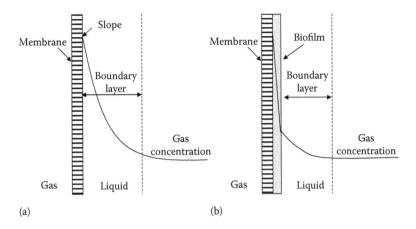

FIGURE 7.4 Effect of the biofilm on gas transfer rate in MABR: (a) without biofilm and (b) with biofilm.

to the slope of concentration profile. Therefore, mass transfer from the lumen to shell can be greatly enhanced by the existence of the biofilm. Biofilms are known to increase mass transfer by a factor of 2 to 10 in general depending on gas component, biofilm activity, biofilm structure, etc. (Cussler 2009).

Membrane humidifiers have been used for medical purposes to humidify oxygen for patients with pulmonary disease (Figure 7.3c). Hydrophobic porous hollow fibers or nonporous hollow fibers are submerged in water, and sweeping air is supplied to one end of the lumen to obtain wet air in the other end. This method eliminates the aerosol production from bubbling humidifiers, which potentially carries pathogens. Larger systems have been promoted commercially for humidifying rooms and buildings, but the high capital costs compared with the conventional humidifier are the hurdle. Membrane humidifiers have also been used to prehumidify the inlet air of fuel cells to prevent the loss of electrolytes through the polymer electrolyte membrane by the evaporation (Oko and Kralick 2001; Chen and Peng 2005). As shown in Figure 7.3d, dehumidification is an opposite of humidification, where dry sweeping gas is flowed through the shell side to extract moisture from the lumen through nonporous membranes (Zaw et al. 2013). Membrane dehumidification does not have an advantage over the conventional dehumidifier if the sweeping air should be regenerated by removing the moisture from it. Therefore, this system targets an air/gas stream for dehumidifying using the other dry air/gas stream available from somewhere else. No prominent commercial applications are known for this process.

There are many different types of membranes available in the market for membrane contactors. Proper membranes must be chosen depending on the type of application because each membrane has its own strength and weakness for specific application.

- *Membranes with nonporous polydimethylsiloxane layer*—The nonporous silicone layer coated on the porous membrane provides high gas permeance. SuperPhobic® membranes from Membrana, Permaselect® membranes from Medarray, and SEPAREL® membranes from DIC are microporous hollow fiber membranes coated with proprietary silicone layer. The nonporous silicone layer provides high gas permeance.
- *Membranes with nonporous polyurethane layer*—Mitsubishi Rayon developed nonporous hollow fiber membranes with a multilayer structure called Sterapore® (Takemura et al. 1987). This membrane has a triple layer structure in which an ultrathin nonporous polyurethane layer is sandwiched between two porous polyethylene layers. The inner diameter and the outer diameter are known to be 0.28 and 0.42 mm, respectively. The 1 to 2 µm thick nonporous layer fundamentally prevents membranes from leaking by eliminating the pore wetting problem. The ultrathin nonporous layer enables a high mass transfer rate and the porous layer provides mechanical strength to the membrane.

- *Single layer nonporous membranes*—GE's Zeelung™ is a nonporous membrane made of a single-layer polymethyl pentene (PMP). PMP membranes have been originally used for blood oxygenation in artificial lung and are known to have a high gas permeance. The hollow fiber membrane with around 50 µm diameter provides very large surface-to-volume ratio and is suitable for dissolving oxygen in wastewater. Zeelung® aims for the economical aeration of the activated sludge by obtaining high OTE at low air pressure (Buer et al. 2008; Stricker et al. 2011). However, the intrinsic lumen condensation issue hampered the overall aeration efficiency, and the challenges in biofilm control diminished the attractiveness of the process. No commercial operations have been reported since the product was introduced. Membrana's Accurel® is also made of a single-layer PMP and used for various degassing processes.
- *Porous hydrophobic membranes*—Flat sheet hydrophobic microporous membranes made of polypropylene (PP), polyvinylidene difluoride (PVDF), and polytetrafluoroethylene (PTFE) with 0.2 to 1.0 µm pore sizes are used for MD (Camacho et al. 2013). Hydrophobic hollow fiber membranes such as Liqui-Cel® are also used for water degassing. This type of membrane is intrinsically prone to pore wetting because of the lowering contact angle by the organic and inorganic deposits around the pore entrance. Periodic membrane cleaning can help mitigate the issue to some extent.

7.2.2 FUNDAMENTAL ISSUES ASSOCIATED WITH MEMBRANE CONTACTORS

7.2.2.1 Lumen Condensation

When gases pass the hollow fiber membranes submerged in water, mass exchanges occur across the membrane because of the partial pressure difference in between the lumen and the shell. Depending on the condition applied, water vapor that intruded the lumen can cause condensation and plug the fibers. Lumen condensation is an inherent drawback in membrane oxygenation and in the MSBR developed for BOD, nitrate, and perchlorate removal. On the contrary, lumen condensation is not an issue in the degassing application, where the water vapor pressure in the lumen is always lower than the water vapor pressure.

Figure 7.5 illustrates the mechanisms of lumen condensation in the oxygenation process. In this example, hollow fiber membrane is submerged in the activated sludge that treats COD using the oxygen supplied through the membrane. The mechanisms of lumen condensation are as follows:

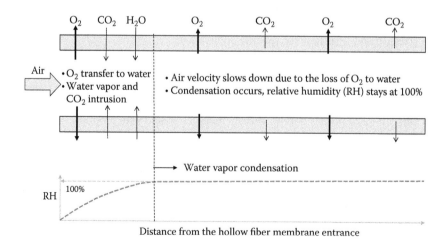

FIGURE 7.5 Conceptual diagram showing water condensation issue in hollow fiber lumen.

- As soon as air enters the membrane lumen, oxygen starts to diffuse through the membrane because of the partial pressure difference. Near the entrance, the partial oxygen pressure is approximately 0.2 bar in the lumen, but it is 0.02 to 0.04 bar in the shell if DO is 1 to 2 mg/L.
- Simultaneously, water vapor and dissolved CO_2 in the activated sludge diffuse into the lumen because the partial pressure of these gases is lower in the lumen than that in the activated sludge.
- After the air proceeds a certain distance (or transitional length), the H_2O vapor pressure in the lumen reaches a saturation point. No further water vapor intrusion occurs from this point on.
- Likewise, CO_2 also reaches the equilibrium pressure in the lumen. The membrane length required for CO_2 to reach the equilibrium is dependent on the permeance of the gas, the initial partial pressure, etc., and is different from that for O_2.
- Although no H_2O and CO_2 intrude after the equilibrium points, O_2 is continuously transferred to the activated sludge. This causes gas volume contraction and slows down the gas velocity in the lumen.
- Because of the loss of O_2 to the activated sludge and a consequent gas volume shrinkage, water vapor pressure in lumen exceeds the saturated vapor pressure whereas CO_2 pressure exceeds the CO_2 pressure in the shell side. Water and CO_2 reach their maximum pressure in different locations in the fiber, but, as soon as the maximum pressures are reached, water vapor starts to condense and CO_2 starts to diffuse out from the lumen.

Water condensation in the lumen is unavoidable in membrane oxygenation and MABR. Once one fiber is plugged up by condensate drops, inlet gas is redistributed among other intact fibers. As more and more fibers are plugged up, the apparent mass transfer rate decreases and the pressure required to flow air increases.

The condensate formed in hollow fiber membrane can be removed by making one end of the porous hydrophobic membrane hydrophilic and seal the end, as shown in Figure 7.6. The remaining hydrophobic areas are coated with silicone-based polymers to make it nonporous (Semmens 1991; Ahmed and Semmens 1992). If pure oxygen is supplied from the open end with a slight pressure, the condensate drop can be driven to the hydrophilic end by the pressurized pure oxygen. Once the condensate reaches the hydrophilic end, it is pushed out by the oxygen pressure.

7.2.2.2 Pore Wetting

When dry hydrophobic membranes are contacted with water, water does not intrude pores unless its pressure exceeds the bubble point, as shown in Figure 1.29. The pressure required to wet the pore can be calculated using Equation 2.12. According to the equation, bubble point is directly proportional to the cosine of the contact angle, and it should be larger than 90° to keep pores dry. Therefore, it is crucial to maintain the contact angle stable and large by keeping the membrane surface hydrophobic.

However, keeping the membrane surface hydrophobic is extremely hard in long-term operation because of membrane fouling. Inorganic and organic precipitates from feed water and microbial attachment/growth contaminate the pore entrance eventually. If the deposition of inorganic and organic material occurs near the pore entrance, it compromises the local contact angle that in turn

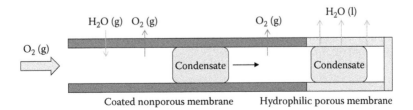

FIGURE 7.6 Nonporous hollow fiber membrane with hydrophilic porous dead end.

decreases the bubble point. Once the bubble point drops below the water pressure, water intrudes the lumen and plugs the fiber. Pore wetting will become more significant if biological activity occurs in water phase because of the production of biological surfactants. According to Lu et al. (2008), pore wetting was more prevalent when membranes have high porosities, feed water contains organics, and water temperature was high. The contaminated membrane surface can be cleaned using acid/base, oxidizing agents, surfactants, etc., but the complete removal of such contaminants are always challenging. In addition, the residual surfactants on membrane and the oxidation of the membrane surface can compromise the hydrophobicity of the membrane at least to some extent.

7.2.2.3 Biofilm Control Issue

In vacuum degassing, humidification/dehumidification, and MD, the biofilm can hamper the mass transfer through the membrane by covering up the surface. Thus, periodic membrane cleaning may be necessary to keep the membrane surface clean. If porous membranes are used for such applications, membranes potentially lose the integrity by incomplete membrane cleaning in long-term operation. However, nonporous membranes are free of pore wetting, although the mass transfer rates are slower.

In membrane oxygenation, MSBR, etc., the biofilm enhances the mass transfer. As shown in Figure 7.4b, the biofilm consumes oxygen on the immediate surface of the membrane, which accelerates oxygen transfer from the gas phase by making the concentration profile steeper. However, microorganisms eventually die in the biofilm and act as a mass transfer barrier. The concentration profile in the biofilm becomes less steep with the dead microorganisms in the biofilm. In addition, parts of the biofilm slough off randomly if old biofilm loses its ability to attach on the membrane surface. The random loss of the biofilm can cause a performance fluctuation in MABR process (Semmens 2005). If the biofilm grows excessively on hollow fiber membrane, it can merge with the other biofilm grown on other hollow fiber in the vicinity. This eventually forms a giant biofilm bundle through which hollow fibers pass. Biofilms must be renewed periodically at a controlled manner to prevent the biofilm aging, random detachment, and outgrowth.

7.2.3 Water Vapor Condensation Modeling

Water vapor condensation fundamentally limits the performance of certain membrane contactors such as membranes oxygenation and MABR for nitrate and perchlorate removal. The rigorous mathematical modeling of water vapor transfer is considerably complicated, but the following simple model provides the ideas on how quickly the water condensation occurs in hollow fiber membranes. It is assumed that only water vapor transfers through the membrane and no humidity gradient exists in the radial direction of hollow fiber.

The driving force of water vapor transfer is the water vapor pressure difference across the membrane wall. Therefore, the water vapor transfer rate can be assumed proportional to the vapor pressure difference in between the lumen (P, Pa) and the shell (P_{wv}^*, Pa). The number of moles of water vapor added to the sweeping gas flow (Q_g, m³/s) in a small segment of a hollow fiber membrane can be calculated by multiplying the sweeping gas flow rate by the changes in water vapor pressure in the same segment (dP, Pa), as shown by the term in the left side of the Equation 7.1. The amount of water vapor added to the same segment of the hollow fiber can also be calculated by the term in the right side of the equation, where gas permeance (K, mol·m/m²/s/Pa) is multiplied by the driving force $(P_{wv}^* - P)$ and membrane surface area (dA, m²) of the segment and divided by the thickness of the membrane wall (δ, m).

$$Q_g = \frac{dP}{RT} = \frac{K(P_{wv}^* - P)}{\delta} dA. \tag{7.1}$$

In the meantime, the membrane surface area of the small segment in interest can be calculated by Equation 7.2, as follows:

$$dA = \pi \phi n dx. \tag{7.2}$$

By combining the two equations previously mentioned, Equation 7.3 is obtained as follows:

$$\frac{dP}{P_{wv}^* - P} = \frac{\pi \phi n K R T}{\delta Q_g} dx. \tag{7.3}$$

The previously mentioned equation can be integrated with two boundary conditions, for example, $(x = 0, P = P_0)$ and $(x = x, P = P)$:

$$\int_{P_0}^{P} \frac{dP}{P_{wv}^* - P} = \frac{\pi \phi n K R T}{\delta Q_g} \int_0^x dx. \tag{7.4}$$

The water vapor pressure along the fiber is expressed as Equation 7.5, but it can be normalized by dividing both sides by saturated water vapor pressure (P_{wv}^*). Finally, Equation 7.6 is obtained, where water vapor pressure is expressed as relative humidity (ξ) along the fiber:

$$P = P_{wv}^* - \left(P_{wv}^* - P_0\right) \exp\left[-\frac{\pi \phi n K R T x}{\delta Q_g}\right] \tag{7.5}$$

$$\xi = 1 - (1 - \xi_0) \exp\left[-\frac{\pi \phi n K R T x}{\delta Q_g}\right] \tag{7.6}$$

In silicone-coated nonporous membranes, water vapor has 60-folds higher permeance than oxygen, for example, 36,000 versus 600 Barrer (Robb 1968). Therefore, when the silicone-coated membranes are used to dissolve oxygen into water, it is not hard to imagine air is saturated with water vapor while only a small fraction of oxygen is transferred out to water. The permeance unit of Barrer can be converted to mol·m/m^2/s/Pa to be used with the previously mentioned equations by multiplying a conversion factor of 3.348×10^{-16}.

For the conditions listed in Table 7.3, relative humidity profile can be calculated using Equation 7.6. The result in Figure 7.7 shows that the air entering into the lumen is saturated with water vapor before it proceeds 5 cm. According to the literature (Fang et al. 2004), when similar modeling studies were performed for several different conditions, water vapor saturation occurred within 1.5 to 3.0 cm from the hollow fiber entrance depending on the condition. Once the air is saturated with water vapor in the membrane lumen, condensation starts to occur because of the oversaturated water vapor as oxygen is lost to the water phase and the gas volume shrinks.

7.2.4 OXOANION REMOVAL

Hollow fiber contactors can be used to remove oxoanions, for example, bromate $\left(BrO_3^-\right)$, perchlorate $\left(ClO_4^-\right)$, and nitrate $\left(NO_3^-\right)$, from contaminated water streams. In this process, hydrogen gas is supplied to the lumen side of nonporous or porous hydrophobic hollow fiber membranes submerged in contaminated water. In the shell side surface, naturally occurring autotrophic microorganisms grow

TABLE 7.3

Conditions Used for Water Vapor Transfer Modeling

Parameters	Symbol	Value
Gas flow rate, total	Q_g	0.001 m³/s
Fiber inner diameter	ϕ	0.0002 m
No. of fiber	n	50,000 ea
Water vapor permeability of silicone	K	1.21×10^{-5} mol·m/(m²·s·Pa)
Thickness of nonpermeable layer	d	1.00×10^{-5} m
Water vapor pressure, inlet air	P_0	500 Pa
Water vapor pressure, saturated at 20°C	P_{wv}^*	2338 Pa
Gas constant	R	8.314 J/mol/K
Gas velocity in the lumen	v	0.637 m/s

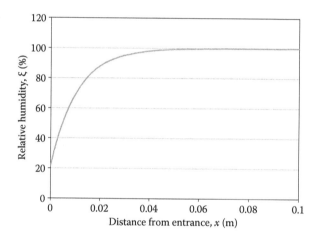

FIGURE 7.7   Relative humidity profile in hollow fiber lumen under the condition given in Table 7.3.

consuming the hydrogen diffused from the lumen as an energy source. The nitrate and perchlorates are used as oxygen source to which the electrons obtained from hydrogen are dumped. Meanwhile, dissolved CO_2 is used as a carbon source for the autotrophic microorganisms growing on membrane surface. The final reaction products are mainly water, nitrogen gas, bromide, and chloride ions. The reaction equations for nitrate and perchlorate ion removal are as follows:

$$NO_3^- + 2.5H_{2(g)} \rightarrow 1/2N_{2(g)} + 2H_2O + OH^-,$$

$$ClO_4^- + 4H_{2(g)} \rightarrow Cl^- + 4H_2O.$$

Nonporous hollow fiber membranes have an advantage over porous membranes because of the lack of pore wetting issues, for example, silicone-coated membranes, PMP membranes, and Mitsubishi's Sterapore membranes with nonporous polyurethane layer sandwiched by porous polyethylene layers. When nonporous Mitshubishi membrane was used to treat 5 mg/L nitrate, 99% of nitrate was removed at an HRT of 25 min by the naturally occurring autotrophic microorganisms (Nerenberg and Rittmann 2004). Perchlorate concentration was reduced from 55 to 4 μg/L at 93% efficiency (Nerenberg and Rittman 2003). In another study, 16.4 mg/L of NO_3-N was completely removed by a hollow fiber contactor made of silicone-coated nonporous membranes (Haugen et al. 2002) (Figure 7.8).

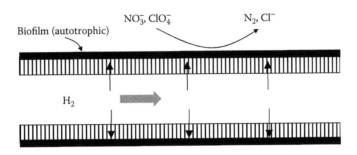

FIGURE 7.8 Hydrogen-based, hollow-fiber membrane biofilm reactor for reduction of oxoanions.

Although any level of oxoanion removal efficiency can be readily achieved by extending the contact time, questions remain for this process with regard to the mass transfer efficiency in long-term operation. The membrane areas required to treat oxoanions to the level desired are mainly dependent on the specific hydrogen consumption rate per membrane area, but the kinetics has not been analyzed thoroughly beyond the pilot or full-scale demonstrations. As discussed in Section 7.2.2, lumen condensation is inevitable in this application because water vapor intrudes the membrane lumen and is supersaturated as hydrogen is depleted. If a membrane fiber is plugged up by condensate, hydrogen gas can be hardly supplied to it, and hence the biofilm associated with the fiber cannot function properly. As more fibers are plugged up, the overall degradation kinetics is hampered. In addition, it is likely that the process performance fluctuates as aged biofilm sloughs off irregularly and new biofilm grows (Semmens 2005).

The actual specific H_2 transfer rates in long-term operation appear less than 1/10 of the specific H_2 transfer rate suggested by the gas permeance of new membrane in field condition. The much lower actual H_2 transfer rate than the permeance indicates that the H_2 permeance is not a likely bottleneck even for nonporous membranes with a low gas permeance. The low gas transfer rates in MSBR can be better attributed to lumen plugging by condensate, the reduced specific activity of the biofilm by dead cells, biofilm loss from the membrane surface, etc.

7.2.5 MEMBRANE DISTILLATION

MD is a separation process relying on vapor pressure difference in between the two sides of a hydrophobic porous membrane. In this process, warm water contacts with one side of the membrane, and the water vapor diffused to the other side of the membrane through micropores is condensed on a cold surface. If the vapor passes a membrane and condenses on the surface of the other side of the same membrane, the process is called direct contact MD (DCMD). If the vapor passes a membrane and travels a distance before being condensed on a cooling plate, it is called air gap MD (AGMD). If vacuum is applied to AGMD to facilitate vapor permeation through the membrane, it is called vacuum MD (VMD). If the vapor passes a membrane and is swept by other carrier gas and condenses elsewhere, it is called sweep gas MD (SGMD). The heat efficiency of DCMD is low because heat easily leaks through the thin membrane in between warm water and condensate by conduction. SGMD is complex because additional devices are required to condense the swept water vapor. As a result, AGMD and VMD are more commonly used for commercial processes.

MD runs at a lower temperature than other distillation methods such as multistage flashing (MSF) and multieffect distillation (MED), for example, 40°C to 80°C versus 70°C to 120°C (Wikipedia 2013b). Capital costs can be saved with MD by using less expensive plastics, for example, PP and PVDF, rather than stainless steel. In addition, solar energy or low-grade heat, such as power plant cooling water, can be used as a heat source instead of high-grade heat at a high temperature. Currently, the solar-powered MD system is commercialized for small-scale applications (<10 m³/day) to produce potable water in remote areas by Fraunhofer ISE of Germany. Brackish or seawater is circulated through a solar collector to be heated before it is fed to MD, and the electricity can be generated by optional photovoltaic cells placed on the surface of the solar collector.

Pore wetting is the intrinsic challenge of MD, as discussed in Section 7.2.2.2. Organic and inorganic deposits gradually increase the hydrophilicity of the membrane and make the membrane vulnerable to pore wetting. High water temperature is essential to improve the productivity of MD, but it increases pore wetting potential by reducing the surface tension of water and increasing scaling potential. Oil and surfactant contained in feed water as contaminants also increase the chance of pore wetting (Zhao et al. 2011). Once pores are wet, saltwater can pass the membrane and permeate conductivity surges. Large membrane porosity and pore size help reduce the resistance of vapor transport through the membrane, but pore wetting becomes easier. Commercial MD membranes are known to have 0.2 to 1.0 μm pores. The maintenance issues associated with membrane fouling and pore wetting limit the applicability of MD only for small-scale applications today (Camacho et al. 2013).

PP, PVDF, and PTFE are the most common membrane materials. PTFE has the highest surface energy essential for maintaining membrane integrity, but it has relatively high heat conductivity that reduces heat efficiency of the process (Camacho et al. 2013). Flat sheet membranes are most commonly used in the form of flat panel (Memstill, Memsys, etc.) and spiral wound (Fraunhofer ISE) configurations. Flux can be 20 to 30 LMH in laboratory tests at high feed water temperatures, for example, >90°C, simultaneously with well-managed hydrodynamics in small modules (Sirkar and Song 2009; Song et al. 2013; Suárez et al. 2010; Winter et al. 2013), but it is much lower in large-scale systems at 3 to 12 LMH at 60°C to 80°C feed water (Jansen et al. 2013; Kesieme et al. 2013). Hollow fiber membranes provide large surface-to-volume ratio, but flux is only 1 to 4 LMH at 40°C to 60°C. The poor flux performance is attributed to the poor flow distribution among fibers and the low heat efficiency because of the difficulties in controlling heat leaks among fibers.

Commercial MD processes are mostly based on AGMD, and the process diagram is shown in Figure 7.9. The AGMD consists of multiple stacked panels with a condenser foil on one side and a hydrophobic porous membrane on the other side. Feed water is preheated in the brine side of the module by heat exchanger. Additional heating is performed to meet the target feed temperature right before the feed enters the process. In the first stage, water vapor passes the porous membrane and travels through the air gap to be condensed on the condense foil. The latent heat stored in the vapor is transferred to the condenser foil and heats up the feed water behind it. Once the water in the

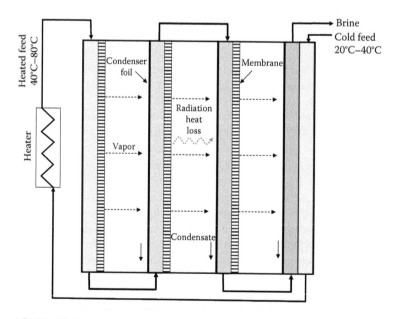

FIGURE 7.9   AGMD with three stages.

second stage is heated up sufficiently, water vapor passes the second membrane and travels to the next condenser foil in the third stage. The same processes repeat until the end of the stages. As the stage goes to the end, the temperature difference between two neighboring stages decreases and the amount of permeate produced decreases. The maximum number of stages should be decided by comparing the additional capital costs and the additional permeate obtainable when extra stages exist.

In AGMD, the dissolved air in feed water evaporates with water and accumulates in the air gap because it does not condense under the condition. The air accumulated in air gap compromises the vapor transportation through the air gap and increases heat leakage by conveying the heat from the membrane to the condenser in the next stage. Vacuum can be applied to the air gap to drive the accumulated air out. Alternatively, feed water can be deaerated to remove dissolved air (Winter et al. 2012; Jansen et al. 2013). In this method, deaerated feed water absorbs the air in air gap by equilibrium reaction while supplying steam to the air gap.

In an ideal condition, heat transfer between the two neighboring stages occurs only by water vapor. However, the heat leak occurs from the membrane to the next stage condenser foil through radiation and air convection. The heat leak can be reduced by increasing the air gap, but it increases the resistance for vapor transportation. Therefore, some extents of mass transfer and heat efficiencies are inevitably compromised to obtain the optimum performance. Most AGMD systems have air gaps bigger than 2 mm, but some MD systems have very small air gap of around 1 mm, for example, Scarab AB. On the contrary, the heat leak can be almost completely prevented by insulating two neighboring stages in conventional distillation processes, for example, MSF and MED.

The heat efficiency of distillation processes is measured by the gain-to-output ratio (GOR). It is a ratio between the total latent heat of the water obtained by MD and the actual heat spent to obtain the water. Thereby, 628 kWh/m³ corresponds to a GOR of 1. In MSF and MED, insulation among vessels/stages are relatively easy so that heat is carried among vessels through water vapor. In MD, however, heat can transfer directly from one membrane fiber/plate to other fiber/plate through radiation and convection instead of being carried by water vapor. Because of the heat leak in the closely packed membranes, the GOR of MD ranges realistically at 2 to 6, whereas the GOR of MSF and MED ranges at 8 to 12 with a maximum at around 20 (Winter et al. 2012; Camacho et al. 2013; Khayet 2013).

The low operating temperature enables MD to be built with lower-cost materials when it is compared with MSF and MED, but the low temperature also causes a bulky system. The water vapor transport rate is dictated not only by the temperature difference between two stages but also by the water temperature and the associated vapor pressure. Thus, the size of the distillation system is inversely correlated with the operating temperature in a given process. As shown in Figure 7.10,

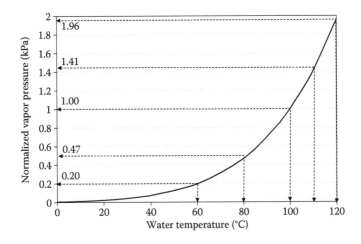

FIGURE 7.10 Normalized water vapor pressure against 100°C as a function of temperature.

water vapor pressure sharply declines as temperature declines. At the median feed water temperature used by MD, for example, 60°C, vapor pressure is only 20% of that at 100°C according to the figure. Because of the disadvantage in the mass transfer rate and the potential membrane integrity issue by pore wetting, MD can have difficulty in competing with MSF and MED for large-scale desalination.

7.3 MEMBRANE PROCESS FOR WATER RECLAMATION

7.3.1 NONPOTABLE REUSE

MBR produces a high-quality effluent that can be discharged to river, ocean, or environmentally sensitive areas, but it can also be recycled for recreational field irrigation, agriculture, artificial waterfall and recreational pond makeup, nonpotable domestic use, cooling tower makeup, vehicle washing, fire protection, dust control, construction, etc. For the existing CAS process, the secondary effluent can be filtered by membrane to remove suspended solids, but the effluent quality is not as good as the MBR effluent in terms of organic contaminants, for example, BOD, COD, and total organic carbon (TOC).

When the MBR effluent is used for irrigation, the salts contained in the recycled water can accumulate in the soil. The salinity of water is determined by electrical conductivity and total dissolved solid (TDS). It has been considered that if the conductivity is below 0.7 mS/cm and TDS is below 450 mg/L, recycled water does not negatively affect the soil. In addition to the salinity, the ratio of monovalent ions and divalent ions also affects soil erosion. In fact, divalent ions such as calcium and magnesium improve granule sizes by bridging soil particles. On the contrary, excess monovalent ions reduce the granule size by depriving divalent ions from soil particles through the ion exchange. The small granule size causes poor soil permeability and makes the soil vulnerable to erosion, which eventually leads to poor tilth. The sodium absorption ratio (SAR) shown in Equation 7.7 is used as a parameter indicating the ion exchange potential. In this equation, all the ionic concentrations are measured as millimole per liter. The SAR should be carefully monitored along with electrical conductivity in recycled water. Equation 7.7 is presented as follows:

$$\text{SAR} = \frac{[\text{Na}^+]}{\sqrt{[\text{Ca}^{2+}] + [\text{Mg}^{2+}]}} . \tag{7.7}$$

The allowed SAR for irrigation is dependent on electrical conductivity, but it is considered safe in general if it is below 3 with a conductivity of 1000 to 1500 μS/cm for the municipal MBR effluent. If the value of SAR is 3 to 6, care must be taken for the sensitive crops such as citrus, nuts, and deciduous fruits. Recycled water with an SAR of 6 to 12 can be used for beans, but conductivity should be higher than 1900 μS/cm (USEPA 2012). The land application of gypsum and lime can be considered if the SAR of recycled water is excessive. Depending on SAR and conductivity, the alert levels vary, as summarized in Table 7.4.

For surface irrigation, most tree crops (fruits) and woody plants (nuts) are sensitive to sodium and chloride, but most annual crops (rice, wheat, oats, barley, tomatoes, etc.) are not sensitive. With overhead sprinkler irrigation and low humidity (<30%), sodium and chloride may be absorbed through the leaves of sensitive crops. The residual chlorine in the recycled water should be below 1 mg/L in general, but some sensitive crops may be damaged at the level as low as 0.05 mg/L. Most plants can be severely damaged by the residual chlorine greater than 5 mg/L (USEPA 2012).

The recommended maximum allowable levels of trace metals for irrigation are also available (USEPA 2012), for example, aluminum (5 mg/L), arsenic (0.1 mg/L), boron (0.75 mg/L), cadmium (0.01 mg/L), chromium (0.1 mg/L), cobalt (0.05 mg/L), copper (0.2 mg/L), fluoride (1.0 mg/L), iron

TABLE 7.4
Guidelines for Interpretation of Water Quality for Irrigation

Potential Irrigation Problem		Units	Degree of Restriction on Irrigation		
			None	Slight to Moderate	Severe
Salinity (Affects Crop Water Availability)					
Conductivity		mS/cm	<0.7	0.7–3.0	>3.0
TDS		mg/L	<450	450–2000	>2000
Infiltration (Affects Infiltration Rate of Water into the Soil; Evaluate Using ECw and SAR Together)					
SAR	0–3	Conductivity	>0.7	0.7–1.2	<0.2
	3–6	(mS/cm)	>1.2	1.2–0.3	<0.3
	6–12		>1.9	1.9–0.5	<0.5
	12–20		>2.9	2.9–1.3	<1.3
	20–40		>5.0	5.0–2.9	<2.9
Specific Ion Toxicity (Affects Sensitive Crops)					
Sodium (Na^+)	Surface irrigation	SAR	<3	3–9	>9
	Sprinkler irrigation	mEq/L	<3	>3	
Chloride (Cl^-)	Surface irrigation	mEq/L	<4	4–10	>10
	Sprinkler irrigation	mEq/L	<3	>3	
Boron		mg/L	<0.7	0.7–3.0	>3.0
Miscellaneous Effects (Affects Susceptible Crops)					
Nitrate		mg/L	<5	5–30	>30
Bicarbonate		mEq/L	<1.5	1.5–8.5	>8.5
pH		–	Normal range 6.5–8.4		

Source: USEPA, *Guidelines for Water Reuse.* EPA/600/R-12/618, 2012.

(5.0 mg/L), lithium (2.5 mg/L), manganese (0.2 mg/L), molybdenum (0.01 mg/L), nickel (0.2 mg/L), selenium (0.02 mg/L), vanadium (0.1 mg/L), and zinc (2.0 mg/L).

There is no water quality criteria established for the use of recycled water for the cooling tower makeup. It is not only because there are several different types of cooling tower built with various materials but also because most cooling towers can accommodate a wide range of water quality with the modification of operating conditions and chemical treatment programs as long as the quality of makeup water does not change abruptly. The cycle of concentration should be adjusted by reflecting the recycled water quality. Proper biocides, antiscalants, and corrosion inhibitors should be selected and dosed depending on the water quality and the building materials of cooling tower. The MBR effluent is advantageous over CAS effluent when it is used for cooling tower makeup because it does not contain suspended solids that can deposit on surface and increase corrosion and biofouling.

The USEPA guideline suggests that regardless of the purpose, reclaimed water should be disinfected, but the target removal rate varies depending on the purpose. If the reclaimed water is used for the purposes without human contact, it should be disinfected to achieve an average fecal coliform of less than 200/100 mL. If the reclaimed water is likely or expected to contact with human or if there is a potential for cross-connects with potable water lines, disinfection should be performed to achieve zero coliforms per 100 mL. Limits for parasite and viruses are not included in the guideline because those are deemed acceptable for nonhuman contact purposes as long as the reclaimed water meets the fecal coliform limits.

7.3.2 CAS Microfiltration or Ultrafiltration versus MBR as a Pretreatment of Reverse Osmosis

When recycled water is used for boiler feed water, process water, nonpotable domestic water with human contact, etc., the filtered effluent may need further treatment using RO to remove dissolved organics and ions. There are two process options, as shown in Figure 7.11. If there is an existing CAS process, the secondary effluent is filtered by microfiltration (MF) or ultrafiltration (UF) followed by RO and disinfection using ultraviolet (UV), chlorine, ozone, etc. For the greenfield projects starting from no existing processes/equipment, MBR-RO can be used instead of CAS-MF/UF-RO. This MBR-RO process has been known to be not only more compact but also more economical than CAS-MF/UF-RO because of the lack of a secondary clarifier and the savings from the footprint (Côté et al. 2004).

In the NEWater project performed in the mid-2000s in Singapore, the MBR-RO system was compared with the CAS-MF-RO system in a municipal wastewater plant. To inhibit microbial growth and to reduce the subsequent membrane biofouling, 2 mg/L of chlorine was added to the RO feed tank. The feed water was declorinated, and 2 mg/L of antiscalant (PC-191T, Nalco) was injected to the RO feed water pipe. As summarized in Table 7.5, MBR produced RO feed water with slightly lower TOC than CAS-MF. In addition, the effluent TOC fluctuated less in MBR-RO than in CAS-MF-RO. It was concluded that MBR-RO produced not only better quality water but also the effluent quality was more consistent. The low NH_4-N and NO_3-N in the MBR-RO effluent was attributed to the superiority of MBR to CAS in removing biodegradable contaminants. The slightly greater inorganic contents in the MBR effluent than in CAS-MF might be caused by the intrinsically lower excess biosolid production that curtailed the amount of inorganic ions assimilated to the excess biosolids (Qin et al. 2006; Tao et al. 2008). Similar observations were made in an industrial wastewater reclamation project. MBR-RO produced consistently better water qualities than CAS-RO when the wastewater from LCD manufacturing process was treated (Lee et al. 2011).

RO membranes are fouled less with MBR permeate than with the CAS-MF/UF effluent, although both processes produce RO feed waters with a sufficiently low silt density index (SDI), for example, <3 (Hosseinzadeh et al. 2013). In a long-term side-by-side test in a municipal wastewater treatment plant (WWTP), the MBR effluent had consistently lower TOC than the CAS-UF hybrid system (Kent and Farahbakhsh 2011). As a consequence of the better feed water quality, the rate of permeability loss of RO membrane was around one-half in MBR-RO compared with MBR-UF, that is, 0.27 LMH/bar/month versus 0.60 LMH/bar/month.

MBR-RO has been used to reclaim water in various industries, for example, electronics, paper, mining, tannery, and brewing (Mortazavi 2008; Hadler and Kullmann 2012). When pilot studies were performed using the wastewater produced from LCD manufacturing process containing dimethyl sulfoxide, tetramethyl ammonium hydroxide, and ethanolamine, the initial TOC of 500 to 800 mg/L was reduced to 10 to 30 mg/L by MBR. The TOC was further reduced to 0.14 to

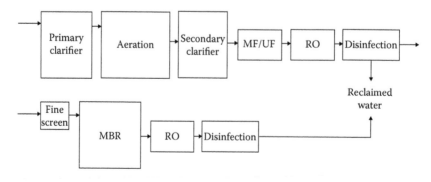

FIGURE 7.11   Water reclamation processes by either CAS-MF-RO or MBR-RO.

TABLE 7.5

Comparison of Effluent Qualities of CAS-MF-RO and MBR-RO in Municipal WWTP

Parameter	RO Feed (mg/L)		RO Permeate	
	From CAS-MF	From MBR	CAS-MF-RO	MBR-RO
TOC	7	4.8–5.0	0.033–0.053	0.024–0.033
B	0.08–0.10	0.071–0.104	0.059–0.074	0.051–0.078
Ca	24.8–30.0	30.0–33.5	0.001–0.004	0.002–0.006
Na	123–169	149–185	2.83–3.86	2.85–3.84
F	1.29–1.42	1.51–1.55	0.012–0.029	0.015–0.033
Cl	146–193	164–214	1.51–2.51	1.75–2.69
NH_4-N	1.94–5.54	1.0–1.7	0.07–0.42	0.05–0.22
NO_3-N	26.2–60.5	6.37–28.4	1.62–3.86	0.53–1.92
SO_4	66.2–71.7	71.9–82.1	<0.010–0.034	0.029–0.48
Hardness as $CaCO_3$	104–129	125–136	0.011–0.018	0.008–0.042
Total silica as SiO_2	7.43–8.34	9.225–9.82	0.140–0.160	0.151–0.201
TDS	573–703	557–749	9.8–13.3	8.9–13.1
pH	6.6–6.8	6.7–6.9	5.4–5.5	4.7–5.2

Source: Qin, J.-J. et al., *J. Membrane Sci.* 272(1–2), 70–77, 2006.

0.34 mg/L by RO when the flux was maintained at 13 to 20 LMH (Lee et al. 2009). Meanwhile, the initial conductivity was reduced from 1050 to 20.6 μS/cm. On the basis of the observations from the pilot study, a full-scale RO system was built with a capacity of 12,000 m^3/day to produce reclaimed water from the MBR effluent.

MBR is a more promising option than the hybrid process of CAS and MF/UF in greenfield projects. Although it can be a more economical option to produce high-quality RO feed water out of wastewater, it has one disadvantage in terms of integrity monitoring. Because of the difficulty of applying a high air pressure in submerged membranes in situ, it is capable of detecting only the defects larger than few micrometers, as discussed in Section 2.7.1.

In the MBR-RO hybrid process, antiscalants and acids are added to the MBR effluent to reduce the scale formation in RO unless the water recovery is below 30% to 50%. Bleach (NaOCl) is also dosed to the MBR effluent to prevent biological growth in the holding tank, but sodium bisulfite (NaHSO$_3$) needs to be injected to remove the residual-free chlorine before RO system. In general, RO/nanofiltration (NF) membranes are very sensitive to free chlorine, and the tolerances of most membranes are known to be only 200 to 1000 ppm·h. Alternatively, monochloramine (NH$_2$Cl) can be added at 2 to 5 mg/L (Stanford 2010; Peck 2011) instead of chlorine or hypochlorite. Because the tolerance of the membrane against chloramine is very high at 150,000 to 300,000 ppm·h before salt passage doubles, dechlorination steps are not required before RO (Bates 1990; DOW 2013). The water recovery of RO ranges from 60% to 85% depending on the initial raw wastewater quality, the MBR operating condition, the RO system design, etc. (Lozier and Fernandez 2001; Bartels et al. 2004; Yang et al. 2009). The average flux of RO in such application is 17 to 20 LMH or somewhat lower. The flux of RO in water reclamation is slightly lower than those in typical surface water filtration because of the higher fouling potential by the organics. In a large municipal WWTP in Sulaibiya, Kuwait, 80% to 83% of wastewater recovery was realized using CAS-UF-RO at a product flow rate of approximately 350,000 m^3/day. In a large-scale municipal plant (groundwater replenishment system [GWRS]) in California, close to 85% recovery is realized using CAS-MF-RO system without recycling RO concentrate.

Overall water recovery can be raised by recycling RO concentrate to the upstream of MBR (Joss et al. 2011). In a long-term pilot study performed in municipal WWTP, overall water recovery could be maintained higher than 90% by recycling the RO concentrate to MBR. The conductivity in MBR could be controlled within the tolerance limit by discharging a portion of the RO concentrate. As a consequence of the partial salt accumulation in MBR, the conductivity of reclaimed RO permeate was somewhat high at 280 μS/cm. Figure 7.12 shows a flow diagram as an example of MBR-RO process. In another study, 92% of industrial wastewater was recovered in a pharmaceutical plant using MBR with nanofiltration membrane (MBR-NF) hybrid system by recycling a portion of NF concentrate to MBR. The removal efficiency of spiramycin was measured at 95%, and the COD and the TOC of the reclaimed water were 35 and 5.7 mg/L, respectively (Wang et al. 2014a). In another long-term study on MBR-NF, 100% of NF concentrate was recycled to MBR to obtain nearly 100% water recovery, except the water lost with excess sludge (Kappel 2014). Under this condition, monovalent ion concentration reaches a steady state in which the amount incoming through feed is balanced with the amount lost through the NF permeate. The accumulated calcium ions in the system was supposed to form mainly hydroxyapatite ($(Ca_{10}(PO_4)_6(OH)_2)$) in mixed liquor and deposit on the membrane surface under the given condition. The same scale was also observed on the NF membrane surface. The slightly elevated membrane fouling tendencies in MF and NF were mainly due to the elevated nonbiodegradable COD in the system because of the recycle of NF concentrate.

Nearly complete removals of pharmaceutical compounds have been obtained by MBR-RO except for charge neutral nonbiodegradable compounds at neutral pH, for example, benzotriazoles and propranolols. The high removal efficiencies of charged trace organics by RO is due to the rejection mechanism of RO, as discussed in Section 1.2.10.

The following conditions should be considered in the MBR-RO system design:

- Chlorine (1–3 mg/L) is injected into MBR permeate pipes or holding tanks to minimize microbial activities. If biofilm is formed and sloughed off in holding tank, it expedites the cartridge filter fouling and potentially plugs up the RO membrane channels by being caught by mesh spacers.
- Minimizing the pipe length between MBR and RO is beneficial to minimize the chance of biofilm growth. However, minimum pipe length should be secured to enable a sufficient mixing of chemicals injected into the pipe.
- $NaHSO_3$ should be added in between the cartridge filter and the RO to minimize the cartridge filter fouling/plugging by microbial activities. The mixing of NaOCl with RO feed water should be secured by static mixer or multipoint radial injection.
- Antiscalants are required to prevent scaling of RO membrane. The exact dosages and chemistries of antiscalants should be consulted by chemical suppliers. Alternatively, acids or CO_2 can be added to RO feed to lower the pH, but this method is primarily for $CaCO_3$ scale. Acids are not effective for $CaSO_4$ scale because its solubility is hardly affected by the pH. The pH adjustment is considered more costly than using antiscalant in general.

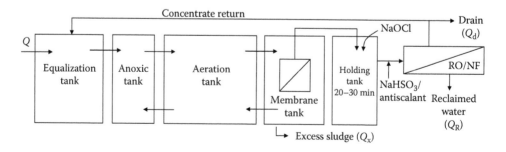

FIGURE 7.12   Water reclamation processes.

In the MBR-RO system, the OLRs to the biological system not only affect the membrane performance of MBR but also affect the RO membrane performance. In a study, RO membrane performance was compared in two parallel MBR-RO systems with different organic loadings. When the F/M ratios of MBR were maintained at 0.17 g COD/g MLSS/day and 0.50 g COD/g MLSS/day, RO membrane was fouled faster when the F/M ratio was high. It was attributed to the soluble polysaccharides (PS) and soluble transparent exopolymer particles (TEP) contained in the MBR effluent (Kitade et al. 2013).

CAS-MF-RO is being used to produce 265,000 m^3/day of reclaimed water for GWRS since 2008 in Orange County in California (Shu et al. 2014). The GWRS is in the middle of expansion to 379,000 m^3/day and is planned to be inline by 2015. The main purpose of this project is to produce potable quality water required to recharge the aquifer. The recharged water raises the groundwater level, preventing the seawater from intruding into the underground aquifer. Eventually, the recharged water becomes a source of groundwater. The average flux of RO is known to be around 20 LMH at a recovery of 80% to 85%.

7.3.3 INDIRECT POTABLE REUSE

7.3.3.1 Concept

Because of the water shortages caused by climate change, industrialization, and population growth, the need of reclaiming wastewater has been growing in many geographical areas, for example, Southern California, Florida, Texas, North Africa, Middle East, and Singapore. In the United States, water flows have significantly reduced in many large rivers particularly in semiarid regions in the southwest. The abstraction of groundwater is often unsustainable in such dry areas resulting in declining water tables (Rodriguez et al. 2009). As a consequence, wastewater has been reclaimed for the activities that do not involve direct human contact in many geographical areas. Recently, reclaimed water has started to be used for all human activities, including potable purposes in some areas.

Historically, wastewater has been reclaimed for potable purposes without being recognized even before wastewater treatment was introduced. Freshwater taken from the upstream of river is used for human activity, and the wastewater produced is discharge back to the river with or without treatment. The effluent is diluted in the river by a factor of hundreds or thousands. While the diluted wastewater is flowing through the river, it is further treated biologically, physically, and photochemically. The diluted wastewater can be taken again in the downstream and used for human activities. These cycles can repeat multiple times until the river meets with the ocean. The concept of unplanned indirect potable reuse (IPR) is illustrated in Figure 7.13.

The unplanned IPR can be better controlled by storing the treated effluent in reservoirs. In this method, the secondary effluent filtered by MF/UF can be further treated by RO followed by advanced oxidation before being stored in reservoirs. While the tertiary effluent is stored in the reservoir for the predetermined period, the treated effluent is diluted by fresh river water, rainwater, groundwater, etc., and is further treated biologically, physically, and photochemically, just as what happens in the river in unplanned IPR. According to the local regulation, the treated water should be stored in the reservoir for a minimum of 12 months while being blended with the incoming natural water within the reservoir (City of San Diego 2005). Alternatively, the reclaimed water can be injected into the well/aquifer, where the injected water is further treated by soil filtration, adsorption, biological degradation, etc., while being diluted by the groundwater. The environmental buffer is beneficial for providing further opportunities to treat residual contaminants in the reclaimed water, but it can also introduce industrial, municipal, and agricultural contaminants depending on the local condition. The water taken from reservoirs or wells can be treated in a drinking water plant and supplied as tap water. IPR has been performed in the Orange County and West Basin Municipal Water Districts in California, the City of Scottsdale, Arizona, El Paso Water Utilities in Texas, the Upper Occoquan Service Authority in Virginia, etc., in the United States.

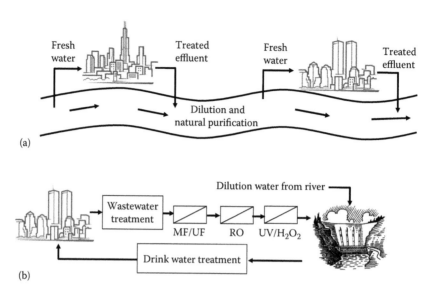

FIGURE 7.13 Example of (a) unplanned and (b) planned IPR of reclaimed wastewater.

7.3.3.2 Reclaimed Water Quality

Reverse osmosis (RO) of MF permeate followed by advanced oxidation process (AOP) provides sufficient water quality that meets the drinking water standards. In NEWater project, it was observed that the 190+ drinking water standards set by the USEPA and the World Health Organization were consistently met. Similar observations were also made in the GWRS project performed by the Orange County Water District (OCWD) of California. Although this compliance is essential to protect the public from the unknown health risks, it does not necessarily guarantee the safety of the recycled water when it is used in potable purposes. It is because the list of water quality parameters only addresses the known concerns, whereas the municipal wastewater is a result of complex human and natural activities that cannot be defined holistically. Therefore, concerns are still remaining despite the rigorous compliances of reclaimed water with all regulations and guidelines. If planned IPR is compared with unplanned IPR, pollutants are more thoroughly treated in planned IPR by more extensive processes. However, the low dilution ratio of planned IPR may impose higher risks from unknown artificial compounds. With an additional hurdle from the public resistance, the reclaimed water has been primarily used for IPR rather than direct potable reuse (DPR) despite its excellent quality. Only few cases have known for DPR to date, for example, direct blend of 6800 m³/day of treated municipal wastewater with raw water in the Colorado River Municipal Water District in Texas since April 2013 and direct blend of 18,900 m³/day of reclaimed water with raw water in Wichita Falls, Texas, since July 2014, using virtually the same processes shown in the next section (Sloan 2014).

The log removal values (LRVs) required for IPR vary depending on the state and the country, but the minimum values are 8 log for *Cryptosporidium*, 9.5 to 10 log for enteric viruses, and 8 log for *Campylobacter* (Rodriguez et al. 2009). For example, California requires 12-log removal for enteric virus, 10 log for *Giardia cyst*, and 10 log for *Cryptosporidium oocyst*. In addition, the target LRV should be met using at least three serially connected treatment processes regardless of the LRV performance of each unit process to prevent catastrophic failure when one unit process fails. The LRV credits for each unit process are awarded by the state government in the United States, depending on the capability of removing pathogens and the existence of the verification methods. The free chlorine or chloramine added to the CAS effluent disinfects the effluent microorganisms, and its efficacy maintains until reducing agents such as bisulfite are added before RO process. MF is able to remove all parasites and most bacteria and viruses, as discussed in Section 2.6.4.1, but membrane integrity must be tested periodically using approved integrity test methods. RO is capable of removing all leftover microorganisms leaked through MF, but the LRV credit allowed is typically low at 1.5 to 2.0 because of the lack of more sophisticated and practical integrity test method than conductivity.

In a pilot test, it was demonstrated that RO was effective in removing all metal ions and 29 pharmaceutical compounds and personal care products from the filtered secondary effluent (City of San Diego 2005). Studies for nonregulated contaminants such as endocrine disrupters, pharmaceuticals, and personal care products are ongoing in many projects as part of regulatory requirements or research interest. As discussed in Section 7.3.2, RO is proficient to remove most nonbiodegradable organics not treated by CAS as long as they are charged, but it is not effective in removing charge neutral small contaminants, for example, N-nitrosodimethylamine (NDMA). Therefore, in the GWRS in Orange County, high-dose UV reactors were used along with H_2O_2 to enhance the oxidation reaction. Instead of a regular UV dose (50 mJ/cm^2) used for disinfection, a high dose at 500 to 750 mJ/cm^2 is required to destroy NDMA. With a high-dose UV and H_2O_2, estrone (E1), 17-α-ethinyl estradiol, and 17-β-estradiol were removed below the detection limit of 10 ng/L, whereas caffeine was destroyed below 100 ng/L (Rodriguez et al. 2009). A similar observation was made in another study targeting estrone, 17-β-estradiol, ethinyl estradiol, 2,4-dichlorophenoxyacetic acid, mecoprop, 2,4,5-trichlorophenoxyacetic acid, atrazine, terbutryn, metaldehyde, and NDMA (James et al. 2014). In this study, most of the target trace organics were removed at well more than 90% efficiencies from the filtered CAS effluent when 3 mg/L of H_2O_2 was dosed along with approximately 700 mJ/cm^2 of UV. However, metaldehyde was removed only by 45%, whereas 2,4-dichlorophenoxyacetic acid, 2,4,5-trichlorophenoxyacetic acid, and atrazine were removed by 85% to 90%. When the same method was applied for the water treated by CAS-MF-RO with a raised UV dose to approximately 2000 mJ/cm^2, removal efficiencies of all compounds exceeded 99%, but metaldehyde removal efficiency was slightly lower at 98%.

7.3.3.3 Groundwater Replenishment System

In the United States, IPR has been performed in Southern California for more than 40 years (Patel 2012; GWRS 2014). IPR is also performed in other states such as Arizona, Colorado, Texas, Florida, and Virginia. The Water Factory 21 project performed by the OCWD was the oldest project known to the public. The initial capacity of 19,000 m^3/day (or 5 MGD) was upgraded through the GWRS project. Since January 2008, 265,000 m^3/day (or 70 MGD) of reclaimed water is being produced. The treated water is transported through a 20 km long pipeline that runs through the cities of Fountain Valley, Santa Ana, Orange, and Anaheim. Half of the reclaimed water is being pumped into the injection wells to raise the barrier for seawater intrusion, and the other half is pumped to recharge basins in Anaheim, California. The water injected percolates through the sand and gravel in the basins and is naturally filtered into the groundwater basin. This groundwater is pumped from more than 400 wells operated by local water agencies, cities, and other groundwater users. The capacity of GWRS is planned to increase to 378,000 m^3/day (or 100 MGD) by 2015 and again increase to 492,000 m^3/day (or 130 MGD) ultimately.

In GWRS, the secondary effluent from the CAS process is filtered by MF before it is fed to RO. To further remove trace organics not removed by CAS and RO, for example, NDMA, 1,4-dioxane, etc., AOP based on UV and H_2O_2 is performed before the treated water is fed to injection wells. The net LRV target for enteric virus is 12 using a hybrid process of CAS-MF-RO-AOP. No log-removal credits were awarded to the process with less than 1 log removal performance, and the maximum credit given to one process was limited at 6 log.

To produce 325,000 m^3/day of filtrate out of 363,000 m^3/day of the secondary effluent at approximately 90% recovery efficiency, 15,808 MF membrane modules (MEMCOR® CS, Evoqua) with 0.2 μm pore sizes are submerged in 26 basins/cells as of 2010. Each basin/cell contains 19 membrane racks that consist of 36 CS modules in each rack. MF system produces a permeate with <0.2 NTU and an SDI below 3. MF concentrate is recycled back to the equalization tank and fed to the CAS. Thus, the effective water recovery of MF becomes 100%. Chlorine is injected to the CAS effluent to protect MF membranes from biofouling. Maintenance cleaning is performed every 22 min by backwashing the MF membranes using filtrate with a simultaneous air scouring. Recovery cleaning is performed every 21 days for 4 h using NaOCl, citric acids, and proprietary cleaners. Under this condition, membrane life is expected to be 5 to 7 years.

MF permeate is pretreated before it is fed to RO. To reduce scaling potential, pH is lowered using sulfuric acid before antiscalant is added. The low pH is also crucial to secure high ammonia removal efficiency in RO (>90%) by moving the equilibrium toward NH_4^+ from NH_3. Bisulfite is added before the MF filtrate reaches the subsequent RO process. Cartridge filters with a 10 µm rating are located upstream of RO to remove the debris that might be introduced with the chemicals or present on pipe walls. The RO with three stages (78–48–24 array with 7 module vessels) are used to obtain close to 85% water recovery as permeate without concentrate recycle. The concentrate is directed to the ocean, and hence salt accumulation in CAS does not occur. Low-pressure hydranautics RO membranes (ESPA2) are used at 10.3 bar (or 150 psi) at a design flux of 20 LMH (or 12 gfd). There are 15 banks of RO that produce 18,900 m³/day of permeate each. Each bank consists of 150 pressure vessels with seven 8-inch modules each in a 78–48–24 array. The total number of membrane modules installed is 15,750. Each bank has a dedicated feed pump equipped with a 746 kW (or 1000 hp) motor. Membrane life is expected to be 5 years at the operating condition used.

To destroy trace organics while further sanitizing the RO effluent, advanced oxidation is performed using UV and H_2O_2. There are nine trains of UV system (TrojanUVPhox™, TrojanTechnologies Inc.) consisting of three vessels in a series stacked on the top of one another. Each UV train treats 33,100 m³/day of RO permeate. Each of the UV/H_2O_2 system is aimed to obtain 4-log removal (or 99.99%) for viruses and 1.2 log removal (or 94%) for NDMA with 3 mg/L of H_2O_2. UV alone without H_2O_2 removes only 43% to 66% of NDMA (Plumlee et al. 1998). The UV systems are equipped with low-pressure high output amalgam lamps. The primary water quality targets are TOC below 0.5 mg/L, total nitrogen below 5 mg/L, TDS, and NDMS below 10 ng/L (or 10 ppt).

Around 75% of treated water is sent to the decarbonation tower packed with media to remove excess carbon dioxides and adjust pH. While the water is falling down through the media, carbon dioxides are stripped out by the air flowing upward. The large surface area of media helps the water contact with the air and increase the mass transfer rate. Hydrate lime (20 mg/L) is added to the reclaimed water at a wet well to minimize the corrosiveness of the water in the pipeline. The total energy consumption is estimated at 1.1 kWh/m³ excluding the CAS process. RO consumes a largest amount of energy at approximately 48% of total energy or 0.53 kWh/m³. MF, effluent pumps, and AOP consume approximately 25%, 18%, and 7% of total energy, respectively.

7.3.4 MEMBRANE INTEGRITY TEST/MONITORING

7.3.4.1 Overview

Maintaining membrane integrity is a crucial part of DPR and IPR of wastewater to ensure the safety of water from pathogenic contamination. The integrity of MF and UF membranes treating the secondary effluent can be monitored based on pressure decay tests, vacuum decay tests, etc., as discussed in Section 2.7. These tests allow up to 4-log virus removal credits for UF and up to 0.5-log for MF depending on the jurisdiction. On the contrary, because of the lack of sensitive and reliable integrity test methods, much less credits (0–2 log) have been given to RO despite its superior virus removal capabilities. According to a study, RO is capable of at least 5.4 log (or 99.9996%) with a virgin membrane and much greater log removals with a fouled membrane, for example, 6.0 to 7.9 log or higher (Mi et al. 2004). Various integrity test methods have been investigated for RO, for example, conductivity, turbidity, particle counts, TOC, sulfate, and fluorescent dye. However, each method has its own strengths and limitations, as will be discussed in the following sections.

7.3.4.2 Method of Using Conductivity, TOC, and Sulfate

Because conductivity is always present (or not zero) in permeate and is variable depending on the flux of membrane, water temperature, ionic composition of feed water, etc., subtle changes in permeate conductivity are not a definitive indication of integrity breach. For example, the theoretical LRV can be calculated for a hypothetical condition, where feed and permeate conductivities are 500 and 10 µS/cm, respectively. If the rise of permeate conductivity from 10 to 11 µS/cm can be measured reliably, the corresponding LRV becomes 2.7 (= $\log_{10}500/(11 - 10)$). However, such small changes in permeate conductivity can occur without an integrity breach when any of feed water quality, water temperature, recover, etc., changes. If a rise of 20 µS/cm in permeate conductivity is larger than any conductivity change anticipated by the fluctuating operating condition, the corresponding LRV traceable becomes 1.4 (= $\log_{10}(500/20)$) under the condition.

TOC has the same limitation as conductivity in RO integrity tests because of similar reasons. The background TOC in RO permeate is not zero, but it is 0.02 to 0.1 mg/L without integrity breaches in municipal wastewater recycle (Tao et al. 2008; GWRS 2014). Like conductivity, permeate TOC varies depending on RO operating condition and feed water quality. Therefore, although very sensitive inline TOC analyzers that can detect down to 0.5 µg/L exist, for example, Sivers 800 model of GE, the maximum LRV traceable based on the TOC rejection efficiency is not high. For example, if TOC of RO feed (or MF/UF permeate) is 5 mg/L and permeate TOC is 0.05 mg/L, the rise of TOC by 0.01 mg/L does not necessarily indicate an integrity breach because the same can happen by the natural fluctuations caused by the dynamic operating condition. If the change by 0.05 ppm is larger than expected based on the natural fluctuations, integrity breach can be called when TOC exceeds 0.10 mg/L. The corresponding LRV becomes 2 in this case (= $\log_{10}(5/(0.10 - 0.05))$).

Turbidity and particle counts are simply not sensitive enough to be used for integrity monitoring in RO. In particular, when RO feed water is pretreated by MF/UF, turbidity and particle counts are not useful in monitoring RO integrity because there are very little particles present from the beginning. Sulfate can be easily measured, but similar issues exist as for TOC.

7.3.4.3 Method of Using Fluorescent Dye

More recently, the use of fluorescent dye has been proposed as a surrogate molecule as shown in Figure 7.14 (Zeiher et al. 2004; Frenkel and Cohen 2014). The measured LRV can be considered a conservative indicator of membrane integrity because the dyes used are much smaller than actual viruses. In this method, fluorescent markers injected into the RO feed continuously or intermittently while the dynamic changes of the marker concentration are monitored in the permeate in real time using inline fluorometers. The maximum LRV traceable is calculated using the dye concentrations in feed and permeate, as shown in Equation 2.11, when no integrity breaches exist. For example, if the dye rejection efficiency is 99.9999%, the maximum LRV traceable becomes 6. However, the actual LRV traceable is 0.5 to 1.0 units lower than the maximum because of the necessity of the margins between no leaks and leaks. Many things need to be considered when choosing a proper dye, listed as follows:

- *Molecular weight and charge*—In general, rejection efficiency increases as the dye's molecular weight and charge increases. The bulkiness of the molecule is another factor affecting the efficiency. If dyes are rejected at 6-log efficiency (or 99.9999%), the maximum

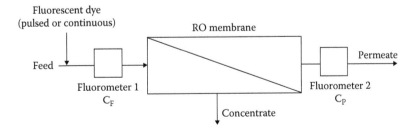

FIGURE 7.14 Membrane integrity monitoring using fluorescent dye.

trackable LRV will be somewhat lower at 5 to 5.5 because the rejection efficiency in the field varies depending on operating condition, and some margins are required to decisively call the dye leakage.

- *pH dependence of the charge*—If carboxyl and amino groups exist in the molecule, the charges of dye are variable depending on pH. The molecular charge affects not only the rejection efficiency by RO membrane but also the fluorescence strength.
- *Fluorescent strength and the emission wavelength*—The stronger the fluorescence response (or the lower the detection limit), less chemicals are necessary or higher LRV is trackable using same amount of dye. In addition, the emission spectra should not overlap with the background spectra to avoid interferences. When specific fluorescent strengths were compared with uranine, eosin, rhodamine WT, and fluorescein, uranine showed the strongest response and was detectable down to very low level at 0.01 µg/L (Frenkel et al. 2014).
- *Chlorine resistance*—The dye fed to feed water might be exposed to free chlorine. Depending on dye chlorine resistance varies.
- *Temperature dependence*—Fluorescence strength tends to decrease as temperature increases. The sensitivity to temperature varies depending on dye.
- *Toxicity*—Some dye inevitably passes membrane and mixed with permeate. During the test, the maximum dye in permeate should not exceed the level allowed for drinking water.

When rhodamine WT was used as a marker, the maximum equivalent LRV to the rejection efficiency was around 5.5-log in the beginning of dye injection, but it was stabilized at around 3.6 log after 45 min (Kitis et al. 2003). It is considered that the hydraulic delays occurring in feed and permeate channels caused a high apparent dye rejection efficiency in the beginning of the test. The delays are also caused by the time required to stabilize the concentration polarization (CP) layer on membranes, which determines the effective (or actual) dye concentration on the membrane surface. The time taken until adsorption and desorption kinetics are settled in the feed and permeate sides of the membrane is another potential cause of the delay. In another study (Frenkel et al. 2014), uranine was injected to the feed side of a flat sheet test cell with 21.6 cm^2 of surface area. It was reported that uranine was capable of tracing up to 4.2-log at a rejection efficiency of 99.994%. However, the actual LRV trackable appears somewhat lower than the reported value because the rejection efficiency is obtained while pulsing uranine in the laboratory-scale test rig only for 60 s. The permeate produced during the pulsing period is estimated at 1.5 mL at 42 LMH, and that a small amount of permeate is diluted by the permeate preexisting in the permeate channel. The underestimated uranine concentration in permeate causes an overestimated dye rejection efficiency.

The low dye rejection efficiencies for the high molecular weights (300–800 Da) and the multiple charges are partly caused by the high dye concentration in the CP layer. According to the film theory, the actual dye concentration on the membrane surface is greater than the concentration in the feed. Because dye molecules have much smaller diffusivities than inorganic salts because of the high molecular weight and the bulkiness, the actual dye concentration in CP is supposed to be much greater than in bulk. In fact, the diffusivities of dye range at 3×10^{-6} to 5×10^{-6} cm^2/s, whereas the diffusivities of Na$^+$ and Cl$^-$ are 1.3×10^{-5} cm^2/s and 2.0×10^{-5} cm^2/s, respectively. However, no guidelines have been developed to take the CP effect into consideration in membrane integrity tests.

In a field trial performed in a municipal WWTP, the fluorescent dye-based method resulted in the most accurate leak detection in RO when it was compared with the methods based on conductivity, turbidity, and particle count (CDM 2010). In this field trial, undisclosed fluorescence marker provided by Ecolab (formerly Nalco) was used, and the marker was detected by handheld fluorometers. Although insensitive handheld fluorometers with a detection limit of 10 µg/L were used, fluorescence dye method showed the greatest LRV performance. Table 7.6 summarizes the test results, where fluorescence dye successfully demonstrates the continuous monitoring of 3.2-log removal (= log(14.9/0.01). In other field study (MWH 2007), the monitoring of 6-log removal was demonstrated using the same dye, but the details of the empirical procedure were not disclosed. In

TABLE 7.6
Sensitivity of Various Integrity Test Methods

	Fluorescence Dye (mg/L)		Conductivity (μS/cm)	Turbidity (NTU)	Particle Count (2–5 μm)
	Feed	Permeate			
Baseline[a]	14.9	<0.01	470	0.016	1.98
Pinched O-ring[b]	14.3	0.1	532	0.016	0.78
Cut O-Ring[c]	17.8	1.0	2361	0.0185	24.3
Membrane puncture[d]	18.1	0.3	2271	0.020	62.3

Source: CDM, Test technology innovations and optimize systems in the city of Santa Cruz pilot plant. Report to SCWD2, 2010.

[a] Values when membrane is functioning without integrity breach.
[b] Values after a small slit is made in the pinched O-ring between the first and the second elements.
[c] Values after a roughly 1/8-inch section was cut out of the same O-ring.
[d] Values after the membrane surface is punctured with a needle.

this study, the more advanced track leak detection system provided by Ecolab was used to detect down to 0.1 μg/L of dye. The frequency of the integrity test is depending on the economics and the regulatory requirement.

The integrity test/monitoring methods available for RO are all based on detecting the surrogate molecules/ions that are much smaller than the actual viruses having 20 to 80 nm effective diameters. Partial oxidation of RO membranes and micro pinholes on the membrane skin layer may cause rises of conductivity, TOC, and dye concentration in permeate, but those integrity breaches are not necessarily large enough to leak viruses. Therefore, all the integrity test methods except virus challenge tests are considered conservative, which project the worst possible cases. The challenge tests using MS2 phage gives the most accurate LRV by definition since it is an actual measurement of virus leakage. MS2 phage is not a human enterovirus, but it is commonly used as a model virus because its shape and size (24–27 nm) are similar to those of polio and hepatitis viruses. In addition, MS2 is one of the smallest viruses that can provide rather conservative results. Although accurate results are anticipated, challenge tests are not a real-time method and are costly because of the requirement of substantial resources, for example, virus inoculum, labor, and test equipment.

7.4 ANAEROBIC MBR

7.4.1 OVERVIEW

Anaerobic digestion (AD) was first introduced as a means to reduce excess biosolids from sewage treatment plant in the early 20th century. Later, the process was modified for industrial wastewater treatment in the 1950s. In the beginning, AD was a single-stage process, but two-stage processes were developed in the 1970s by separating acidogenesis and methanogenesis. With a dominance of the activated sludge process, AD has been used mainly for treating high-strength wastewater that could not be treated economically by the activated sludge. AD received an attention during the oil shock in the 1970s because of its capability of producing methane gas, but it did not last long with a stabilization of the situation. Currently, AD is mainly used for organic waste treatment, for example, excess biosolids from the activated sludge process, agricultural by-products, and human and animal manures, but it is also used to treat high-strength organic wastewater from the industry.

In AD, microorganisms obtain energy by disproportioning electrons among carbons. As shown in the following chemical equation, the oxidation states of six carbons in glucose are zero, but it splits to three carbons with −4 and there carbons with +4 after the anaerobic disproportioning. Because the energy available from the anaerobic disproportioning reaction is much less than the aerobic oxidation reaction, biosolid yields are an order of magnitude lower in AD than in CAS. Because of the low sludge yield, much less nutrients are required to remove the same amount of COD. In addition, because methane escapes from the system without being oxidized, the COD removal rate from wastewater is greater in AD than that in CAS in general:

$$C_6H_{12}O_6 \rightarrow 6CH_4 + 6CO_2.$$

Low sludge yields, for example, 0.03 to 0.06 g MLSS/g COD, are advantageous with respect to excess sludge treatment, but the slowly growing methanogens and the delicate balances among the microorganisms make the AD operation difficult (Visvanathan and Abeynayaka 2012). As a result, the development efforts have been focused on balancing the process while retaining high MLVSS without having microbial washout problems.

Traditionally, batch reactors were used, where waste or wastewater mixed with anaerobic seed sludge is filled up in a reactor and sufficient time is given until the anaerobic reaction completes. This process is simple, but large reactors are required to compensate the slow overall reaction rate caused by the induction period in the early part of the batch cycle. To overcome the kinetic limitation of batch reactor, continuously stirred tank reactor (CSTR) was developed for high-strength wastewater. In this process, high-strength wastewater is fed to the reactor, and the treated wastewater is discharged with suspended microorganisms at continuous or semicontinuous modes. At a steady state, the microbial population is stabilized at the level microbial production rate equals to the microbial discharge rate. This method is valid only for very high-strength wastewater that can naturally produce sufficient amount of suspended microorganisms using the organics in the wastewater. Clarifier can be added to the CSTR to enrich MLVSS by settling sludge and returning it to the reactor. This method improves the stability of AD, but the sludge settling in a clarifier is often challenging because of the sludge floatation by gas bubbles and the inherently small floc particles compared with the activated sludge. In addition, capital and operating costs are high for CSTR because of the necessity of sealed reactors and the mixing mechanism. To improve solid retention in the reactor without excessive costs, an upflow anaerobic sludge blanket (UASB) reactor was devised. In this process, wastewater is passed upward through a sludge blanket located in the middle of the reactor, as shown in Figure 7.15. The solids floating with gas bubbles are captured by the gas–liquid–solid separator in the top of the reactor and settled back to the sludge blanket/bed. The filtered wastewater by the sludge blanket is partially recycled back to the feed; thereby, wastewater passes the sludge blanket two to three times before it permanently exits the reactor.

Because anaerobic microorganisms do not form large flocs compared with aerobic microorganisms, the effluent often contains high-level suspended solids. Untreated organic acids, alcohols,

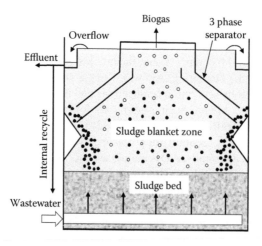

FIGURE 7.15 Upflow anaerobic sludge blanket (UASB) reactor.

hydrolysis by-products, etc., also exist in the effluent at varying concentrations depending on the process condition. Thus, the effluent quality of AD is typically not sufficient to meet the discharge limit, and hence further treatment is required by CAS. This is a reason why AD is viewed as an economical pretreatment method before high-strength wastewater is treated by CAS. The relative advantages of CAS and AD over each other are summarized in Table 7.7.

The hybrid process of AD-CAS can be competitive enough over the CAS process for high-strength wastewater treatment because of the low-energy costs for aeration. However, the biogas produced from AD is generally not economical or reliable to be used as a regular heat source, if more reliable heat sources are available such as natural gas from the grid, gasoline, kerosene, etc. Although the global natural gas prices range at $3 to $16 per million BTU depending on location as of 2013, biogas is not always used as heat source even in the areas with the highest natural gas prices. The use of biogas as an auxiliary heat source has been practiced, especially when specific benefits exist. For example, biogas can be used as an auxiliary fuel for combined heat and power (CHP) system that supplies electricity and heat for large buildings. By supplementing CHP with the renewable methane gas, the CHP efficiency based on the nonrenewable natural gas increases, and the system can be qualified for Energy Star label. In one site with a small CHP, biogas produced from AD boosted the energy efficiency of CHP by 5% or 85 kW at 1700 kW output. The estimated heat savings by the biogas was only 0.73 MBtu/h at a generator efficiency of 40%. Although the incremental efficiency gain is sufficient for the building to obtain the Energy Star label, the natural gas savings are estimated only at $90/day, assuming a natural gas price of $5/MBtu.

Anaerobic MBR (AnMBR) is the combination of CSTR and membrane separation. Just like aerobic MBR, membranes remove solids from the effluent and enrich microorganisms in the digester; thereby, SRT is decoupled from HRT. In its original configuration, sidestream cross-flow tubular membranes were used in the late 1960s, and plate and frame modules were introduced later for the process. In the late 1980s, AnMBR was studied in a large scale for water reuse and energy recovery through the "Aqua Renaissance 90" project sponsored by the Japanese government (Kimura 1991). A few tens of membrane manufactures, engineering firms, and universities participated, and various types of membranes and reactors were studied. Hollow fibers, tubular, flat sheet, and rotating membranes made of polymeric or ceramic materials were tested in sidestream, submerged, or airlift configurations. Although the project contributed immensely to the progress of modern MBR technologies, it was realized that combining membranes with AD was not economically feasible in the vast majority of the cases primarily because of the prohibitively high capital and operating costs compared with the benefits of biogas production at the time. In fact, the tubular membranes used at the time are still much more expensive than the submerged membranes used today. The specific energy consumption ranges from 4 to 10 kWh/m^3 even in an optimized system, as discussed in Section 2.4.

In the meantime, a great progress was made in the membrane technology with regard to aerobic MBR. The newly developed submerged membranes demonstrated nearly two orders of magnitude lower specific energy consumption than cross-flow membranes simultaneously with a lower specific membrane costs. With the development of economical submerged membrane systems, the hope

TABLE 7.7
Relative Advantages of Activated Sludge and AD

Activated Sludge	AD
Better effluent quality	Lower excess sludge production
More robust biology	Lower energy consumption
Easier operation	Fuel gas production
Less odor issue	Less nutrient demand
Nutrient removal (N & P)	Smaller footprint
Better sludge settling property	

for AnMBR had been growing. Eventually, AnMBR received a new attention when the gas and oil prices surged in the later part of 2000s and, as a result, several AnMBR based on flat sheet membranes have been commissioned in North America by the early 2010s. However, despite the more favorable environment for AnMBR than in the previous decades, AnMBR has not been proliferated yet because of some intrinsic limitations discussed in the following sections. As a result, there are no fully established standard design protocols or operating practices.

7.4.2 Biological Aspect of AD

AD consists of two major stages, that is, acid formation and methane formation, but the acid formation can be further broken down into three substeps, that is, (1) hydrolysis of large/complex organics to small organics, (2) acidogenesis of the small organics to organic acids, and (3) acetogenesis of higher organic acids to H_2, CO_2, and acetate. Fermentative acetogens, homoacetogens, hydrogenotrophic methanogens, and aceticlastic methanogens are responsible for the degradation of complex organics. The bacteria involving in anaerobic process include *Clostridium* spp., *Peptococcus anaerobius*, *Bifidobacterium* spp., *Desulfovibrio* spp., *Corynebacterium* spp., *Lactobacillus*, *Actinomyces*, *Staphylococcus*, and *Escherichia coli*. Fermentative and acetogenesis bacteria are facultative and grow much faster than methanogenic bacteria in general (Visvanathan and Abeynayaka 2012).

In the hydrolysis step, fats are hydrolyzed by lipase to 1 mol of glycerol and 3 mol of fatty acids. Glycerols are further broken down by alcoholic pathway, whereas fatty acids are broken down to short chain fatty acids. Cellulose needs to be hydrolyzed to smaller molecules and solubilized to be digested to sugars. If a large amount of substrates exist as solids, hydrolysis can be a rate-limiting step because the accessibility of solids is limited for microorganisms. In the acidogenesis step, the long-chain organics formed by the hydrolysis of large/complex organics are broken down further to propionic/butyric acids, but some are directly converted to acetate, H_2, and CO_2. Hydrolytic and nonhydrolytic fermentative bacteria are responsible for the hydrolysis and acidogenesis steps. The higher organic acids produced in the acidogenesis step are further broken down to acetate and H_2 in acetogenesis step by syntrophic acetogens.

Finally, H_2 and acetate are converted to methane in the methanogenesis step, but some H_2 are converted to acetate before they become methane. There are four morphologically different bacterial species responsible for the methanogenesis, that is, *Methanobacterium*, *Methanococcus*, *Methanosarcina*, and *Methanospirillum*, but these bacteria can use only acetate, H_2, CO_2, formic acid, and methanol. Methane formers not only grow slowly but also are highly sensitive to environmental factors such as pH, temperature, oxygen, nutrients, and trace metals. Therefore, methanogenesis can be easily disrupted by the fluctuations in operating condition and is the most common rate-limiting step. It is crucial to balance acidogenesis and methanogenesis to avoid acid accumulation and the drop of pH. The metabolic pathway of AD is summarized in Figure 7.16. The groups of microorganisms responsible for each step are also marked in the figure.

The optimum pH of the acidogens and acetogens is known to be low at 5.5 to 7.0, but the optimum pH for methanogens is high at 6.8 to 7.8. Because acidogenesis and acetogenesis are much more robust reactions than methanogenesis, pH is set high at 6.8 to 7.6 to encourage the growth of methanogens in a single-stage AD. In a two-stage AD, the pH in each tank can be maintained at its optimum to maximize the overall process efficiency. Methane formation occurs only at strict anaerobic conditions without DO. Nitrates and sulfates should be mostly reduced to N_2, NH_3, and H_2S before methane formation occurs at an ORP less than −300 mV. Microbial activities of anaerobic bacteria are affected by any abrupt environmental changes as mentioned earlier. The temperature of the digester should not change more than 0.6°C to 1.2°C per day to prevent the performance drop (WPCF 1987). Not only the toxic components such as cyanide, some chlorinated hydrocarbons, and heavy metals but also NH_3, sulfides, excess fatty acids, etc., can inhibit the AD. It has been known that excessive NH_3 is the most common cause of inhibition in AD, but it is difficult to avoid the problem especially with the wastewater from meat packing, dairy, soy bean processing, etc., where NH_3

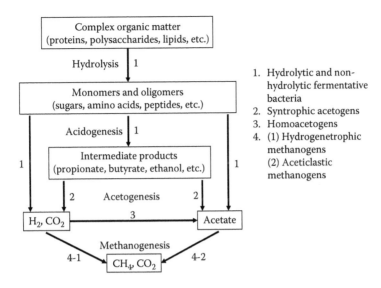

FIGURE 7.16 Metabolic pathway of anaerobic degradation. (Reproduced from del Real Olvera, J. et al., Biogas production from anaerobic treatment of agro-industrial wastewater. In *Biogas*, ed. Kumar, S. p. 94, 2012. Available at http://www.intechopen.com.)

is contained as ammonium, amino acids, proteins, etc. It has been known that halogenated organics and tannins can also be a significant factor in the treatment of paper mill wastewater.

Depending on the reaction temperature, AD is divided into mesophilic digestion (30°C–38°C) and thermophilic digestion (50°C–57°C). The degradation of general organics is faster in the thermophilic condition than in the mesophilic condition, and this may enable a smaller digester size. However, the biological reaction becomes more prone to fail in thermophilic digestion when subtle changes occur in organic loading, feed characteristics, oxygen intrusion, temperature, pH, etc., because of the narrowing microbial diversities. In fact, it is plausible to consider that more diverse organisms have evolved for the average environment than for the rare hot environment in nature. It was observed that phenols and phthalic acids were easily degraded at 37°C, but not at 55°C. The subsequent phylogenetic analysis of the 16S rRNA genes from the archaeal and bacterial communities revealed a narrower microbial diversity at higher temperatures (Levén 2006). In addition to the vulnerability to environmental factors, thermophilic AD often requires extra fuel beyond the biogas generated in situ because of the fast heat loss. As a result, mesophilic AD is dominantly used over thermophilic AD in the industry.

7.4.3 Advantages and Disadvantages over Conventional AD

AD has been primarily used to reduce wastewater COD from very high to moderate levels without consuming the energy for biological system oxygenation. The effluent from AD typically has much higher TSS and COD than the effluent from aerobic wastewater treatment processes. As a result, the AD effluent typically needs further treatment by the activated sludge process before being discharged to nature or reclaimed. Sludge washout is a common issue in AD because sludge yield is an order of magnitude lower than that in CAS. The high microbial sensitivity to the environmental factors contributes to the sludge washout by causing deflocculation. The produced biogas can be used to generate heat for the AD itself or to generate electricity. However, it is common to combust biogas in flare stack because of the lack of economic advantages. In fact, the costs of the equipment required to recycle the low-quality biogas tend not to justify the savings in fuel costs, especially in small-scale systems. The biogas may be required to be pretreated to remove sulfides before it is recycled or flared, depending on the regulation imposed in the area the plant is located.

With an addition of membrane filtration, AD can be operated more stably without sludge wash-out issues even at very low influent COD, for example, less than 500 mg/L. This makes AnMBR suitable for treating municipal wastewater using suspended culture. AnMBR can produce a particle-free effluent with a fairly low COD, for example, 50 mg/L, if wastewater contains predominantly readily biodegradable compounds and the OLR is low enough (Gao et al. 2013). However, the odor-ous nature of the effluent may still prevent it from being recycled or being discharged to nature in many cases. The effluent can be further filtered by RO before being recycled or discharged, but the odor-causing sulfides are not readily rejected by RO unless the feed pH is 9 or higher. In one study (Hall and Bérubé 2006), when AnMBR was used to treat municipal wastewater, the BOD and COD removal efficiency was 65% and 72%, respectively. However, the permeate needed further aerobic treatment to meet the discharge limit in terms of BOD and COD. From the perspective of water reclamation, AnMBR does not provide definitive advantages over conventional AD in many cases unless the AnMBR effluent can be either directly recycled for specific purposes or discharged to nature. In another pilot study, full-scale submerged membranes (Puron®, Koch Membrane Systems) were used to treated municipal wastewater at a specific bubbling rate of 0.23 $Nm^3/m^2/h$ (Robles et al. 2012). The F/M ratio was maintained very low at 0.1 to 0.13 g COD/g MLTS/day, where MLTS stands for mixed liquor total solids and is only slightly higher than the MLSS because of the low dissolved matter contents in the mixed liquor. MLTS was around 30 g/L and SRT was 70 days dur-ing the experiment. COD was reduced from 445 to 77 mg/L on average at 83% efficiency. Under this condition, sustainable flux was observed at around 10 LMH. The exact cause of this much higher flux than those observed in industrial wastewater treatment (5 LMH or so) is not clear, but high sus-pended solids that originated from the raw wastewater might contribute to the membrane scouring under the condition. In fact, the raw municipal wastewater used for the study contained high-level TSS (186 mg/L average), which accumulated in the reactor at the long SRT condition.

Membrane pore size is not a factor affecting permeate quality in AnMBR just like in aerobic MBR as discussed in Section 2.6.4.3. It is partly because separation occurs by the cake layer ini-tially formed on the membrane surface, which is also known as "dynamic membrane," instead of the membrane itself, as discussed in Section 1.2.4. As a result, the permeate quality of AnMBR is decided by the biological condition that determines the concentrations of organic acids, alcohols, and other small organics that can easily pass the cake layer.

Conceptually, the size of the digester can be reduced by enriching microorganisms in AnMBR compared with the conventional AD. There is no sufficient knowledge on the optimum MLSS in AnMBR, but it is typically set at equal or lower than the levels commonly found in aerobic MBR (6–15 g/L), because membrane fouling is much more severe in AnMBR. MLSS can be raised fur-ther, but larger membrane areas are required and more energy must be used to bubble the larger membrane surfaces. In the meantime, the F/M ratio of AnMBR should be kept lower than those in conventional AD because membrane fouling is in a positive correlation with it, as discussed in Section 5.2.3.3 (Gao et al. 2011; Liu et al. 2012b; Yeo and Lee 2013). Although MLSS of AnMBR may be higher than that of the comparable CSTR-based AD, reactor size is not necessarily smaller due to the lower F/M ratio required for low membrane fouling. If the reactor size of the AnMBR is compared with that of UASB, there are little or no benefits. The average MLSS of UASB can be much higher than that of AnMBR based on the total reactor volume, that is, 15 to 30 g/L versus 6 to 15 g/L. Therefore, AnMBR does not provide a definitive benefit with respect to reactor size.

AnMBR can produce more biogas than AD by retaining suspended solids at longer SRT com-pared with conventional AD for the same wastewater (Yeo et al. 2013). When biogas production rates were compared at different SRT in side-by-side tests of AnMBR with a common feed source, it was observed that nearly 45% more biogas was produced when SRT increased from 20 to 60 days. Less sludge production was also observed simultaneously. The high biogas production at long SRT was attributed to the endogenous respiration. The amount of additional biogas deliverable by AnMBR varies depending on environment, but the additional value created by using the membrane would not be sufficient enough to justify the capital costs for membrane and biogas recycle systems given the fact that the value of biogas does not justify even the equipment required to recycle it, especially in small-scale system.

From the technical perspective, AnMBR can treat low-strength wastewater without microbial washout issues as mentioned previously. However, several financial barriers prevent it from being a viable option despite its advantage of not requiring biological aeration. First, permeate quality is not sufficient to be qualified for the secondary effluent because of its odorous nature perhaps with high BOD and nutrients contents. If subsequent aerobic treatment is required, AnMBR can hardly be justified against aerobic treatment options because aerobic treatment can produce qualified water directly from the wastewater. Second, the capital and operating costs of AnMBR are many folds greater than those of aerobic MBR because of the low sustainable flux, for example, 3 to 10 LMH for submerged membranes and 50 to 100 LMH for sidestream tubular membranes, and the necessity of air-tight reactors. Third, the low-strength wastewater produces only a small amount of methane gas, but substantial amount is lost through permeate (Yeo and Lee 2013). For example, if a municipal wastewater has 400 mg/L COD, the maximum theoretical methane gas yield is 100 mg/L considering the conversion factor of 0.25 kg CH_4/kg COD (or 0.35 m^3 CH_4/kg COD at STP). In the meantime, the methane gas dissolved in the effluent is as much as 14 mg/L if the biogas contains 60% methane at 25°C, according to Equation 7.8, where T is the temperature (°C), C_{CH4} (mg/L) is the methane concentration in water, and P_{CH4} (atm) is the partial pressure of methane gas in headspace. Therefore, the lost methane gas through the effluent is at least 14% of the total amount produced:

$$C_{CH4} = (2.046 \times 10^{-6}\,T^4 - 3.674 \times 10^{-4}x^3 + 2.860 \times 10^{-2}\,T^2 - 1.270\,T + 41.53) \cdot P_{CH4}. \quad (7.8)$$

7.4.4 MEMBRANE FOULING

In AnMBR, the SMP content of mixed liquor is in a similar range with that of aerobic MBR, but membrane fouling is much more severe (Hall and Bérubé 2006). Small floc sizes and the unique chemical properties of SMP are considered the cause. The sustainable flux of submerged membranes has been known to be only approximately 20% to 40% of those in typical aerobic MBR at 5 to 10 LMH versus 25 LMH in aerobic MBR at a comparable bubbling rate. The low flux can be a fatal disadvantage of AnMBR because the membrane surface area and the energy required to scour the membrane are inversely proportional to the flux. In a more recent study (Dagnew et al. 2014), it was claimed that a stable flux of 6 to 10 LMH was obtained at 38.5°C in treating refinery, petrochemical, meat processing, or potato processing wastewaters using laboratory- to pilot-scale MBR systems equipped with submerged hollow fiber membranes (ZW500d®, GE), as shown in Figure 7.17. Depending on the wastewater source and the purpose of the tests, MLSS was varied between 4.2 and 17.5 g/L. The applicable F/V ratio was higher than 6 COD/m^3/day for food wastewaters, but it was less than that for refinery and petrochemical wastewaters.

In sidestream cross-flow membrane filtration, flux is known to be only 50 to 100 LMH despite the high cross-flow velocity of 1.5 to 3 m/s in laboratory-scale studies (Yoon 1994; Kang et al. 2002).

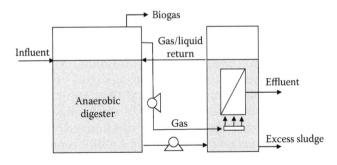

FIGURE 7.17 Process diagram of AnMBR with submerged membranes.

In the same study, flux was mainly affected by the cross-flow velocity rather than TMP used because the filtration is in a mass transfer–controlled region, as discussed in Section 1.2.7.1. The effluent quality in terms of COD, BOD, and TSS was hardly affected by the type of membrane and the pore size just like in aerobic MBR. At such filtration condition dominated by the cake layer (or dynamic membrane), pore sizes and membrane permeability are not a primary factor affecting the steady-state flux, as discussed in detail in Section 1.4.4.

In thermophilic AnMBR operated at 55°C using alcohol distillery wastewater, sidestream tubular ceramic membranes with 0.45 µm pore size and 6 mm inner diameter suffered from a quick and severe flux loss despite the high cross-flow velocity at 1.5 m/s and low TMP at 0.5 bar (Yoon et al. 1999b; Kang et al. 2002). The high initial, high flux at approximately 120 LMH decreased to approximately 50 LMH within a week. Membrane autopsy revealed that there was no cake layer built on the membrane surface. The loss of flux was mainly attributed to the pore plugging by struvite ($MgNH_4PO_4$). However, the flux was recovered well by backflushing the membrane using 1N H_2SO_4. The abundant struvite precipitation potential at the operating condition was attributed to the lowest solubility at the operating pH of 7.5. In the same experiment, much greater flux was observed with tubular polymeric membranes because of the much greater membrane porosity that delayed pore plugging. It appeared that struvite formation was less significant in polymeric membrane, but the exact causes were not known. In another study performed using municipal wastewater, membrane foulants were found to be very different from the previously mentioned experiment (Hall and Bérubé 2006). When municipal wastewater was treated by AnMBR at the mesophilic condition (32°C–36°C), membrane fouling occurred mainly by organic deposits rather than inorganic struvite in both submerged and cross-flow membranes. Membrane autopsy revealed that carbon contents were around 60% in the cake layer, whereas magnesium contents were only 1% or so.

The reduction of membrane fouling has been attempted by using floating biocarriers that directly contact with membrane (Yoo et al. 2012, 2014; Bae et al. 2013). Similar methods discussed in Section 5.4.1 have been tested, but biocarriers are fluidized by either recirculating headspace gas through diffusers or generating upflow by recirculating the mixed liquor through external loops in AnMBR. Biocarriers potentially reduces SMP concentrations in mixed liquor by adsorbing them and providing longer contact time with microorganisms. In a laboratory study, municipal wastewater was treated using a tall fluidized AnMBR column with 50 cm height and 2.5 cm diameter. Eight 45 cm hollow fibers with 0.0215 m^2 total surface area were submerged in the center of the column, and 30 g of 10 × 30 mesh granular activated carbon (GAC) was added as a biocarrier. Filtered municipal wastewater by 2 mm screen was fed to the column, and HRT was maintained at 2.3 h. The anaerobic broth in the column was circulated through a sidestream loop at 0.75 L/min to generate cross-flow on the membrane surface and to fluidize GAC particles. It was observed that the filtration was sustainable for more than 200 days at a flux of 9 LMH at 10°C to 25°C, which was the high end of the commonly known sustainable flux in AnMBR equipped with submerged membranes (5–10 LMH). In the same experiment, the BOD and COD removal efficiencies were 94% and 89%, respectively. Sludge yield was measured low at 0.01 to 0.03 kg VSS/kg COD.

7.4.5 Design of Mesophilic AnMBR

Many sidestream membranes are compatible with thermophilic AD at 50°C to 60°C, but commercial submerged membranes are compatible with only mesophilic AD at below 40°C. Therefore, all AnMBR equipped with submerged membranes adopt mesophilic temperatures or below. There are no universally applicable design parameters for AnMBR primarily because anaerobic kinetics are heavily dependent on the wastewater source and the slowly growing anaerobic microorganisms are much more sensitive to the substrate characteristics than aerobic counterparts. The lack of universal design parameter also stems from the rarity of AnMBR because of the lack of convincing value proposition relative to the conventional AD. As a consequence, the design parameters introduced in this section are based on the anecdotal observations from the field and the literature.

AnMBR with submerged membranes are strongly preferred to be designed with separate membrane tanks. If any maintenance work is necessary for membrane tanks, mixed liquor can be pumped to the anaerobic digester in this configuration. Nonetheless, because it is much harder to access the membrane system in AnMBR than in aerobic MBR, wastewater pretreatment should be performed much more rigorously to prevent any trouble unfiltered debris can cause in the membrane system. Membrane system also needs to be designed with more redundancies to minimize membrane cleaning than for aerobic MBR. The diagram of mesophilic AnMBR with submerged membranes is shown in Figure 7.17. The head space gas is pumped to the membrane tank to scour the membrane surface while the mixed liquor in the digester is recycled through the membrane tank at $2.5Q$ to $4Q$ to mitigate the solids accumulation in the membrane tank.

The following steps can be taken to design AnMBR for the wastewater with 20 g/L COD at a flow rate of 100 m³/day:

- Calculate the total COD to be treated: 20 kg COD/m³ × 100 m³/day = 2000 kg/day.
- Calculate the reactor volume assuming MLSS and F/M ratio. In this example, MLSS is assumed at 10 g/L in the digester and the F/M ratio at 0.3 g COD/g MLSS/day. The applicable F/M ratio varies depending on the source of wastewater, but it may be 30% to 50% lower than the F/M ratio known for conventional AD for the given wastewater to prevent excessive membrane fouling. The realistic F/M ratio should be found out from pilot tests:

$$V = \frac{2000 \, \text{kg COD} \cdot \text{d}^{-1}}{0.3 \, \text{kg COD kg MLSS}^{-1} \cdot \text{d}^{-1} \cdot 10 \, \text{kg MLSS m}^{-3}} = 667 \, \text{m}^3.$$

- The F/V ratio is calculated at 3.0 kg COD/m³/day, which is acceptable because it is lower than the maximum OLR shown in Table 7.8:

$$\text{OLR} = \frac{2000 \, \text{kg COD} \cdot \text{d}^{-1}}{667 \, \text{m}^3} = 3.0 \frac{\text{kg COD}}{\text{m}^3 \cdot \text{d}^{-1}}.$$

- Assuming the sludge yield is 0.05 kg MLSS/kg COD/day, sludge production (Q_x) and SRT can be calculated. Because the sludge yield is low from the beginning, it does not change much regardless of the SRT:

TABLE 7.8
Design Parameters of Mesophilic AnMBR for Petrochemical Wastewater

	Range	Note
Temperature	30°C–40°C	• Abrupt change should be avoided
pH	6.8–7.6	• NaOH is used to control pH
F/M	<0.4 kg COD/MLVSS/day	• Varies depending on the source of wastewater
F/V	3–6 kg COD/m³/day	• Inversely correlated with membrane fouling rate
MLSS	6–15 g/L	• Same as those for aerobic MBR
Y_{obs}	0.04–0.05 g MLSS/g COD	• Observed biosolid yield
Flux	3–7.5 LMH	• 15–25 LMH for aerobic MBR
Gas sparging	0.2 m³/m²/h for HF	• Gross gas flow for HF is 0.4 m³/m²/h at 10–10 s
	0.3 m³/m²/h for FS	intermittent mode
		• Same as those summarized in Table 3.4
Specific energy demand	0.5–1.0 kWh/m³ for HF	• Fivefold higher than those for aerobic MBR
	0.75–1.5 kWh/m³ for FS	

$$Q_x = 2000 \frac{\text{kg COD}}{\text{d}} 0.05 \frac{\text{kg MLSS}}{\text{kg COD}} = 100 \frac{\text{kg MLSS}}{\text{d}},$$

$$\text{SRT} = \frac{10 \text{ kg MLSS m}^3 \cdot 667 \text{ m}^3}{100 \text{ kg MLSS d}^{-1}} = 67 \text{ d}.$$

- Calculate the membrane area required. Here sustainable flux is assumed at 7.5 LMH or 0.18 m/day. Safety factors of 1.2 to 2.0 can also be multiplied. The membranes can be submerged in two different membrane tanks to avoid shut down of the process during the maintenance of the membrane tank. If the safety factor is 1.5, the membrane flux will be 5 LMH during the normal time, but it will increase to 10 LMH temporarily when the other membrane tank is under maintenance work:

$$A_m = \frac{100 \text{ m}^3 \text{d}^{-1}}{0.18 \text{ m d}^{-1}} \times 1.5 = 833 \text{ m}^3.$$

- Calculate the mixed liquor recycle pump capacity by multiplying 2.5 to 4.0 to the influent flow rate. In this case, it is 250 to 400 m³/day or 10.4 to 16.7 m³/h.
- Calculate in-line gas compressor capacity by multiplying the specific aeration demand per membrane surface area (SAD_m) with the membrane area. Head pressure loss is dependent on membrane height, but it ranges from 0.2 to 0.4 bar:

$$Q_{gas} = 833 \text{ m}^2 \times 0.2 \frac{\text{m}^3}{\text{m}^2 \text{ h}} = 167 \frac{\text{m}^3}{\text{h}}$$

- Specific biogas production per COD removed varies widely depending on substrate and operating condition. The more exact values must be found from the respirometry, but it can be approximately calculated using an assumption of 0.35 m³ methane/kg COD or 0.25 g methane/g COD (Kale and Singh 2014). If the effluent COD is 100 mg/L in this example, the total treated COD will be 1990 kg COD/day from which methane production is calculated at 697 m³/day (= 1990 × 0.35). If the methane content in the biogas is 65%, the total biogas flow becomes 45 Nm³/h (= 697/0.65/24).

7.5 HIGH-RETENTION MEMBRANE BIOREACTOR

7.5.1 MBR WITH NANOFILTRATION MEMBRANE

NF has been mostly used to soften surface water with low TDS. In terms of molecular weight cutoff (MWCO), it is located in between UF and RO, as shown in Figure 2.1. It can reject multivalent ions at high efficiency at 70% to 99%, but reject monovalent ions at low efficiency at 20% to 70%, depending on the tightness of the membrane. NF is also known to reject most organic compounds larger than a few hundred Daltons, but the lower limit of MWCO decreases for the molecules with charges. If NF is applied for MBR, the high-quality effluent with very low hardness and TOC can be directly recycled for various purposes without further treatment processes.

The widely available NF membranes in spiral wound configuration are economical options for surface water treatment, but they are not compatible with the mixed liquor with high MLSS in MBR. NF membranes in tubular or plate and frame configuration would be suitable to handle high MLSS, but capital and operating costs are prohibitively expensive in the vast majority of the cases, as discussed in Section 2.1.3. As a result, submerged NF becomes a more realistic option for NFMBR. But, NFMBR has a potentially fatal limitation in the filtration driving force because the maximum TMP available is no more than 0.8–0.9 bar in practical situation. If the osmotic pressure that acts against TMP is deducted from the maximum TMP available, the effective TMP for filtration can be too small. Therefore, if ionic strength is too high in feed water, submerged NF would not be about to draw permeate at all.

Osmotic pressure of wastewater can be calculated using Equation 7.9 (Wolfe 2003). In this equation, Π is the osmotic pressure (kPa), C_{fb} is the TDS as NaCl (ppm), and T is the water temperature (K). If the TDS in feed water is 500 ppm as NaCl at 20°C (or 293 K) and the average ion rejection efficiency of submerged NF membrane is 45%, C_{fb} in the membrane tank will be 909 ppm (= 500/(1 − 0.45)). In this case, C_{fb} in permeate should be the same as C_{fb} in feed, that is, 500 ppm, according to the mass conservation law. The net difference across NF membrane is 409 ppm, which translates to the steady-state osmotic pressure of 32 kPa. If the maximum available vacuum pressure in practical situation is 85 kPa, the maximum effective TMP becomes only 53 kPa. Because NF membranes have high filtration resistance compared with MF/UF, the operating flux should be low at the effective TMP. If the wastewater has a C_{fb} of 1000 mg/L as NaCl or higher, osmotic pressure approaches to the maximum vacuum pressure available, and the effective suction pressure becomes negligible:

$$\Pi = \frac{0.2654 C_{fb} T}{1000 - C_{fb}/1000}. \tag{7.9}$$

In a laboratory study, municipal wastewater was treated by NF-MBR equipped with submerged hollow fiber NF membranes based on cellulose acetate chemistry (Toyobo Co., Japan) (Choi et al. 2006). The pure water permeability of the membrane was 0.0028 LMH/kPa. The DOC of the permeate was measured at 0.5 to 2.0 mg/L, and the ion removal was 40% to 60% for monovalent ions and 70% to 90% for divalent ions. At the very low operating flux at 0.021 to 0.029 LMH, no sign of membrane fouling was observed. Consequently, much lower SAD_m was sufficient in the NF-MBR compared with the conventional MBR. Overall, considering the packing density and the flux, the specific aeration demand per permeate volume (SAD_p) appears a few hundred times greater in NF-MBR. In another study, side-by-side comparisons were performed using two MBRs equipped with polyamide-based submerged hollow fiber NF membranes and polyethylene-based submerged hollow fiber MF membranes, respectively (Choi et al. 2006). The pure water permeability of NF membrane was 0.0033 LMH/kPa and was operated at 0.021 LMH at 70 to 80 kPa. NF-MBR produced the permeate with much less TOC than MF-MBR, for example, 0.5 to 2 mg/L versus 4 to 6 mg/L for most of the time, as shown in Figure 7.18.

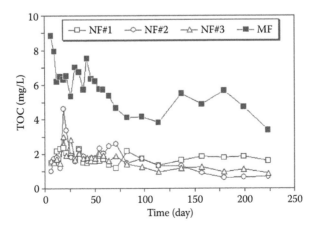

FIGURE 7.18 Profile of TOC concentrations in the permeate of NF-MBR and MF-MBR. (From Choi, J.H. et al., *Water Sci. Technol.* 53(6), 131–136, 2006.)

NF-MBR is essentially a competing technology with the hybrid system of MBR and NF/RO, as discussed in Section 7.3. Although it is a one-step process with a very simple configuration, it has significant disadvantages mainly because of the low flux that causes excessive capital and operating costs. Another limitation is that the NF-MBR based on submerged membrane cannot treat the wastewater with greater than certain salt concentration because of the limitation in TMP. Although the limitation in the applicable salt concentration is inherent, it can still be a viable option for the wastewater with low salt concentration, if the permeability of NF membrane can be dramatically improved to 10–50 LMH/bar from the current level of less than 0.01 LMH/bar, which allows 3–5 LMH at 0.1–0.3 bar of effective TMP.

7.5.2 Application of Forward Osmosis to MBR

7.5.2.1 Principle of Forward Osmosis

Forward osmosis (FO) relies on the osmotic pressure difference across a semipermeable membrane that has much greater water permeability than salt permeability. In this process, a feed water with low salt concentration flows from one side of a membrane and a draw solution with high salt concentration flows from the other side of the same membrane. Permeate is spontaneously drawn from the feed water to the draw solution to balance the chemical potential in both sides of the membrane. No additional hydraulic pressure is required beyond the level required to drive the feed water through the FO module. If inorganic salts were used for the draw solution, the diluted draw solution is regenerated by concentrating it with RO. The deionized water obtained as RO permeate is the final product. If volatile solutes were used for the draw solution, the diluted draw solution can be heated to remove/recover the solutes.

As any other membrane process, FO suffers from the performance loss by the CP. However, FO is unique that the CP occurs in both sides of the membrane unlike any other filtration processes. In the feed side, CP occurs when water permeates through the membrane, leaving solutes on the membrane surface (Figure 7.19). The rate of solute back transport to the bulk feed is dependent on the diffusivity of the solute molecules and the extent of turbulence on the membrane surface. The CP is settled, where the convective solute transport toward the membrane and the back transport toward the bulk are balanced. The CP in the feed side is not as significant as that in RO because the flux of FO is much lower, for example, 5 to 10 LMH (FO) versus 17 to 22 LMH (RO). Considering the low CP factor in RO at less than 1.5 for NaCl, it is plausible to consider that the CP in the feed side of

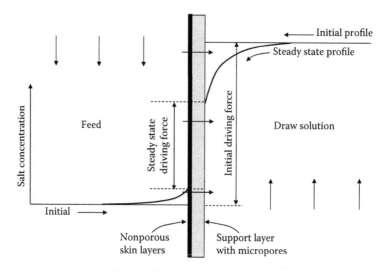

FIGURE 7.19 The effect of CP in both sides of the membrane on driving force in forward osmosis.

FO is not significant, if the smooth side of membrane faces the feed just like in RO. Meanwhile, the CP in the draw solution can significantly hamper the system performance by directly compromising the driving force. The clean permeate emerging from the membrane surface dilutes the draw solution right on the membrane surface. Although the salts in the bulk are diffused to the membrane surface to balance the salt concentration, the dilution of the draw solution on the membrane surface proportionally cuts down the driving force. Because high flux causes stronger CP in both sides of membrane, flux does not increase proportionally to the apparent driving force based on salt concentration difference. This self-limiting flux behavior is similar to that found in MF and UF as discussed in Section 1.6.2. To mitigate the CP effect in the draw solution, strong turbulent condition should be maintained by using high flow velocity and spacers with tortuous water paths. However, the fundamental imitation exists in reducing CP effect in the draw solution because of the porous support layer structure as discussed in the next few sections.

7.5.2.2   Challenges of FO

The structures of RO and FO membranes are similar. To minimize the filtration resistance, ultrathin nonporous skin layers with a depth of no more than a few microns are coated on the porous UF membrane casted on nonwoven fabric sheets. In RO, the skin layer with smooth surface faces feed water to minimize membrane fouling by scaling, particle deposition, etc. The support layer with a microporous structure faces the clean permeate without a concern on fouling or the CP. On the contrary, in FO, the smooth skin layer can face either feed water or draw solution depending on application. Depending on the membrane orientation, FO suffers from varying degrees of internal concentration polarization (ICP) and pore plugging in the microporous support layer in addition to the CP in the skin layer.

If feed water contains suspended solids and high hardness, facing the skin layer to the feed water is beneficial (Figure 7.20a). It is because if the porous support fabric faces feed water, the convective flow toward the membrane carries particles into pores and plug them. In addition, scaling can easily occur in the pore because the ICP developed inside the pore cannot be sufficiently controlled by the turbulence in the bulk. The high salt concentration in the pore also diminishes the flux by counteracting the filtration driving force. These drawbacks can be mitigated by facing the support layer to the draw solution that does not contain solids and hardness. In this configuration, however, the clean permeate emerging from the internal spaces of pores in the support layer dilutes the draw solution and directly compromises the filtration driving force. The ICP can be reduced by raising the turbulence in the bulk, but it has only a limited impact on mitigating the ICP problem inside the micropores. Hypothetically, this problem can be eliminated by eliminating the support layer, but new ways of assembling the ultrathin membrane as a module are required.

Although drawbacks exist regardless of the membrane orientation, it has been known for forward osmosis MBR (FOMBR) that facing the skin layer to the feed side results in greater flux than the other way around. In a study using submerged FO membranes based on hollow fiber configuration,

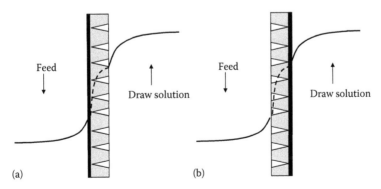

FIGURE 7.20   Comparison of the two configurations of forward osmosis: (a) active layer facing feed and (b) active layer facing the draw solution.

flux was nearly doubled from approximately 4 LMH to approximately 8 LMH when membrane orientation was reversed (Zhang et al. 2012). In this experiment, a draw solution with a conductivity of 47 mS/cm was used, which was equivalent to an osmotic pressure of approximately 25 atm. The tiny flux compared with the high driving force may hint the intrinsic inefficiency of FO because of the ICP.

On the contrary, facing the skin layer to the draw solution can be beneficial for the clear feed water with low or no suspended solids and scaling potential (Figure 7.20b). In this orientation, the ICP problem does not exist in the draw solution, and as a result, the large loss of driving force can be minimized. At the same time, the risk of plugging the micropores in the feed side remains low because of the clear feed water. In a laboratory-scale study, particle-free deionized water was used as feed, and NaCl, dextrose, and sucrose were used as a draw solution at various concentrations (Gary et al. 2006). It was observed that the flux was much higher when the skin layer was facing the draw solution. By maintaining sufficient turbulence on the skin layer, the flux could be maintained at close to the theoretical maximum, for example, 3 to 25 LMH depending on the draw solution strength. By contrast, when support layer faced the draw solution, a much lower flux was obtained at the same turbulence because of the inability of controlling ICP. The ICP effect was more significant when the large molecules with low diffusivity were used for the draw solution because of the slow back transport from the pores. When support layer faced draw solution, highest and lowest fluxes were obtained with NaCl and sucrose, respectively, although both were at the same molar concentration.

The driving force for FO can be readily raised by raising salt concentration in the draw solution. For example, although each mole per liter of species generates an osmotic pressure of 24.5 atm, 1 mol/L of NaCl is sufficient to generate nearly 49 atm of osmotic pressure because NaCl splits to two oppositely charged ions. The diluted draw solution is concentrated again by RO and recycled back to FO. Alternatively, the ammonium bicarbonate (NH_4HCO_3) solution can be used as a draw solution. It can be easily recovered by raising the temperature of the spent draw solution to 40°C or higher using a low-grade waste heat. Ammonium bicarbonates are decomposed to ammonia and carbon dioxide that easily escape from the draw solution as gas. The recovered ammonia and carbon dioxide are recycled for the draw solution (McCutcheon et al. 2005). However, the high solubility of ammonia in water at 40°C (320 mg/L) suggests a substantial amount of ammonia will remain in treated water unless the solution pH is raised substantially. The residual ammonia after the stripping process can be recovered/removed using adsorption columns filled with a form of zeolite called clinoptilolite (Kirts 2009), but the draw solution should be cooled down to maximize the efficiency of adsorption process. Despite various attempts, ammonium bicarbonate-based FO has not been proliferated because of the costs associated with the technical challenges in removing ammonia completely from the draw solution.

One potential drawback of FO is that the concentrated solutes in the draw solution diffuse to the feed water through membrane because of the concentration gradient, just like the solutes diffuse from concentrate to permeate in RO. Because the solutes diffuse to the opposite direction from the desired direction, it is called "reverse solute (or salt) diffusion (RSD)." The very high solute concentration in the draw solution causes a significant RSD in typical FO condition. It not only causes a loss of solute but also contaminates the feed water if the concentrate is the final product. In FOMBR, the solute lost to the feed is accumulated in the system diminishing the driving force, as will be discussed in Section 7.5.2.4. If the sodium sulfate (Na_2SO_4) solution is used as the draw solution instead of the sodium chloride (NaCl) solution, RSD can decrease to some extent because of the low permeability of sulfate ion (Tang and Ng 2014).

NaCl is considered the most economical and effective compound with high rejection efficiency by FO membrane, but it still suffers from the RSD. Even if only 0.01% of solutes are lost in one pass of the draw solution, the total loss can add up high as the draw solution recirculates through FO repeatedly. The accumulated solutes in feed water (or mixed liquor in FOMBR) diminish the driving force for the process and may affect the biology. In the meantime, the necessity of making up the lost solute negatively impacts the economics of the process. In addition, the loss of solute to feed

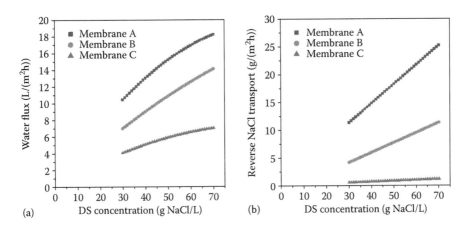

FIGURE 7.21 Relation between water flux and reverse salt diffusion (or transport) rate: (a) water flux and (b) reverse salt diffusion rate. (From Achilli A. et al., *Desalination* 239(1–3), 10–21, 2009.)

potentially deteriorates the feed quality when feed is the product of the FO process, for example, in juice or chemical concentration processes.

Macromolecules with multiple charges can be used to reduce the RSD, but they not only limit the maximum osmotic pressure deliverable but also exacerbate CP and ICP problems because of the low back transport velocity. The RSD can be also reduced by using tighter membranes, but permeability decreases as membranes become tighter in general. Thereby, a larger membrane surface area may be required to treat a given flow. As shown in Figure 7.21, the flux and the RSD of the three membranes based on same cellulose triacetate chemistry are positively correlated with each other (Achilli et al. 2009). The ideal FO membranes should have high water permeability and low salt permeability, which is also true for RO membranes. The currently available commercial membranes based on cellulose triacetate chemistry (HTI Inc., USA) have a clean water permeability of 0.5 to 0.7 LMH/bar (Lay et al. 2012). It is noticeable that the permeability of FO membrane is much lower than those of RO membranes, that is, 1 to 4 LMH/bar, which suggests a lot of room to improve the FO membranes.

FO membranes suffer from membrane fouling just like any other membrane primarily because of the accumulation of contaminants in the CP or ICP layer depending on the membrane orientation. However, if the skin layer faces the feed in FOMBR, the extent of membrane fouling has been reported not as significant as in other comparable membrane filtrations (Achilli et al. 2009). In principle, the low fouling in FO is not an inherent feature of the process, but it is a consequence of the low flux. In fact, the amount of particle deposition and the extent of the cake layer compaction are directly related with the flux as discussed in Section 1.2. At low flux, less cake layer is formed because of the slower convective flow toward the membrane. In addition, less cake layer compaction occurs because of the lower pressure loss through cake layer. In terms of capital costs, the membrane surface area required to treat a given flow is inversely proportional to the flux and this is a significant factor affecting the economic feasibility of FO.

7.5.2.3 Energetics

If waste heat can be used to regenerate the spent draw solution based on volatile solutes, the energy costs of the FO are not a critical factor in justifying the process. However, because of the lack of the volatile solutes suitable to produce the deionized water with satisfactory quality, inorganic salt solutions are used as draw solution and hence RO is necessary to regenerate the draw solution. From the energetics stand point, the first stage (permeate withdrawal) is a spontaneous reaction that occurs naturally by osmosis, where the free energy of the system drops when the permeate is drawn. The second stage (draw solution regeneration) is a forced reaction that requires external energy input

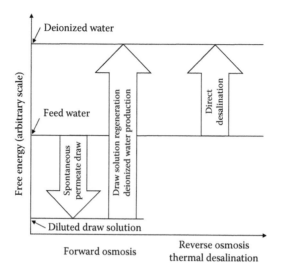

FIGURE 7.22 Conceptual comparision of free energy changes of feed water during the FO and direct desalinations, where external energy is required to compensate the free energy gains by water.

because the deionized water is in a higher free energy status than the diluted draw solution. As illustrated in Figure 7.22, the free energy of the feed water drops in the FO and increases in the subsequent draw solution regeneration process by RO/thermal desalination. Overall, the drop in the free energy in the FO process should be compensated by external energy in the deionization process. In the meantime, the free energy of the feed water moves straight up without being decreased in the conventional desalination processes such as RO, MSF, MED, etc. Because of the lack of the free energy drop that must be compensated later, the less external energy is required in those conventional processes.

According to an energy analysis of a hypothetical FO process for seawater desalination, FO followed by draw solution regeneration requires more energy than a single step RO process (McGovern and Lienhard 2014). From the energetics stand point, FO can be competitive to the single-stage processes such as RO, MED, and MSF, only when the draw solution does not need to be regenerated. For example, wastewater concentration can be performed to reduce the volume using abundant seawater as a draw solution. In this process, the diluted seawater can be discharged to the ocean without being regenerated. Similar applications can be considered for concentrating process streams, for example, juice, milk, and chemicals. However, the high capital costs caused by low flux are still a constraint in justifying FO.

7.5.2.4 FOMBR

The concept of applying FO membranes for MBR was proposed relatively recently (Achilli et al. 2009). FO membranes in flat sheet or hollow fiber configuration are submerged in the aeration tank, and aeration is performed to scour the membrane surface as shown in Figure 7.20. Draw solutions with high salt concentration, for example, 0.5 to 1.0 M NaCl, are circulated through the submerged FO membranes to obtain permeate. The diluted draw solution is concentrated by RO from which deionized water is obtained as permeate. The RO concentrate is returned back to the submerged FO membrane unit. The characteristics/chemistry of FO membranes is basically the same as RO membrane because both are to remove ions, but the support layer of FO membrane needs to be optimized to minimize the performance loss from ICP. Cellulose triacetate and polyamide are the known membrane chemistries, but they are also the predominant chemistries for RO membranes (Zhang et al. 2012). The basic separation mechanism of FO is same as those of RO, as discussed in Section 1.2.10. Thereby, solution diffusion mechanisms are still valid (Lay et al. 2012).

The hybrid process of FOMBR-RO produces water with an excellent quality simply because water must pass the double barriers imposed by the two tight membranes, that is, FO and RO. In terms of inorganic ion content in treated water, the effluent quality of FOMBR-RO is superior to that of MBR-RO except NaCl contents if NaCl is used for the draw solution. For example, if 0.5 N or 29.2 g/L NaCl is used as a draw solution and the average NaCl rejection efficiency of the RO membrane was 99%, the NaCl content in the permeate will be around 300 mg/L. FOMBR-RO process is also known to produce the effluent with much lower organic contents. In a laboratory test, the FOMBR-RO process removed more than 99% of organic carbon (from 1325 to 2.5 mg/L) and 98% of NH_4–N (from 65 to 0.4 mg/L) from the raw wastewater (Achilli et al. 2009).

In FOMBR-RO hybrid process, the organics passing the submerged FO membrane can accumulate in the draw solution and eventually cause organic fouling or biofouling in the subsequent RO membrane. If seawater or other saline water is readily available as the draw solution, the diluted draw solution from FOMBR can be discharged without concentration. In a study (Achilli et al. 2009), the soluble TOC in the draw solution was measured at 3 mg/L. But the effective TOC might be higher if the attached TOC in the internal loop was considered. During the course of the experiment, the draw solution that contained 50 g/L of NaCl initially was diluted gradually because of RSD. The lost NaCl to mixed liquor and RO permeate was compensated periodically. Salt concentration in mixed liquor gradually increased as soon as the experiment was commissioned. A steady state was reached at slightly more than 4 g NaCl/L in the aeration tank after 12 days of commissioning. At the steady state, the amount of salts brought into the mixed liquor by feed wastewater and draw solution was balanced with the amount removed through the excess sludge. Only a mild membrane fouling was observed during the experiment, but the operating flux gradually dropped to 9.4 LMH from the initial flux of 11 LMH because of the salt accumulation in mixed liquor. Meanwhile, RSD was estimated at 5.0 to 6.5 g NaCl/m^2/h depending on the extent of membrane fouling.

SRT (or the F/M ratio) is a primary factor affecting membrane fouling in aerobic MBR, as discussed in Section 5.2.3. However, the effect of SRT appears much less in FOMBR. When the fluxes of three FOMBR with 3-, 5-, and 10-day SRT were compared in a side-by-side test, the initial flux of 4 LMH ended up at around 2.5 LMH in all reactors. The flux loss was attributed mainly to the salt accumulation in mixed liquor as a consequence of RSD along with some membrane fouling (Zhang 2011). In fact, the salt accumulation in mixed liquor by RSD increases as SRT increases because less mixed liquor is removed from the system, and hence the net driving force for permeate drawing decreases. However, it must be noticed that the no or little influence of SRT on FO membrane performance is not an inherent feature of FO but a consequence of low flux. In fact, the flux observed with FO membrane in this study (2.5–4 LMH) is well below the sustainable flux measured in MBR, for example, 20–30 LMH. At such low flux, membrane fouling occurs very slowly because most fine particles in mixed liquor are transported back to the bulk by the particle back transport phenomena, as discussed in Chapter 1. Not much has been known about the optimization of FOMBR, but the SAD_m can be maintained much lower than that in MBR (0.2–0.4 m^3/m^2/h) because the operating flux is low. Although FOMBR is expected to require approximately threefold to fivefold larger submerged membrane surface areas compared with MBR, the net airflow rate required is less than the membrane surface area suggests.

FO can also be combined with AnMBR (Chen et al. 2014). Flat sheet FO membranes from HTI, LLC (USA) were submerged in the anaerobic digester, and the head space gas was recycled to scour the membrane while circulating the draw solution of 0.5 N or 29.2 g/L NaCl, through the submerged FO membranes. Low strength feed with 460 mg/L COD was fed, whereas SRT and HRT were maintained at 90 days and 15 to 40 h, respectively. Because of membrane fouling and revere salt diffusion, flux decreased from 9.5 LMH to 3.5 LMH. Alongside, the conductivity in the digester rose from 1.0 to 20.5 mS/cm, which corresponded to approximately 13 g/L NaCl. Despite the salt accumulation in the digester, no apparently negative impacts on anaerobic biology were observed. Gas composition and production rate were all in the normal range when they were compared with regular AD.

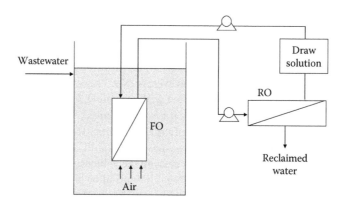

FIGURE 7.23 Process diagram of forward osmosis MBR followed by RO (FOMBR-RO).

Because the FOMBR concept is relatively new and no full-scale demonstration has been performed, the economics of the process has not been known in detail. However, if RO is required to concentrate the draw solution and obtain final permeate, it is apparent that the FOMBR-RO process is not as economical as the MBR-RO process. It is because FOMBR requires much larger membrane surface areas submerged in aeration tank because of its low submerged membrane flux, for example, 5 to 10 LMH or less versus 20–30 LMH. The larger membrane surface in turn requires more aeration for membrane scouring. In addition, the subsequent RO process used to regenerate the draw solution should operate at much higher operating pressure than the RO in MBR-RO process because of the high salt concentration in draw solution, for example, 25 to 50 bar versus 7 to 15 bar. No external energy is required to draw permeate in FOMBR, but it is not a significant benefit because only a very small amount of energy is required for the permeate suction performed at <0.3 bar in the regular MBR, as discussed in Section 6.15.6. In addition, as discussed in the previous section using Figure 7.22, much more energy needs to be added in the drawing solution recovery process to compensate the free energy loss in the permeate drawing (Figure 7.23).

7.6 MBR WITH WOVEN OR NONWOVEN TEXTILE FILTER

7.6.1 Overview

Nonwoven textile filters are the material composed of random networks of overlapping fibers. The typical cartridge filters used for air and water are based on nonwoven fabrics. Woven textile filters are manufactured by weaving microfibers using a loom according to a predetermined pattern, and the most fabrics used for clothes are woven. Various woven or nonwoven textiles have been used to fabricate submerged flat sheet filters or tubular filters to be used for MBR.

Initially, those textile filters were considered as an alternative of submerged membrane in the 1990s to reduce the capital costs of MBR. The high porosity and the excellent mechanical and chemical strengths were expected to result in high flux and long service life. In the initial studies, it was observed that permeate quality was almost equivalent to that of the regular MBR as the cake layer acted as a dynamic membrane despite the large nominal pore sizes of woven and nonwoven filters (Seo et al. 2002). However, the sustainable flux turned out to be lower than that of MF/UF membranes in the initial attempts primarily because of the large amount of suspended solids captured on the rough filter surfaces. As a result of the low flux, a larger surface area was required, and the filtration system became much bulker than regular MF/UF membrane systems. Thereby, textile filters failed to provide decisive economic advantages over regular MF/UF membranes in MBR. In addition, with the progress of MF/UF membrane manufacturing, the service life of the membrane was prolonged beyond 6 to 10 years, and the advantage of using fabric filter decreased. With rapidly declining membrane prices, the interest on using textile filters for MBR had faded out gradually.

In recent years, however, the flux of textile filters has been raised dramatically from 200 to 500 LMH with a subtle change of the filtration goal (Chen and Pang 2006; Bai et al. 2010). Instead of targeting the high-quality effluent comparable with that of the regular MBR, it was aimed to obtain the effluent with a compromised water quality yet sufficient for specific purposes, for example, irrigation and discharging to nature. Textile filters with coarse pores, for example, 5 to 100 μm, are sufficient to achieve this goal, although the effluent contains suspended solids at the level typically found in the CAS process. As a result of the high flux, the filtration system became much more compact than the regular membrane-based MBR.

Although there have been tremendous progresses in using textile filters in MBR, the textile filter–based MBR is still in the initial development stage. The factors/mechanisms affecting the sustainable flux are still not well understood. Thereby, the sustainable flux varies widely at 10 to 500 LMH depending on the filter media selected, module configurations, operating method and condition, etc. In addition, module and system designs, aeration strengths and patterns, backwashing frequencies, cleaning methods, and other operating guidelines are yet to be fully established. Finally, the overall economics of textile filter–based MBR concept needs to be proven in conjunction with the system optimization, for example, the optimum MLSS in the aeration tank, dewatering of the low MLSS sludge, filter cleaning and operation, and long-term system maintenance.

7.6.2 Factors Affecting the Performance

One fundamental difference between textile filters and membranes is the surface roughness. Because textile materials have much rougher surfaces than regular membranes, suspended solids are easily caught on the filter surface and form a cake layer. The cake layer thickness can be controlled by frequent backwashing with continuous or intermittent aeration. The rate of cake layer formation is more directly correlated with MLSS than in MBR, and hence MLSS is maintained lower, for example, 2 to 6 g/L. Air and permeate can be used to perform backwashing, but no general practices have been established with respect to the optimum backwashing frequency and backwashing fluids. Some of the particles captured inside the filter medium can be dislodged during the backwashing, but the remaining particles can eventually plug the internal pores and reduce backwashing efficiency. Because the cake layer formed on the filter surface acts as a dynamic membrane that interferes particle intrusion into the pores as discussed in Section 1.2.4, maintaining a certain degree of the cake layer may be beneficial in controlling further filter fouling and obtaining good permeate quality. As shown in Figure 7.24, as scouring air flow increases, filtration resistance tends to increase due to the overly thinning cake layer that allows particles to intrude internal pores of the filter medium. In the meantime, in a typical membrane filtration process, high airflow mitigates cake layer formation and almost certainly improves membrane performance because internal pore plugging rarely occurs. Because of the complex response of filtration resistance to aeration rate, the optimum airflow rate, which enables the high flux and good permeate quality, must be estimated in a given condition empirically (Figure 7.24).

Because the bubble point pressure is not high for the pores of 5 to 100 μm on morphologically rough surfaces, backwashing can be performed at a low pressure, such as 50 kPa or below. Despite the air scouring and the periodic backwashing, textile filters are still prone to excess cake layer formation. Thereby, the optimum MLSS of textile filter–based MBR tends to be lower than that of the regular MBR, for example, 2 to 6 g/L versus 6 to 15 g/L. The low MLSS may cause bulky biological tanks, but if the filter fouling rate is not sensitive to the OLR (or the F/M ratio), biological tanks can be sized comparably with those of the regular MBR. No sufficient knowledge exists to date regarding the relation between filter fouling rate and OLR in fabric filter–based MBR. Various textiles based on polyethylene (PE), nylon, PP, polyester (PES), polyethylene terephthalate (PET), polyamide (Nylon), etc., are widely available. The compatibility of such textiles with acid, base, and oxidants are affected not only by the base material but also by the additives used and the manufacturing method. In general, however, PE, PP, PES, PVDF, and PET are known to be resistant

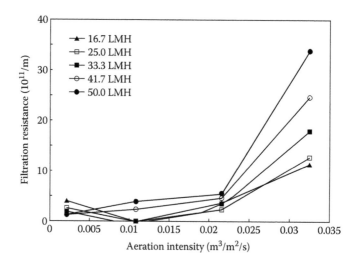

FIGURE 7.24 Effect of aeration intensity on filtration resistance at different fluxes. Nonwoven filters with 25.2 μm was used at an MLSS of 5 g/L. (From Chang, M.-C. et al., *Desalination* 191(1–3), 8–15, 2006b.)

to most acid, base, and oxidants at the typical concentrations used for membrane cleaning at surrounding temperatures. Polyamides are stable at high pH but may be vulnerable to low pH below 2. Polyamide-based filters may also be vulnerable to oxidants such as hydrogen peroxide and sodium hypochlorite.

Permeate quality is known to be comparable with MBR if the textile filters with small enough pores such as <5 μm are used and the filters are operated with a sufficient amount of cake layer that acts as a dynamic membrane. Sustainable fluxes are low under this condition, for example, <20 LMH, because of the high filtration resistance. As a result, large filter surface areas are required, and this system is not as economical as the regular MBR. By contrast, if thin textile filters with large pores (5–100 μm) are used with continuous or intermittent air scouring along with periodic backwashing, cake layer depth can be controlled, and thereby very high fluxes such as 200 to 500 LMH can be obtained. Under this condition, the effluent contains a moderate level of suspended solids, for example, 10 to 30 mg/L. In terms of the effluent-soluble COD, little differences exist regardless of the filter pore size or cake layer depth because biological COD degradation is rarely a rate determining step in biological wastewater treatment.

7.6.3 Operating Experiences in Literature

In an early study, PP nonwoven filters with varying depths were used, that is, 0.26, 0.36, and 0.42 mm, respectively (Seo et al. 2002). In this pilot test, a filter unit with 2 m² total area was submerged in the aeration tank of A/O type reactor, and permeate was drawn using the gravity. MLSS in the aeration tank was maintained low at 1.3 to 2.5 g/L, and no backwashing was performed. The initial TSS in the effluent was high at more than 100 mg/L, but it gradually dropped with the stabilization of the cake layer. The effluent TSS was 3.2 mg/L on average once the filtration was stabilized. There were no substantial differences from the regular MBR in terms of soluble COD and nutrient removal efficiencies. Permeate quality tended to improve slightly with thick filters, but the filtration resistance grew quicker to the higher level. In nonwoven filter, large particles are rejected by the filter surface by the size exclusion mechanism, but fine particles intrude the filter. The fine particles that intruded the filter are partially captured on the walls while traveling through the tortuous pores. The uncaptured fine particles are discharged from the filter with permeate.

In another study, PP-based nonwoven fabrics with 0.6 mm depth and 20 μm pore size were used (Chang et al. 2007). The nonwoven filters were manufactured as tubular configuration with 37 mm diameter and 230 mm length and fabricated as a module with 0.72 m² of total filtration area. The OLR was 2 to 4 kg COD/m³/day, which was comparable with that of the regular MBR. At a high MLSS at 15 g/L, a stable flux of 7.5 LMH was obtained at <5 kPa for 50 days without backwashing or chemical cleaning. The effluent TSS was below 10 mg/L for most of the time during the experiment. When MLSS was lower at 10 g/L, sustainable fluxes of the three filter units with different pore sizes were approximately 22 LMH (25.2 μm), 14 LMH (38.8 μm), and approximately 7 LMH (13.1 μm) at a specific airflow of 0.6 m³/m²/min (Chang et al. 2006b). Compared with the regular MBR, fluxes are comparable or lower, but specific airflow rates are greater by more than an order of magnitude. In another study, textiles were wrapped on porous tubular stainless steel pipes with 28.6 mm outer diameter and 280 mm length (Zahid and El Shafai 2011). The tubular filter units were vertically mounted in aeration tank. Aeration was performed at 20 L/m²/min to scour the filter surface. Considering the short filter length, aeration strength is in line with regular submerged membranes. When MLSS was 5.3 to 5.5 g/L and F/M ratio was 0.32 g COD/g MLSS/day, flux was maintained at 16.3 LMH. The effluent TSS and COD were 2 to 3 mg/L and 30 to 37 mg/L, respectively.

The flux of textile filters has been dramatically improved recently by fabricating filters with more optimum textile materials and operating them at more optimum conditions with respect to aeration rate, MLSS, air scouring pattern, backwashing intensity and frequency, etc. Xiong et al. (2014) used PET meshes with 300 mesh size (or 53 μm opening) in the configurations of flat sheet (270 mm × 95 mm) and tubular (φ 6 mm × 290 mm). PET nonwoven fabric with 75 μm pores (No. 2) was also fabricated as tubular configuration to compare with PET meshes. Flat sheet modules (No. 1) were fabricated similarly to the flat sheet submerged membranes. Tubular membranes were fabricated by welding mesh or nonwoven fabric around coil spring–like supports (Nos. 3 and 4). All filter units were submerged in the same tank, and filtration was performed at an identical water head of 70 mm. MLSS was maintained at 3.8 g/L and HRT was 8 h. Municipal wastewater was fed to the system at an OLR of 0.2 to 0.25 g COD/g VSS/day. SRT was at 40 days. If the flux decreased below 75% of the initial, air scouring was performed for 5 min at 8 L/min. As shown in Figure 7.25, filtrate turbidity was high in the beginning, but it decreased to <5 NTU within several minutes as the cake layer built up. No significant cake layer compaction occurred because the filtration was performed at a low water head of 70 mm, and thereby nearly complete flux recovery was achieved by air souring alone. Flat sheet submerged modules with PET mesh showed the highest flux at around 160 LMH, and tubular modules with PET mesh were the next at around 140 LMH for 800 to 1000 h. The tubular module with nonwoven fabric (No. 2) showed the lowest flux at around 100 LMH in the beginning,

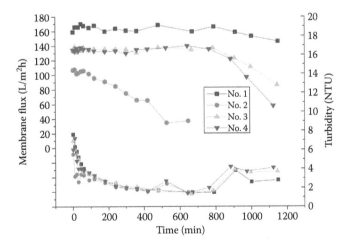

FIGURE 7.25 Flux and effluent turbidity at 70 mm H₂O head loss, where No. 1, No. 2, and No. 3 are PET meshes in flat sheet (No. 1) or tubular (Nos. 2 and 3) configuration and No. 4 is nonwoven PET filter in tubular configuration. (From Xiong, J. et al., *J. Membrane Sci.* 471, 308–318, 2014.)

and it declined rapidly. The large performance difference between meshes and nonwoven fabrics suggests that selecting efficient filter materials is critically important in obtaining high fluxes in textile filter–based MBR.

In another study (Chen and Pang 2006), using nonwoven textile filters with nominal pores larger than 5 μm, a steady-state flux was obtained at 62.5 LMH at an MLSS of less than 5 g/L. In this study, daily backwashing was performed along with continuous aeration. The effluent turbidity, TSS, and COD were less than 1.5 NTU, 3 mg/L, and 35 mg/L, respectively. When the similar filters with larger pores (>10 μm) were used for MBR treating saline sewage, an average flux of 250 LMH was maintained for 270 days (Bai et al. 2010, 2011). The total COD and TSS in the effluent were 27 mg/L and 15 mg/L on average, respectively. MLSS was 2.2 to 6 g/L during the pilot testing. The SAD_p was 15, whereas average flux was 200 LMH. For comparison, the typical design flux of hollow fiber membranes with 0.03 to 0.45 μm pore sizes is approximately 25 LMH at a similar SAD_p in field condition. In another study, nylon woven fabrics with nominal pore size of 55 μm were used as filter (Guan et al. 2014). While continuous aeration was performed, daily backwash was also performed for 1 min every day. A flux of 546 LMH was maintained for 74 days at an MLSS of 3.5 g/L, and the effluent TSS was at 28.8 mg/L on average. Meanwhile, the total COD was reduced from 593 to 86 mg/L on average.

7.7 THERMOPHILIC MBR

Thermophilic aerobic biological treatment has been used widely to stabilize the excess biosolids from the activated sludge process for many decades. One prominent process that falls into this category is the autothermal thermophilic aerobic digestion (ATAD) that operates at 50°C to 65°C using the heat generated from the biological degradation of sewage sludge. ATAD enables 30% to 50% biosolid reductions at a retention time of 5 to 10 days. According to a study, autothermal reaction is plausible even for wastewater, if the wastewater has a COD higher than 20 to 40 g/L and the OTE is 10% to 20% (LaPara and Alleman 1999). Thermophilic aerobic wastewater treatment can be a promising option especially when preserving the heat in the wastewater is crucial for overall economics, for example, hot white water recycle in paper mill.

It has been often considered that thermophilic reactions are faster than mesophilic reactions because of the high enzymatic reaction rates. When it is used for wastewater treatment, excess biosolid production has also been deemed low because of the high maintenance energy requirement at high temperature. Such notions might be true when the same reactor is experiencing short-term moderate temperature rises because activities of the microorganisms in the reactor rise as temperature rises and microbial decay rate accelerates. However, if the elevated temperature sustains for a sufficiently long period, the microbial consortia will change with the species that fit in the hot environment, and hence the microbial activity returns near to the original level. It has been found from rigorous experimental setups that biodegradation rates are relatively constant at 25°C to 65°C and the biosolid yields were also nearly constant at the same temperature range (LaPara et al. 2001a,b). When complex organic medium was used as a feed, the maximum specific substrate utilization rates remained nearly constant at 0.64 ± 0.04 g COD/g MLSS/h at a temperature range of 25°C to 65°C (LaPara et al. 2000b). The biosolid decay constant (k_d) was found at 0.046 per day at 55°C (LaPara et al. 2000a,b) and was similar to the k_d observed in CAS, as summarized in Table 6.3. Depending on the temperature (25°C–65°C), distinctively different microbial consortia were confirmed based on DNA analysis. It is considered that the different microbial consortia have almost same metabolic rates at the temperature they are adapted to.

Similar to the observations made in anaerobic processes as discussed in Section 7.4.2, the microbial diversities in thermophilic aerobic conditions are much narrower than those in mesophilic conditions. Denaturing gradient gel electrophoresis (DGGE) studies have revealed narrowing genetic diversities as temperature increases. As a result of the narrow genetic diversities of the microbial community, treatment efficiency tends to decrease as temperature increases, especially when feed water contains diverse organics that require complex biological steps to be decomposed. For instance, when feed water with a complex growth medium was used, the efficiency of substrate removal declined at elevated temperatures particularly for gelatin and α-lactose (LaPara et al. 2000b).

In addition to the potentially lower degradation efficiencies, poor floc formation is commonly observed in thermophilic processes (de Sousa et al. 2011). At 35°C, compact and dense flocs more than 500 μm were observed along with abundant type 021N filamentous microorganisms, but floc size decreased below 150 μm at 45°C with a proliferation of type 0581 filamentous microorganisms. At 55°C, larger flocs with a size of 150 to 500 μm were observed, but those were weak at shear field perhaps because of the lack of filamentous bacteria that can provide structural strength to the floc. Meanwhile, clarifiers of thermophilic processes also suffer from more vertical mixing than those of mesophilic counterparts. It is because the strong cooling effect on the water surface and the clarifier wall causes water density gradient in the clariier. As a consequence, the effluent quality of thermophilic processes is not as good as those of mesophilic processes.

The operating costs of thermophilic process are sensitive to the OTE at the high temperature, but the OTE is not sensitive to the temperature in the range between 20°C and 55°C (Vogelaar et al. 2000). It is because oxygen molecules diffuse quicker through the air–water interface at a high temperature, although the declining oxygen solubility cut down the driving force. In fact, the oxygen dissolution rate is proportional to the mass transfer coefficient (k_{La}) and the magnitude of the difference between the saturation DO (C_0^*) and the current DO (C), as described as Equation 6.24. According to a study, k_{La} and C_0^* move to the opposite direction as temperature increases, as shown in Figure 7.26. It is noticeable that k_{La} became nearly twofolds, whereas C_0^* drops by less than one-half when temperature rises from 20°C to 55°C. Therefore, if the DO is low at below 1 mg/L, overall oxygen dissolution rate can increase slightly as temperature increases based on the Equation 6.24. Meanwhile, in the empirical equation shown in Equation 6.20, the increase of kinetic term, θ^{T-20}, is more than enough to compensate the declining thermodynamic term, where C_0^* is calculated as $0.0017T^2$ to $0.2409T + 13.21$. Therefore, aeration capacities for thermophilic processes can be designed according to the design parameters developed for mesophilic processes, which was discussed in Section 6.5.

Not much has been known for thermophilic MBR, but the treatment efficiency and excess biosolid production are deemed similar to the regular MBR because most biological characteristics are common for both processes, except the potentially lower biological stabilities due to the narrow genetic diversity. In fact, removal efficiencies were reciprocally related with the temperature in a

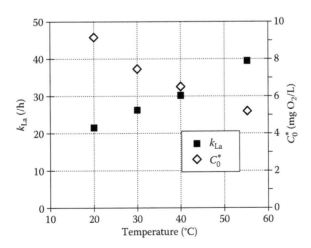

FIGURE 7.26 Mass transfer coefficient (k_{La}) and saturation DO (C_0^*) as a function of temperture in tap water. (From Vogelaar, J.C. et al., *Water Res.* 34(3), 1037–1041, 2000.)

side-by-side test performed with three parallel MBRs equipped with submerged hollow fibers (de Sousa et al. 2011). The three MBRs were maintained at 35°C, 45°C, and 55°C, respectively, and paper mill wastewater was fed at an identical food-to-volume ratio (F/V) ratio of 2.57 kg COD/m³/day. It was observed that COD removal efficiency decreased gradually as temperature increased, that is, 95.5% to 94.2% to 91.9%, respectively.

Meanwhile, the sustainable flux of thermophilic MBR is somewhat lower than that of normal MBR despite the low water viscosity at high water temperature. When flat sheet submerged PVDF membranes were used to treat paper mill wastewater at 51°C, flux was not able to be raised higher than 12.5 LMH (Qu et al. 2012). The MLSS and SRT used for the laboratory-scale experiment were 8 to 12 g/L and 10 days, respectively. Membrane autopsy revealed that the cake layer was responsible for the low flux. The small and weak flocs appeared as the major components of the cake layer in the thermophilic condition. In another study, pet food wastewater containing high FOG contents (6 g/L) and high COD (51 g/L) was treated using two laboratory-scale MBRs at 20°C and 45°C (Kurian et al. 2005). Submerged membranes were used for the 20°C MBR, but sidestream membranes with circulation pump were used for 45°C MBR because of the membrane compatibility issue at the high temperature. Removal efficiencies at 20°C were mostly more than 90%, but they varied at 75% to 98% in the thermophilic MBR. Biosolid yield was much lower in 45°C MBR, but it might be caused by the repeated exposure of microorganisms to the high shear stress in circulation pump. It has been known that the circulation pump expedites cell decay by partially disintegrating the microorganisms and eventually causes low biosolid yield, as discussed in Section 5.2.10.2.

In another study (Ramaekers et al. 2001), a thermophilic MBR equipped with sidestream tubular membranes (12 m³/h) was piloted at 50°C–55°C. F/M and F/V ratios were maintained at 0.1 to 0.6 kg COD/kg MLSS/day and 3.75 kg COD/m³/day, respectively, by feeding paper mill wastewater with 4390 mg/L COD. When the thermophilic MBR effluent was recycled to bleaching process, 10% more chemical was required to obtain the same performance in the bleaching process because of the organics contained in the MBR effluent. No significant changes were noticed in the product quality with the excess chemical consumption.

As discussed previously, most of thermophilic MBRs tested in the field have been performed using sidestream tubular membranes because most of commercially available submerged membranes are not compatible with the high temperature, for example, >40°C. Despite the satisfactory technical performance, thermophilic MBR has not been proliferated because of the high capital and operating costs of the sidestream membranes. Substantial cost reduction is essential to make thermophilic MBR competitive. Perhaps the success of thermophilic MBR hinges on the availability of economical submerged membranes compatible with the high temperature just as the MBR technology became popular with the introduction of submerged membranes.

References

Abdullah, S., Bérubé, P. 2013. Assessing the effects of sodium hypochlorite exposure on the characteristics of PVDF based membranes. *Water Res.* 47:5392–5399.

Achilli, A., Cath, T.Y., Marchand, E.A., Childress, A.E. 2009. The forward osmosis membrane bioreactor: A low fouling alternative to MBR processes. *Desalination* 239(1–3):10–21.

ACWA Ltd. 2007. Slide for SAWEA conference. Available at http://www.sawea.org/pdf/2007/DEC%205 /MBR%20Challenge/ACWA.pdf (accessed June 30, 2011).

Adham, S.S., Jacangelo, J.G., Laine, J.M. 1995. Low-pressure membranes: Assessing integrity. *J. AWWA* 3:62–75.

Adham, S.S., DeCarolis, J.F., Pearce, W. 2004. Optimization of Various MBR Systems for Water Reclamation—Phase III. Desalination and Water Purification Research and Development Program Final Report No. 103. US Department of Interior.

Ahmed, T., Semmens, M.J. 1992. Use of sealed end hollow fibers for bubbleless membrane aeration: Experimental studies. *J. Membrane Sci.* 69(1–2):1–10.

Ahmed, Z., Cho, J., Lim, B.-R., Song, K.-G., Ahn, K.-H. 2007. Effects of sludge retention time on membrane fouling and microbial community structure in a membrane bioreactor. *J. Membrane Sci.* 287:211–218.

Ahn, K.H., Park, K.Y., Maeng, S.K. et al. 2002. Ozonation of wastewater sludge for reduction and recycling. *Water Sci. Technol.* 46(10):71–77.

Akamatsu, K., Lu, W., Sugawara, T., Nakao, S. 2010. Development of a novel fouling suppression system in membrane bioreactors using an intermittent electric field. *Water Res.* 44(3):825–830.

Al-Halbouni, D., Lyko, S., Janot, A., Tacke, D., Dott, W., Hollender, J. 2007. Impact of solids retention time on EPS and fouling tendency in membrane bioreactors. *Proceedings of the 2nd IWA National Young Water Professionals Conference*, June 4–6. Berlin, Germany.

Amiri, S., Mehrnia, M.R., Azami, H., Barzegari, D., Shavandi, M., Sarrafzadeh, M.H. 2010. Effect on heavy metals on fouling behavior in membrane bioreactors. *Iran. J. Environ. Healt.* 7(5):377–384.

APHA. 1998. *Standard Method for the Examination of Water and Wastewater*, 20th ed. Washington, DC: American Public Health Association.

Arkhangelsky, E., Kuzmenko, D., Gitis, V. 2007. Impact of chemical cleaning on properties and functioning of polyethersulfone membranes. *J. Membrane Sci.* 305(1–2):176–184.

Arora, N., Davis, R.H. 1994. Yeast cake layers as secondary membranes in dead-end microfiltration of bovine serum albumin. *J. Membrane Sci.* 92(3):247–256.

ASCE. 1991. A standard for the measurement of oxygen transfer in clean water. Method 02-84. Reston, VA: American Society of Civil Engineers.

ASTM. 2010. Standard practice for integrity testing of water filtration membrane systems. D 6908-03. West Conshohocken, PA: ASTM International.

Atkins, P., de Paula, J. 2009. *Physical Chemistry*, 9th ed. Oxford, UK: Oxford University Press.

Ayala, D.F., Ferre, V., Judd, S.J. 2011. Membrane life estimation in full-scale immersed membrane bioreactors. *J. Membrane Sci.* 378:95–100.

Bae, J.-H., Kim, J.-H., McCarty, P.L. 2013. Fluidized membrane bioreactor. US Patent No. 8,404,111.

Baker, R.W. 2004. *Membrane Technology and Applications*, 2nd ed. Chichester, West Sussex, England: Wiley.

Banu, J.R., Uan, D.K., Chung, I.-J., Kaliappan, S., Yeom, I.-T. 2009. A study on the performance of a pilot scale A2/O-MBR system in treating domestic wastewater. *J. Environ. Biol.* 30(6):959–963.

Banu, R., Uan, K., Kaliappan, S., Yeom, I.T. 2011. Effect of sludge pretreatment on the performance of anaerobic/anoxic/oxic membrane bioreactor treating domestic wastewater. *Int. J. Environ. Sci. Tech.* 82:281–290.

Barillon, B., Martin Ruel, S.M., Langlais, C., Lazarova, V. 2012. Energy optimization in membrane bioreactors. *Proceedings of the IWA World Congress on Water, Climate and Energy*, May 13–18. Dublin, Ireland.

Barillon, B., Martin Ruel, S., Langlais, C., Lazarova, V. 2013. Energy efficiency in membrane bioreactors. *Water Sci. Technol.* 67(12):2685–2691.

Barnett, J., Richardson, D., Stack, K., Lewis, T. 2012. Addition of trace metals and vitamins for the optimization of pulp and paper mill activated sludge wastewater treatment plant. *Appita J.* 65(3):237–243.

Bartels, C., Franks, R., Furukawa, R., Murkute, P., Papukchiev, U. 2004. *Integrated Membrane System for Low Fouling RO: Desalting of Municipal Wastewater*. Oceanside, CA: Hydranautics Inc.

Bassin, J.P., Pronk, M., Muyzer, G., Kleerebezem, R., Dezotti, M., van Loosdrecht, M.C. 2011. Effect of elevated salt concentrations on the aerobic granular sludge process: Linking microbial activity with microbial community structure. *Appl. Environ. Microbiol.* 77(22):7942–7953.

Bates, W. 1990. Reducing the fouling rate of surface and wastewater RO systems. *Proceedings of the International Water Conference (IWC)*, October 22–24. Pittsburgh, PA.

Beck, R.W. 2010. San Antonio water system—Brackish groundwater desalination facility—Enhanced recovery alternatives evaluation and pilot test report. Texas Water Development Board Contract #0704830718.

Beier, S.P., Jonsson, G. 2009. A vibrating membrane bioreactor VMBR: Macromolecular transmission-influence of extracellular polymeric substances. *Chem. Eng. Sci.* 64(7):1436–1444.

Belfort, G., Davis, R.H., Zydney, A.L. 1994. The behavior of suspensions and macromolecular solutions in crossflow microfiltration. *J. Membrane Sci.* 96:1–58.

Bendick, J.A., Modise, C.M., Miller, C.J., Neufeld, R.D., Vidic, R.D. 2004. Application of crossflow microfiltration for the treatment of combined sewer overflow wastewater. *J. Environ. Eng.* 130:1442–1449.

Bendick, J.A., Miller, C.J., Kindle, B.J., Shan, H., Vidic, R.D., Neufeld, R.D. 2005. Pilot scale demonstration of crossflow ceramic membrane microfiltration for treatment of combined and sanitary sewer overflows. *J. Environ. Eng.* 131:1532–1539.

Bentzen, T.R., Ratkovich, N., Madsen, S., Jensen, J.C., Bak, S.N., Rasmussen, M.R. 2012. Analytical and numerical modeling of Newtonian and non-Newtonian liquid in a rotational cross-flow MBR. *Water Sci Technol.* 66(11):2318–2327.

Berghof GmbH. 2014. Available at http://www.berghof.com/ (accessed August 2014).

Bérubé, P.R., Lei, E. 2006. The effect of hydrodynamic conditions and system configurations on the permeate flux in a submerged hollow fiber system. *J. Membrane Sci.* 271(1–2):29–37.

Bérubé, P.R., Afonso, G., Taghipour, F., Chan, C.C.V. 2006. Quantifying the shear at the surface of submerged hollow fiber membranes. *J. Membrane Sci.* 279(1–2):495–505.

Bérubé, P.R., Lin, H., Watai, Y. 2008. Fouling in air sparged submerged hollow fiber membranes at sub- and super-critical flux conditions. *J. Membrane Sci.* 307(2):169–180.

Bian, R., Yamamoto, K., Watanabe, Y. 2000. The effect of shear rate on controlling the concentration polarization and membrane fouling. *Proceedings of the Conference on Membranes in Drinking and Industrial Water Production*, October 3–6. L'Aquila, Italy, 1:421–432.

Bilad, M.R., Mezohegyi, G., Declerck, P., Vankelecom, I.F.V. 2012. Novel magnetically induced membrane vibration (MMV) for fouling control in membrane bioreactors. *Water Res.* 46(1):63–72.

BKT Inc. 2013. Available at http://www.bkt21.com/ (accessed January 24, 2013).

Böhm, L., Drews, A., Prieske, H., Bérubé, P.R., Kraume, M. 2012. The importance of fluid dynamics for MBR fouling mitigation. *Bioresour. Technol.* 122:50–61.

Boonyaroj, V., Chiemchaisri, C., Chiemchaisri, W., Yamamoto, K. 2012. Removal of organic micro-pollutants from solids waste landfill leachate in membrane bioreactor operated without excess sludge discharge. *Water Sci. Technol.* 66(8):1774–1780.

Bougrier, C., Delgenès, J.-P., Carrère, H. 2006. Combination of thermal treatment and anaerobic digestion to reduce sewage sludge quantity and improve biogas yield. *Process Saf. Environ. Prot.* 84(4):280–284.

Brannock, M., Wang, Y., Leslie, G. 2010. Mixing characterization of full-scale membrane bioreactor: CFD modelling with experimental validation. *Water Res.* 44(10):3181–3191.

Brepols, C. 2011. *Operating Large Scale Membrane Bioreactors for Municipal Wastewater Treatment.* London: IWA Publishing.

Brepols, C.H., Schäfer, H., Engelhardt, N. 2010. Considerations on the design and financial feasibility of full-scale membrane bioreactors for municipal applications. *Water Sci. Technol.* 61(10):2461–2468.

Bretscher, U. 2014. Cleaning clogged aerators in place. Available at http://www.musketeer.ch/sewage/formic_acid.html (accessed June 2014).

Brockmann, M., Seyfried, C.F. 1996. Sludge activity and crossflow microfiltration: A non-beneficial relationship. *Water Sci. Technol.* 34(9):205–213.

Buer, T., Cumin, J. 2010. MBR module design and operation. *Desalination* 250:1073–1077.

Buer, T., Adams, N., Hong, Y. 2008. High efficiency oxygen transfer membrane supported biofilm reactor for wastewater treatment. *Proceedings of the IWA North American Membrane Research Conference*, August 10–13. Amherst, MA.

Burgess, J.E., Quarmby, J., Stephenson, T. 1999. Micronutrient supplements for optimization of the treatment of industrial wastewater using activated sludge. *Water Res.* 33(18):3707–3714.

Burgess, J.E., Harkness, J., Longhurst, P.J., Stephenson, T. 2000. Nutrient balancing for enhanced activated sludge reactor performance: UK perspective. *Water Sci. Technol.* 41(12):223–231.

Cabassud, C., Ducom, G., Laborie, S. 2003. Measurement and comparison of wall shear stresses in a gas/liquid two-phase flow for two module configurations. *Proceedings of the 5th International Membrane Science and Technology (IMSTEC)*, November 10–14. Sydney, Australia.

Camacho, L.M., Dumée, L., Zhang, J. et al. 2013. Advances in membrane distillation for water desalination and purification applications. *Water* 5(1):94–196.

Cantor, J., Sutton, P.M., Steinheber, R., Novachis, L. 2000. Industrial biotreatment plant capacity expansion and upgrading through application of membrane biomass-effluent separation. *Proceedings of WEFTEC*, October 14–18. Anaheim, CA.

Cao, Z.P., Zhang, J.L., Zhang, H.W. 2008. Influence of solids retention time on sludge characteristics and effluent quality in immersed membrane bioreactor. *Chinese Sci. Bull.* 53(24):3942–3950.

Carletti, G., Fantone, F., Bolzonella, D., Cecchi, F. 2008. Occurrence and fate of heavy metals in large wastewater treatment plants treating municipal and industrial wastewaters. *Water Sci. Technol.* 57(9):1329–1336.

CDM. 2010. Test technology innovations and optimize systems in the city of Santa Cruz pilot plant. Report to SCWD2.

Cha, G.C., Hwang, M.G., Chung, H.K., Kim, D.J., Yoo, I.K., Han, S.K. 2003. Characteristics of cell-growth and behavior of SMP in the MBR process. *J. Korean Soc. Environ. Eng.* 25(2):155–162 (in Korean).

Chan, C.C.V., Bérubé, P.R., Hall, E.R. 2007. Shear profiles inside gas sparged submerged hollow fiber membrane modules. *J. Membrane Sci.* 297:104–120.

Chan, C.C.V., Bérubé, P.R., Hall, E.R. 2011. Relationship between types of surface shear stress profiles and membrane fouling. *Water Res.* 45:6403–6416.

Chang, S. 2011. Application of submerged hollow fiber membrane in membrane bioreactors: Filtration principles, operation, and membrane fouling. *Desalination* 283:31–39.

Chang, I.S., Lee, C.H. 1998. Membrane filtration characteristics in membrane coupled activated sludge system: The effect of physiological states of activated sludge on membrane fouling. *Desalination* 120(3):221–233.

Chang, S., Fane, A.G. 2000. Filtration of biomass with axial inter-fibre upward slug flow: Performance and mechanisms. *J. Membrane Sci.* 180(1):57–68.

Chang, S., Fane, A.G. 2001. The effect of fibre diameter on filtration and flux distribution-relevance to submerged hollow fibre modules. *J. Membrane Sci.* 184:221–231.

Chang, S., Fane, A.G. 2002. Filtration of biomass with laboratory-scale submerged hollow fibre membrane module: Effect of operational conditions and module configuration. *J. Chem. Technol. Biotechnol.* 77:1030–1038.

Chang, I.-S., Kim, S.-N. 2005. Wastewater treatment using membrane filtration-effect of biosolids concentration on cake resistance. *Proc. Biochem.* 40(3–4):1307–1314.

Chang, I.-S., Choo, K.-H., Lee, C.-H. et al. 1994. Application of ceramic membrane as a pretreatment in anaerobic digestion of alcohol-distillery wastes. *J. Membrane Sci.* 90:131–139.

Chang, I.S., Lee, C.H., Ahn, K.H. 1999. Membrane filtration characteristics in membrane coupled activated sludge system: The effect of floc structure of activated sludge on membrane fouling. *Sep. Sci. Technol.* 34:1743–1758.

Chang, S., Fane, A.G., Vigneswaran, S. 2002a. Experimental assessment of filtration of biomass with transverse and axial fibres. *Chem. Eng. J.* 87:121–127.

Chang, S., Fane, A.G., Vigneswaran, S. 2002b. Modelling and optimizing submerged hollow fiber membrane modules. *AIChE J.* 48:2203–2212.

Chang, W.C., Jou, S.J., Chien, C.C., He, J.A. 2004. Effect of chlorination bulking control on water quality and phosphate release/uptake in an anaerobic-oxic activated sludge system. *Water Sci. Technol.* 50(8):177–183.

Chang, S., Fane, A.G., Waite, T.D. 2006a. Analysis of constant permeate flow filtration using dead-end hollow fiber membranes. *J. Membrane Sci.* 268:132–141.

Chang, M.-C., Horng, R.-Y., Shao, H., Hu, Y.-J. 2006b. Performance and filtration characteristics of non-woven membranes used in a submerged membrane bioreactor for synthetic wastewater treatment. *Desalination* 191(1–3):8–15.

Chang, W.-K., Hu, A.Y.-J., Horng, R.-Y., Tzou, W.-J. 2007. Membrane bioreactor with nonwoven fabrics as solid-liquid separation media for wastewater treatment. *Desalination* 202(1–3):122–128.

Chang, S., Fane, A.G., Waite, T.D., Yeo, A. 2008. Unstable filtration behaviour with submerged hollow fiber membranes. *J. Membrane Sci.* 308:107–114.

Chang, I.-S., Field, R., Cui, Z. 2009. Limitations of resistance-in-series model for fouling analysis in membrane bioreactors: A cautionary note. *Desalin. Water Treat.* 8(1–3):31–36.

Chang, I.S., Lee, S., Joung, S.Y., Lee, C.K. 2011. Advanced treatment system of wastewater having plasma discharging vessel. US Patent No. 8,574,435.

Chen, G.H., Pand, S.K., 2006. Development of a low-cost dynamic filter immersed in activated sludge syatem. *Water. Sci. Technol: Water Suppl.* 6(6):111–117.

Chen, D., Peng, H. 2005. A thermodynamic model of membrane humidifiers for PEM fuel cell humidification control. *J. Dyn. Sys. Meas. Control.* 127:424–432.

Chen, G.H., Saby, S., Djafer, M., Mo, H.K. 2001. New approaches to minimize excess sludge in activated sludge systems. *Water Sci. Technol.* 44(10):203–208.

Chen, D., Weavers, L.K., Walker, H.W. 2006. Ultrasonic control of ceramic membrane fouling by particles: Effect of ultrasonic factors. *Ultrason. Sonochem.* 13(5):379–387.

Chen, J., Ng, W., Luo, R. et al. 2012a. Membrane bioreactor process modeling and optimization: Ulu Pandan water reclamation plant. *J. Environ. Eng.* 138(12):1218–1226.

Chen, J., Zhang, M., Wang, A., Lin, H., Hong, H., Lu, X. 2012b. Osmotic pressure effect on membrane fouling in a submerged anaerobic membrane bioreactor and its experimental verification. *Bioresour. Technol.* 125:97–101.

Chen, L., Gu, Y., Cao, C., Zhang, J., Ng, J.-W., Tang, C. 2014. Performance of a submerged anaerobic membrane bioreactor with forward osmosis membrane for low-strength wastewater treatment. *Water Res.* 50:114–124.

Cherchi, C., Onnis-Hayden, A., El-Shwabkeh, I., Gu, A.Z. 2009. Implication of using different carbon sources for denitrification in wastewater treatments. *Water Env. Res.* 81(8):788–799.

Cho, B.D., Fane, A.G. 2002. Fouling transients in nominally subcritical flux operation of a membrane bioreactor. *J. Membrane Sci.* 209:391–403.

Choi, J.-H., Ng, H.-Y. 2007. Influence of membrane material on performance of a submerged membrane bioreactor. *Proceedings of the IWA Conference*, May 15–17. Harrogate, UK.

Choi, J.H., Fukushi, K., Ng, H.Y., Yamamoto, K. 2006. Evaluation of a long-term operation of a submerged nanofiltration membrane bioreactor (NF MBR) for advanced wastewater treatment. *Water Sci. Technol.* 53(6):131–136.

Choi, J.-S., Hwang, T.-M., Lee, S., Hong, S. 2009. A systematic approach to determine the fouling index for a RO/NF membrane process. *Desalination* 238(1–3):117–127.

Choi, B.G., Cho, J., Song, K.G., Maeng, S.K. 2013. Correlation between effluent organic matter characteristics and membrane fouling in a membrane bioreactor using advanced organic matter characterization tools. *Desalination* 309:74–83.

Chon, D.-H., Rome, M., Kim, H.-S., Park, C. 2010. Investigating the mechanism of sludge reduction in activated sludge with an anaerobic side-stream reactor. *Water Sci. Technol.* 63(1):93–99.

Choo, K.-H., Lee, C.-H. 1996. Membrane fouling mechanisms in the membrane-coupled anaerobic bioreactor. *Water Res.* 30(8):1771–1780.

Ciccotelli, J. 2012. Efficiency of MBR with a giant leap forward. *Proceedings of WEFTEC*, September 27–October 1. New Orleans, LA.

Cicek, N., Macomber, J., Davel, J., Suidan, M.T., Audic, J., Genestet, P. 2001. Effect of solids retention time on the performance and biological characteristics of a membrane bioreactor. *Water Sci. Technol.* 43(11):43–50.

City of San Diego. 2005. Water reuse study phase 1. Project No. 491,098.

Clarke, A., Blake, T.D., Carruthers, K., Woodward, A. 2002. Spreading and imbibition of liquid droplets on porous surfaces. *Langmuir* 18:2980–2984.

Codianne, B. 2012. Energy optimization in membrane bioreactors (MBR) through proper design and operations. *Proceedings of WEFTEC*, September 27–October 1. New Orleans, LA.

Conner, W.G. 2011. Oily wastewater reuse technologies. *Proceedings of the Water Arabia Conference*, January 31–February 2. Manama, Bahrain.

Cornel, P., Wagner, M., Krause, S. 2003. Investigation of oxygen transfer rates in full scale membrane bioreactors. *Water Sci. Technol.* 47(11):313–319.

Cosenza, A., Di Bella, G., Mannina, G., Torregrossa, M. 2013. The role of EPS in fouling and foaming phenomena for a membrane bioreactor. *Bioresour. Technol.* 147:184–192.

Côté, P. 2007. Future of membrane technology in worldwide sanitation. *Proceedings of IWA Young Water Professionals Conference*, June 4–6. Berlin, Germany.

Côté, P. 2010. The future of MF/UF in water treatment. *Proceedings of the WETSUS Conference*, October 19. Leeuwarden, The Netherlands.

Côté, P., Thompson, D. 2000. Wastewater treatment using membranes: The North American experience. *Water Sci. Technol.* 41(10–11):209–215.

Côté, P., Liu, M. 2004. MBR beats tertiary filtration for indirect water reuse. *Desalin. Water Reuse* 13(4):32–37.

Côté, P., Janson, A., Rabie, H., Singh, M. 2001. Cyclic aeration system for submerged membrane modules. US Patent No. 6,245,239.

Côté, P., Cadera, J., Adams, N., Best, G. 2002. Monitoring and maintaining the integrity of immersed ultrafiltration membranes used for pathogen protection. *Water Supply* 2(5–6):307–311.

Côté, P., Masini, M., Mourato, D. 2004. Comparison of membrane options for water reuse and reclamation. *Desalination* 167:1–11.

Côté, P., Brink, D., Adnan, A. 2006. Pretreatment requirements for membrane bioreactors. *Proceedings of WEFTEC*, October 21–25. Dallas.

Côté, P., Alam, Z., Penny, J. 2012. Hollow fiber membrane life in membrane bioreactor (MBR). *Desalination* 288(1):145–151.

Crawford, G., Daigger, G., Erdal, Z. 2006. Enhanced biological phosphorus removal within membrane bioreactors. *Proceedings of WEFTEC*, October 21–25. Dallas.

Crozes, G.F., Sethi, S., Mi, B., Curl, J., Mariñasc, B. 2002. Improving membrane integrity monitoring indirect methods to reduce plant downtime and increase microbial removal credit. *Desalination* 149:493–497.

Cui, Z.F., Bellara, S.R., Homewood, P. 1997. Airlift crossflow membrane filtration—A feasibility study with dextran ultrafiltration. *J. Membrane Sci.* 128(1):83–91.

Cui, Z.F., Chang, S., Fane, A.G. 2003. The use of gas bubbling to enhance membrane processes. *J. Membrane Sci.* 221(1–2):1–35.

Culkin, J.B. 1989. Method and device for separation of colloidal suspensions. US Patent No. 4,872,988.

Cumin, J., Behmann, H., Hong, Y., Bayly, R. 2011. Gas sparger for an immersed membrane. US Patent Application No. 2011/0049047.

Cussler, E.L. 2009. *Diffusion: Mass Transfer in Fluid Systems*, 3rd ed. Cambridge, UK: Cambridge University Press.

Dagnew, M., Hong, Y., Adams, N., Fonseca, N., Cumin, J. 2014. Zeeweed AnMBR: Advancing anaerobic digestion for industrial application. *Proceedings of WEFTEC*, September 26–30. New Orleans, LA.

Daigger, G.T., Adams, C.D., Steller, H.K. 2007. Diffusion of oxygen through activated sludge flocs: Experimental measurement, modeling, and implications for simultaneous nitrification and denitrification. *Water Environ. Res.* 79(4):375–387.

Daigger, G.T., Crawford, G.V., Johnson, B.R. 2009. Achieving low effluent nutrient concentrations using membrane bioreactor technology: Documenting "how low you can go." *Proceedings of WEFTEC*, October 17–21. Orlando, FL.

Daigger, G.T., Crawford, G.V., Johnson, B.R. 2010. Full-scale assessment of the nutrient removal capabilities of membrane bioreactors. *Water Environ. Res.* 82(9):806–818.

de Bruin, B., Giesen, A., de Kreuk, M., Power, S. 2006. A breakthrough in biological wastewater treatment: Aerobic granules. *Proceedings of the Biennial Water Institute of Southern Africa Conference*, May 21–25. Durban, South Africa.

DeCarolis, J.F., Adham, S. 2007. Performance investigation of membrane bioreactor systems during municipal wastewater reclamation. *Water Environ. Res.* 79(13):2536–2550.

DeCarolis, J., Adham, S., Pearce, W., Hirani, Z., Lacy, S., Stephenson, R. 2007. Cost trends of MBR systems for municipal wastewater treatment. *Proceedings of WEFTEC*, September 26–30. New Orleans, LA.

de la Torre, T., Lesjean, B., Drews, A., Kraume, M. 2008. Monitoring of transparent exopolymer particles (TEP) in a membrane bioreactor (MBR) and correlation with other fouling indicators. *Water Sci. Technol.* 58(10):1903–1909.

del Real Olvera, J., Lopez-Lopez, A. 2012. Biogas production from anaerobic treatment of agro-industrial wastewater. In *Biogas*, ed. S. Kumar. Rijeka, Croatia: Intech. Available at http://www.intechopen.com.

Denecke, M., Liebig, T. 2003. Effect of carbon dioxide on nitrification rates. *Bioprocess Biosyst. Eng.* 25(4):249–253.

de Sousa, C.A., Silva, C.M., Vieira, N.M. et al. 2011. Thermophilic treatment of paper machine white water in laboratory-scale membrane bioreactors. *Desalin. Water Treat.* 27:1–7.

de Temmerman, L., Maere, T., Temmink, H., Zwijnenburg, A., Nopens, I. 2014. Salt stress in a membrane bioreactor: Dynamics of sludge properties, membrane fouling and remediation through powdered activated carbon dosing. *Water Res.* 63:112–124.

De Wever, H., Brepols, C., Lesjean, B. 2009. Decision tree for full-scale submerged MBR configurations. *Proceedings of the Final MBR-Network Workshop*, March 31–April 1. Berlin, Germany.

De Wilde, W., Richard, M., Lesjean, B., Tazi-Pain, A. 2008. Towards standardization of the MBR technology? *Desalination* 231:156–165.

Diamantis, V.I., Antoniou, I., Melidis, P., Aivasidis, A. 2009. Direct membrane filtration of sewage using aerated flat sheet membranes. *Proceedings of the 11th International Conference on Environmental Science and Technology*, September 3–5. Chania, Crete, Greece.

Diamantis, V.I., Antoniou, I., Athanasoulia, E., Melidis, P., Aivasidis, A. 2010. Recovery of reusable water from sewage using aerated flat-sheet membranes. *Water Sci. Technol.* 62(12):2769–2775.

Diamond, J. 2010. The home stretch: Approaching of a startup of a state of the art, sustainable MBR. *77th Annual Conference & Exhibition of Pacific Northwest Clean Water Association*, October 27. Bend, OR.

Díaz, H., Azócar, L., Torres, A., Lopes, S.I.C., Jeison, D. 2014. Use of flocculants for increasing permeate flux in anaerobic membrane bioreactors. *Water Sci. Technol.* 69(11):2237–2242.

Diya'uddeen, B.H., Daud, W.M., Aziz, A.R. 2011. Treatment technologies for petroleum refinery effluents: A review. *Process Saf. Environ. Prot.* 89(2):95–105.

Dizge, N., Koseoglu-Imer, D.Y., Karagunduz, A., Keskinler, B. 2011. Effects of cationic polyelectrolyte on filterability and fouling reduction of submerged membrane bioreactor MBR. *J. Membrane Sci.* 377(1–2):175–181.

Dojlido, J. 1979. Investigation of biodegradability and toxicity of organic compounds. USEPA Report 600/2–79–163.

DOW. 2013. Filmtec™ product information. Form No. 609-22010-604.

Drews, A. 2010. Membrane fouling in membrane bioreactors: Characterization, contradictions, cause and cures. *J. Membrane Sci.* 363:1–28.

Drews, A., Vocks, M., Iversen, V., Lesjean, B., Kraume, M. 2006. Influence of unsteady membrane bioreactor operation on EPS formation and filtration resistance. *Desalination* 192(1–3):1–9.

Drews, A., Mante, J., Iversen, V., Vocks, M., Lesjean, B. 2007. Impact of ambient conditions on SMP elimination and rejection in MBRs. *Water Res.* 41(17):3850–3858.

Drews, A., Vocks, M., Bracklow, U., Iversen, V., Kraume, M. 2008. Does fouling in MBRs depend on SMP? *Desalination* 231:141–149.

Dubois, M., Gilles, K.A., Hamilton, J.K., Rebers, P.A., Smith, F. 1956. Colorimetric method for determination of sugar and related substances. *Anal. Chem.* 28(3):350–356.

Durden, K., Dodson, D. 2004. Show me the money! *AMTA News* 22:10–13.

Eckstein, E.C., Bailey, D.G., Shapiro, A.H. 1977. Self-diffusion of particles in shear flow of a suspension. *J. Fluid Mech.* 79:191–208.

Electric Choice website. 2013. Electricity prices by state: Updated for 2013. Available at http://www.electricchoice.com/electricity-prices-by-state.php (accessed March 21, 2014).

El Hadidy, A.M., Peldszus, S., Van Dyke, M.I. 2014. Effect of hydraulically reversible and hydrolically irreversible fouling on the removal of MS2 and φX174 bacteriophage by an ultrafiltration membrane. *Water Res.* 61:297–307.

Elimelech, M., Childress, A. 1996. Zeta potential of reverse osmosis membranes: Implications for membrane performance. US Department of Interior, Water Treatment Technology Program Report No. 10. US Department of Interior.

Engelhardt, N. 2008. Long-term experience with Europe's largest membrane bioreactor plant for municipal waste water treatment. *Proceedings of IFAT*, September 23–25. Shanghai, China.

EUROMBRA. 2006. Membrane bioreactor technology (MBR) with an EU perspective for advanced municipal wastewater treatment strategies for the 21st century. D1—Data acquisition and compilation. Project No. 018480.

Evenblij, H. 2006. Filtration characteristics in membrane bioreactors. PhD dissertation, Delft University of Technology, The Netherlands.

Fabiyi, M.E., Novak, R.A. 2007. System and method for eliminating sludge via ozonation. US Patent No. 7,309,432.

Fakhru'l-Razi, A., Pendashteh, A., Abdullah, L.C., Radiah, D. et al. 2009. Review of technologies for oil and gas produced water treatment. *J. Hazard. Mater.* 170(2–3):530–551.

Fan, F., Zhou, H., Husain, H. 2006. Identification of wastewater sludge characteristics to predict critical flux for membrane bioreactor processes. *Water Res.* 40(2):205–212.

Fan, F., Zhou, H., Husain, H. 2007. Use of chemical coagulants to control fouling potential for wastewater membrane bioreactor processes. *Water Environ. Res.* 79(9):952–957.

Fane, A.G. 2005. Factors affecting the membrane performance in MBRs. *Proceedings of the MBR Workshop*, August 29. Tsinghua University, Beijing, China.

Fane, A.G. 2007. Sustainability and membrane processing of wastewater for reuse. *Desalination* 203:53–58.

Fane, A.G. 2009. MBR and the TMP jump—Self accelerating fouling phenomena. *Proceedings of the Final MBR-Network Workshop*, March 31–April 1. Berlin, Germany.

Fane, A.G., Chang, S., Chardon, E. 2002. Submerged hollow fibre membrane module—Design options and operational considerations. *Desalination* 146(1–3):231–236.

Fane, A.G., Yeo, A., Law, A., Parameshwaran, K., Wicaksana, F., Chen, V. 2005. Low pressure membrane process—Doing more with less energy. *Desalination* 185(1–3):159–165.

Fang, Y., Novak, P.J., Hozalski, R.M., Cussler, E.L., Semmens, M.J. 2004. Condensation studies in gas permeable membrane. *J. Membrane Sci.* 231(1–2):47–55.

Farahbakhsh, K., Smith, D.W. 2004. Estimating air diffusion contribution to pressure decay during membrane integrity tests. *J. Membrane Sci.* 237:203–212.

Farahbakhsh, K., Adham, S.S., Smith, D.W. 2003. Monitoring the integrity of low-pressure membranes. *J. AWWA* 95(6):95–106.

Fenu, A., Roels, J., Wambecq, T. et al. 2010. Energy audit of full-scale MBR. *Desalination* 26(1–3):121–128.

Fenu, A., de Wilde, W., Gaertner, M., Weemaes, M., de Gueldre, G., Van De Steene, B. 2012. Elaborating the membrane life concept in a full scale hollow fibers MBR. *J. Membrane Sci.* 421–422:349–354.

Ferré, V., Trepin, A., Giminez, T., Lluch, S. 2009. Design and performance of full-scale MBR plants treating winery wastewater effluents in Italy and Spain. *Proceedings of the 5th International Specialized Conference on Sustainable Viticulture: Winery Waste and Ecological Impacts Management by IWA*, April 2. Trento and Verona, Italy.

Field, R.W., Wu, D., Howell, J.A., Gupta, B.B. 1995. Critical flux concept for microfiltration fouling. *J. Membrane Sci.* 100(3):259–272.

Fitzgerald, C.M., Camejo, P., Oshlag, J.Z., Noguera, D.R. 2015. Ammonia-oxidizing microbial communities in reactors with efficient nitrification at low-dissolved oxygen. *Water Res.* 70:38–51.

Fluxxion. 2007. Fluxxion Microsieves. Available at http://www.mikrocentrum.nl/assets/Themadagen/SIG /FluXXion-Thijs-Bril.pdf (accessed August 28, 2011).

Foley, G., Malone, D.M., MacLoughlin, F. 1995. Modelling the effects of particle polydispersity in crossflow filtration. *J. Membrane Sci.* 99(1):77–88.

Franken, A.C., Nolten, J.A., Mulder, M.H., Bargeman, D., Smolders, C.A. 1987. Wetting criteria for the applicability of membrane distillation. *J. Membrane Sci.* 33:315–328.

Frechen, F.-B., Schier, W., Linden, C. 2008. Pretreatment of municipal MBR applications. *Desalination* 231:108–114.

Frechen, F.-B., Schier, W., Exler, H., Ohme, M. 2010. Mechanical pre-treatment (MPT) on municipal MBR plants. *Proceedings of the MBR Asia 2010*, April 27. Bangkok, Thailand.

Frenkel, V.S., Cohen, Y. 2014. *New Techniques for Real-Time Monitoring of Membrane Integrity for Virus Removal: Pulsed-Marker Membrane Integrity Monitoring*. Alexandria, VA: WateReuse Research Foundation.

Fritzsche, A.K., Cruse, C.A., Kesting, R.E., Murphy, M.K. 1990. Polysulfone hollow fiber membranes spun from Lewis acid: Base complexes. II. The effect of Lewis acid to base ratio on membrane structure. *J. Appl. Polym. Sci.* 39(9):1949–1956.

Frølund, B., Palmgren, R., Keiding, K., Nielsen, P.H. 1996. Extraction of extracellular polymers from activated sludge using a cation exchange resin. *Water Res.* 30(8):1749–1758.

Fulton, B., Bérubé, P.R. 2010. Optimal module configuration and sparging scenario for a submerged hollow fiber membrane system. *Proceedings of the WEF Membrane Applications Conference*, June 6–9. Anaheim, CA.

Futselaar, H., Schonewille, H., Vente, D., Broens, L. 2007. Norit AirLift MBR: Side-stream system for municipal wastewater treatment. *Desalination* 204:1–7.

Gadala-Maria, F., Acrivos, A. 1980. Shear-induced structure in a concentrated suspension of solid spheres. *J. Rheol.* 24:799–814.

Gao, W.J., Lin, H.J., Leung, K.T., Schraft, H., Liao, B.Q. 2011. Structure of cake layer in a submerged anaerobic membrane bioreactor. *J. Membrane Sci.* 374(1–2):110–120.

Gao, W.J., Liao, B.Q., Dagnew, M., Cumin, J. 2013. Comparison of performance between flat-sheet and hollow fiber submerged anaerobic membrane bioreactors for the treatment of synthetic petrochemical wastewater. *Proceedings of WEFTEC*, October 5–9. Chicago.

Gary, G.T., McCutcheon, J.R., Elimelech, M. 2006. Internal concentration polarization in forward osmosis: Role of membrane orientation. *Desalination* 197(1–3):1–8.

GE Water. 2011. Mobile ZeeWeed membrane bioreactor (MBR) system. Available at http://www.gewater.com /content/pdf/Fact%20Sheets_Cust/Americas/English/TS-MOB-WW-1206%20NA%20GE%20Logo.pdf (accessed August 2, 2011).

GE Water. 2013. Low energy costs, increase productivity, and lower your footprint with our latest MBR. Available at http://www.gewater.com/products/leap-mbr.html (accessed December 10, 2013).

Geider, R.J., La Roche, J. 2002. Redfield revisited: Variability of C:N:P in marine microalgae and its biochemical basis. *Eur. J. Phycol.* 37:1–17.

Geilvoet, S.P. 2010. The Delft Filtration Characterisation method assessment. PhD dissertation, Delft University of Technology, the Netherlands.

Germain, E., Nelles, F., Drews, A. et al. 2007. Biomass effects on oxygen transfer in membrane bioreactors. *Water Res.* 41(5):1038–1044.

Ghosh, R., Cui, Z.F. 1999. Mass transfer in gas-sparged ultrafiltration: Upward slug flow in tubular membranes. *J. Membrane Sci.* 162(1–2):91–102.

Giesen, A., van Bentem, A., Gademan, G., Erwee, H. 2008. Lessons learnt in facility design, tendering and operation of MBR's for industrial and municipal wastewater treatment. *Proceedings of the Water Institute of South Africa (WISA) Biennial Conference and Exhibition*, May 18–22. Sun City, South Africa.

Giesen, A., van Loosdrecht, M., de Bruin, B., van der Roest, H., Pronk, M. 2013. Full-scale experiences with aerobic granular biomass technology for treatment of urban and industrial wastewater. *Proceedings of the International Water Week Conference*, November 4–8. Amsterdam, The Netherlands.

Gildemeister, R. 2003. Survey of membrane bioreactor (MBR). Berlin, Germany: Technical University of Berlin.

Gitis, V., Haught, R.C., Clark, R.M., Gun, J., Lev, O. 2006. Nanoscale probes for the evaluation of the integrity of ultrafiltration membranes. *J. Membrane Sci.* 276(1–2):199–207.

Grelier, P., Rosenberger, S., Tazi-Pain, A. 2006. Influence of sludge retention time on membrane bioreactor hydraulic performance. *Desalination* 192:10–17.

Grélot, A., Grelier, P., Tazi-Pain, A., Lesjean, B., Brüss, U., Grasmick, A. 2010. Performances and fouling control of a flat sheet membrane in a MBR pilot-plant. *Water Sci. Technol.* 61(9):2185–2192.

Gresch, M., Braun, D., Gujer, W. 2010. The role of the flow pattern in wastewater aeration tanks. *Water Sci. Technol.* 61(2):407–414.

Gresch, M., Armbruster, M., Braun, D., Gujer, W. 2011. Effects of aeration patterns on the flow field in wastewater aeration tanks. *Water Res.* 45(2):810–818.

Guan, D., Fung, W.C., Lau, F. et al. 2014. Pilot trial study of a compact macro-filtration membrane bioreactor process for saline wastewater treatment. *Water Sci. Technol.* 70(1):120–126.

Güell, C., Czekaj, P., Davis, R.H. 1999. Microfiltration of protein mixtures and the effects of yeast on membrane fouling. *J. Membrane Sci.* 155:112–122.

Guibert, D., Ben Aim, R., Rabie, H., Côté, P. 2002. Aeration performance of immersed hollow-fiber membranes in a bentonite suspension. *Desalination* 148:395–400.

Guillen, G.R., Pan, Y., Li, M., Hoek, E.M. 2011. Preparation and characterization of membranes formed by nonsolvent induced phase separation: A review. *Ind. Eng. Chem. Res.* 50(7):3798–3817.

Gujer, W. 2010. Nitrification and me—A subjective review. *Water Res.* 44(1):1–19.

Günder, B. 2000. *The Membrane Coupled-Activated Sludge Process in Municipal Wastewater Treatment.* Lancaster, PA: Technomic Publishing Co.

Guo, W.S., Vigneswaran, S., Ng, H.H., Kandasamy, J., Yoon, S. 2008. The role of a membrane performance enhancer in a membrane bioreactor: A comparison with other submerged membrane hybrid systems. *Desalination* 231:305–313.

GWRS website. 2014. Available at http://www.gwrsystem.com/about-gwrs.html (accessed February 28, 2014).

Hadler, J., Kullmann, C. 2012. Integrated membrane bioreactors (MBR) and reverse osmosis (RO) for water reuse. *Filtr. Sep.* 2:40–44.

Hall, E.R., Bérubé, P.R. 2006. Membrane bioreactors for anaerobic treatment of wastewaters: Phase I. WERF Report No. 02-CTS-4a.

Haugen, K.S., Semmens, M.J., Novak, P.J. 2002. A novel in situ technology for the treatment of nitrate contaminated groundwater. *Water Res.* 36(14):3497–3506.

Henshaw, W.J., Mahendran, M., Behmann, H. 1998. Vertical cylindrical skein of hollow fiber membranes and method of maintaining clean fiber surfaces. US Patent No. 5,783,083.

Henze, M., Grady Jr., C.P.L., Gujer, W., Marais, G.V.R., Matsuo, T. 1986. A model for single-sludge wastewater treatment systems. *Water Sci. Technol.* 18(6):47–61.

Henze, M., Gujer, W., Mino, T., van Loosdrecht, M. 2000. *Activated Sludge Models ASM1, ASM2, ASM2d and ASM3.* London: IWA Publishing.

Herath, G., Yamamoto, K., Urase, T. 2000. The effect of suction velocity on concentration polarization in microfiltration membranes under turbulent flow conditions. *J. Membrane Sci.* 169(2):175–183.

Hermanowicz, S.W. 2004. Membrane filtration of biological solids: A unified framework and its application to MBR. *Proceedings of the Water Environment-Membrane Technologies Conference by IWA*, June 7–10. Seoul, Korea.

Hernandez Rojas, M.E., Van Kaam, R., Schetrite, S., Albasi, C. 2005. Role and variations of supernatant compounds in submerged membrane bioreactor fouling. *Desalination* 179:95–107.

Herold, D. 2011. Energy efficient MBR design: Rabigh Refinery, Saudi Arabia. *Proceedings of the Water Arabia Conference*, February 2. Manama, Bahrain.

Hirani, Z.M., DeCarolis, J.F., Adham, S.S., Jacangelo, J.G. 2010. Peak flux performance and microbial removal by selected membrane bioreactor systems. *Water Res.* 44(8):2431–2440.

Hirani, Z.M., Bukhari, Z., Oppenheimer, J., Jjemba, P., LeChevallier, M.W., Jacangelo, J.G. 2013. Characterization of effluent water qualities from satellite membrane bioreactor facilities. *Water Res.* 47(14):5065–5075.

Ho, C.-C., Zydney, A. 2006. Overview of fouling phenomena and modeling approaches for membrane bioreactors. *Sep. Sci. Technol.* 41:1231–1251.

Ho, J., Smith, S., Patamasank, J., Tontcheva, P., Kim, K.D., Roh, H.K. 2013. Development of alternative energy saving MBR using reciprocating vibration in place of membrane air scouring. *Proceedings of WEFTEC*, October 5–9. Chicago.

Hoang, T., Ang, H.-M., Rohl, A. 2007. Effects of temperature on the scaling of calcium sulphate in pipes. *Powder Technol.* 179:31–37.

Hobbs, C., Hong, S.K., Taylor, J. 2000. Effect of membrane properties on fouling in RO/NF membrane filtration of surficial groundwater. *Proceedings of the ACS Biannual Conference*, Washington, DC.

Hobbs, C., Hong, S.K., Taylor, J. 2006. Effect of surface roughness on fouling of RO and NF membranes during filtration of a high organic surficial groundwater. *J. Water Supply Res. Technol.* 55(7–8):559–570.

Honey, H.C., Pretorius, W.A. 2000. Laminar flow pipe hydraulics of pseudoplastic-thixotropic sewage sludges. *Water SA* 26(1):19–26.

Hong, S.P., Bae, T.H., Tak, T.M., Hong, S., Randall, A. 2002. Fouling control in activated sludge submerged hollow fiber membrane bioreactor. *Desalination* 143:219–228.

Hosseinzadeh, M., Bidhendi, G.N., Torabian, A., Mehrdadi, N. 2013. Evaluation of membrane bioreactor for advanced treatment of industrial wastewater and reverse osmosis pretreatment. *J. Environ. Health Sci. Eng.* 11:34.

Huber, S.E. 2013. Huber vacuum rotation membrane VRM® bioreactor. Available at http://www.huber.de /fileadmin/01_Produkte/06_Membranbelebung_MBR//mbr_uebersicht_en.pdf (accessed November 15, 2013).

Hugaboom, D.A., Sethi, S. 2004. An evaluation of MF/UF membrane integrity monitoring strategies as proposed in LT2. *AMTA Conference*, Washington, DC.

Hung, C.H., Boyle, W.C. 2001. The effect of acid cleaning on a fine pore ceramic diffuser aeration system. *Water Sci. Technol.* 44(2–3):211–218.

Hwang, B.-K., Lee, C.-H., Lee, W.-N., Park, P.-K., Lee, C.-H., Chang, I.-S. 2007. Effect of membrane fouling reducer on cake structure and membrane permeability in membrane bioreactor. *J. Membrane Sci.* 288:149–156.

Hwang, B.-K., Kim, J.-H., Ahn, C.-H., Lee, C.-H., Song, J.-Y., Ra, Y.-H. 2010a. Effect of disintegrated sludge recycling on membrane permeability in a membrane bioreactor combined with a turbulent jet flow ozone contactor. *Water Res.* 44:1833–1840.

Hwang, B.-K., Son, H.-S., Ahn, C.-H., Lee, C.-H., Song, J.-Y., Ra, Y.-H. 2010b. Decomposition of excess sludge in a membrane bioreactor using a turbulent jet flow ozone contactor. *J. Ind. Eng. Chem.* 16:602–608.

Igunnu, E.T., Chen, G.Z. 2012. Produced water treatment technologies. *Int. J. Low-Carbon Technol.* 0:1–21.

Imasaka, T., Kanekuni, N., So, H., Yoshino, S. 1989. Crossflow filtration of methane fermentation broth by ceramic membrane. *J. Ferment. Bioeng.* 68(3):200–206.

Imasaka, T., So, H., Matsushita, K., Furukawa, T., Kanekuni, N. 1993. Application of gas-liquid two-phase crossflow filtration to pilot scale membrane fermentation. *Drying Technol.* 11(4):769–785.

Irizar, I., Zambrano, J.A., Montoya, D., De Garacia, M., García, R. 2009. Online monitoring of OUR, K_{La} and OTE indicators: Practical implementation in full-scale industrial WWTPs. *Water Sci. Technol.* 60(2):459–466.

Ishak, S., Malakahmad, A., Isa, M.H. 2012. Refinery wastewater biological treatment: A short review. *J. Sci. Ind. Res.* 71:251–256.

Ishigami, T., Nii, Y., Ohmukai, Y., Rajabzadeh, S., Matsuyama, H. 2014. Solidification behavior of polymer solution during membrane preparation by thermally induced phase separation. *Membranes* 4(1):113–122.

Iversen, V., Mohaupt, J., Drews, A., Lesjean, B., Kraume, M. 2008. Side effects of flux enhancing chemicals in membrane bioreactors MBRs: Study on their biological toxicity and their residual fouling propensity. *Water Sci. Technol.* 57(1):117–123.

Iversen, V., Koseoglu, H., Yigit, N.O. et al. 2009. Impacts of membrane flux enhancers on activated sludge respiration and nutrient removal in MBRs. *Water Res.* 43(3):822–830.

Jahangir, D., Oh, H.-S., Kim, S.-Y., Park, P.-K., Lee, C.-H., Lee, J.-K. 2012. Specific location of encapsulated quorum quenching bacteria for biofouling control in an external submerged membrane bioreactor. *J. Membrane Sci.* 411–412:130–136.

James, C.P., Germain, E., Judd, S. 2014. Micropollutant removal by advanced oxidation of microfiltered secondary effluent for water reuse. *Sep. Purif. Technol.* 127:77–83.

Jang, N., Ren, X., Choi, K., Kim, I.S. 2006. Comparision of membrane biofouling in nitrification and denitrification for the membrane bioreactor (MBR). *Water Sci. Technol.* 53(6):43–49.

Jansen, A.E., Assink, J.W., Hanemaaijer, J.H., van Medevoort, J., van Sonsbeek, E. 2013. Development and pilot testing of full-scale membrane distillation modules for deployment of waste heat. *Desalination* 323:55–65.

Jenkins, D., Richard, M., Daigger, G. 2004. *Manual on the Causes and Control of Activated Sludge Bulking, Foaming, and Other Solids Separation Problems.* London: CRC Press LLC.

Jia, X.S., Furumai, H., Fang, H.P. 1996. Yields of biomass and extracellular polymers in four anaerobic sludges. *Environ. Technol.* 17(3):289–291.

Jin, Y.-L., Lee, W.N., Lee, C.-H., Chang, I.-S., Huang, X. 2006. Effect of DO concentration on biofilm structure and membrane filterability in submerged membrane bioreactor. *Water Res.* 40:2829–2836.

Jin, W., Guo, W., LÜ, X., Han, P., Wang, Y. 2008. Effect of the ultrasound generated by flat plate transducer cleaning on polluted polyvinylidenefluoride hollow fiber ultrafiltration membrane. *Chinese J. Chem. Eng.* 16(5):801–804.

Jin, L., Ong, S.L., Ng, H.Y. 2013. Fouling control mechanism by suspended biofilm carriers addition in submerged ceramic membrane bioreactors. *J. Membrane Sci.* 427:250–258.

Johannessen, E., Samstag, R.W., Stensel, H.D. 2006. Effect of process configurations and alum addition on EBPR in membrane bioreactors. *Proceedings of WEFTEC*, October 21–25. Dallas.

Johnson, W.T. 1998. Predicting log removal performance of membrane systems using in-situ integrity testing. *Filtr. Sep.* 35(1):26–29.

Johnson, B.R., Daigger, G.T. 2009. Integrated nutrient removal design for very low phosphorus levels. *Water Sci. Technol.* 60(9):2455–2462.

Johnson, D.W., Semmens, M.J., Gulliver, J.S. 1998. Unconfined membranes: Transfer performance and module design. *J. Membrane Sci.* 140:13–25.

Joss, A., Baenninger, C., Foa, P. et al. 2011. Water reuse: >90% water yield in MBR/RO through concentrate recycling and CO_2 addition as scaling control. *Water Res.* 45(18):6141–6151.

Jouanneau, S., Recoules, L., Durand, M.J. et al. 2014. Methods for assessing biochemical oxygen demand (BOD): A review. *Water Res.* 49:62–82.

Judd, S. 2004. A review of fouling of membrane bioreactors in sewage treatment. *Water Sci. Technol.* 49(2):229–235.

Judd, S. 2006. *MBR Book.* Amsterdam, The Netherlands: Elsevier Ltd., pp. 165–205.

Judd, S. 2008. The status of membrane bioreactor technology. *Trend Biotechnol.* 262:109–116.

Judd, S. 2011. *The MBR Book: Principles and Applications of Membrane Bioreactors for Water and Wastewater Treatment*, 2nd ed. Amsterdam, The Netherlands: Elsevier.

Judd, S.J., Le-Clech, P., Taha, T., Cui, Z.F. 2001. Theoretical and experimental representation of a submerged membrane bio-reactor system. *Membrane Technol.* 135:4–9.

Kabsch-Korbutowicz, M., Majewska-Nowak, K., Winnicki, T. 1999. Analysis of membrane fouling in the treatment of water solutions containing humic acids and mineral salts. *Desalination* 126:179–185.

Kaempfer, H., Daigger, G., Adams, C. 2000. Characterization of the floc micro-environment in dispersed growth systems. *Proceedings of WEFTEC*, October 14–18. Anaheim, CA.

Kale, M.M., Singh, K.S. 2014. Performance of novel sludge-bed anaerobic membrane bioreactor (SB-AnMBR) treating prehydrolysis liquor. *Water Sci. Technol.* 69(4):796–802.

Kang, I.-J., Yoon, S.-H., Lee, C.H. 2002. Comparison of the filtration characteristics of organic and inorganic membranes in a membrane-coupled anaerobic bioreactor. *Water Res.* 37(7):1803–1813.

Kang, I.-J., Lee, C.-H., Kim, K.-J. 2003. Characteristics of microfiltration membranes in a membrane coupled sequencing batch reactor system. *Water Res.* 37:1192–1197.

Kang, S., Subramani, A., Heok, E., Deshusses, M., Matsumoto, M. 2004. Direct observation of biofouling in crossflow microfiltration mechanisms of deposition and release. *J. Membrane Sci.* 244:151–165.

Kappel, C. 2014. An integrated membrane bioreactor—Nanofiltration concept with concentrate recirculation for wastewater treatment and nutrient recovery. PhD dissertation, University of Twente, The Netherlands.

Katsou, E., Malamis, S., Lozidou, M. 2011. Performance of a membrane bioreactor used for the treatment of wastewater contaminated with heavy metals. *Bioresource Technol.* 102(6):4325–4332.

Kaya, D., Yagnur, E.A., Yigit, K.S., Kilic, F.C., Eren, A.S., Celik, C. 2008. Energy efficiency in pumps. *Energ. Convers. Manage.* 49:1662–1673.

Keller, J., Giesen, A. 2010. Advancements in aerobic granular biomass processes. *Neptune and Innowatech End User Conference*, January 27. Gent, Belgium.

Kent, F.C., Farahbakhsh, K. 2011. Addressing reverse osmosis fouling within water reclamation: A side-by-side comparison of low-pressure membrane pretreatments. *Water Environ. Res.* 83(6):515–526.

Kesieme, U.K., Milne, N., Aral, H., Cheng, C.Y., Duke, M. 2013. Economic analysis of desalination technologies in the context of carbon pricing, and opportunities for membrane distillation. *Desalination* 323:66–74.

Ketrane, R., Saidani, B., Gil, O., Leleyter, L., Baraud, F. 2009. Efficiency of five scale inhibitors on calcium carbonate precipitation from hard water: Effect of temperature and concentration. *Desalination* 249(3):1397–1404.

Khayet, M. 2013. Solar desalination by membrane distillation: Dispersion in energy consumption analysis and water production costs. *Desalination* 308:89–101.

Khirani, S., Ben Aim, R., Manero, M.-H. 2006. Improving the measurement of the modified fouling index using nanofiltration membrane (NF-MFI). *Desalination* 191(1–3):1–7.

Kim, Y.-K., Boyle, W.C. 1993. Mechanisms of fouling in fine pore diffuser aeration. *J. Environ. Eng.* 119(6):1119–1138.

Kim, K.J., Fane, A.G., Ben Aim, R. et al. 1994. A comparative study of techniques used for porous membrane characterization: Pore characterization. *J. Membrane Sci.* 87:35–46.

Kim, J.-S., Lee, S.-J., Yoon, S.-H., Lee, C.-H. 1996. Competitive adsorption of trace organics on membranes and powdered activated carbon in powdered activated carbon-ultrafiltration system. *Water Sci. Technol.* 34(9):223–229.

Kim, J., Lee, C.-H., Chang, I.-S. 2001. Effect of pump shear on the performance of a crossflow membrane bioreactor. *Water Res.* 35(9):2137–2144.

Kim, J.-H., Choi, D.-C., Yeon, K.-M., Kim, S.-R., Lee, C.-H. 2011a. Enzyme-immobilized nanofiltration membrane to mitigate biofouling based on quorum quenching. *Environ. Sci. Technol.* 45(4):1601–1607.

Kim, J.-Y., Lee, J.-H., Chang, I.-S., Lee, J.-H., Yi, C.-W. 2011b. High voltage impulse electric fields: Disinfection kinetics and its effect on membrane bio-fouling. *Desalination* 283:111–116.

Kim, M., Guerra, P., Shah, A., Parsa, M., Alaee, M., Smyth, A. 2014. Removal of pharmaceuticals and personal care products in a membrane bioreactor wastewater treatment plant. *Water Sci. Technol.* 69(11):2221–2229.

Kimura, S. 1991. Japan's Aqua Renaissance '90 project. *Water Sci. Technol.* 23(7–9):1573–1582.

Kimura, K., Kurita, T., Watanabe, Y. 2013. Energy saving in operation of submerged MBRs by insertion of baffles and introduction of granular materials. *Proceedings of the IWA Membrane Technology Conference*, August 26–29. Toronto, Canada.

Kirts, R.E. 2009. Method and apparatus for producing potable water from seawater using forward osmosis. US Patent Application No. 2009/0308727.

Kitade, T., Wu, B., Chong, T.H., Fane, A.G., Uemura, T. 2013. Fouling reduction in MBR-RO processes: The effect of MBR F/M ratio. *Desalin. Water Treat.* 51:4829–4838.

Kitis, M., Lozier, J.C., Kim, J.-H., Mi, B., Marinas, B. 2003. Microbial removal and integrity monitoring of RO and NF membranes. *J. AWWA* 95(12):105–119.

Kloosterman, J., van Wassenaar, P.D., Slater, N.K.H., Baksteen, H. 1988. The effect of anti-foam agents on the ultrafiltration of a protease solution. *Biopress Eng.* 3(4):181–185.

KMS. 2013. Spec sheet of Puron® hollow fiber modules. Available at http://www.kochmembrane.com/PDFs/Data-Sheets/Hollow-Fiber/UF/KMS_Puron_Hollow_Fiber_PSH300_PSH600_PSH1800_Modul.aspx.

Knoell, T., Safarik, J., Cormack, T., Riley, R., Lin, S.W., Ridgway, H. 1999. Biofouling potentials of microporous polysulfone membranes containing a sulfonated polyether-ethersulfone/polyethersulfone block copolymer: Correlation of membrane surface properties with bacterial attachment. *J. Membrane Sci.* 157:117–138.

Kochkodan, V.M., Hilal, N., Goncharuk, V.V., Al-Khatib, L., Levadna, T.I. 2006. Effect of the surface modification of polymer membranes on their microbiological fouling. *Colloid J.* 68(3):267–273.

Kola, A., Ye, Y., Ho, A., Le-Clech, P., Chen, V. 2012. Application of low frequency transverse vibration on fouling limitation in submerged hollow fiber membranes. *J. Membrane Sci.* 409–410:54–65.

Kola, A., Ye, Y., Le-Clech, P., Chen, V. 2014. Transverse vibration as novel membrane fouling mitigation strategy in anaerobic membrane bioreactor applications. *J. Membrane Sci.* 455:320–329.

Komesli, O.T., Teschner, K., Hegemann, W., Gokcay, C. 2007. Vacuum membrane applications in domestic wastewater reuse. *Desalination* 215:22–28.

Kondo, M. 1998. Airlift pump apparatus and method. US Patent No. 6,162,020.

Kose, B., Ozgun, H., Ersahin, M.E. et al. 2012. Performance evaluation of a submerged membrane bioreactor for the treatment of brackish oil and natural gas field produced water. *Desalination* 285:295–300.

Koseoglu, H., Yigit, N.O., Iversen, V. et al. 2008. Effects of several different flux enhancing chemicals on filterability and fouling reduction of membrane bioreactor (MBR) mixed liquors. *J. Membrane Sci.* 320:57–64.

Krampe, J., Krauth, K. 2003. Oxygen transfer into activated sludge with high MLSS concentrations. *Water Sci. Technol.* 47(12):297–303.

Kraume, M., Wedi, D., Schaller, J., Iversen, V., Drews, A. 2009. Fouling in MBR: What use are lab investigations for full scale operation? *Desalination* 236(1–3):94–103.

Krause, S., Cornel, P., Wagner, M. 2003. Comparison of different oxygen transfer testing procedures in full scale membrane bioreactors. *Water Sci. Technol.* 47(12):169–176.

Krzeminski, P., van der Graaf, J., van Lier, J. 2012. Specific energy consumption of membrane bioreactor (MBR) for sewage treatment. *Water Sci. Technol.* 65(2):380–392.

Kubota website. Available at http://www.kubota-mbr.com/resources/Kubota%20SMU%20Brochure_2010.pdf (accessed June 2011).

Kuo, C.-Y., Lin, H.-N., Tsai, H.-A., Wang, D.M., Lai, J.-Y. 2008. Fabrication of a high hydrophobic PVDF membrane via nonsolvent induced phase separation. *Desalination* 233:40–47.

Kurian, R., Acharya, C., Nakhla, G., Bassi, A. 2005. Conventional and thermophilic aerobic treatability of high strength oily pet food wastewater using membrane-coupled bioreactors. *Water Res.* 39(18):4299–4308.

Kwon, S., Sullivan, E.J., Katz, L. et al. 2008. Pilot scale test of a produced water treatment system for organic compound removal from CBM water. *SPE Annual Technical Conference and Exhibition*. Denver, CO.

Lackner, S., Terada, A., Horn, H., Henze, M., Smets, B.F. 2010. Nitritation performance in membrane-aerated biofilm reactors differs from conventional biofilm systems. *Water Res.* 44(20):6073–6084.

Lala, J., Lebumfacil, R., Mira, E., Grant, S.R., Christian, S.J., Xie, K. 2014. Full scale MBR operation experience for the treatment of high salt concentration wastewater from carrageenan production. *Proceedings of WEFTEC*, September 26–30. New Orleans, LA.

Lamminen, M.O., Walker, H.W., Weavers, L.K. 2004. Mechanisms and factors influencing the ultrasonic cleaning of particle-fouled ceramic membranes. *J. Membrane Sci.* 237(1–2):213–223.

LaPara, T.M., Alleman, J.E. 1999. Thermophilic aerobic biological wastewater treatment. *Water Res.* 33(4):895–908.

LaPara, T.M., Konopka, A., Nakatsu, C.H., Alleman, J.E. 2000a. Thermophilic aerobic wastewater treatment in continuous-flow bioreactors. *J. Environ. Eng.* 126(8):739–744.

LaPara, T.M., Konopka, A., Nakatsu, C.H., Alleman, J.E. 2000b. Effects of elevated temperature on bacterial community structure and function in bioreactors treating a synthetic wastewater. *J. Ind. Microbiol. Biotechnol.* 24(2):140–145.

LaPara, T.M., Nakatsu, C.H., Pantea, L.M., Alleman, J.E. 2001a. Aerobic biological treatment of a pharmaceutical wastewater: Effect of temperature on COD removal and bacterial community development. *Water Res.* 35(18):4417–4425.

LaPara, T.M., Konopka, A., Nakatsu, C.H., Alleman, J.E. 2001b. Thermophilic aerobic biological wastewater treatment of a synthetic wastewater in a membrane-coupled bioreactor. *J. Ind. Microbiol. Biotechnol.* 26(4):203–209.

Larrea, A., Rambor, A., Fabiyi, M. 2014. Ten years of industrial and municipal membrane bioreactor (MBR) systems—Lessons from the field. *Water Sci. Technol.* 70(2):279–288.

Lawrence, N., Perera, J.M., Iyer, M., Hickey, M.W., Stevens, G.W. 2006. The use of streaming potential measurements to study the fouling and cleaning of ultrafiltration membranes. *Sep. Purif. Technol.* 48:106–112.

Lay, W.C.L., Liu, Y., Fane, A.G. 2010. Impacts of salinity on the performance of high retention membrane bioreactors for water reclamation: A review. *Water Res.* 44:21–40.

Lay, W.C., Zhang, J., Tang, C., Wang, R., Fane, A.G. 2012. Analysis of salt accumulation in a forward osmosis system. *Sep. Sci. Technol.* 47(13):1837–1848.

Le Roux, I., Krieg, H.M., Yeates, C.A., Breytenbach, J.C. 2005. Use of chitosan as an antifouling agent in a membrane bioreactor. *J. Membrane Sci.* 248(1–2):127–136.

Lebegue, J., Heran, M., Grasmick, A. 2008. Membrane bioreactor: Distribution of critical flux throughout an immersed HF bundle. *Desalination* 231:245–252.

Le-Clech, P., Chen, V., Fane, A.G. 2006. Fouling in membrane bioreactors used in wastewater treatment. *J. Membrane Sci.* 284:17–53.

Lee, W.-N. 2002. A study on the filtration characteristics in membrane-coupled moving bed biofilm reactor. MS thesis, Seoul National University, Seoul, Korea.

Lee, J., Ahn, W.-Y., Lee, C.-H. 2001a. Comparison of the filtration characteristics between attached and suspended growth microorganisms in submerged membrane bioreactor. *Water Res.* 35:2435–2445.

Lee, J.C., Kim, J.S., Kang, I.J., Cho, M.H., Park, P.K., Lee, C.H. 2001b. Potential and limitations of alum or zeolite addition to improve the performance of a submerged membrane bioreactor. *Water Sci. Technol.* 43(11):59–66.

Lee, H.S., Kim, C.G., Yoon, T.I. 2002. Comment on "Comparison of the filtration characteristics between attached and suspended microorganisms in submerged membrane bioreactor." *Water Res.* 36:4938–4939.

Lee, W.-N., Kang, I.-J., Lee, C.-H. 2006. Factors affecting filtration characteristics in membrane coupled moving bed biofilm reactor. *Water Res.* 40:1827–1835.

Lee, W.-N., Chang, I.-S., Hwang, B.-K., Park, P.-K., Lee, C.-H., Huang, X. 2007. Changes in biofilm architecture with addition of membrane fouling reducer in a membrane bioreactor. *Process Biochem.* 42:655–661.

Lee, C.H., Park, P.K., Lee, W.N. et al. 2008. Correlation of biofouling with the bio-cake architecture in an MBR. *Desalination* 231:115–123.

Lee, C.-H., Yeon, K.-M., Kim, J.-H. 2009. Quorum quenching: A new biofouling control paradigm in MBR. *Proceedings of the Final MBR-Network Workshop*, March 31–April 1. Berlin, Germany.

Lee, Y.-W., Lee, J., Rittmann, B.E., Chung, J. 2011. Wastewater recycling at an electronics company using a combined system of membrane bioreactor and reverse osmosis membrane processes. *Canadian J. Civil Eng.* 38(7):762–771.

Lees, E.J., Noble, B., Hewitt, R., Parsons, S.A. 2001. The impact of residual coagulant on the respiration rate and sludge characteristics of an activated microbial biomass. *Process Saf. Environ. Prot.* 79(5):283–290.

Leighton, D., Acrivos, A. 1987. The shear-induced migration of particles in concentrated suspensions. *J. Fluid Mech.* 181:415–439.

Leiknes, T., Semmens, M.J. 2001. Vacuum degassing using microporous hollow fiber membrane. *Sep. Purif. Technol.* 22–23(1):287–294.

Leiknes, T., Ødegaard, H. 2005. The development of a biofilm membrane bioreactor. *Desalination* 202:135–143.

Lesjean, B., Huisjes, E.H. 2008. Survey of the European MBR market: Trends and perspectives. *Desalination* 231:71–81.

Lesjean, B., Rosenberger, S., Laabs, C., Jekel, M., Gnirss, R., Amy, G. 2005. Correlation between membrane fouling and soluble/colloidal organic substances in membrane bioreactors for municipal wastewater treatment. *Water Sci. Technol.* 51(6–7):1–8.

Lesjean, B., Tazi-Pain, A., Thaure, D., Moeslang, H., Buisson, H. 2011. Ten persistent myths and the realities of membrane bioreactor technology for municipal applications. *Water Sci. Technol.* 63(1):32–39.

Leu, S.Y., Rosso, D., Jiang, P., Larson, L.E., Stenstrom, M.K. 2008. Real-time efficiency monitoring for wastewater aeration systems. *Water Pract. Technol.* 3(3):1–7.

Leu, S.-Y., Rosso, D., Larson, L., Stenstrom, M.K. 2009. Real-time monitoring aeration efficiency in the activated sludge process and methods to reduce energy consumption. *Water Environ. Res.* 81(12):2471–2481.

Leu, S.-Y., Chan, L., Stenstrom, M.K. 2010. Toward long SRT of activated sludge processes: Benefits in energy saving, effluent quality, and stability. *Proceedings of WEFTEC*, October 2–6. New Orleans, LA.

Levén, L. 2006. Anaerobic digestion at mesophilic and thermophilic temperature. PhD dissertation, Uppsala University, Sweden.

Levisky, I., Duek, A., Arkhangelsky, E. et al. 2011. Understanding the oxidative cleaning of UF membranes. *J. Membrane Sci.* 377:206–213.

Li, X.Y., Yang, S.F. 2007. Influence of loosely bound extracellular polymeric substances (EPS) on the flocculation, sedimentation and dewaterability of activated sludge. *Water Res.* 41(5):1022–1030.

Li, H., Fane, A.G., Coster, H.G.L., Vigneswaran, S. 1998. Direct observation of particle deposition on the membrane surface during crossflow microfiltration. *J. Membrane Sci.* 149(1):83–97.

Li, X., Li, J., Wang, J., Wang, H., Cui, C., He, B., Zhang, H. 2014. Direct monitoring of sub-critical flux fouling in a horizontal double-end submerged hollow fiber membrane module using ultrasonic time domain reflectometry. *J. Membrane Sci.* 451:226–233.

Lin, H., Zhang, M., Wang, F. et al. 2014. Experimental evidence for osmotic pressure-induced fouling in a membrane bioreactor. *Bioresour. Technol.* 158:119–126.

Liu, H., Fang, H.P. 2002. Extraction of extracellular polymeric substances EPS of sludges. *J. Biotechnol.* 95:249–256.

Liu, C., Wachinski, A. 2009. Integrity testing for low-pressure membrane systems. *Proceedings of the NEWWA Spring Joint Regional Conference and Exhibition*, April 1. Worcester, MA.

Liu, R., Huang, X., Wang, C., Chen, L., Qian, Y. 2000. Study on hydraulic characteristics in a submerged membrane bioreactor process. *Process Biochem.* 36:249–254.

Liu, F., Hashim, A., Liu, Y., Moghareh Abed, M.R., Li, K. 2011. Progress in the production and modification of PVDF membranes. *J. Membrane Sci.* 375(1–2):1–27.

Liu, L., Liu, J., Gao, B., Yang, F., Chellam, S. 2012a. Fouling reduction in a membrane bioreactor using an intermittent electric field and cathodic membrane modified by vapor phase polymerized pyrrole. *J. Membrane Sci.* 394–395:202–208.

Liu, Y., Liu, H., Cui, L., Zhang, K. 2012b. The ratio of food-to-microorganism (F/M) on membrane fouling of anaerobic bioreactors treating low-strength wastewater. *Desalination* 297:97–103.

Low, E.W., Chase, H.A., Milner, M.G., Curtis, T.P. 2000. Uncoupling of metabolism to reduce biomass production in the activated sludge process. *Water Res.* 34(12):3204–3212.

Lowry, O.H., Rosebrough, N.J., Farr, A.L., Randall, R.J. 1951. Protein measurement with the folin phenol reagent. *J. Biol. Chem.* 193:265–275.

Lozier, J., Fernandez. A. 2001. Using a membrane bioreactor/reverse osmosis system for indirect potable reuse. *Water Supply* 1(5–6):303–313.

Lu, J.-G., Zheng, Y.-F., Cheng, M.-D. 2008. Wetting mechanism in mass transfer process of hydrophobic membrane gas absorption. *J. Membrane Sci.* 308(1–2):180–190.

Lüdicke, C., Stüber, J., Gnirss, R., Lesjean, B., Kraume, M. 2009. Operational experience of MBR demonstration plant with post denitrification in Berlin-Magretenhöhe (ENREM-Project). *Proceedings of the Final MBR-Network Workshop*, March 31–April 1. Berlin, Germany.

Luo, J., Zhu, Z., Ding, L. et al. 2013. Flux behavior in clarification of chicory juice by high-shear membrane filtration: Evidence for threshold flux. *J. Membrane Sci.* 435:120–129.

Luxmy, B.S., Yamamoto, K. 2003. Investigation of microorganisms associated with the foam of a submerged membrane bioreactor in Japan. *Microbes Environ.* 18(2):62–68.

Luxmy, B.S., Nakajima, F., Yamamoto, K. 2000. Predator grazing effect on bacterial size distribution and floc size variation in membrane-separation activated sludge. *Water Sci. Technol.* 42(3):211–217.

Lv, W., Zheng, X., Yang, M., Zhang, Y., Liu, Y., Liu, J. 2006. Virus removal performance and mechanism of a submerged membrane bioreactor. *Process Biochem.* 41(2):299–304.

Macomber, J., Cicek, N., Suidan, M., Davel, J., Ginestet, P., Audic, J. 2005. Biological kinetic data evaluation of an activated sludge system coupled with an ultrafiltration membrane. *J. Environ. Eng.* 131(4)579–586.

Madsen, S., Bin, Y. 2013. Filtration apparatus. US Patent Application No. US 2013/0175208.

Maeng, S.K., Choi, B.G., Lee, K.T., Song, K.G. 2013. Influences of solids retention time, nitrification and microbial activity on the attenuation of pharmaceuticals and estrogens in membrane bioreactors. *Water Res.* 47(9):3151–3162.

Manson, S., Ewert, J., Ratsey, H., Sears, K., Beale, J. 2010. Flat sheet membrane bioreactors operational experiences—A New Zealand perspective. *Proceedings of the NZWWA Conference*, September 22–24. Christchurch, Canterbury, New Zealand.

Martinelli, L., Guigui, C., Line, A. 2010. Characterisation of hydrodynamics induced by air injection related to membrane fouling behavior. *Desalination* 250:587–591.

Matas, J.-P., Morris, J.F., Guazzelli, E. 2004. Inertial migration of rigid spherical particles in Poiseuille flow. *J. Fluid Mech.* 515:171–195.

Matsuyama, H., Okafuji, H., Maki, T., Teramoto, M., Tsujioka, N. 2002. Membrane formation via thermally induced phase separation in polypropylene/polybutene/diluent system. *J. Appl. Polym. Sci.* 84(9):1701–1708.

Mayhew, M., Stephenson, T. 1998. Biomass yield reduction: Is biochemical manipulation possible without affecting activated sludge process efficiency? *Water Sci. Technol.* 38(8–9):137–144.

MBR-Network website. 2006. Available at http://www.mbr-network.eu/ (accessed September 11, 2011).

McCabe, W.L., Smith, J.C., Harriot, P. 2005. *Unit Operations of Chemical Engineering*, 7th ed. New York: McGraw-Hill, 188–189.

McCarthy, A.A., Conroy, H., Walsh, P.K., Foley, G. 1998. The effect of pressure on the specific resistance of yeast filter cakes during dead-end filtration in the range 20–500 kPa. *Biotechnol. Tech.* 12(12):909–912.

McCutcheon, J.R., McGinnis, R.L., Elimelech, M. 2005. A novel ammonia-carbon dioxide forward (direct) osmosis desalination process. *Desalination* 174(1):1–11.

McGovern, R.K., Lienhard, J.H. 2014. On the potential of forward osmosis to energetically outperform reverse osmosis desalination. *J. Membrane Sci.* 469:245–250.

Melcer, H., Keucken, A., Hackner, T. 2009. Pilot testing of new vacuum rotating membrane bioreactor technology. *Proceedings of WEFTEC*, October 17–21. Orlando, FL.

Melin, T. 2011. Energy consumption, a critical success factor in membrane filtration. *Proceedings of the 6th IWA Special Conference on Membrane Technology for Water and Wastewater*, October 4–7. Aachen, Germany.

Meng, F., Zhang, H., Yang, F., Li, Y., Xiao, J., Zhang, X. 2006. Effect of filamentous bacteria on membrane fouling in submerged membrane bioreactor. *J. Membrane Sci.* 272:161–168.

Menniti, A., Morgenroth, E. 2010. The influence of aeration intensity on predation and EPS production in membrane bioreactors. *Water Res.* 44:2541–2553.

Menniti, A., Kang, S., Elimelech, M., Morgenroth, E. 2009. Influence of shear on the production of extracellular polymeric substances in membrane bioreactors. *Water Res.* 43:4305–4315.

Merlo, R.P., Trussell, R.S., Hermanowicz, S.W., Jenkins, D. 2004. Physical, chemical and biological properties of submerged membrane bioreactor and conventional activated sludges. *Proceedings of WEFTEC*, October 2–6. New Orleans, LA.

Mezohegyi, G., Bilad, M.R., Vankelecom, I.F.J. 2012. Direct sewage up-concentration by submerged aerated and vibrated membranes. *Bioresour. Technol.* 118:1–7.

Mi, B., Eaton, C.L., Kim, J.-H., Colvin, C.K., Lozier, J.C., Mariñas, B.J. 2004. Removal of biological and non-biological viral surrogates by spiral-wound reverse osmosis membrane elements with intact and compromised integrity. *Water Res.* 38(18):3821–3832.

Mi, B., Mariñas, B.J., Curl, J. 2005. Microbial passage in low pressure membrane elements with compromised integrity. *Environ. Sci. Technol.* 39:4270–4279.

Mishima, I., Nakajima, J. 2009. Control of membrane fouling in membrane bioreactor process by coagulant addition. *Water Sci. Technol.* 59(7):1255–1262.

Mori, M., Seyssiecq, I., Roche, N. 2006. Rheological measurements of sewage sludge for various solids concentrations and geometry. *Process Biochem.* 41(7):1656–1662.

Mortazavi, S. 2008. Application of membrane separation technology to mitigation of mine effluent and acidic drainage. Report 3.15.1. The Mining Association of Canada.

Moslehi, P., Shayegan, J., Bahrpayma, S. 2008. Performance of membrane bioreactor in removal of heavy metals from industrial wastewater. *Iran. J. Chem. Eng.* 5(4):33–38.

Moustafa, M.A.E. 2011. Effect of the pre-treatment on the performance of MBR, Berghausen WWTP, Germany. *Alexandria Eng. J.* 50(2):197–202.

Mueller, J.A., Boyle, W.C., Popel, H.J. 2002. *Aeration: Principles and Practice.* Boca Raton, FL: CRC Press.

MWH. 2007. City of San Diego advanced water treatment research studies. Report to City of San Diego.

Myung, S.-W., Choi, I.-H., Lee, S.-H., Kim, I.-C., Lee, K.-H. 2005. Use of fouling resistant nanofiltration and reverse osmosis membranes for dyeing wastewater effluent treatment. *Water Sci. Technol.* 51(6–7):159–164.

Nagasaki, Y., Nakazawa, H. 1996. Method for treatment of wastewater by activated sludge process. US Patent No. 5,573,670.

Nanda, M., Sharma, D., Kumar, A. 2010. Removal of heavy metals from industrial effluent using bacteria. *Int. J. Environ. Sci.* 2(2):781–787.

Natvik, O., Zaghi, E., Todd, R. 2009. WAS thickening using ultrafiltration membranes—Preliminary results at Oxford water pollution control plant. *Proceedings of WEFTEC*, October 10–14. Orlando, FL.

Ndinisa, N.V. 2006. Experimental and CFD simulation investigations into fouling reduction by gas–liquid two-phase flow for submerged flat sheet membranes. PhD dissertation, The University of New South Wales, Sydney, Australia.

Ndinisa, N.V., Fane, A.G., Wiley, D.E. 2006a. Fouling control in a submerged flat sheet membrane system: Part I—Bubbling and hydrodynamic effects. *Sep. Sci. Technol.* 41:1383–1409.

Ndinisa, N.V., Fane, A.G., Wiley, D.E. 2006b. Fouling control in a submerged flat sheet membrane system: Part II—Two-phase flow characterization and CFD simulations. *Sep. Sci. Technol.* 41:1411–1445.

Neethling, J.B. 2008. Tertiary phosphorus removal. Water Environment Research Foundation (WERF) Nutrient Compendium.

Nerenberg, R., Rittmann, B.E. 2003. Perchlorate reduction using a hollow-fiber membrane biofilm reactor: Kinetics, microbial ecology, and pilot-scale studies. *Proceedings of the Seventh International In Situ and On-Site Bioremediation Symposium*, June 2–5. Orlando, FL.

Nerenberg, R., Rittmann, B.E. 2004. Hydrogen-based, hollow-fiber membrane biofilm reactor for reduction of perchlorate and other oxidized contaminants. *Water Sci. Technol.* 49(11–12):223–230.

Ng, H.Y., Tan, T.W., Ong, S.L. 2006. Membrane fouling of submerged membrane bioreactors: Impact of mean cell residence time and the contributing factors. *Environ. Sci. Technol.* 40:2706–2713.

Nghiem, L.D., Tadkaew, N., Sivakumar, M. 2009. Removal of trance organic contaminants by submerged membrane bioreactors. *Desalination* 236:127–134.

Nguyen Cong Duc, E., Fournier, L., Levecq, C., Lesjean, B., Grelier, P., Tazi-Pain, A. 2008. Local hydrodynamic investigation of the aeration in a submerged hollow fibre membrane cassette. *J. Membrane Sci.* 321:264–271.

Nguyen, L.N., Hai, F.I., Kang, J., Price, W.E., Nghiem, L.D. 2012. Removal of trace organic contaminants by a membrane bioreactor-granular activated carbon (MBR-GAC) system. *Bioresour. Technol.* 113:169–173.

Nielsen, P.H., Kragelund, C., Nielsen, J.L. et al. 2005. Control of *Microthrix parvicella* in activated sludge plants by dosage of polyaluminium salts: Possible mechanisms. *Acta Hydrochem. Hydrobiol.* 33(3):255–261.

Nishimori, K., Tokushima, M., Oketani, S., Churchose, S. 2010. Performance and quality analysis of membrane cartridges used in long-term operation. *Water Sci. Technol.* 62(3):518–524.

Norberg, D., Hong, S., Taylor, J., Zhao, Y. 2006. Surface characterization and performance evaluation of commercial fouling resistant low-pressure RO membranes. *Desalination* 202:45–52.

Nunes, S.P., Peinemann, K.-V. 2001. *Membrane Technology in the Chemical Industry*. Weinheim, Germany: Wiley-VCH Verlag GmbH.

Nyström, M., Pihlajamaki, A., Ehsani, N. 1994. Characterization of ultrafiltration membranes by simultaneous streaming potential and flux measurements. *J. Membrane Sci.* 87(3):245–256.

Ognier, S., Wisniewski, C., Grasmick, A. 2004. Membrane bioreactor fouling in sub-critical filtration conditions: A local critical flux concept. *J. Membrane Sci.* 229:171–177.

Oh, Y.-K., Lee, K.-R., Ko, K.-B., Yeom, I.-T. 2007. Effects of chemical sludge disintegration on the performances of wastewater treatment by membrane bioreactor. *Water Res.* 41(12):2665–2671.

Oko, U.M., Kralick, J.H. 2001. Fuel cell system having humidification membranes. US Patent No. 6,284,399.

Prakasam, T.B.S., Soszynski, S., Zenz, D.R., Lue-Hing, C., Blyth, L., Sernel, G. 1990. Effect of recycling thermophilic sludge on the activated sludge process. USEPA Report. EPA/600/S2–90/037.

Parameshwaran, K., Fane, A.G., Kim, K.J. 2001. Analysis of microfiltration performance with constant flux processing of secondary effluent. *Water Res.* 35(18):4349–4358.

Park, C. 2002. Cations and activated sludge floc structure. MS thesis, Virginia Polytechnic Institute, Blacksburg, VA.

Park, K.Y., Ahn, K.-H., Maeng, S.K., Hwang, J.H., Kwon, J.H. 2003. Feasibility of sludge ozonation for stabilization and conditioning. *Ozone Sci. Eng.* 251:73–80.

Park, P.K., Lee, C.H., Lee, S. 2006. Permeability of collapsed cakes formed by deposition of fractal aggregates upon membrane filtration. *Environ. Sci. Technol.* 40(8):2699–2705.

Patel, M.V. 2012. Groundwater replenishment system—Energy usage implications. In *Water Energy Interaction of Water Reuse*, eds. Lazarova, V., Choo, K.-H., Cornel, P., London: IWA Publishing, pp. 183–192.

Patsios, S.I., Karabelas, A.J. 2011. An investigation of the long-term filtration performance of a membrane bioreactor MBR: The role of specific organic fractions. *J. Membrane Sci.* 372:102–115.

Pearce, G. 2008. Introduction to membranes—MBRs: Manufacturers' comparison: Part 3—Supplier review. *Filtr. Sep.* 45(4):23–25.

Peck, S. 2011. Reclaiming Las Vegas municipal wastewater by integrating membrane bioreactor and reverse osmosis technologies: Recent pilot test results. *Proceedings of 26th Water Reuse Symposium*, September 11–14. Phoenix, AZ.

Pentair Inc. 2013. X-flow procut: Compact 33V-5.2 mm. Available at http://www.x-flow.com/Engineered Product_Compact52_Airlift.aspx (accessed January 15, 2014).

Philips, S., Rabaey, K., Verstraete, W. 2003. Impact of iron salts on activated sludge and interaction with nitrite or nitrate. *Bioresour. Technol.* 88(3):229–239.

Pilutti, M., Nemeth, J.E. 2003. Technical and cost review of commercially available MF/UF membrane products. *Proceedings of the IDA Conference*, September 28–October 3. Paradise Island, Bahamas.

Plumlee, M.H., López-Mesas, M., Heidberger, A., Ishida, K.P., Reinhard, M. 2008. N-nitrosodimethylamine (NDMA) removal by reverse osmosis and UV treatment and analysis via LC-MS/MS. *Water Res.* 42(1–2):347–355.

Poorasgari, E., Bugge, T.V., Christensen, M.L., Jørgensen, M.K. 2015. Compressibility of fouling layers in membrane bioreactors. *J. Membrane Sci.* 475:65–70.

Porter, M.C. 1972. Concentration polarization with membrane ultrafiltration. *Ind. Eng. Chem. Prod. Res. Dev.* 11(3):234–248.

Prieske, H., Böhm, L., Drews, A., Kraume, M. 2010. Optimized hydrodynamics for membrane bioreactors with immersed flat sheet membrane modules. *Desalin. Water Treat.* 18:270–276.

Qaisrani, T.M., Samhaber, W.M. 2011. Impact of gas bubbling and backflushing on fouling control and membrane cleaning. *Desalination* 266:154–161.

Qasim, S.R. 1999. *Wastewater Treatment Plants: Planning, Design, and Operation*. Boca Raton, FL: CRC Press.

Qin, J.-J., Kekre, K.A., Tao, G. et al. 2006. New option of MBR-RO process for production of NEWater from domestic sewage. *J. Membrane Sci.* 272(1–2):70–77.

Qu, X., Gao, W.J., Han, M.N., Chen, A., Liao, B.Q. 2012. Integrated thermophilic submerged aerobic membrane bioreactor and electrochemical oxidation for pulp and paper effluent treatment-towards system closure. *Bioresour. Technol.* 116:1–8.

Rabie, H., Côté, P., Singh, M., Janson, A. 2001. Cyclic aeration system for submerged membrane modules. US Patent No. 6,550,747.

Rabinowitz, B., Stephenson, R. 2006. Effect of microsludge on anaerobic digester performance and residuals dewatering at LA County's JWPCP. *Proceedings of WEFTEC*, October 21–25. Dallas.

Ramaekers, J.P., van Dijk, L., Lumpe, C., Verstraeten, E., Joore, L. 2001. Application of thermophilic membrane bioreactors in the paper industry—A successful key to in mill water treatment. *Paper Technol.* 32–40.

Ratkovich, N., Chan, C.C., Bérubé, P.R., Nopens, I. 2011. Analysis of shear stress and energy consumption in a tubular airlift membrane system. *Water Sci. Technol.* 64(1):189–198.

Ratkovich, N., Horn, W., Helmus, F.P. et al. 2013. Activated sludge rheology: A critical review on data collection and modelling. *Water Res.* 47(2):463–482.

Rautenbach, R., Albrecht, R. 1989. *Membrane Processes*. New York: John Wiley & Sons.

Redmon, D.T., Boyle, W.C., Ewing, L. 1983. Oxygen transfer efficiency measurements in mixed liquor using off-gas techniques. *J. WPCF* 55:1338–1347.

Reid, E., Liu, X., Judd, S.J. 2006. Effect of high salinity on activated sludge characteristics and membrane permeability in an immersed membrane bioreactor. *J. Membrane Sci.* 283:164–171.

Robb, W.L. 1968. Thin silicone membranes—Their permeation properties and some applications. *Ann. N. Y. Acad. Sci.* 146:119–137.

Robles, A., Durán, F., Ruano, M.V., Ribes, J., Ferrer, J. 2012. Influence of total solids concentration on membrane permeability in a submerged hollow fibre anaerobic membrane bioreactor. *Water Sci. Technol.* 66(2):377–384.

Rodriguez, C., van Buynder, P., Lugg, R. et al. 2009. Indirect potable reuse: A sustainable water supply alternative. *Int. J. Environ. Res. Public Health* 6:1174–1209.

Roels, T., Dauwe, F., van Damme, S., de Wilde, K., Roelandt, F. 2002. The influence of PAX-14 on activated sludge systems and in particular on Microthrix parvicella. *Water Sci. Technol.* 46(1–2):487–490.

Rosenberger, S. 2003. Charakterisierung von belebtem Schlamm in Membranbelebungsreaktoren zur Abwasserreinigung. Dissertation, Technical University of Berlin, Fortschr.-Ber. VDI Reihe 3 Nr. 769, VDIVerlag, Düsseldorf.

Rosenberger, S., Kubin, K., Kraume, M. 2002. Rheology of activated sludge in membrane bioreactors. *Eng. Life Sci.* 2(9):269–275.

Rosenberger, S., Evenblij, H., te Poele, S., Wintgens, T., Laabs, C. 2005. The importance of liquid phase analyses to understand fouling in membrane assisted activated sludge processes—Six case studies of different European research groups. *J. Membrane Sci.* 263(1–2):113–126.

Rosenberger, S., Laabs, C., Lesjean, B. et al. 2006. Impact of colloidal and soluble organic material on membrane performance in membrane bioreactors for municipal wastewater treatment. *Water Res.* 40(4):710–720.

Rosso, D., Stenstrom, M.K. 2007. Energy-saving benefits of denitrification. *Env. Eng.: Appl. Sci. Pract.* 3:1–11.

Rosso, D., Iranpour, R., Stenstrom, M.K. 2005. Fifteen years of off-gas transfer efficiency measurements on fine-pore aerators: Key role of sludge age and normalized air flux. *Water Environ. Res.* 77(3):266–273.

Rosso, D., Leu, S.-Y., Stenstrom, M.K. 2007. Energy-conservation in fine pore diffuser installations in activated sludge processes: Final report to California government. Southern California Edison Project 500-03-001.

Rosso, D., Lothman, S.E., Stone, A.L. et al. 2010. Comparative analysis of parallel IFAS and ASP reactors: Oxygen transfer and uptake, nutrient removal, carbon and energy footprint. Research Report of Hazen and Sawyer Co.

Rosso, D., Lothman, S.E., Stone, A.L., Howard, D., Gellner, W.J., Pitt, P. 2011. Comparative analysis of parallel IFAS and ASP reactors: Oxygen transfer and uptake, nutrient removal, carbon and energy footprint. *Proceedings of the WEF Nutrient Recovery and Management Conference*, January 9–12. Miami, FL.

Russotti, G., Goklen, K. 2001. Crossflow membrane filtration of fermentation broth. In *Membrane Separations in Biotechnology*, ed. W.K. Wang, 85–159. New York: Marcel Dekker.

Saby, S., Djafer, M., Chen, C.H. 2002. Feasibility of using a chlorination step to reduce excess sludge in activated sludge process. *Water Res.* 36(3):656–666.

Sachtler, J., Schmidt, W. 1989. System for exchanging a substance between fluids. US Patent No. 4,859,331.

Sahar, E., Ernst, M., Godehardt, M. et al. 2011. Comparison of two treatments for the removal of selected organic micropollutants and bulk organic matter: Conventional activated sludge followed by ultrafiltration versus membrane bioreactor. *Water Sci. Technol.* 63(4):733–744.

Saini, G. 2010. Bacterial hydrophobicity: Assessment techniques, applications and extension to colloids. PhD dissertation, Oregon State University, Corvallis, OR.

Sakai, Y., Fukase, T., Yasui, H., Shibata, M. 1997. An activated sludge process without excess sludge production. *Water Sci. Technol.* 36(11):163–170.

Santos, A., Judd, S.J. 2010a. The commercial status of membrane bioreactors for municipal wastewater. *Sep. Sci. Technol.* 45(7):850–857.

Santos, A., Judd, S.J. 2010b. The fate of metals in wastewater treated by the activated sludge process and membrane bioreactors: A brief review. *J. Environ. Monit.* 12(1):110–118.

Santos, A., Ma, W., Judd, S.J. 2011. Membrane bioreactors: Two decades of research and implementation. *Desalination* 273:148–154.

Schier, W., Frechen, F.-B., Fischer, St. 2009. Efficiency of mechanical pre-treatment on European MBR plants. *Desalination* 236:85–93.

Schippers, J.C., Verdouw, J. 1980. The modified fouling index, a method of determining the fouling characteristics of water. *Desalination* 32:137–148.

Schoichet, M.S., Sefton, M.V. 1999. Immunoisolation. In *Handbook of Biomaterials Evaluation*, eds. A.F. Von Recum. Boca Raton, FL: CRC Press, pp. 411–424.

Semmens, M.J. 1991. Bubbless gas transfer device and process. US Patent No. 5,034,164.

Semmens, M.J. 2005. Membrane technology: Pilot studies of membrane-aerated bioreactors. WERF Report. Project 00-CTS-11.

Seo, G.T., Moon, B.H., Lee, T.S., Lim, T.J., Kim, I.S. 2002. Non-woven fabric filter separation activated sludge reactor for domestic wastewater reclamation. *Water Sci. Technol.* 47(1):133–138.

Seyfried, C.F. 1988. Influence of sludge from chemical biological wastewater treatment on nitrification and digestion. *Proceedings of the Pretreatment in Chemical Water and Wastewater Treatment Conference*, June 1–3. Gothenburg, Germany, pp. 307–317.

Shammas, N. 2007. Fine pore aeration of water and wastewater. In *Advanced Physic-Chemical Treatment*, eds. L.K. Wang, Y.-T. Hung, N.K. Shammas. Totowa, NJ: Humana Press.

Shao, Y.J., Starr, M., Kaporis, K., Kim, H.S., Jenkins, D. 1997. Polymer addition as a solution to Nocardia foaming problems. *Water Environ. Res.* 69(1):25–27.

Sharghi, E.A., Bonakdarpour, B. 2013. The study of organic removal efficiency and halophilic bacterial mixed liquor characteristics in a membrane bioreactor treating hypersaline produced water at varying organic loading rates. *Bioresour. Technol.* 149:486–495.

Sharghi, E.A., Bonakdarpour, B., Roustazade, P. et al. 2013. The biological treatment of high salinity synthetic oilfield produced water in a submerged membrane bioreactor using a halophilic bacterial consortium. *Chem. Technol. Biotechnol.* 88(11):2016–2026.

Sharp, R.R., Heslin, G., Dolphin, M. 2006. Evaluation of a novel membrane bioreactor system for water reuse applications in urban environments. In *Water Pollution VIII—Modelling, Monitoring, and Management*, eds. C.A. Brebbia, A. do Carmo. Southamton, UK: WIT Press, pp. 479–488.

Shaw, D.J. 1969. *Electrophoresis*. London: Academic Press.

Sheng, G.-P., Yu, H.-Q. 2006. Characterization of extracellular polymeric substances of aerobic and anaerobic sludge using three-dimensional excitation and emission matrix fluorescence spectroscopy. *Water Res.* 40(6):1233–1239.

Shu, J., Majamaa, K., Knoell, T. 2014. Advancing reverse osmosis membrane performance in a water reuse application: A collaborative approach. *Proceedings of the Membrane Technology Conference*, March 10–14. Las Vegas, NV.

Singh, K.S., Mi, Z., Grant, S.R. 2008. Stress effects on fouling of flat sheet membrane bioreactor treating biodegradable wastewater. *Proceedings of the World Environmental and Water Resources Congress*, May 12–16. Honolulu, HI.

Sirkar, K.K., Song, L. 2009. Pilot-scale studies for direct contact membrane distillation-based desalination process. Report to US Department of Interior. DWPR Report No. 134.

Sloan, D. 2014. A new spring in big spring. *Proceedings of WEFTEC*, September 26–30. New Orleans, LA.

Smith, S., Takács, I., Murthy, S., Daigger, G.T., Szabó, A. 2008. Phosphate complexation model and its implications for chemical phosphorus removal. *Water Environ. Res.* 80(5):428–438.

Smolders, G.J.F., van der Meij, J., van Loosdrecht, M.C.M., Heijnen, J.J. 1994. Model of the anaerobic metabolism of the biological phosphorus removal process: Stoichiometry and pH influence. *Biotechnol. Bioeng.* 43(6):461–470.

Sofia, A., Ng, W.J., Ong, S.L. 2004. Engineering design approaches for minimum fouling in submerged MBR. *Desalination* 160:67–74.

Solsona, F., Méndez, J.P. 2003. *Water Disinfection*. World Health Organization (WHO) PAHO/CEPIS/PUB/03.89.

Sombatsompop, K., Visvanathan, C., Ben Aim, R. 2006. Evaluation of biofouling phenomenon in suspended and attached growth membrane bioreactor systems. *Desalination* 201(1–3):138–149.

Song, K.-G., Choung, Y.-K., Ahn, K.-H., Cho, J., Yun, H. 2003. Performance of membrane bioreactor system with sludge ozonation process for minimization of excess sludge production. *Desalination* 157 (1–3):353–359.

Song, K.-G., Kim, Y., Ahn, K.-H. 2008. Effect of coagulant addition on membrane fouling and nutrient removal in a submerged membrane bioreactor. *Desalination* 221(1–3):467–474.

Song, Z., Li, L., Wang, H., Li, B., Wang, S. 2013. DCMD flux curve characteristics of crossflow hollow fiber membrane. *Desalination* 323:107–113.

Sridang, P.C., Heran, M., Crasmick, A. 2005. Influence of module configuration and hydrodynamics in water clarification by immersed membrane systems. *Water Sci. Technol.* 51(6–7):135–142.

Stanford, B. 2010. Pilot scale evaluation of an MBR-ozone-RO system for water reuse. *Proceedings of the 14th Water Reuse & Desalination Research Conference*, May 23–24. Tampa, FL.

Stefanski, M., Kennedy, S., Judd, S. 2011. The determination and origin of fibre clogging in membrane bioreactors. *J. Membrane Sci.* 375:198–203.

Stenstrom, M.K., Poduska, R.A. 1980. The effect of dissolved oxygen concentration on nitrification. *Water Res.* 14(6):643–649.

Stenstrom, M.K., Masutani, G. 1990. Fine pore diffuser fouling: The Los Angeles Studies, A final report to the ASCE and the USEPA, UCLA ENG 90–02. University of California, Los Angeles.

Stenstrom, M.K., Leu, S.-Y., Jiang, P. 2006. Theory to practice: Oxygen transfer and the new ASCE standard. *Proceedings of WEFTEC*, October 22–26. Dallas.

Stephenson, R.J., Dhaliwal, H.S. 2000. Method of liquefying microorganisms derived from biological wastewater treatment process. US Patent No. 6,013,183.

Stevens, J. 2008. Cleaning of diffusers at Edgeworth wastewater treatment plant. *Proceedings of 2nd Annual WIOA NSW Water Industry Engineers & Operators Conference*, April 8–10. Newcastle, New Zealand.

STOWA. 2009. Experience with the hybrid MBR in Ootmarsum. Report No. 36 (in Dutch).

Stricker, A.E., Lossing, H., Gibson, J.H., Hong, Y., Urbanic, J.C. 2011. Pilot scale testing of a new configuration of the membrane aerated biofilm reactor (MABR) to treat high-strength industrial sewage. *Water Environ. Res.* 83(1):3–14.

Suárez, F., Tyler, S.W., Childress, A.E. 2010. A theoretical study of a direct contact membrane distillation system coupled to a salt-gradient solar pond for terminal lakes reclamation. *Water Res.* 44(15):4601–4615.

Suk, D.E., Park, J.S., Hwang, B.K. et al. 2012. Development of high performance membrane fibre/module/aerator for a membrane bioreactor. *Proceedings of the IWA Biannual Conference*, September 16–21. Busan, Korea.

Sumitomo Inc. 2011. Poreflon™ module. Available at http://www.sei-sfp.co.jp/english/products/poreflon-module_1.html (accessed January 2, 2011).

Sutton, P.M., Melcer, H. Schraa, O.J., Togna, A.P. 2011. Treating municipal wastewater with the goal of resource recovery. *Water Sci. Technol.* 63(1):25–31.

Syed, W., Zhou, H., Sheng, C., Mahendraker, V., Adnan, A., Theodoulou, M. 2009. Effects of hydraulic and organic loading shocks on sludge characteristics and its effects on membrane bioreactor performance. *Proceedings of WEFTEC*, October 17–21. Orlando, FL.

Szabó, A., Takács, I., Murthy, S., Daigger, G.T., Licskó, I., Smith, S. 2008. Significance of design and operational variables in chemical phosphorus removal. *Water Environ. Res.* 80(5):407–416.

Tadkaew, N., Sivakumar, M., Khan, S.J., McDonald, J.A., Nighie, L.D. 2010. Effect of mixed liquor pH on the removal of trace organics contaminants in a membrane bioreactor. *Bioresour. Technol.* 101:1494–1500.

Tadkaew, N., Hai, F.I., McDonald, J.A., Khan, S.J., Nghiem, L.D. 2011. Removal of trace organics by MBR treatment: The role of molecular properties. *Water Res.* 45(8):2439–2451.

Tajima, F., Yamamoto, T. 1986. Apparatus for filtering water containing radioactive substances in nuclear power plants. US Patent No. 4,756,875.

Takács, I., Murthy, S., Smith, S., McGrath, M. 2006. Chemical phosphorus removal to extremely low levels: Experience of two plants in the Washington, DC area. *Water Sci. Technol.* 53(12):21–28.

Takemura, T., Itoh, H., Kamo, J., Yoshida, H. 1987. Multilayer composite hollow fibers and method of making same. US Patent No. 4,713,312.

Takizawa, S., Fujita, K., Kim, H.S. 1996. Membrane fouling decrease by microfiltration with ozone scrubbing. *Desalination* 106:423–426.

Tang, M.K.Y., Ng, H.Y. 2014. Impacts of different draw solutions on a novel anaerobic forward osmosis membrane bioreactor (AnFOMBR). *Water Sci. Technol.* 69(10):2036–2042.

Tao, G., Kekre, K., Qin, J., Oo, M.H., Viawanath, B., Seah, H. 2008. MBR-RO process for water reclamation and purification form used water. *Water Conditioning and Purification*, March.

Tao, J., Wu, S., Sun, L., Tan, X., Yu, S., Zhang, Z. 2012. Composition of waste sludge from municipal wastewater treatment plant. *Procedia Environ. Sci.* 12:964–971.

Tarabara, V., Koyuncu, I., Wiesner, M. 2004. Effect of hydrodynamics and solution ionic strength on permeate flux in crossflow filtration: Direct experimental observation of filter cake cross-sections. *J. Membrane Sci.* 241:65–78.

Tardieu, E., Grasmick, A., Geaugey, A., Manem, J. 1998. Hydrodynamic control of bioparticle deposition in a MBR applied to wastewater treatment. *J. Membrane Sci.* 147:1–12.

Tarnacki, K., Lyko, S., Wintgens, T., Melin, T., Natau, F. 2005. Impact of extra-cellular polymeric substances on the filterability of activated sludge in membrane bioreactors for landfill leachate treatment. *Desalination* 179(1–3):181–190.

Tarre, S., Green, M. 2004. High-rate nitrification at low pH in suspended- and attached-biomass reactors. *Appl. Environ. Microbiol.* 70(11):6481–6487.

Tazi-Pain, A., Schrotter, J.-C., Gaid, K. 2006. How to select a membrane for water application? The experience of Veolia water. *Desalination* 199:310–311.

Tchobanoglous, G., Burton, F.L., Stensel, H.D. 2003. *Wastewater Engineering: Treatment and Reuse*. Boston: McGraw-Hill.

Teksoy Başaran, S., Aysel, M., Kurt, H. et al. 2012. Removal of readily biodegradable substrate in super fast membrane bioreactor. *J. Membrane Sci.* 423–424:477–486.

Teli, A., Antonelli, M., Bonomo, L., Malpei, F. 2012. MBR fouling control and permeate quality enhancement by polyaluminum chloride dosage online: A case study. *Water Sci. Technol.* 66(6):1289–1295.

Thiemig, C. 2011. The importance of measuring the sludge filterability at MBR: Introduction of a new method. *Proceedings of the IWA Membrane Conference*, October 4–7. Aachen, Germany.

Tian, J.-Y., Xu, Y.-P., Chen, Z.-L., Nan, J., Li, G.-B. 2010. Air bubbling for alleviating membrane fouling of immersed hollow-fiber membrane for ultrafiltration of river water. *Desalination* 260:225–230.

Till, S.W., Judd, S.J., McLoughlin, B. 1998. Reduction of fecal coliform bacteria in sewage effluents using microporous polymeric membrane. *Water Res.* 32(4–5):373–382.

Tiller, R.M. 1953. The role of porosity in filtration. *Chem. Eng. Prog.* 49(9):467–479.

Todar, K. 2005. Online textbook of bacteriology. Available at http://textbookofbacteriology.net/nutgro.html.

Tomescu, I.J., Simon, R. 2013. Formed sheet membrane element and filtration system. US Patent Application No. 2013/0101739.

Torpey, W.N., Andrews, J., Basilico, J.V. 1984. Effect of multiple digestion on sludge. *J. Water Pollut. Control Fed.* 56:62–68.

Torretta, V., Urbini, G., Raboni, M. et al. 2013. Effect of powdered activated carbon to reduce fouling in membrane bioreactors: A sustainable solution. *Sustainability* 5(4):1501–1509.

Trimboli, P., Lozier, J., Johnson, W. 2001. Demonstrating the integrity of a large scale microfiltration plant using a Bacilus spore challenge test. *Water Sci. Technol: Water Supply* 1(5–6):1–12.

Trinh, T., van den Akker, B., Stuetz, M., Coleman, H.M., Le-Clech, P. 2012. Removal of trace organic chemical contaminants by a membrane bioreactor. *Water Sci. Technol.* 66(9):1856–1863.

Trivedi, H. 2004. Flat plate microfiltration membrane bioreactor designed for ultimate nutrient removal (UNR™). *Proceedings of WEFTEC*, October 2–6. New Orleans, LA.

Trussell, R.S., Merlo, R.P., Hermanowicz, S.W., Jenkins, D. 2006. The effect of organic loading on process performance and membrane fouling in a submerged membrane bioreactor treating municipal wastewater. *Water Res.* 40:2675–2683.

Ueda, T., Hata, K., Kikuoka, Y., Seino, O. 1997. Effects of aeration on suction pressure in a submerged membrane bioreactor. *Water Res.* 31(3):489–494.

Um, M.-J. 1996. Flux enhancement with gas injection in crossflow ultrafiltration of oily wastewater. M.S. Thesis. Seoul National University. Seoul, Korea.

UNC website. 2014. Available at http://www.unc.edu/~shashi/TablePages/totalfecalcoliforms.html (accessed January 30, 2014).

Urase, T., Yamamoto, K., Ohgaki, S. 1993. Evaluation of virus removal in membrane separation processes using coliphage. *Water Sci. Technol.* 28(7):9–15.

USEPA. 1994. A plain English guide to the EPA Part 503 biosolids rule. EPA/832/R-93/003.

USEPA. 2001a. Low-pressure membrane filtration for pathogen removal: Application, implementation, and regulatory issues. EPA 815-C-01–001.

USEPA. 2001b. Specific oxygen uptake rate in biosolids. PEA-821-R-01-014.

USEPA. 2005. Membrane filtration guidance manual. EPA/815-R-06-009.

USEPA. 2010. Evaluation of energy conservation measures for wastewater treatment facilities. EPA/832-R-10-005.

USEPA. 2012. Guidelines for water reuse. EPA/600/R-12/618.

van den Brink, P., Satpradit, O.A., van Bentem, A., Zwijnenburg, A., Temmink, H., van Loosdrecht, M. 2011. Effect of temperature shocks on membrane fouling in membrane bioreactors. *Water Res.* 45(15):4491–4500.

van den Broeck, R., Krzeminski, P., van Dierdonck, J. et al. 2011. Activated sludge characteristics affecting sludge filterability in municipal and industrial MBRs: Unraveling correlations using multi-component regression analysis. *J. Membrane Sci.* 378:330–338.

van den Broeck, R., van Dierdonck, J., Nijskens, P. et al. 2012. The influence of solids retention time on activated sludge bioflocculation and membrane fouling in a membrane bioreactor (MBR). *J. Membrane Sci.* 401–402:48–55.

van der Marel, P., Zwijnenburg, A., Kemperman, A., Wessling, M., Temmink, H., van der Meer, W. 2010. Influence of membrane properties on fouling in submerged membrane bioreactors. *J. Membrane Sci.* 348:66–74.

van Haandel, A.C., van der Lubbe, J.G.M. 2012. *Handbook of Biological Wastewater Treatment.* London: IWA Publishing.

Visvanathan, C., Abeynayaka, A. 2012. Developments and future potentials of anaerobic membrane bioreactors (AnMBRs). *Membrane Water Treat.* 3(1):1–23.

Vogelaar, J.C., Klapwijk, A., van Lier, J.B., Rulkens, W.H. 2000. Temperature effect on the oxygen transfer rate between 20°C and 55°C. *Water Res.* 34(3):1037–1041.

Vrijenhoek, E.M., Elimelech, M., Hong, S.K. 2000. Influence of membrane properties, solution chemistry and hydrodynamics on colloidal fouling of reverse osmosis and nanofiltration membranes. *Proceedings of the ACS Biannual Conference*, June 17–19. Washington, DC.

Vrijenhoek, E.M., Elimelech, M., Hong, S.K. 2001. Influence of membrane surface properties on initial rate of colloidal fouling of reverse osmosis and nanofiltration membranes. *J. Membrane Sci.* 188:115–128.

Wagner, M., Cornel, P., Krause, S. 2002. Efficiency of different aeration systems in full scale membrane bioreactors. *Proceedings of WEFTEC*, September 28–October 2. Chicago.

Wang, J., Wei, Y., Li, K., Cheng, Y., Li, M., Xu, J. 2014a. Fate of organic pollutants in a pilot-scale membrane bioreactor-nanofiltration membrane system at high water yield in antibiotic wastewater treatment. *Water Sci. Technol.* 69(4):876–881.

Wang, Y., Tng, K.H., Wu, H., Leslie, G., Waite, T.D. 2014b. Removal of phosphorus from wastewaters using ferrous salts—A pilot scale membrane bioreactor study. *Water Res.* 57:140–150.

Wang, Z., Ma, J., Tnag, C.Y., Kimura, K., Wang, Q., Han, X. 2014c. Membrane cleaning in membrane bioreactors: A review. *J. Membrane Sci.* 468:276–307.

WEF. 2009. Energy conservation in water and wastewater treatment facilities. Manual of Practice No. 32.

WEF. 2012. Membrane bioreactors. Manual of Practice No. 36. WEF Press.

Wei, C.-H., Huang, X., Wang, C.-W., Wen, X.-H. 2006. Effect of a suspended carrier on membrane fouling in a submerged membrane bioreactor. *Water Sci. Technol.* 53(6):211–220.

Weiss, R. 1970. The solubility of nitrogen, oxygen, and argon in water and seawater. *Deep-Sea Res.* 17:721–735.

Wicaksana, F., Fane, A.G., Chen, V. 2006. Fibre movement induced by bubbling using submerged hollow fibre membranes. *J. Membrane Sci.* 271:186–195.

Wienk, I.M., Boom, R.M., Beerlage, M.A., Bulte, A.M., Smolders, C.A., Strathmann, H. 1996. Recent advances in the formation of phase inversion membranes made from amorphous or semi-crystalline polymers. *J. Membrane Sci.* 113(2):361–371.

Wiesner, M.R., Chellam, S. 1992. Mass transport considerations for pressure-driven membrane processes. *J. AWWA* 84:88–95.

Wijnbladh, E. 2007. Ozone technology for sludge bulking control. PhD dissertation, Uppsala University, Uppsala, Sweden.

Wikipedia. 2012. Non-Newtonian fluid. Available at http://en.wikipedia.org/wiki/Non-Newtonian_fluid (accessed December 3, 2012).

Wikipedia. 2013a. Defoamer. Available at http://en.wikipedia.org/wiki/Defoamer (accessed January 1, 2014).

Wikipedia. 2013b. Fraunhofer Institute for Solar Energy Systems ISE. Available at http://en.wikipedia.org /wiki/Fraunhofer_Institute_for_Solar_Energy_Systems_ISE (accessed February 28, 2014).

Williams, R., Schuler, P., Comstock, K., Pope, R. 2008. Large membrane bioreactors of Georgia, a guide and comparison. *Proceedings of the Water Environment Federation Membrane Technology Conference*, January 28. Atlanta, GA.

Winter, D., Koschikowski, J., Ripperger, S. 2012. Desalination using membrane distillation: Flux enhancement by feed water deaeration on spiral-wound modules. *J. Membrane Sci.* 423–424:215–224.

Wolfe, T.D. 2003. Membrane process optimization technology. Desalination and water purification research and development program. US Dept. Interior. Report No. 100.

Won, Y.-J., Lee, J., Choi, D.-C. et al. 2012. Preparation and application of patterned membranes for wastewater treatment. *Environ. Sci. Technol.* 46:11021–11027.

WPCF. 1987. Anaerobic sludge digestion. Manual of Practice No. 16.

WRF. 2011. New concepts of UV/H$_2$O$_2$ oxidation. BTO 2011.046.

Wu, J., Huang, X., Li, H., Wei, C., Wang, J. 2011. Seasonal variation of activated sludge mixed liquors in a long-term steadily-operating membrane bioreactor. *Proceedings of the IWA Membrane Conference,* October 4–7. Aachen, Germany.

Wu, B., Kitade, T., Chong, T.H., Lee, J.Y., Uemura, T., Fane, A.G. 2013. Flux-dependent fouling phenomena in membrane bioreactors under different food to microorganisms F/M ratio. *Sep. Sci. Technol.* 48(6):840–848.

Xia, L., Law, A., Fane, A.G. 2013. Hydrodynamic effects of air sparging on hollow fiber membranes in a bubble column reactor. *Water Res.* 47(11):3762–3772.

Xie, R.J., Wang, J., Hu, J.Y. et al. 2013. Field studies of UV/H$_2$O$_2$ and ozone/H$_2$O$_2$ systems for removal of taste and odor and selected emerging trace organic compounds under tropical conditions. *Proceedings of the IOA-IUVA World Congress,* September 22–26. Las Vegas, NV.

Xiong, J., Fu, D., Singh, R.P. 2014. Self-adaptive dynamic membrane module with a high flux and stable operation for the municipal wastewater treatment. *J. Membrane Sci.* 471:308–318.

Yamamoto, K., Hiasa, M., Mahmood, T., Matsuo, T. 1989. Direct solid–liquid separation using hollow fiber membrane in an activated sludge aeration tank. *Water Sci. Technol.* 21(4–5):43–54.

Yamanoi, I., Kageyama, K. 2010. Evaluation of bubble flow properties between flat sheet membranes in membrane bioreactor. *J. Membrane Sci.* 360:102–108.

Yang, Q., Chen, J., Zhang, F. 2006. Membrane fouling control in a submerged membrane bioreactor with porous, flexible suspended carriers. *Desalination* 189:292–302.

Yang, Y.F., Huang, S.S., Takabatake, H., Hanada, S. 2009. Development of MBR-RO integrated system for wastewater reclamation. *J. EICA* 14(2–3):87–90.

Yang, W., Syed, W., Zhou, H. 2014. Comparative study on membrane fouling between membrane-coupled moving bed biofilm reactor and conventional membrane bioreactor for municipal wastewater treatment. *Water Sci. Technol.* 69(5):1021–1027.

Yasui, H. 2000. Apparatus and method for ozone-treating biosludges. US Patent No. 6,146,521.

Yasui, H., Shibata, M. 1994. An innovative approach to reduce excess sludge production in the activated sludge process. *Water Sci. Technol.* 30(9):11–20.

Yasui, H., Nakamura, K., Sakuma, S., Iwasaki, M., Sakai, Y. 1996. A full-scale operation of a novel activated sludge process without excess sludge production. *Water Sci. Technol.* 34(3–4):395–404.

Yasui, H., Komatsu, K., Goel, R., Li, Y.Y., Noike, T. 2005. Full-scale application of anaerobic digestion process with partial ozonation of digested sludge. *Water Sci. Technol.* 52(1–2):245–252.

Ye, F.X., Shen, D.S., Li, Y. 2003. Reduction in excess sludge production by addition of chemical uncouplers in activated sludge batch cultures. *J. Appl. Microbiol.* 95(4):781–786.

Ye, Y., Chen, V., Fane, A.G. 2006. Modeling long term sub-critical filtration of model EPS solutions. *Desalination* 191(1–3):318–327.

Yeo, A.P.S., Fane, A.G. 2005. Performance of individual fibers in a submerged hollow fiber bundle. *Water Sci. Technol.* 51(6–7):165–172.

Yeo, H., Lee, H.S. 2013. The effect of solids retention time on dissolved methane concentration in anaerobic membrane bioreactors. *Environ. Technol.* 34(13–16):2105–2112.

Yeo, H.-T., Lee, S.-T., Han, M.J. 2000. Role of a polymer additive in casting solution in preparation of phase inversion polysulfone membranes. *J. Chem. Eng. Japan* 33(1):180–184.

Yeo, A.P.S., Law, W.K., Fane, A.G. 2006. Factors affecting the performance of a submerged hollow fiber bundle. *J. Membrane Sci.* 280:969–982.

Yeo, A.P.S., Law, W.K., Fane, A.G. 2007. The relationship between performance of submerged hollow fibers and bubble-induced phenomena examined by particle image velocimetry. *J. Membrane Sci.* 304:125–137.

Yeom, I.T., Lee, K.R., Choi, Y.G. et al. 2005. A pilot study on accelerated sludge degradation by a high-concentration membrane bioreactor coupled with sludge pretreatment. *Water Sci. Technol.* 52(10–11):201–210.

Yeon, K.-M., Cheong, W.-S., Oh, H.-S. et al. 2009a. Quorum sensing: A new biofouling control paradigm in a membrane bioreactor for advanced wastewater treatment. *Environ. Sci. Technol.* 43(2):380–385.

Yeon, K.-M., Lee, C.-H., Kim, J. 2009b. Magnetic enzyme carrier for effective biofouling control in the membrane bioreactor based on enzymatic quorum quenching. *Environ. Sci. Technol.* 43(19):7403–7409.

Yigit, N.O., Harman, I., Civelekoglu, G., Koseoglu, H., Cicek, N., Kitis, M. 2008. Membrane fouling in a pilot-scale submerged membrane bioreactor operated under various conditions. *Desalination* 231:124–132.

Yoo, R., Kim, J., McCarty, P.L., Bae, J. 2012. Anaerobic treatment of municipal wastewater with a staged anaerobic fluidized membrane bioreactor (SAF-MBR) system. *Bioresour. Technol.* 120:133–139.

Yoo, R.H., Kim, J.H., McCarty, P.L., Bae, J.H. 2014. Effect of temperature on the treatment of domestic waste-water with a staged anaerobic fluidized membrane bioreactor. *Water Sci. Technol.* 69(6):1145–1150.

Yoon, S.-H. 1994. Filtration characteristics of ceramic membranes in membrane coupled anaerobic digester. MS thesis, Seoul National University, Korea.

Yoon, S.-H. 2003. Important operational parameters of membrane bioreactor-sludge disintegration (MBR-SD) system for zero excess sludge production. *Water Res.* 87:1921–1931.

Yoon, S.-H., Lee, S. 2005. Critical operational parameters for zero sludge production in biological wastewater treatment processes combined with sludge disintegration. *Water Res.* 39:3738–3754.

Yoon, S.-H., Collins, J.H. 2006. A novel flux enhancing method for membrane bioreactor (MBR) process using polymer. *Desalination* 191:52–61.

Yoon, S.-H., Lee, C.-H., Kim, K.-J., Fane, A.G. 1998. Effect of calcium ion on the fouling of nanofilter by humic acid in drinking water production. *Water Res.* 32(7):2180–2186.

Yoon, S.-H., Lee, C.-H., Kim, K.-J., Fane, A.G. 1999a. Three-dimensional simulation of the deposition of multi-dispersed charged particles and prediction of resulting flux during cross-flow microfiltration. *J. Membrane Sci.* 161(1–2):7–20.

Yoon, S.-H., Kang, I.-J., Lee, C.H. 1999b. Fouling of inorganic membrane and flux enhancement in membrane-coupled anaerobic bioreactor. *Sep. Sci. Technol.* 34(5):709–724.

Yoon, S.-H., Collins, J.H., Musale, D. et al. 2004a. Application of membrane performance enhancer MPE for full scale membrane bioreactors. *IWA's Water Environment Membrane Technology (WEMT) Conference*, Seoul, Korea.

Yoon, S.-H., Kim, H.-S., Lee, S. 2004b. Incorporation of ultrasonic cell disintegration into a membrane biore-actor for zero sludge production. *Process Biochem.* 39(12):1923–1929.

Yoon, T.I., Lee, H.S., Kim, C.G. 2004c. Comparison of pilot scale performance between membrane bioreactor and hybrid conventional wastewater treatment systems. *J. Membrane Sci.* 242(1–2):5–12.

Yoon, S.-H., Kim, H.-S., Yeom, I.-T. 2004d. Optimization model of submerged hollow fiber membrane mod-ules. *J. Membrane Sci.* 234(1–2):147–156.

Yoon, S.-H., Collins, J.H., Musale, D., Sundararajan, S., Tsai, S.-P. 2005. Effects of flux enhancing polymer on the characteristics of sludge in membrane bioreactor process. *Water Sci. Technol.* 51(6)151–157.

Yoon, Y., Westerhoff, P., Snyder, S.A., Wert, E.C., Yoon, J. 2007. Removal of endocrine disrupting compounds and pharmaceuticals by nanofiltration and ultrafiltration membranes. *Desalination* 202(1–3):16–23.

Yoon, S.-H., Yeom, I.-T., Lee, S. 2008. Experimental verification of pressure drop models in hollow fiber mem-brane. *J. Membrane Sci.* 310:7–12.

You, S.J., Sue, W.M. 2009. Filamentous bacteria in a foaming membrane bioreactor. *J. Membrane Sci.* 342(1–2):42–49.

Yukawa, H., Shimura, K., Suda, A., Maniwa, A. 1983. Cross flow electro-ultrafiltration for colloidal solution of protein. *J. Chem. Eng. Japan* 164:305–311.

Zahid, W.M., El-Shafai, S.A. 2011. Use of cloth-media filter for membrane bioreactor treating municipal wastewater. *Bioresour. Technol.* 102:2193–2198.

Zaw, K., Safizadeh, M.R., Luther, J., Ng, K.C. 2013. Analysis of a membrane based air-dehumidification unit for air conditioning in tropical climates. *Appl. Therm. Eng.* 59(1–2):370–379.

Zeiher, K.E., Ho, B., Hoots, J.E. 2004. Method of monitoring membrane separation process. US Patent No. 6,838,001.

Zenon Environmental Inc. 2003. A Zenon design and pilot report. Available at http://www.gov.mb.ca /conservation/eal/registries/brandonwastewater/eia/append-b.pdf (accessed November 30, 2010).

Zhang, J. 2011. Forward osmosis membrane bioreactor for water reuse. MS thesis, National Singapore University, Singapore.

Zhang, X., Bishop, P.L., Kinkle, B.K. 1999. Comparison of extraction methods for quantifying extracellular polymers in biofilms. *Water Sci. Technol.* 39(7):211–218.

Zhang, J., Chua, H.C., Zhou, J., Fane, A.G. 2006. Factors affecting the membrane performance in submerged membrane bioreactors. *J. Membrane Sci.* 284:54–66.

Zhang, K., Cui, Z., Field, R.W. 2009. Effect of bubble size and frequency on mass transfer in flat sheet MBR. *J. Membrane Sci.* 332:30–37.

Zhang, Y., Gao, B., Lu, L., Yue, Q., Wang, Q., Jia, Y. 2010a. Treatment of produced water from polymer flood-ing in oil production by the combined method of hydrolysis acidification-dynamic membrane bioreactor-coagulation process. *J. Petrol. Sci. Eng.* 74(1–2):14–19.

Zhang, J., Zhou, J., Liu, Y., Fane, A.G. 2010b. A comparison of membrane fouling under constant and variable organic loadings in submerge membrane bioreactors. *Water Res.* 44:5407–5413.

Zhang, J., Loong, W.L., Chou, S., Tang, C., Wang, R., Fane, A.G. 2012. Membrane biofouling and scaling in forward osmosis membrane bioreactor. *J. Membrane Sci.* 403–404:8–14.

Zhang, Z., Bligh, M.W., Wang, Y. et al. 2015. Cleaning strategies for iron-fouled membranes from submerged membrane bioreactor treatment of wastewater. *J. Membrane Sci.* 475:9–21.

Zhao, Q., Guan, W. 2011. Effect of salinity on biological phosphorus removal by anaerobic/aerobic/anoxic process in sequencing batch reactor. *Proceedings of the International Symposium on Water Resource and Environmental Protection (ISWREP)*, May 20–22. Xian, Shaanxi, China.

Zilverentant, A., van Nieuwkerk, A., Vance, I., Watlow, A., Rees, M. 2009. Pilot-scale membrane bioreactor treats produced water. *World Oil* 230(8):21–27.

Zsirai, T., Wang, Z.-Z., Gabarrón, S. et al. 2014. Biological treatment and thickening with a hollow fibre membrane bioreactor. *Water Res.* 58(1):29–37.

Zuehike, S., Duennbier, U., Lesjean, B., Gnirss, R., Buisson, H. 2007. Long-term comparison of trace organics removal performances between conventional and membrane activated sludge processes. *Water Environ. Res.* 78(13):2480–2486.

Zydney, A.L., Colton, C.K. 1986. A concentration polarization model for the filtrate flux in crossflow microfiltration of particulate suspensions. *Chem. Eng. Commun.* 47:1–21.

Abbreviations

AD	Anaerobic digestion
ADF	Average daily flux
AFM	Atomic force microscopy
AGMD	Air gap membrane distillation
AnMBR	Anaerobic MBR
AOP	Advanced oxidation process
API	American Petroleum Institute
ASM	Activated sludge model
ATAD	Autothermal aerobic digestion
ATP	Adenosine triphosphate
BF-MBR	Biofilm membrane bioreactor
BNR	Biological nitrogen removal
BOD	Biological oxygen demand
BPT	Bubble point test
BSA	Bovine serum albumin
CA	Cellulose acetate
CAPEX	Capital expenditure
CAS	Conventional activated sludge
CFD	Computational fluid dynamics
CHP	Combined heat and power
CLSM	Confocal laser scanning microscopy
COD	Chemical oxygen demand
CP	Concentration polarization
CPI	Corrugated plate interceptor
CST	Capillary suction time
CSTR	Continuously stirred tank reactor
Da.	Dalton
DAF	Dissolved air floatation
DAM	Diffusive air flow monitoring
DBP	Disinfection byproducts
DCMD	Direct contact membrane distillation
DFCm	Delft filtration characterization method
DGGE	Denaturing gradient gel electrophoresis
DI water	Deionized water
DMAEA	Dimethylaminoethylacrylate
DMF	Direct membrane filtration
DMSO	Dimethyl sulfoxide
DNA	Deoxyribonucleic acid
DO	Dissolved oxygen
DOTM	Direct observation through membrane
DPR	Direct potable reuse
EBPR	Enhanced biological phosphorous removal
EBS	Ethylene *bis* stearamide
EDC	Endocrine disrupting compounds
EDI	Electrodeionization reversal
EPDM	Ethylene propylene diene monomer

Epi-DMA	Epichlorohydrin-dimethylamine
EPS	Extracellular polymeric substances
F/M ratio	Food-to-microorganism ratio
F/V ratio	Food-to-volume ratio
FO	Forward osmosis
FOG	Fat, oil, and grease
FOMBR	Forward osmosis membrane bioreactor
FS	Flat sheet
GAC	Granular activated carbon
GOR	Gain–output ratio
GTE	Gas transfer efficiency
GWRS	Ground water replenishing system
HAA	Haloacetic acids
HF	Hollow fiber
HFMBfR	Hydrogen-based hollow-fiber membrane biofilm reactor
HFO	Hydrous ferric oxides
HRT	Hydraulic retention time
ICP	Internal concentration polarization
ID	Inner diameter
IFAS	Integrated fixed film sludge
IPR	Indirect potable reuse
IWA	International water association
LMH	$L/m^2/h$ (flux unit)
LPM	Liters per minute
LRV	Log removal value
MABR	Membrane assisted biofilm reactor
MATH	Microbial adhesion to hydrocarbons
MBBR	Moving bed biofilm reactor
MBR	Membrane bioreactor
MCE	Mixed cellulose acetate
MD	Membrane distillation
MED	Multieffect distillation
MEK	Methyl ethyl ketone
MF	Microfiltration
MFI	Modified fouling index
MLE process	Modified Ludzack–Ettinger process
MLSS	Mixed liquor suspended solids
MLVSS	Mixed liquor volatile suspended solids
MPN	Most probable number
MSBR	Membrane supported biofilm reactor
MSF	Multistage flashing
MWCO	Molecular weight cutoff
NDMA	N-nitrosodimethylamine
NF	Nanofiltration
NFMBR	Nanofiltration membrane bioreactor
NIPS	Nonsolvent induced phase separation
NTU	Nephelometric turbidity unit
OD	Outer diameter
OLR	Organic loading rate
OPEX	Operating expenditure
ORP	Oxidation–reduction potential

OTE	Oxygen transfer efficiency
OTR	Oxygen transfer rate
OUR	Oxygen uptake rate
PAC	Powdered activated carbon
PACl	Polyaluminum chloride
PAH	Polycyclic aromatic hydrocarbons
PAN	Polyacrylonitrile
PAO	Phosphorus-accumulating organisms
PC	Polycarbonate
pDADMAC	Polydiallyldimethylammonium chloride
PDMS	Polydimethylsiloxane
PDT	Pressure decay test
PE	Polyethylene
PEG	Polyethylene glycol
PEM	Polymer electrolyte membrane
PES	Polyethersulfone
PETE	Polyester
PF	Plate and frame module
PFR	Plug flow reactor
PHA	Polyhydroxyalkanoates
PHB	Polyhydroxybutyrates
PMMA	Polymethyl methacrylate
PMP	Polymethyl pentene
PP	Polypropylene
PPCP	Pharmaceutical and personal care products
PPI	Parallel plate interceptor
PS	Polysaccharides
PSU	Polysulfone
PTFE	Polytetrafluoroethylene
PVA	Polyvinyl alcohol
PVC	Polyvinyl chloride
PVDF	Polyvinylidene difluoride
PVP	Polyvinylpyrrolidone
RAS	Return activated sludge
RDS	Rate determining step
RH	Relative hydrophobicity
RNA	Ribonucleic acid
RO	Reverse osmosis
RSD	Reverse solute diffusion
$\textbf{SAD}_\textbf{m}$	Specific aeration demand per membrane surface area
$\textbf{SAD}_\textbf{p}$	Specific aeration demand per permeate volume
SAR	Sodium absorption ratio
SBR	Sequencing batch reactor
SDI	Silt density index
SDN	Sludge disintegration number
SED	Specific energy demand per permeate volume
SEM	Scanning electron microscopy
SFI	Sludge filterability index
SGMD	Sweep gas membrane distillation
SMP	Soluble microbial products
SNDR	Specific denitrification rates

SOD	Specific oxygen demand
SOTE	Standard oxygen transfer efficiency
SOTR	Standard oxygen transfer rate
SOUR	Specific oxygen uptake rate
SRT	Solids retention time
ST	Sonic test
SW	Spiral wound module
TB	Tubular
TDS	Total dissolved solids
TEP	Transparent exopolymer particles
THM	Trihaloacetic acid
TIPS	Thermally induced phase separation
TKN	Total Kjeldahl nitrogen
TMP	Transmembrane pressure
TN	Total nitrogen
TOC	Total organic carbon
TP	Total phosphorus
TPH	Total petroleum hydrocarbon
TTF	Time to filter
UASB	Upflow sludge blanket reactor
UCT process	University of Cape Town process
UF	Ultrafiltration
USEPA	United States Environmental Protection Agency
UV	Ultra violet
VDT	Vacuum decay test
VFA	Volatile fatty acids
VFD	Variable frequency drive
VIP process	Virginia initiative plant process
VMD	Vacuum membrane distillation
VOC	Volatile organic carbon
WHO	World Health Organization
WWTP	Wastewater treatment plant

Unit Conversion

	X	Y	Relation
Temperature	°C	°F	$Y = 1.8X + 32$
	°C	K	$Y = X + 273.15$
Length/area	m	in	$Y = X/0.0254$
	m	ft	$Y = X/0.3048$
	m^2	ft^2	$Y = X/0.3048^2$
Mass	kg	lb	$Y = X/0.453592$
Volume	m^3	ft^3	$Y = X/0.3048^3$
	m^3	L	$Y = 1000X$
	L	gal	$Y = X/3.785$
Flux	L/m²/h (LMH)	gal/ft²/d (gfd)	$Y = X/1.698$
	L/m²/h (LMH)	m/s	$Y = X/3.6 \times 10^6$
	L/m²/h (LMH)	m/day	$Y = 0.024X$
Permeability	LMH/bar	gfd/psi	$Y = 0.0406X$
	LMH/bar	LMH/kPa	$Y = X/100$
Viscosity	cP	kg/m/s	$Y = 0.001X$
Force	N	kg·m/s²	$Y = X$
Pressure	Pa	kg/m/s² (or N/m²)	$Y = X$
	psi	lb/in²	$Y = X$
	bar	kg/m²/s (or Pa)	$Y = 1 \times 105X$
	bar	kPa	$Y = 100X$
	atm	bar	$Y = 1.01325X$
	atm	mm Hg	$Y = 760X$
	bar	psi	$Y = 14.504X$
	bar	m H_2O	$Y = 10.197X$
	psi	ft H_2O	$Y = X/0.4335$
Power	kW	hp	$Y = X/0.7457$
	W	N·m/s	$Y = X$
Energy	kWh	Joule	$Y = 3.6 \times 10^6X$
	J	N·m	$Y = X$
	J	cal	$Y = X/4.184$
Flow rate	m^3/day	MGD	$Y = X/3785$
	m^3/day	gal/min (gpm)	$Y = 0.1835X$
	m^3/h	gal/min (gpm)	$Y = 4.403X$
	m^3/min	ft³/min (CFM)	$Y = 35.315X$
Conductivity	S/cm	1/Ω/cm (or mho/cm)	$Y = X$
	mS/cm	μS/cm	$Y = 1000X$
	μS/cm	μmho/cm	$Y = X$

Index

Page numbers followed f and t indicate figures and tables, respectively.

Milton Keynes UK
Ingram Content Group UK Ltd.
UKHW051942071024
449327UK00026B/2132